Biomolecular NMR Spectroscopy

Biomolecular NMR Spectroscopy

JEREMY N. S. EVANS
Department of Biochemistry and Biophysics
Washington State University

Oxford   New York   Tokyo
OXFORD UNIVERSITY PRESS

Oxford University Press, Walton Street, Oxford OX2 6DP

Oxford   New York
Athens   Auckland   Bangkok   Bombay
Calcutta   Cape Town   Dar es Salaam   Delhi
Florence   Hong Kong   Istanbul   Karachi
Kuala Lumpur   Madras   Madrid   Melbourne
Mexico City   Nairobi   Paris   Singapore
Taipei   Tokyo   Toronto

and associated companies in
Berlin   Ibadan

Oxford is a trade mark of Oxford University Press

Published in the United States
by Oxford University Press Inc., New York

© Jeremy N. S. Evans, 1995

First printed in 1995
Reprinted 1996 (with corrections)

A catalogue record for this book is available from the British Library

Library of Congress Cataloging in Publication Data
Evans, Jeremy N. S.
Biomolecular NMR spectroscopy/Jeremy N. S. Evans.
Includes bibliographical references and index.
1. Nuclear magnetic resonance spectroscopy.   2. Biomolecules—
Analysis.   I. Title.
QP519.9.N83E94   1995   574.19′285′028—dc20   94-34841

ISBN 0 19 854767 6 (Hbk)
ISBN 0 19 854766 8 (Pbk)

Printed in Great Britain by
The Bath Press, Somerset

Dedicated to the memory of my Mother, Billa Evans, who would have laughed at all this gobbledegook

Preface

I was sitting in a shiny new laboratory, largely empty and woefully bereft of NMR spectrometers, and preparing my lecture notes on a new graduate level course on 'biomolecular NMR spectroscopy', when I became aware of the serious lack of modern, up-to-date textbooks in the area. In what I have since realized was rather a weak moment, I resolved to write the textbook that I felt was needed. Of course, what I did not realize was that it would be at least two years before I got around to thinking seriously about the project, and a further year before I really started on the writing. And, like most people, I procrastinated as long as I could, relying on the gentle but firm reminders from Howard Stanbury (Senior Chemistry Editor at Oxford University Press) to needle me into action.

The result is a somewhat personal view of biological NMR spectroscopy. It differs substantially from those biological NMR texts that are currently available, even though most of them are now very outdated. First, it does not fight shy of mathematical equations. But for those readers who feel that they are not in the least mathematical, do not be too put off! I have never regarded myself as being particularly mathematical, and yet the mathematical equations used here, particularly the choice of product operator formalism over vector formalism for explaining NMR pulse sequences, are all relatively simple, provided you don't lose your head over the amount of algebra. Except in a few places, the mathematics does not exceed the level of a British first-year undergraduate 'mathematics for the physical sciences' course, or a 300-level undergraduate course in the United States. Furthermore, I have attempted to provide physical explanations for the majority of the equations used, so it is possible to skip over them and simply read the text, without getting too bogged down in integrals and summations.

The second major difference from previous biological NMR texts is that on a number of occasions this book discusses the importance of solid-state NMR as well as solution-state NMR. This arises in part because my own research interests are equally divided between the two states of matter, but also because I firmly believe that solid-state NMR is about to undergo the kind of revolution that solution-state NMR underwent ten years ago.

Third, the book attempts to span the relatively diverse uses of biological NMR by including detailed discussions on protein structures, enzyme mechanisms and stereochemistry, DNA, RNA and carbohydrate structures, and NMR of membranes. I have limited the book to biomolecular applications, with only a brief discussion of *in vivo* NMR, leaving this topic and MRI to the more clinically oriented texts.

I have deliberately adopted a pedagogical approach for Chapters 1–3, and a highly selective literature review for the remainder of the book. This is because I believe that we learn best through example, and I have chosen examples that I think exemplify the theoretical principles outlined in the early part of the book. Furthermore, because of the diversity of biological problems, the ways in which NMR spectroscopy may be applied are limitless. The survey of the ways in which it has been applied tries to give some flavour of the breadth of the technique in its ability to address key structural and mechanistic problems.

Pullman, WA, USA   J.N.S.E.
May 1994

Acknowledgements

I have benefited from (and probably vastly overstretched) the goodwill and help of a number of colleagues. I would like to thank my former mentors, A. Ian Scott, Mary F. Roberts, and Iain D. Campbell for encouraging me into this field and teaching me an enormous amount. I would like to thank Raymond Dwek for introducing me to Oxford University Press. I would also like to mention the following for reading chapters, and/or providing me with guidance, discussion, and preprints: Richard Appleyard, Bill Bachovchin, Ad Bax, (the late) Andy Derome, Keith Dunker, Jim Frye, Bob Griffin, Erwin Hahn, Toshiko Ichiye, Malcolm Levitt, Yan Li, Dan Raleigh, Cecilia Ramilo, Jake Schaefer, Wendy Shuttleworth, and Jaroslav Zajicek. I am particularly grateful to David Neuhaus who waded through the entire manuscript, pointed out errors, and made numerous helpful suggestions. I would also like to single out Jaroslav Zajicek and Wendy Shuttleworth, who provided invaluable assistance in obtaining the two-dimensional spectra shown in Appendices 2 and 3; Yan Li, who obtained the glycine solid-state spectra shown in Fig. 1.40 and who devised the product operator analogy for the REDOR experiment (Appendix 1); and Richard Appleyard who also assisted with some of the graphics. A special word of thanks to Cindy Neal, who typed in all the tables and who, with the help of Cecilia Ramilo, typed in a large number of the figure legends and finally carried out the onerous task of photocopying the completed manuscript.

I should like to acknowledge the following for permission to reproduce their illustrations. Firstly, all those individuals whose work has been reproduced, including those who granted formal approval: Ian Armitage (Figs 7.27 and 7.28); Robert Baxter (Figs 6.25 and 6.26); Mildred Cohn (Fig. 6.8); Colin Fyfe (Figs 1.46, 1.48, 1.49, and 1.57); Peter Kim (Fig. 5.20(a)); Eric Oldfield (Fig. 10.11); Mary Roberts (Fig. 6.18); Heinrich Roder (Fig. 5.19), and David Wemmer (Fig. 9.24). Secondly, all those publishers who have granted us formal approval: Figs 1.38, 1.39, 1.53, 1.54, 2.5, 2.28, 2.41, 4.12–4.18, 4.22, 4.23, 4.40–4.45, 5.1, 5.2, 5.4–5.7, 5.13, 6.6, 6.7, 6.10, 6.11, 6.27, 7.26, 7.36, 8.5, 8.6, 9.22, 9.25, 10.3, 10.5–10.7, 10.12, 10.12 © Academic Press; Figs 4.46, 4.47, 5.14, 7.4, 9.32, 9.33, 10.10, 10.14, 10.16 © American Association for the Advancement of Science; Figs 1.1, 1.51, 2.30, 2.32–2.34, 2.43, 2.47, 2.51–2.53, 2.55, 3.9, 3.10, 4.5, 4.10, 4.11, 4.19, 4.20, 4.21, 4.24, 4.25, 4.27–4.29, 4.31, 4.32, 4.34–4.39, 4.48(b), 5.15, 5.18, 6.1, 6.5, 6.9, 6.12, 6.13, 7.6, 7.8, 7.9, 7.11, 7.12, 7.14–7.22, 7.25, 7.31, 7.37, 8.7–8.9, 8.11, 8.13, 8.16, 8.23, 8.24, 8.26, 9.12, 9.15, 9.18, 9.19, 9.21, 9.26, 9.30, 9.31, 9.38, 10.2, 10.4, 10.8, 10.9, 10.17–10.22 © American Chemical Society; Figs 3.7, 5.3, 5.8, 10.1, 10.15 © Annual Reviews Inc.; Figs 8.2, 8.3 © Biophysical Society; Figs 2.21–2.23, 3.13, 4.30, 9.6–9.9, 9.11, 9.29 © Cambridge University Press; Figs 3.4, 4.4 © CRC Press Inc.;

Figs 4.9, 6.29, 7.33, 7.38, 7.39 © Elsevier Science Publishers B.V.; Figs 1.31, 1.37, 2.2–2.4, 2.9, 2.10, 2.13, 2.26, 2.45, 2.49, 6.28 with kind permission of Elsevier Science Ltd., The Boulevard, Langford, Langford Lane, Kidlington OX5 1GB, UK; Fig. 2.39 © Escom Publisher B.V.; Figs 2.16, 4.26, 4.48(a), 5.9–5.12, 5.16, 5.17, 9.13, 9.23 © Macmillan & Sons; Figs 1.17, 1.52, 2.7, 2.8, 2.19, 2.20, 7.29, 8.21, 8.22 © Oxford University Press; Fig. 1.56 © Royal Society; Figs 6.23, 6.24, 8.15, 9.14 © Royal Society of Chemistry; Figs 4.6–4.8, 7.28, 7.32, 9.17, 9.20, 9.27, 9.28 © Springer-Verlag Inc.; Fig. 1.2, with kind permission of Varian Associates; Figs 3.1, 3.2, 4.3, 9.1–9.5, 9.36, 9.37 © VCH Publishers Inc.; Figs 1.32, 1.50, 4.1, 4.2, 7.30, 7.40, 9.10, 9.34, 9.35, Table 4.1 © John Wiley & Sons.

Although every effort has been made to trace and contact copyright holders, in a few instances this has not been possible. If notified, the publishers will be pleased to rectify any omission in future editions.

In a book of this nature, there are bound to be errors, for which I take full responsibility. I would, however, welcome comments and suggestions from readers so that future editions can be improved.

Last, but by no means least, I would like to thank my wife Michèle, without whose love, affection, and support none of this would have ever seen the light of day, and our son, Matthew, whose gestation followed that of this book.

Contents

Part I · *Theory*

Part 1: Theory

1 · Introduction

The first application of nuclear magnetic resonance spectroscopy (NMR, sometimes referred to as n.m.r. in old-fashioned texts) to a biological sample was reported[1] in 1954 by Jacobson, Anderson, and Arnold on the effect of hydration of deoxyribonucleic acid, one year after Watson and Crick's historic discovery. Three years later, Saunders, Wishnia, and Kirkwood obtained the first ^{1}H NMR spectrum of a protein, ribonuclease.[2] This spectrum, obtained at 40 MHz, is shown in Fig. 1.1.

In contrast, the 750 MHz ^{1}H NMR spectrum of the enzyme lysozyme is shown in Fig. 1.2. In the intervening 38 years, the field of NMR spectroscopy has undergone a revolution, culminating in the award of the 1991 Nobel Prize for Chemistry to Richard Ernst, one of the key figures in the development of NMR. However, the recognition given to this field is only just beginning, and the area where the method is likely to have the greatest impact is at the interface between biology, chemistry, and physics. Although there are a number of excellent texts on NMR spectroscopy, they are primarily for chemists, and tend to focus either on the theory of NMR[5-6] or on applications to organic chemistry.[7-8] Those books on biological NMR, while still very useful, tend to be somewhat outdated[9-12] or directed solely towards structure.[13] This book seeks to provide a general outline of basic NMR theory and examine selected examples of applications to

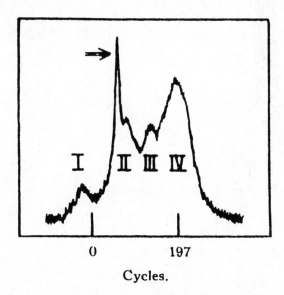

Fig. 1.1 The first ^{1}H NMR spectrum of an enzyme, ribonuclease, at 40 MHz, obtained in 1957. (Reprinted from Ref. 2 with permission.)

biochemical and biophysical problems, with a strong emphasis on the relationship between structure and function.

As a structural method, NMR spectroscopy has many unique features which set it apart from other structural techniques. It is the only structural method which spans three states of matter. Although gas-phase NMR spectroscopy is not particularly common, and of no real interest to biochemists, the ability to correlate structural details obtained from solid-state NMR spectroscopy with those obtained in the liquid state is of the utmost importance, given the, sometimes controversial, debate in recent years between NMR spectroscopy and X-ray crystallographers as to the physiological relevance of their respective biological structures. NMR spectroscopy is also the only structural method that spans many levels of biological organization. The method of magnetic resonance imaging (MRI) is now becoming relatively widespread in hospitals in developed countries, and is simply ^{1}H NMR spectroscopy of water used in a particular fashion to generate images of the human head, or other parts of the body. Other nuclei have been used to follow metabolism *in vivo* in humans and animals, and also in cell cultures. Metabolism has been studied by multinuclear NMR in cell-free extracts, and using purified proteins and enzymes. The interaction of small molecules with isolated enzymes and proteins has been studied by NMR, focusing on active site regions of a biomolecule. Finally, NMR spectroscopy has been used with considerable success to solve the three-

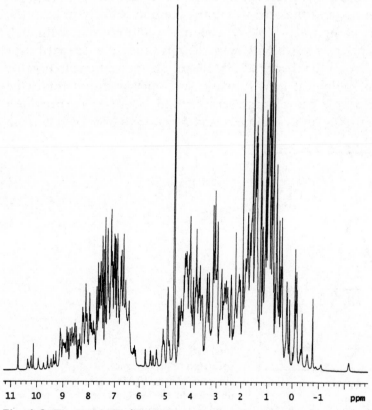

Fig. 1.2 The 750 MHz ^{1}H NMR spectrum of lysozyme.

dimensional structures of individual components in the cell, such as proteins, enzymes, DNA, RNA, and membranes. Therefore, a technique which can obtain an image of a human head on the one hand, and the structure of DNA on the other, must be worth learning about.

The following sections will attempt to provide sufficient basic NMR theory for use in solving biological problems. Throughout this book the emphasis will be on molecular details rather than macroscopic details such as the images obtained with MRI. There are a number of excellent texts on the MRI technique, and the reader is referred to those for more information.[14–15]

1.1 BASIC THEORY OF NMR

This chapter attempts to provide a brief summary of the salient features of basic NMR theory. It does not presume to be exhaustive, and the reader should refer to one or more of the increasing number of texts in this area for more detailed information (*vide supra*). In our treatment of NMR theory we have chosen to introduce both classical vector formalism and also quantum mechanical Cartesian product operator formalism. In our experience vector formalism, while being extremely useful for simple experiments, is not very helpful in understanding multidimensional NMR experiments. Rather than introduce the far more complicated density operator formalism, we make no apology for adopting the product operator formalism where appropriate. The mathematics follows simple rules, and, while sometimes generating lengthy expressions, algorithms are even available[16,17] for computer programs such as Mathematica. The product operator formalism is relatively straightforward and very powerful, and is now the method of choice for evaluating new pulse sequences. It is important for researchers using NMR to be familiar with it.

1.1.1 The NMR phenomenon

The magnetic resonance phenomenon occurs as a result of the quantum mechanical property of spin. This is a source of angular momentum intrinsic to a number of different nuclei. The spin angular momentum confers a magnetic moment on a nucleus and therefore a given energy in a magnetic field. The nuclear spin (I) can have the values $I = 0$, $\frac{1}{2}$, 1, $1\frac{1}{2}$, . . . , etc. (see Table 1.1).

Note that common biological nuclei such as ^{12}C or ^{16}O have $I = 0$ and therefore do not give NMR spectra. The nuclear magnetic moment (μ) is given by:

$$\mu = \gamma I \hbar \tag{1.1}$$

The gyromagnetic ratio (also known as the magnetogyric ratio) γ is the proportionality constant which determines the resonant frequency of the nucleus for a given external field. Typical nuclei of interest in biological NMR are given in Table 1.2. In a magnetic field, a nucleus of spin I has $2I + 1$ possible orientations, given by the value of the magnetic quantum number m_I, which has values of $-I, -i+1, \ldots I-1$ (e.g. for a nucleus of spin $\frac{3}{2}$, $m_I = -\frac{3}{2}, -\frac{1}{2}, \frac{1}{2}, \frac{3}{2}$). We can regard a spin-$\frac{1}{2}$ nucleus as a small bar magnet, which when placed in a static field has an energy which varies with orientation to the field. The possible energies are quantized, with the two possible values of $m_I(\pm\frac{1}{2})$ corresponding to parallel and antiparallel orientations of this small magnet and the external field. As we shall see

Table 1.1

Mass no.	Atomic no.	I
Odd	Even or odd	$\frac{1}{2}, \frac{3}{2}, \frac{5}{2} \ldots$
Even	Even	0
Even	Odd	$1, 2, 3 \ldots$

shortly, the NMR absorption is a consequence of transitions between the energy levels stimulated by applied radiofrequency (RF) radiation.

Although we should bear in mind the quantized nature of nuclear spin, we can describe the motion of a nucleus in a magnetic field in terms of classical mechanics. In the presence of an applied magnetic field $\mathbf{B}_0$, the magnetic moment experiences a torque which is the vector product of the nuclear angular momentum $\mathbf{J}$ and the magnetic moment μ (see Fig. 1.3). Note, however, that the physical picture presented here does not represent reality. The findings of the famous Stern–Gerlach experiment clearly suggest that spin angular momentum is quantized, and therefore classical mechanics is not applicable. Furthermore, the nucleus is not necessarily spinning about its axis—indeed were it to be spinning the radial velocity would exceed the speed of light. Thus although we speak of 'spin' as if the nucleus were actually rotating about its axis, this is an unfortunate choice of words since the source of the magnetic moment is a purely quantum mechanical property which could just as easily be called 'sweetness', 'bitterness', or whatever.

According to Newtonian mechanics, this torque equals the rate of change of angular momentum:

$$\frac{\mathrm{d}\mathbf{J}}{\mathrm{d}t} = \mu \times \mathbf{B}_0 \tag{1.2}$$

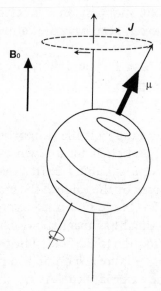

Fig. 1.3 Schematic representation of the motion of a nucleus in a magnetic field.

Table 1.2 Magnetic properties of some biologically useful nuclei

Isotope	Spin	Natural abundance (%)	Quadrupole moment Q (10^{-28} m^2)	Gyromagnetic ratio γ (10^7 rad s^{-1} T^{-1})	Sensitivity rel.[a]	abs.[b]	NMR-frequency (MHz) at a field (T) of 2.3488
^{1}H	1/2	99.98	—	26.7522	1.00	1.00	100.000
^{2}H	1	1.5×10^{-2}	2.87×10^{-3}	4.1066	9.65×10^{-3}	1.45×10^{-6}	15.351
^{3}H	1/2	0	—	28.5350	1.21	0	106.663
^{7}Li	3/2	92.58	-3.7×10^{-2}	10.3976	0.29	0.27	38.863
^{11}B	3/2	80.42	4.1×10^{-2}	8.5847	0.17	0.13	32.084
^{13}C	1/2	1.108	—	6.7283	1.59×10^{-2}	1.76×10^{-4}	25.144
^{14}N	1	99.63	1.67×10^{-2}	1.9338	1.01×10^{-3}	1.01×10^{-3}	7.224
^{15}N	1/2	0.37	—	-2.7126	1.04×10^{-3}	3.85×10^{-6}	10.133
^{17}O	5/2	3.7×10^{-2}	-2.6×10^{-2}	-3.6280	2.91×10^{-2}	1.08×10^{-5}	13.557
^{19}F	1/2	100	—	25.1815	0.83	0.83	94.077
^{23}Na	3/2	100	0.10	7.0704	9.25×10^{-2}	9.25×10^{-2}	26.451
^{25}Mg	5/2	10.13	0.22	-1.6389	2.67×10^{-3}	2.71×10^{-4}	6.1195
^{31}P	1/2	100	—	10.8394	6.63×10^{-2}	6.62×10^{-2}	40.481
^{35}Cl	3/2	75.53	-8.2×10^{-2}	2.6242	4.70×10^{-3}	3.55×10^{-3}	9.798
^{39}K	3/2	93.1	5.5×10^{-2}	1.2499	5.08×10^{-4}	4.73×10^{-4}	4.667
^{43}Ca	7/2	0.145	-5×10^{-2}	-1.8028	6.40×10^{-3}	9.28×10^{-6}	6.728
^{51}V	7/2	99.76	2.17×10^{3}	-5.2×10^{-2}	0.38	0.38	26.289
^{57}Fe	1/2	2.19	—	0.8687	3.37×10^{-5}	7.38×10^{-7}	3.231
^{75}As	3/2	100	0.29[u]	4.5961	2.51×10^{-2}	2.51×10^{-2}	17.126
^{77}Se	1/2	7.58	—	5.1214	6.93×10^{-3}	5.25×10^{-4}	19.067
^{113}Cd	1/2	12.26	—	-5.9609	1.09×10^{-3}	1.33×10^{-3}	22.182

[a] At constant field for equal number of nuclei.
[b] Product of relative sensitivity and natural abundance.

$$\mathbf{J} = I\hbar \tag{1.3}$$

$$\therefore \frac{d\mu}{dt} = \gamma\boldsymbol{\mu} \times \mathbf{B}_0 \quad \text{using} \quad \boldsymbol{\mu} = \gamma I\hbar = \gamma\mathbf{J} \tag{1.4}$$

This equation is analogous to the equation of motion for a body with angular momentum $\mathbf{L}$ in a gravitational field $\mathbf{g}$ with mass m at a distance $\mathbf{r}$ from the fixed point of rotation, if we equate $\mathbf{J}$ to $\mathbf{L}$ and $\mathbf{B}_0$ to $\mathbf{g}$, and regard $\mathbf{r} \times m$ as an intrinsic property of the body analogous to $\gamma\boldsymbol{\mu}$:

$$\frac{d\mathbf{L}}{dt} = \mathbf{r} \times m\mathbf{g} \tag{1.5}$$

Thus this is just like the motion of a gyroscope which in a gravitational field precesses, i.e. its axis of rotation itself rotates about the field direction. In the classical analogy, the same motion occurs for nuclear spins in a magnetic field.

The energy of the interaction is proportional to μ and $\mathbf{B}_0$ (see Fig. 1.4), so

$$E = -\gamma\hbar m_I \mathbf{B}_0 \tag{1.6}$$

and since $\Delta m_I = 1$,

$$\Delta E = \gamma\hbar\mathbf{B}_0 \tag{1.7}$$

and from Planck's law,

$$\Delta E = h\nu \tag{1.8}$$

then

$$\nu = \frac{\gamma\mathbf{B}_0}{2\pi} \text{ (in Hz)} \tag{1.9}$$

or

$$\omega = \gamma\mathbf{B}_0 \text{ (in rad s}^{-1}) \tag{1.10}$$

Note that we have dropped the minus sign between Equations (1.6) and (1.7). This is a convention that has been widely adopted for convenience in the NMR literature, although, as pointed out by Ernst and co-workers,[4] strictly speaking, $\omega = -\gamma\mathbf{B}_0$, which

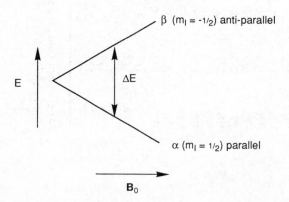

Fig. 1.4 Energy level diagram.

has consequences for the Cartesian representation of vectors and product operators considered in Sections 1.1.2 and 1.1.3. Thus nuclei precess around the $\mathbf{B}_0$ axis at a speed which is called the Larmor frequency (which is its NMR absorption frequency, ω). The rotation may be clockwise or anticlockwise depending on the sign of γ, but is always the same for any particular nucleus.

The two energy states α and β will be unequally populated, the ratio being given by the Boltzmann equation:

$$\frac{N_\beta}{N_\alpha} = e^{-\Delta E/k_B T} \tag{1.11}$$

Another way to look at this is that in a sample containing a large number of spins, all possessing the same Larmor frequency, the parallel orientation of the z component of each spin along the $\mathbf{B}_0$ direction is of lower energy than the antiparallel one. So at thermal equilibrium, we expect the Boltzmann surplus as shown in Fig. 1.5. Thus along the z axis there is a net magnetization of the sample parallel to the field. All the contributing spins have components precessing in the xy plane, but because all have equal energy, the phase of the precession is random. Thus for an ensemble of spins, there is no net magnetization in the xy plane and total magnetization of the sample is stationary and aligned with the z axis (called $\mathbf{M}_0$).

1.1.2 The vector model

Radio frequency (RF) radiation is electromagnetic (see Fig. 1.6) and can be represented as an oscillating magnetic field, which in turn can be represented by magnetization vectors (see Fig. 1.7). This represents the half cycle of the oscillation of the magnetization due to the presence of the RF (also known as the $\mathbf{B}_1$) field. Alternatively, we can represent its as two magnetization vectors of constant amplitude rotating about an axis (x) in opposite directions, with angular frequency = RF. Thus this pair of counter-rotating vectors is a valid way of representing the RF (see Fig. 1.8).

So when the sample and the RF field interact, we have a moving field interacting with a static one (although this, too, causes precessional motion in the sample). Conceptually, the way this rather complex picture is simplified is by rotating the coordinate axes at the same rate as the nuclear precession. Now since there is no precession, it looks as though the applied field (which caused the precession) has disappeared. However, the net magnetization remains along the z axis. Furthermore, the RF can be decomposed into two

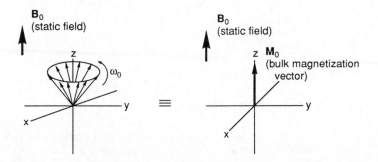

Fig. 1.5 The bulk magnetization vector.

components, one in the xy plane. The other, which was originally moving at an equal speed in the opposite direction, is rotating in the new frame at twice the Larmor frequency. This can now be neglected, and has no effect on the NMR experiment (see Fig. 1.9). When a pulse of RF is applied to the sample, i.e. the $\mathbf{B}_1$ field is switched on and then switched off, in the rotating frame the $\mathbf{M}_0$ and $\mathbf{B}_1$ vectors are static and orthogonal. This generates a torque and the sample magnetization is driven around by the $\mathbf{B}_1$ vector, at a speed dependent on the field strength. Thus we could move it through 90°, as shown in Fig. 1.10. Note that the direction of motion (in this case clockwise relative to the $\mathbf{B}_1$ direction) is to some extent arbitrary, and in general the NMR literature using classical vector formalism adopts the convention used here (although the opposite convention has

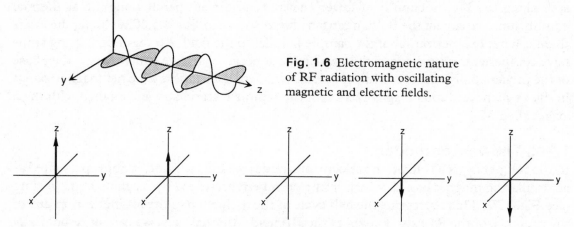

Fig. 1.6 Electromagnetic nature of RF radiation with oscillating magnetic and electric fields.

Fig. 1.7 RF radiation represented by magnetization vectors.

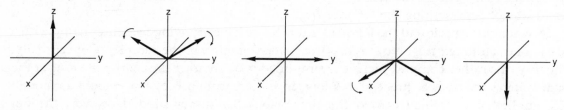

Fig. 1.8 RF radiation represented by two counter rotating vectors.

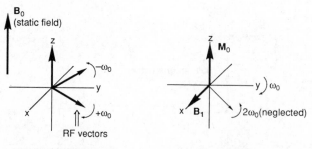

LABORATORY FRAME

ROTATING FRAME

Fig. 1.9 Magnetization vectors shown in the laboratory and rotating frames.

been adopted when using the Cartesian representation of product operators—see Section 1.1.3). Note, however, that the direction is governed by the sign of the gyromagnetic ratio. After the pulse has finished, the sample magnetization remains in the *xy* plane. In the laboratory frame it precesses about the static field, generating radio signals (see Fig. 1.11). These radio signals generate what is called the *free induction decay* (FID), which is a function that decays exponentially with time (see Fig. 1.12). The FID is related to the frequency domain spectrum through *Fourier transformation*. This is as follows:

$$f(\omega) = \int_{-\infty}^{\infty} f(t)e^{i\omega t} \, dt \tag{1.12}$$

$$\mathrm{Re}(f(\omega)) = \int_{-\infty}^{\infty} f(t)\cos \omega t \, dt$$

$$\mathrm{Im}(f(\omega)) = \int_{-\infty}^{\infty} f(t)\sin \omega t \, dt \tag{1.13}$$

since $e^{i\omega t} = \cos \omega t + i \sin \omega t$ (see Fig. 1.13). This is a consequence of the fact that every NMR signal has amplitude, frequency, *and* phase. The NMR signal is detected using a

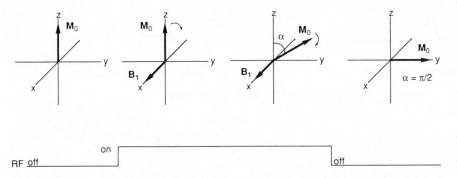

Fig. 1.10 The sample magnetization driven to the *y* axis after a 90° pulse.

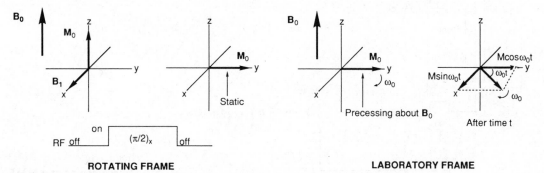

Fig. 1.11 Precession of the sample magnetization about the static field.

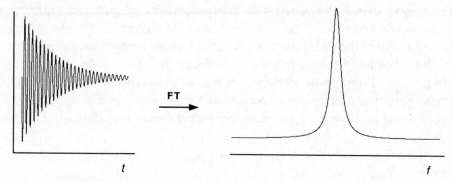

Fig. 1.12 The free induction decay.

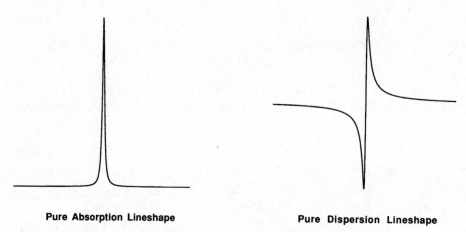

Pure Absorption Lineshape   **Pure Dispersion Lineshape**

Fig. 1.13 Pure absorption lineshape versus pure dispersion lineshape.

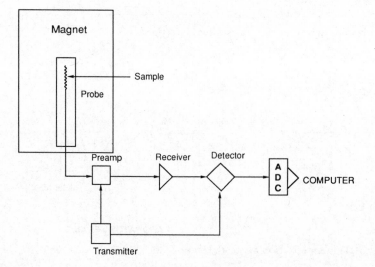

Fig. 1.14 Outline of a NMR spectrometer.

detector, and the basic outline of an NMR spectrometer is given in Fig. 1.14. This represents a greatly simplified picture of a modern NMR spectrometer, which nowadays is commonly equipped with three RF channels and an array of sophisticated equipment, a discussion of which is outside the scope of this book. However, one feature which is particularly important is the way in which the signal is detected. Here a technique known as *quadrature detection* is used. The detectable magnetization can be represented as a vector precessing in the *xy* plane, as we saw in Fig. 1.9, so that a detector aligned along the *x* axis would be insensitive to the direction of rotation of the vector. In other words, the detector cannot distinguish whether the signal frequency is greater or less than the reference frequency in the case where two signals are on opposite sides of the reference by the same frequency difference. The two vectors would be rotating at the same frequency but in opposite directions. In order to distinguish between these we use a detector that can detect the signals along both the *y* and *x* axes simultaneously. Instead of having two coils, however, the signal is manipulated electronically, by having two detectors in which one has had the phase of the reference frequency shifted by 90° and the FIDs stored in separate memory locations in the computer. Since the phase of one of these FIDs is affected by the sign of the frequency, these two FIDs correspond to the real and imaginary components of the signal, and are treated as such in the Fourier transformation.

The question of *phase* both in terms of the transmitter pulses and in terms of the receiver is important. In many pulse sequences, as we shall see later in this chapter and in Chapter 2, elaborate phase cycling of pulses and the receiver are required in order to achieve the particular result desired. This is the means by which desired and undesired signals are separated. The same result can also be achieved using what are called *pulsed field gradients*, which are emerging as an important and fast alternative to lengthy phase cycles.[18] An example of the consequence of changing the phase of the transmitter pulses is summarized in Fig. 1.15, which uses the vector model to represent the four possible phases (*x*, *y*, −*x*, and −*y*).

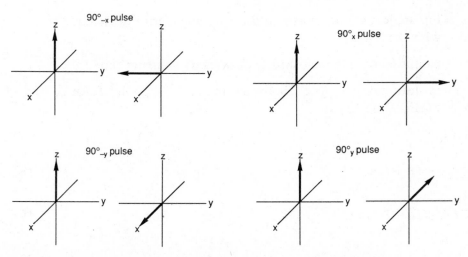

Fig. 1.15 The vector representation of RF pulses of differing phase.

1.1.3 The product operator

There are intrinsic limitations in representing the effects of RF pulses in NMR using vectors, and we need to consider alternative ways of looking at this. In contrast to the vector formalism, in which the macroscopic magnetization rotates under the influences of pulses, and/or precesses under the influence of chemical shift, the entire state of the spins or spin system can be described by the wave function, $\Psi(t)$, or the density operator $\sigma(t)$. Disregarding relaxation, the time evolution of the density operator is described by the Liouville–von Neumann equation:

$$\frac{d\sigma(t)}{dt} = -i[\mathscr{H}(t), \sigma(t)] \tag{1.14}$$

$\mathscr{H}(t)$ is the famous Hamiltonian operator, which includes chemical shift terms, coupling terms, and so on. Instead of getting into the rather complex area of density matrices, we can describe the density operator in terms of Cartesian *product operators*.[19] Thus the ground state of a system at thermal equilibrium is described by a *polarization* (or population–see Section 1.1.7) along the z axis, I_z. After a pulse in the y direction:

$$I_x \xrightarrow{\alpha I_y} I_x \cos \alpha - I_z \sin \alpha \tag{1.15}$$

$$I_y \xrightarrow{\alpha I_y} I_y \tag{1.16}$$

$$I_z \xrightarrow{\alpha I_y} I_z \cos \alpha + I_x \sin \alpha \tag{1.17}$$

These Cartesian operators, I_x, I_y, and I_z transform like magnetizations when a rotation is performed. The properties of a state I_x, I_y, and I_z can be read off directly with such simple operators, since they each consist of only one operator. I_x is associated with an x magnetization, and I_y and I_z accordingly.

We will adopt the following convention:

$\alpha I_{x,y}$ or αI_ϕ = flip angle α about an axis in the xy plane (forming an angle ϕ with the x axis)

$\Omega I_z t$ = evolution of chemical shift Ω of nucleus I during time t

$2\pi \mathscr{J} I_{1z} I_{2z} t$ = evolution of coupling $\mathscr{J}$ between nuclei I_1 (or I) and I_2 (or S) during time t

So for chemical shifts:

$$I_x \xrightarrow{\Omega t I_z} I_x \cos \Omega t + I_y \sin \Omega t \tag{1.18}$$

$$I_y \xrightarrow{\Omega t I_z} I_y \cos \Omega t - I_x \sin \Omega t \tag{1.19}$$

$$I_z \xrightarrow{\Omega t I_z} I_z \tag{1.20}$$

Detection of x magnetization leads to build-up of the signal according to the coefficient $-\sin \Omega t$ of I_x. Fourier transformation leads to a line at $\pm \Omega$. Quadrature detection allows selection at either $+\Omega$ or $-\Omega$. Thus we can summarize the *vector model* and the *product operator model* as in Fig. 1.16 for a simple 90° pulse. The effects of chemical shifts, pulses, and positive scalar couplings on product operators are summarized pictorially in Fig. 1.17, and summarized mathematically in Appendix 1. Note that the sense of rotation for product operators is opposite to the classical description presented in Section 1.1.2. As

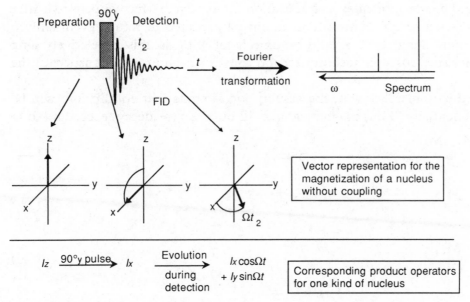

Fig. 1.16 The vector model and the product operator model for a 90° pulse.

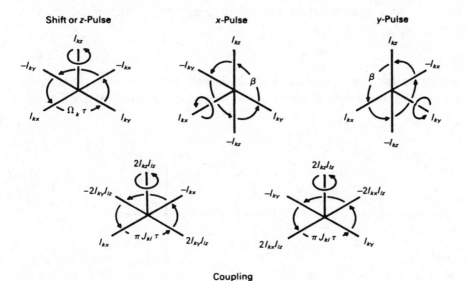

Coupling

Fig. 1.17 Graphic representation of the effect of chemical shifts, pulses, and positive scalar couplings on product operators. (Reprinted from Ref. 4 with permission.)

mentioned in Section 1.1.1, this results from the fact that, strictly speaking, $\omega = -\gamma \mathbf{B}_0$, and when product operator formalism was introduced, the minus sign was retained, thereby changing the sense of the rotation relative to classical formalism. Of course this has nothing to do with the physics, and is simply a consequence of how the equations are written.

1.1.4 Relaxation

Having considered how we obtain xy magnetization, let us turn to what happens to it after the RF pulse is switched off. Naturally, it might be expected to return eventually to thermal equilibrium along the z axis. The time it takes to do this is relatively long compared to other spectroscopic techniques, and constitutes an important aspect of the experiment.

In the Bloch theory of relaxation, the assumption is made that equilibrium will be approached exponentially. Thus magnetization will build up (or decay) according to:

$$\frac{d\mathbf{M}_z}{dt} = \frac{\mathbf{M}_0 - \mathbf{M}_z}{T_1} \tag{1.21}$$

$$\frac{d\mathbf{M}_x}{dt} = -\frac{\mathbf{M}_x}{T_2} \tag{1.22}$$

$$\frac{d\mathbf{M}_y}{dt} = -\frac{\mathbf{M}_y}{T_2} \tag{1.23}$$

Longitudinal relaxation (T_1)

The build-up of $\mathbf{M}_0$ can be represented as shown in Fig. 1.18. Thus if $\mathbf{M}_0$ is the magnetization at thermal equilibrium, then

$$\mathbf{M}_z = \mathbf{M}_0(1 - e^{-t/T_1}) \tag{1.24}$$

T_1 is called the longitudinal (or spin-lattice) relaxation. After a $\pi/2$ pulse, the magnetization precesses about the z axis in the xy plane. With longitudinal relaxation, the

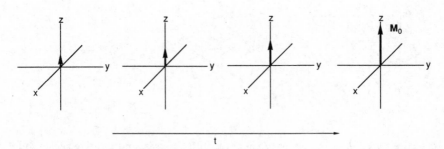

Fig. 1.18 Build-up of magnetization ($\mathbf{M}_0$).

z magnetization reappears (see Fig. 1.19) with the time constant T_1. One common method for the measurement of T_1s is the *inversion-recovery* method, which is shown in Fig. 1.20.

Transverse relaxation (T_2)

The transverse relaxation, or T_2, is not intuitively obvious, and is harder to understand in simple physical terms. When the magnetization is in the xy plane, there is *phase coherence* between the spins in the transverse plane, and loss of this phase coherence due to mutual exchange of spin energies is what gives rise to T_2 relaxation. However, in general, T_1 is always greater than T_2, and this is because there are additional causes of loss of transverse magnetization. Disregarding longitudinal relaxation (T_1) for a moment, the static magnetization in the xy-plane would not remain there because of another simple fact: the static field is not uniform throughout the sample. Thus, if we divide up the sample into small regions such that the field is uniform (regions known as *isochromats*), then the total magnetization is the sum of all these regions, each of which contributes a precessing vector which differs slightly in frequency. This blurs the bulk magnetization vector, and this is due to inhomogeneous broadening, which often dominates the transverse (or spin–spin) relaxation, T_2. Note that the magnetic field inhomogeneity is really an experimental artefact, and when it dominates T_2 the directly observed decay is usually defined as T_2^* whereas T_2 is usually reserved for other causes of loss of order (see Fig. 1.21).

Spin echoes

One of the fascinating things that happens if we apply two RF pulses in quick succession relative to T_1 is the production of a spin echo. Consider the pulse sequence $(\pi/2)-\tau-\pi$ shown in Fig. 1.22. After the $\pi/2$ pulse, the inhomogeneity causes the isochromats to fan

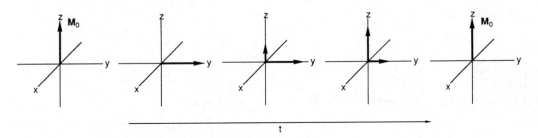

Fig. 1.19 Reappearance of the z-magnetization.

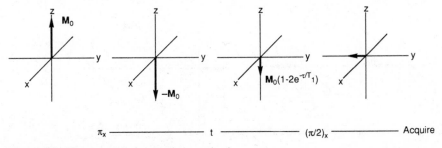

Fig. 1.20 The inversion recovery method.

out. The front edge is labelled '+' for the vectors precessing faster than ω_0, and the back edge is labelled '−' for the slower vectors. The π_x pulse rotates all the isochromats together about the x axis, leaving the vectors on the y axis. Now the '+' sign lags behind the average direction of the vectors and the '−' side is ahead. Thus the faster vectors catch up with the average position and the slow ones fall back to it. Thus after an additional time they refocus to give an echo. In fact, in a sense, *time has been reversed*, since the disorder in the system can be reversed.[20]

Let us also consider the product operator calculation for this pulse sequence:

$$\sigma_0 = I_z$$
$$\downarrow 90_x$$
$$\sigma_1 = -I_y$$
$$\downarrow \Omega\tau I_z$$
$$\sigma_2 = -I_y \cos \Omega\tau + I_x \sin \Omega\tau$$
$$\downarrow 180_y$$
$$\sigma_3 = -I_y \cos \Omega\tau - I_x \sin \Omega\tau$$
$$\downarrow \Omega\tau I_z$$
$$\sigma_4 = -I_y \cos^2 \Omega\tau + I_x \cos \Omega\tau \cdot \sin \Omega\tau - I_x \cos \Omega\tau \cdot \sin \Omega\tau - I_y \sin^2 \Omega\tau$$
$$\downarrow$$
$$\sigma_4^{\text{obs}} = -I_y$$

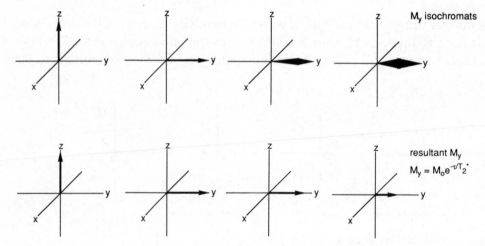

Fig. 1.21 The vector model of transverse relaxation (T_2).

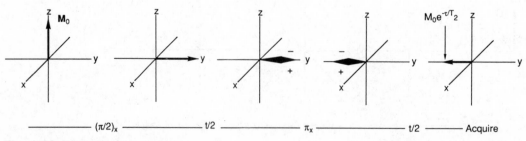

Fig. 1.22 The spin echo method.

So note that we have essentially returned to where we started after the first 90° pulse. Remembering this point greatly simplifies product operator calculations, since many pulse sequences use spin echo sequences to refocus chemical shifts.

Spin echoes play a very important role in modern NMR experiments. For example, two lines of different chemical shifts evolve at different rates in the rotating frame, so if we apply an echo sequence, they diverge before the 180° pulse, but realign after it. Similarly, heteronuclear couplings are refocused by spin echoes, but *not* homonuclear couplings. Spin echoes can be used to measure T_2^*s, although there are inherent difficulties in such measurements. An alternative method is to simply measure the linewidth, since $v_{1/2} = 1/\pi T_2^*$, where $v_{1/2}$ is the linewidth at half-height.

Another spin echo sequence, and the one usually used in practice, is the Carr–Purcell–Meiboom–Gill spin echo sequence, which is:

$$(\pi/2)_x - \tau - \pi_y - 2\tau - \pi_y - 2\tau - \pi_y - \ldots$$

Cumulative errors due to an incorrectly set pulse length do not arise. We can illustrate this if we consider the fate of the fastest moving isochromat only and look at the consequence of the pulse angle being not π but some slightly different value $\pi - \varepsilon$, as shown in Fig. 1.23. Consider a pulse $\theta = \pi - \varepsilon$, which is slightly shorter than π and directed along the y axis. The pulse interchanges the orders of rotation of the slow and fast magnetization isochromats (Fig. 1.23(c)), and leaves them slightly above the xy plane. Refocusing then occurs normally at time 2τ but takes place above the y axis in the zy plane (Fig. 1.23(d)). After the dephasing period τ (Fig. 1.23(e)) the second $\pi - \varepsilon$ pulse rotates the isochromats

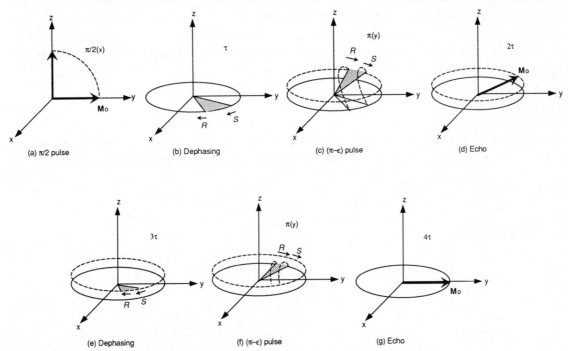

(a) π/2 pulse   (b) Dephasing   (c) (π–ε) pulse   (d) Echo

(e) Dephasing   (f) (π–ε) pulse   (g) Echo

Fig. 1.23 Vectorial representation of the Carr–Purcell–Meiboom–Gill spin echo.

exactly back into the xy plane where they are refocused at time 4τ (Fig. 1.23(g)). Thus all the even-numbered echoes are produced along the right direction, y, whereas all the odd-numbered echoes are only slightly displaced and by a constant angle.

1.1.5 The nuclear Overhauser effect

The Nuclear Overhauser Effect (NOE, η) is the change in the intensity of an NMR resonance when the transitions of another are perturbed. Generally, this means saturation, which is the elimination of a population difference across some transitions (by irradiating one with a weak RF field) while observing the signals from others:

$$\eta_i = (I - I_0)/I_0 \tag{1.25}$$

where I_0 is the thermal equilibrium intensity.

Consider two spin-$\frac{1}{2}$ nuclei I and S, with the same γ but different chemical shifts and not $\mathcal{J}$ coupled where the energy levels are represented as combinations of α and β, and the populations indicated with multiples of N. The system is shown in Fig. 1.24. Chemical shifts are very small compared with the Larmor frequency (i.e. of the order of ppm), so the transitions have nearly equal energy. Assuming we have $4N$ nuclei, then each energy level would have N nuclei of almost equal energy. The population differences are given in Table 1.3.

The first four differences in Table 1.3 are across the normal transitions which give rise to the NMR lines, while the other two involve changes in the quantum number M (which is the total of the individual quantum numbers m for the two nuclei). Although these latter two are not observable, they may contribute to the relaxation pathways. Another way of looking at this is to consider how the system restores itself. The relaxation can take place in several ways and we can assume that relaxation across a single transition is first order, and the rate constant is designated W, with the subscript distinguishing the change in M involved (see Fig. 1.25). T_1 is therefore related to some combination of W_1^I, W_2, and W_0. If W_2 and $W_0 = 0$, then

$$T_1^I = \frac{1}{2W_1} \tag{1.26}$$

Note that if W_2 and $W_0 \neq 0$ then the total relaxation time for spin I involves transitions of spin S. Thus T_1 measurements on multispin systems are not straightforward.

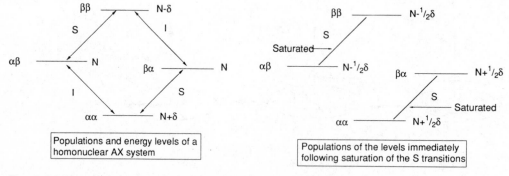

Populations and energy levels of a homonuclear AX system

Populations of the levels immediately following saturation of the S transitions

Fig. 1.24 Populations of levels before and after saturation of the S transitions.

Table 1.3 Population differences for transitions between energy levels in one-dimensional NOE experiment

	At thermal equilibrium		After saturation of S
I transitions:	$\alpha\alpha - \alpha\beta$ $\quad\Big\}\ \delta$ $\beta\alpha - \beta\beta$		δ
S transitions:	$\alpha\alpha - \beta\alpha$ $\quad\Big\}\ \delta$ $\alpha\beta - \beta\beta$		0
$\Delta m = 0$ transition:	$\beta\alpha - \alpha\beta$	0	δ
$\Delta m = 2$ transition:	$\alpha\alpha - \beta\beta$	2δ	δ

If only single-quantum transitions were involved, saturation of spin S would result in no change in intensity for spin I. It is the so-called zero- and double-quantum transitions, W_0 and W_2 that enable us to detect the NOE. If W_0 is the dominant relaxation pathway, saturating S decreases the intensity of the I resonance, and there is a negative NOE at I due to S. Conversely, if W_2 is the dominant relaxation pathway, saturating S increases the intensity of the I resonance, and there is a positive NOE at I due to S.

The steady-state NOE enhancement at spin I on saturation of spin S is described by the Solomon equation:[21]

$$\eta_i = \frac{\gamma_S}{\gamma_I} \frac{W_2 - W_0}{2W_1^I + W_2 + W_0} \tag{1.27}$$

Deriving Equation (1.27) is outside the scope of this book, but interested readers should refer to the book by Neuhaus and Williamson.[22] Note that for homonuclear NOEs $\gamma_S = \gamma_I$ and typically the maximum enhancement is ≈ 50 per cent whereas for heteronuclear NOEs the enhancement is also multiplied by the ratio of the two γs.

Mechanisms for relaxation

The interaction which gives rise to the NOE is the dipolar coupling between nuclei. This is the interaction between nuclear magnetic dipoles, in which the local field at one nucleus is due to the presence of the other. Since this local field is dependent upon the orientation of the whole molecule, it may vary from one molecule to the next. In amorphous or

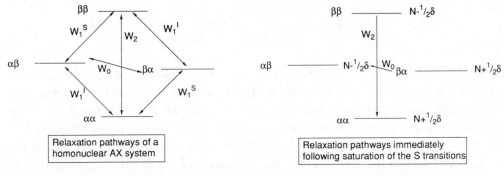

Fig. 1.25 Relaxation pathways before and after saturation of the S transitions.

polycrystalline solids, where the positions of single molecules are fixed, but vary from one molecule to the next, this leads to a range of resonant frequencies and characteristically broad lineshapes. In single crystals, the lineshape is dependent upon the orientation of the crystal with respect to the magnetic field. In solution, rapid molecular motion averages the dipolar interaction, generating fluctuating fields which stimulate longitudinal relaxation. The strength of the dipolar interaction is dependent upon internuclear distance, and can cause the cross-relaxation that gives rise to the NOE.

1.1.6 Dipolar coupling

The relation between dipolar coupling, T_1 and NOE is connected through τ_c, the rotational correlation time. Since the size of the dipolar interaction does not depend on τ_c, but its rate of change does, the total power available from the lattice will be constant (i.e. the area underneath the curve of spectral density versus frequency—see Fig. 1.26), while the upper limit of their frequencies will vary with τ_c. Thus if the strength of the fluctuating field, the spectral density function $\mathcal{J}(\omega)$, is plotted against frequency, equal areas are enclosed, but the upper limits vary (see Fig. 1.26). The spectral density is simply the Fourier transform of a correlation function $g(t)$, and a correlation function establishes a correlation between a parameter at time t and at some time later $(t + \tau)$, for example it correlates the isotropic motion of a molecule with time, and its rate constant for the decay is in fact the rotational correlation time, τ_c.

We are assuming that the spectral density is constant for $\omega \ll 1/\tau_c$. This then allows predictions to be made about the variation of T_1, T_2 and W_S with τ_c. For example, as τ_c decreases, ω_0 also decreases, and therefore T_1 increases. At $1/\tau_c \approx \omega_0$ there is a point of inflexion. What is found is shown in Fig. 1.27. For W_0, W_1 and W_2 relaxation we would predict that in this region at the point of inflexion W_2 would fall off first since it is the sum of two transitions. Without deriving them, the quantitative relaxation rates via dipolar coupling are:

$$W_1^I \propto \frac{3\tau_c}{r^6(1+\omega_I^2\tau_c^2)} \propto \frac{3\tau_c}{r^6} \quad \text{(under extreme narrowing)} \tag{1.28}$$

$$W_0 \propto \frac{3\tau_c}{r^6(1+(\omega_I-\omega_S)^2\,\tau_c^2)} \propto \frac{2\tau_c}{r^6} \quad \text{(under extreme narrowing)} \tag{1.29}$$

$$W_2 \propto \frac{12\tau_c}{r^6(1+(\omega_I+\omega_S)^2\,\tau_c^2)} \propto \frac{12\tau_c}{r^6} \quad \text{(under extreme narrowing)} \tag{1.30}$$

Assuming $1/\tau_c \gg \omega_0$ might also be expressed as $\omega_0^2\tau_c^2 \ll 1$, which is called the *extreme narrowing limit*, then the terms above simplify as shown, with the famous $1/r^6$ dependence for NOE. Note, therefore, that for biological macromolecules, this extreme narrowing limit does not hold, and interpretation of the $1/r^6$ dependence is not straightforward.

Although the dipolar interaction, modulated by molecular motion, is the most important relaxation mechanism in solution for protons and other spin-$\frac{1}{2}$ nuclei such as ^{13}C, there are other mechanisms which should be mentioned. Indeed, the overall T_1 is comprised of many contributing relaxation mechanisms:

$$T_1^{-1} = T_{1DD}^{-1} + T_{1SR}^{-1} + T_{1Q}^{-1} + T_{1SC}^{-1} + T_{1CSA}^{-1} \tag{1.31}$$

In addition to dipolar (T_{1DD}) relaxation, there is paramagnetic relaxation, which is a dipolar interaction with the very large magnetic moment of an unpaired electron; and spin-rotation (T_{1SR}) in which there is segmental motion of an aromatic ring, for example. Also, there is quadrupolar relaxation (T_{1Q}), which is due to the coupling of electric field gradients with nuclei possessing a quadrupole moment, scalar coupling relaxation (T_{1SC}),

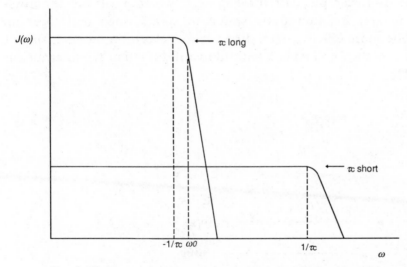

Fig. 1.26 Plot of the spectral density function $\mathcal{J}(\omega)$ against frequency (ω).

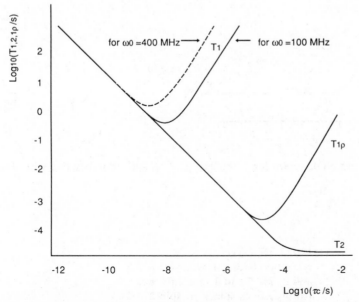

Fig. 1.27 Variations of T_1 and T_2 as a function of τ_c.

which is due to rapidly changing coupling networks; and finally chemical shift anisotropy ($T_{1\text{CSA}}$), which is due to modulation of the chemical shift tensors (σ_{11}, σ_{22}, and σ_{33}) due to molecular motion.

The $^1\text{H}\{^1\text{H}\}$ NOE, which is of prime interest for conformational studies, is $+0.5$ for the extreme motional narrowing situation ($\omega_0^2\tau_c^2 \ll 1$, or $\tau_c \ll \omega_0^{-1}$). For τ_c longer than approximately 1×10^{-9} s it adopts a value of -1.0 (for lower $\mathbf{B}_0$ fields than 11.7 T, the transition between these limiting values will occur at somewhat longer τ_c), as shown in Fig. 1.28. Note that in a network of like spins contained in a macromolecule with $\tau_c \gg \omega_0^{-1}$, *spin diffusion* by two or several subsequent cross-relaxation steps can greatly influence the observed NOE intensities. For example, with three spins, a two-step pathway for cross-relaxation from spin 1 to spin 2 followed by spin 2 to spin 3, may under certain experimental conditions be more efficient than direct cross-relaxation between spins 1 and 3. The NOE on spin 3 is then no longer a faithful manifestation of the internuclear distance $r_{1,3}$ (see Fig. 1.29).

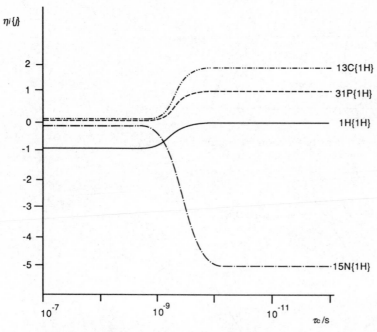

Fig. 1.28 Plots of the maximum NOE versus log τ_c for ^1H, ^{13}C, ^{15}N, and ^{31}P interacting with ^1H.

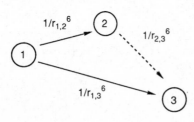

Fig. 1.29 Direct cross-relaxation between two spins 1 and 2, and 1 and 3 spin diffusion pathway from spin 1 via spin 2 to spin 3.

1.1.7 Polarization transfer—the INEPT experiment

In a heteronuclear two-spin AX system, such as ^{13}C—^1H, the energy level diagram looks very similar to that presented in Fig. 1.24. There are four times the population of the ground state for protons as there are for the carbon-13 nuclei, because the protons have four times the Larmor frequency (because of the ratio of the gyromagnetic ratios, γ) and therefore the energy of the transitions is four times greater. Furthermore, the magnetic moment produced by the protons is four times larger than that due to carbon, so that when a 90° pulse is applied to the protons, the transverse magnetization is 16 times that of carbon. Also, the signal precesses at a rate four times faster for protons, leading to a total signal some 64 times greater than for carbon. In general the signal is proportional to γ^3. If this is coupled to the fact that ^{13}C has a low natural abundance (introducing another factor of 100), then it is easy to see why the absolute sensitivities of proton and carbon shown in Table 1.2 are of the magnitude given.

There are two ways in which the polarization of one nucleus can be used to enhance another nucleus, usually by no more than γ (it is not possible to recover the full γ^3). In thermodynamic terms, the *spin temperature* (which is another way of expressing the Boltzmann population) of an abundant spin with a high gyromagnetic ratio can be used to 'warm up' or increase the spin temperature of a rare spin with a low gyromagnetic ratio. In the liquid state the experiment which is used is the 'Insensitive Nuclei Enhanced by Polarization Transfer' or INEPT pulse sequence. In the solid state, as we shall consider in Section 1.2.3, the cross-polarization experiment is used. The basic pulse sequence for the INEPT experiment is shown in Fig. 1.30, where I refers to the source nucleus (e.g. ^1H) and S refers to the destination nucleus (e.g. ^{13}C). Note that the use of the labels I and S is not consistent in the literature, and in the early literature I refers to the 'insensitive' nucleus and S the 'sensitive' nucleus. In this book we will adopt the convention used in Fig. 1.30, although the reader should always be aware of the context when studying the literature on NMR pulse sequences.

The equilibrium density operator is given by:

$$\sigma_0 \approx I_z + (\gamma_S/\gamma_I)S_z \tag{1.32}$$

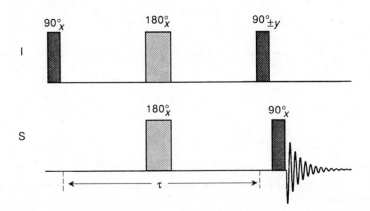

Fig. 1.30 The basic INEPT pulse sequence.

which after the first 90° pulse to the I spins becomes:

$$\sigma_1 \approx -I_y + (\gamma_S/\gamma_I)S_z \tag{1.33}$$

The spins evolve during τ, and the pair of 180° pulses refocuses the chemical shifts, so that the spins undergo the following transformation:

$$-I_y + (\gamma_S/\gamma_I)S_z \xrightarrow{\pi I_x} \xrightarrow{\pi S_x} \xrightarrow{\pi \mathcal{J}_{IS}\tau 2I_zS_z} I_y \cos \pi \mathcal{J}_{IS}\tau - 2I_xS_z \sin \pi \mathcal{J}_{IS}\tau - (\gamma_S/\gamma_I)S_z = \sigma_2 \tag{1.34}$$

Effectively, the 180° I spin pulse removes all chemical shift evolution during the τ period, and the second 180° S spin pulse inverts the S spin labels, ensuring that the I spin vectors continue to precess away from their chemical shift. After the 90° pulse on the I spins (with alternating phase $\pm y$):

$$\sigma_3 \approx I_y \cos \pi \mathcal{J}_{IS}\tau \pm 2I_zS_z \sin \pi \mathcal{J}_{IS}\tau - (\gamma_S/\gamma_I)S_z \tag{1.35}$$

The final 90° pulse on the S spins gives:

$$\sigma_4 \approx I_y \cos \pi \mathcal{J}_{IS}\tau \mp 2I_zS_y \sin \pi \mathcal{J}_{IS}\tau + (\gamma_S/\gamma_I)S_y \tag{1.36}$$

When the two experiments with alternating phase are subtracted, the only term which survives is:

$$\sigma_{\text{obs}} \approx -2I_zS_y \sin \pi \mathcal{J}_{IS}\tau \tag{1.37}$$

If $\tau = 1/2\mathcal{J}_{IS}$, then $\sin \pi \mathcal{J}_{IS}\tau = 1$, and the observed signal is at a maximum. The signal evolves during the detection period, generating antiphase magnetization which is enhanced by a factor of γ_I/γ_S.

The result of this pulse sequence is an antiphase doublet whose intensity is greater than the normal spectrum by a factor of 4 (i.e. by γ_H/γ_C). The result, an example of which is shown in Fig. 1.31, is not perhaps as impressive as one might have predicted. This arises

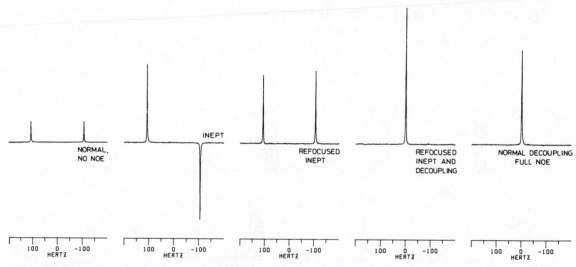

Fig. 1.31 Examples of ^{13}C spectra of chloroform obtained with some variants of the INEPT experiment. (Reprinted from Ref. 7 with permission.)

because heteronuclei intensities are often enhanced by the heteronuclear NOE, which for ^{13}C in the presence of continuous ^1H broadband decoupling is about 3. Therefore, where the INEPT pulse sequence really comes into its own is when the heteronucleus is very insensitive, such as ^{15}N or ^{57}Fe. If a delay is introduced between the last 90° pulse and the start of the acquisition, the antiphase magnetization can be allowed to refocus (i.e. both components of the doublet have the same phase) and this leads to the 'refocused' INEPT, which can also be carried out with broadband decoupling (see Fig. 1.31). The INEPT sequence can also be performed in reverse, in which the population of the ^{13}C for example, is transferred to ^1H. Both variants are the basis for a number of three- and four-dimensional pulse sequences considered in Chapter 2.

1.1.8 $\mathcal{J}$ coupling

When scalar coupling $\mathcal{J}$ occurs between two adjacent spin-$\frac{1}{2}$ nuclei, there is an electron-coupled spin–spin interaction with energy E:

$$E = h\mathcal{J}I_i \cdot I_j \qquad (1.38)$$

The mechanism of this effect is spin pairing of each nucleus to the electrons in the bonds between them, which makes each nucleus sensitive to the possible orientations of the spin of its neighbour. Thus a nucleus coupled to a neighbour with $I = \frac{1}{2}$ will have its line split into a doublet and if $I = 1$, it will be split into a triplet, and so on. The general rule is that the number of lines in a multiplet is $2I + 1$, where I is the spin of the neighbour.

The above rule holds true even if the neighbour is a chemical group with two or more identical nuclei. In this case, the total spin of the group determines the multiplicity, which is given by $2nI + 1$, where n is the number of identical nuclei. The spacing between the lines (in Hz) in a multiplet equals the coupling constant. The intensity of each line in the multiplet is determined by the number of possible combinations of individual spins resulting in a given value of the total spin, and follows a binomial distribution as denoted by Pascal's triangle:

```
              1
            1   1
          1   2   1
        1   3   3   1
      1   4   6   4   1
    1   5  10  10   5   1
```

Note that the equality between the coupling constant and spacing between multiplet lines and the simple intensity relationship apply only when $\mathcal{J} \ll \delta$ (in Hz), which defines a first-order spectrum. If $\mathcal{J} \sim \delta$, more complicated spectral patterns are observed and quantum mechanical analysis of the spectrum is necessary to obtain $\mathcal{J}$ and δ. Unfortunately, like rather a lot of the nomenclature in NMR, there is a somewhat puzzling array of letters used to describe a particular spin system. Each different nucleus is given a letter, e.g. A, B, C, etc. If the nuclei have the same chemical shift, and they are *chemically equivalent nuclei* (as opposed to nuclei with accidental shift degeneracy), subscripts are used, e.g. A_2, B_3, etc. This chemical equivalence can be due to symmetry or rapid rotation or merely by

Table 1.4 Typical ranges of scalar coupling constants

1. ^1H–^1H		
Geminal	^1H—C—^1H	-12 to -15 Hz
Vicinal	^1H—C—C—^1H	$\begin{cases} \text{2–14 Hz} \\ \text{7 Hz (free rotation average)} \end{cases}$
	^1H—C=C—^1H	$\begin{cases} \text{10 Hz (cis)} \\ \text{17 Hz (trans)} \end{cases}$
	^1H—C≡C—^1H	$\sim$2 Hz
	^1H—N—C—^1H	1–10 Hz
Long range		0.5–3 Hz
2. ^1H–^{13}C		
Single bond	^1H—^{13}C (sp^3)	110–130 Hz
Long range	^1H—C—^{13}C	$\sim$5 Hz
	^1H—C=^{13}C	$\sim$2 Hz (ethylene)
3. ^1H–^{15}N		
Single bond	^1H—^{15}N	61 Hz (ammonia)
	^1H—^{15}N	89–95 Hz (peptides)
Long range	^1H—C—^{15}N	15–23 Hz (peptides)
4. ^1H–^{17}O		
Single bond	^1H—^{17}O	17 Hz
Long range	^1H—C—^{17}O	8–10 Hz
5. ^1H–^{31}P		
	^1H—^{31}P	170–230 Hz (trivalent ^{31}P)
		700–900 Hz (pentavalent ^{31}P)
	^1H—O—^{31}P	15–25 Hz
	^1H—C—O—^{31}P	2–20 Hz
6. ^{31}P–^{31}P		
	^{31}P—O—^{31}P	10–30 Hz (pentavalent ^{31}P)
7. ^1H–^{19}F		
	^1H—C—^{19}F	40–50 Hz
	^1H—C–C—^{19}F	5–20 Hz
8. ^1H–^2H		
	^1H—C—^2H	~ -2 Hz
	^2H—C—^2H	$\sim$0.3 Hz

accident. If the nuclei have very different Larmor frequencies, then X, Y are used. Thus, for example, CHCl:CHBr is an AB system, but CHCl:CFBr is an AX system. ClCH$_2$CH$_3$ is an A$_2$B$_3$ system, but ClCH$_2$CF$_3$ is an A$_2$X$_3$ system. When there are more than two well-separated groups of nuclei then the middle letters of the alphabet are used. For example, ^{13}CH$_3$F is an AMX$_3$ system. Examples of calculated AB, AB$_2$, and ABX spectra are given in Fig. 1.32. Typical ranges for scalar couplings are given in Table 1.4.

The magnitude of the three-bond (vicinal) coupling constant $^3\mathcal{J}$ is a function of the dihedral angle θ, originally derived for ^{1}H—C—C—^{1}H coupling from the theoretical calculations of Karplus,[23] and has the form:

$$\mathcal{J} = A + B \cos \theta + C \cos^2 \theta \tag{1.39}$$

where A, B, and C, are coefficients that depend upon the substituent electronegativity. Karplus relations have been used in various forms for calculating bond angles and

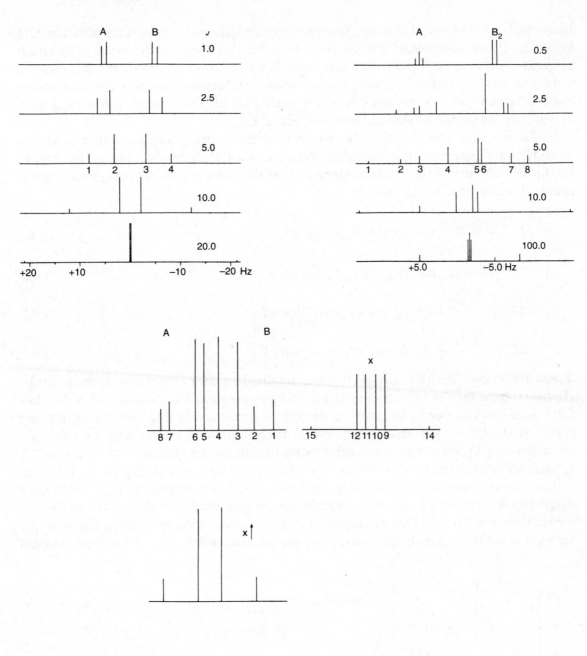

Fig. 1.32 Calculated AB, AB$_2$, and ABX spectra. (Reprinted from R. J . Abraham, J. Fisher, and P. Loftus, *Introduction to NMR Spectroscopy*, John Wiley & Sons, New York (1988) with permission.)

therefore conformations. However, there are two fundamental assumptions which have to be made: (i) that the molecular fragment involved is rigid, and (ii) the limits of precision are unknown (because it is a truncated cosine function, and the coefficients are obtained semi-empirically). Indeed, Karplus himself wrote that 'anyone who uses this relationship for conformational analysis does so at his peril'.[24] With that caveat, it has been used extensively with considerable success (see Fig. 1.33).

So far we have considered scalar coupling without using any particular model to consider what happens to the magnetization under the influence of couplings. In order to do this let us return to the product operator formalism. Spin–spin coupling between two nuclei, 1 and 2, may be described:

$$I_{1x} \xrightarrow{\pi \mathcal{J}_{12} t 2 I_{1z} I_{2z}} I_{1x} \cos \pi \mathcal{J}_{12} t + 2 I_{1y} I_{1z} \sin \pi \mathcal{J}_{12} t \tag{1.40}$$

$$I_{1y} \xrightarrow{\pi \mathcal{J}_{12} t 2 I_{1z} I_{2z}} I_{1y} \cos \pi \mathcal{J}_{12} t - 2 I_{1x} I_{1z} \sin \pi \mathcal{J}_{12} t \tag{1.41}$$

$$2 I_{1x} I_{1z} \xrightarrow{\pi \mathcal{J}_{12} t 2 I_{1z} I_{2z}} 2 I_{1x} I_{1z} \cos \pi \mathcal{J}_{12} t + I_{1y} \sin \pi \mathcal{J}_{12} t \tag{1.42}$$

$$2 I_{1y} I_{1z} \xrightarrow{\pi \mathcal{J}_{12} t 2 I_{1z} I_{2z}} 2 I_{1y} I_{1z} \cos \pi \mathcal{J}_{12} t - I_{1x} \sin \pi \mathcal{J}_{12} t \tag{1.43}$$

These expressions are quite similar to those obtained for the evolution of chemical shifts. However, here for the first time products of two operators are obtained, to which this formalism owes its name. In terms of the vector picture, coupling between spin 1 and spin 2 is shown as two vectors for spin 1, precessing at different angular velocities corresponding to either the α state or the β state of spin 2. The x magnetization of spin 1 is apparently modulated by $\cos \pi \mathcal{J} t$, as is shown in the first part of Equation (1.40). In addition there is a magnetization along the y axis, which is described by $I_{1y}(I_2^\alpha - I_2^\beta)$ (the y magnetization of spin 1 points to $+y$ for the case where spin 2 is in the α state, and to $-y$ for spin 2 in the β state). This y magnetization is not observable, thus, in the macroscopic sense, it is not really a magnetization, since the positive and negative components cancel

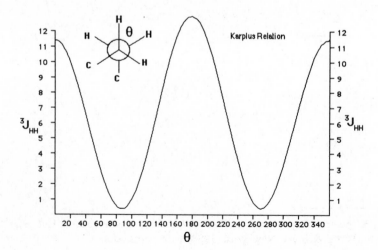

Fig. 1.33 Plot of the three-bond coupling constant $^3\mathcal{J}$ versus the dihedral angle θ.

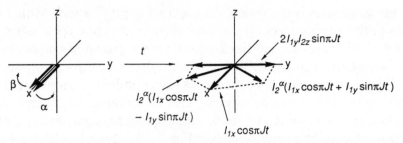

Fig. 1.34 Vectorial representation of the product operator of coupled spins.

exactly. The difference in populations of the α and β levels $(I^{\alpha} - I^{\beta})$ is given by $2I_z$. Hence, $I_{1y}(I_2^{\alpha} - I_2^{\beta})$ is identical to $2I_{1y}I_{1z}$, which is obviously modulated by $\sin \pi \mathcal{J} t$ (see Fig. 1.34).

In contrast to the evolution under chemical shift and/or pulses, the evolution of coupling 'creates' or 'annihilates' operators: $I_{1x} \rightarrow 2I_{1y}I_{2z}$ and $2I_{1x}I_{2z} \rightarrow I_{1y}$. Only coupling and not chemical shift can bring other nuclei into the picture, which is one of the salient features for coherence transfer, since coherence can be transferred only when a second spin is involved in the product operator. Equations (1.40) and (1.41) correspond to the conversion of in-phase magnetization into orthogonal antiphase magnetization, and, starting with antiphase magnetization. Equations (1.42) and (1.43) correspond to the evolution under scalar coupling of in-phase magnetization. This has important consequences for the appearances of multiplets in two-dimensional spectra, which we will consider in Chapter 2.

1.1.9 Populations and coherences

Coherence is a relationship between two states across a single nuclear transition, or multiple states for multiple transitions. A good example of the difference between magnetization and coherence is to consider the case where an NMR transition has been saturated compared with one which has just undergone a 90° pulse. For the saturated transition, there will be no magnetization at all. For the transition which has undergone a 90° pulse, there will be xy components precessing together with the same phase, which they derived from the pulse. In the saturated sample the nuclei are precessing incoherently, and in experiencing a pulse they are precessing with *phase coherence*. For a coherence across one transition this is called a *single-quantum coherence*. When more than one level is involved, then the relationship between the levels is such that *multiple-quantum coherences* may be transferred. However, only single-quantum coherences may be detected by NMR. Thus *coherence* is a generalization of *magnetization*, and in fact corresponds to transverse magnetization. This is in contrast to *polarizations*, which correspond to longitudinal magnetization.

Thus, application of a high-frequency pulse to the z magnetization (spin state I_z) which is initially at its Boltzmann equilibrium produces transverse magnetization (state I_x). During this state the so-called coherence evolves during the detection period time t. This variation of intensity with time can be observed by detection of an induced signal in the receiver coil (I_x, I_y are associated with transverse magnetization). In other words, coherences produced by product operators are displayed at characteristic frequencies.

This concept of the coherence is important. In contrast to populations, which are the occupation of states in the energy level scheme, coherences describe transitions among different energy levels. These are characterized by Δm_I for the quantum number m_I. This change Δm_I is the coherence order p, which is essential for describing phase cycles. Although in general many of the possible coherences occur under the influence of pulses in a pulse sequence, only a few are associated with observable magnetizations, i.e. $\Delta m_I = \pm 1$ (the selection rule for NMR). Of these single-quantum coherences, only a few are detectable under weak coupling conditions (see Fig. 1.35): those in which the state of exactly one spin is changed (i.e. product operators with only one transverse operator).

We can assume that the state of all the spins in the sample (the 'spin ensemble') is a superposition of (i) the populations of the states and (ii) the coherences between the states. These coherences are indexed by their order p. Populations are the same as coherences between a state and itself (e.g. populations are a special type of zero-quantum coherence). The only coherences which induce an observable NMR signal are the (-1)-quantum coherences. Populations are long-lived and do not oscillate; coherences are short-lived and oscillate, and we can refer to coherences + populations as 'normal modes'. The normal modes for a single spin-$\frac{1}{2}$ consist of two energy levels and four normal modes (see Fig. 1.36).

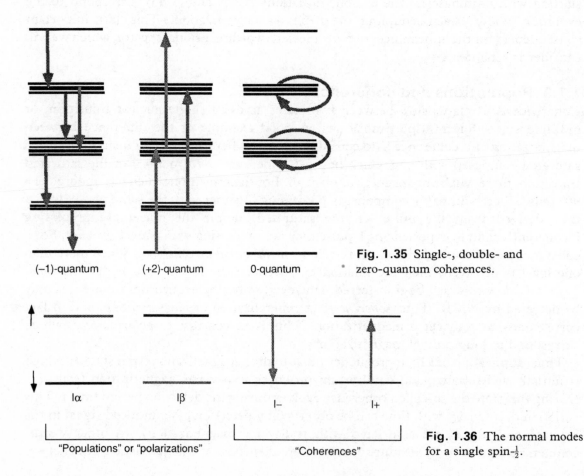

(−1)-quantum   (+2)-quantum   0-quantum

Fig. 1.35 Single-, double- and zero-quantum coherences.

Iα   Iβ   I−   I+

"Populations" or "polarizations"   "Coherences"

Fig. 1.36 The normal modes for a single spin-$\frac{1}{2}$.

The single-quantum coherences can be represented in terms of product operators or vectors, as well as energy levels, as illustrated in Fig. 1.37(a). However, other coherences are also possible in two-spin (or more) systems, and these give rise to zero-quantum and double-quantum coherences, as shown in Fig. 1.37(b). Two-spin coherence terms consist of superpositions of $p=0$ and $p=\pm 2$ quantum coherence. These give rise to the products of two operators which are not observable in the NMR experiment.

1.1.10 Chemical shifts

The chemical shift is one the most basic parameters of NMR, and is generally defined as δ in parts per million (ppm):

$$\delta = \frac{\omega - \omega_0}{\omega_0} \times 10^6 \qquad (1.44)$$

where ω_0 is the Larmor frequency in Hz, and ω is the resonant frequency of the line of

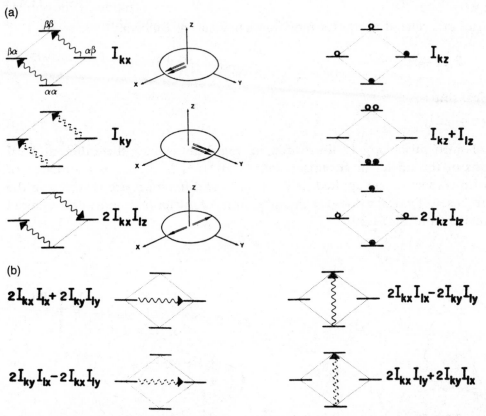

Fig. 1.37 Graphic representation of product operators representing: (a) single-quantum magnetization and longitudinal magnetization; (b) pure zero- and double-quantum coherence, in a two-spin system with $I=\frac{1}{2}$, and each spin operator referred to generally as I_k or I_l. The oscillating x and y magnetization components are represented by wavy lines (with dashed lines for y components) in the energy level diagram, or by the customary vectors in the xy plane of the rotating frame. Populations are represented by open circles for depleted states and filled circles for more populated states (compared with the demagnetized saturated state). (Reprinted from Ref. 19 with permission.)

interest, and note that we have used $\Omega = \omega - \omega_0$ in Section 1.1.3. The origin of the chemical shift is that the moving electric charges of the electron cloud around a nucleus induce a local magnetic field which opposes the applied field. Thus the effective field at the nucleus is:

$$\mathbf{B}_{\text{eff}} = \mathbf{B}_0(1 - \sigma) \tag{1.45}$$

The nucleus is said to be *shielded*, and the extent of the shielding is given by the shielding constant, also called the chemical shift tensor, σ. This is directly related to the electron density ρ at a distance r from the nucleus by Lamb's equation:

$$\sigma = \frac{4\pi e^2}{3mc^2} \int_{-\infty}^{\infty} r\rho(r)\,\mathrm{d}r \tag{1.46}$$

There are three principal components of the shift tensor, σ_{11}, σ_{22}, and σ_{33}, and the isotropic shift tensor σ_{iso} is given by:

$$\sigma_{\text{iso}} = \tfrac{1}{3}(\sigma_{11} + \sigma_{22} + \sigma_{33}) \tag{1.47}$$

The shift tensor σ is related to the Larmor frequency, ω_0, as follows:

$$\omega_0 = \frac{\gamma}{2\pi}\mathbf{B}_0(1 - \sigma) \tag{1.48}$$

and to chemical shift:

$$\delta = 10^6\,(\sigma_{\text{ref}} - \sigma_{\text{sample}}) \tag{1.49}$$

The chemical-shift anisotropy is illustrated in Fig. 1.38, where the different local magnetic fields of the nuclei in an anisotropic C–H bond is shown as a function of orientation with respect to the applied field, $\mathbf{B}_0$. These differences are reflected in the chemical shift tensor. Typical values for chemical shifts of nuclei commonly encountered in biology are shown in Fig. 1.39.

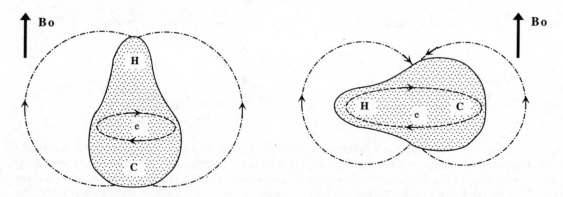

Fig. 1.38 Origins of chemical-shift anisotropy in a C—H bond. The circulation of the electrons in a plane perpendicular to the bond describes a smaller area than circulation in a plane containing the bond. (Redrawn from ref. 12 with permission.)

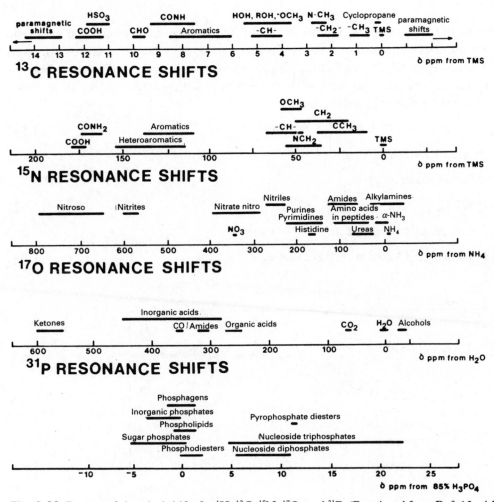

Fig. 1.39 Ranges of chemical shifts for ^1H, ^{13}C, ^{15}N, ^{17}O, and ^{31}P. (Reprinted from Ref. 12 with permission.)

1.2 SOLID-STATE NMR

A number of excellent texts are available in the field of solid-state NMR,[6,25–28] although many are rather dated, somewhat mathematical, and none deal with biological solids.

1.2.1 The source of broad lines in solid samples

Figure 1.40(a) shows the ^{13}C spectrum of crystalline glycine without ^1H decoupling. Given the quality of this spectrum, one might be forgiven for thinking that NMR of solids is not very helpful for obtaining structural information. However, there are a number of methods for obtaining narrow lines in solid-state spectra.

Recall that in its simplest form, the relationship between the wave function Ψ of a quantized particle (i.e. a nucleus) and the energy is given by the Schrödinger equation:

$$\mathcal{H}\Psi = E\Psi \tag{1.50}$$

$\mathcal{H}$ is the Hamiltonian operator, which for a heteronuclear two-spin system in a magnetic field includes a number of contributing Hamiltonians which until now we have expressed in terms of Cartesian product operators. These operators describe all the interactions which the spins experience. In order to understand solid-state NMR, it is more convenient to consider these Hamiltonians:

$$\mathcal{H}_{\text{TOTAL}} = \mathcal{H}_Z + \mathcal{H}_\delta(\theta) + \mathcal{H}_{II}(\theta) + \mathcal{H}_{SS}(\theta) + \mathcal{H}_{IS}(\theta) \tag{1.51}$$

Consider each Hamiltonian in turn:

The Zeeman Hamiltonian, $\mathcal{H}_Z$

$$\mathcal{H}_Z = \gamma \hbar \mathbf{B}_0 I_z \quad \text{where} \quad \omega_0 = \gamma \mathbf{B}_0 \tag{1.52}$$

The chemical-shift Hamiltonian, $\mathcal{H}_\delta(\theta)$

$$\mathcal{H}_\delta(\theta) = \sigma_i \omega_0 I_z + \tfrac{1}{2}(3\cos^2\theta - 1)(\sigma_{33} - \sigma_i)\omega_0 I_z \tag{1.53}$$

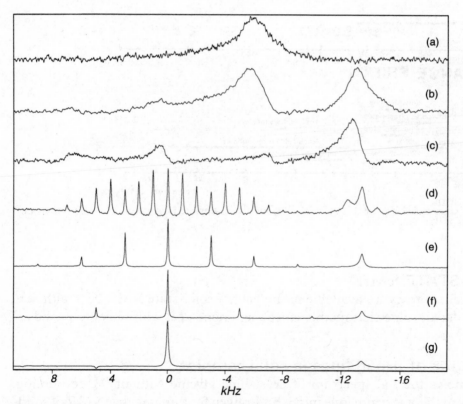

Fig. 1.40 ^{13}C NMR spectra of crystalline glycine: (a) without proton decoupling; (b) with proton decoupling; (c) with proton decoupling and cross-polarization; (d) CPMAS ($\nu_R = 1$ kHz); (e) CPMAS ($\nu_R = 3$ kHz); (f) CPMAS ($\nu_R = 5$ kHz); (g) CPMAS ($\nu_R = 5$ kHz) plus sideband suppression (TOSS).

The homonuclear spin coupling Hamiltonian, $\mathscr{H}_{\mathrm{II}}(\theta)$

$$\mathscr{H}_{\mathrm{II}}(\theta) = \mathscr{J}\mathbf{I}_1 \cdot \mathbf{I}_2 - \tfrac{1}{2}\cos\theta(\mathbf{D}_{x'y'} - \mathbf{D}_{y'x'})(I_{1x}I_{2y} - I_{1y}I_{2x})$$
$$+ \tfrac{1}{2}(3\cos^2\theta - 1)\tfrac{1}{2}(\mathbf{D}_{z'z'} - \mathscr{J})(3I_{1z}I_{2z} - \mathbf{I}_1\mathbf{I}_2) \tag{1.54}$$

The heteronuclear spin coupling Hamiltonian, $\mathscr{H}_{\mathrm{IS}}(\theta)$

$$\mathscr{H}_{\mathrm{IS}}(\theta) = \mathscr{J}I_zS_z + \tfrac{3}{2}(3\cos^2\theta - 1)\tfrac{1}{2}(\mathbf{D}_{z'z'} - \mathscr{J})I_zS_z \tag{1.55}$$

In all these expressions, $\mathbf{D}_{x',y',z'}$ are the principal components of the dipolar coupling tensor in the sample rotation frame, and θ is the angle between the applied magnetic field and the axis of rotation for a pair of dipolar coupled spins (see Fig. 1.41). Note that these Hamiltonians are all time-independent, which means that they are effectively 'stroboscopic', sampling the interactions at one full sample rotation cycle.

1.2.2 Magic-angle spinning

In solution, rapid isotropic motion averages all the θ-dependent terms of the interaction Hamiltonians to zero, leaving the total Hamiltonian as:

$$\mathscr{H}_{\mathrm{TOTAL}} = \mathscr{H}_Z + \mathscr{H}_\delta + \mathscr{H}_{\mathrm{IS}}(\mathscr{J}) \tag{1.56}$$

which is the familiar Hamiltonian that dominates solution-state pulsed NMR. Note that the dipolar coupling D falls to zero.

When the rather interesting condition that $(3\cos^2\theta - 1) = 0$ is met, or $\theta = 54°44'$, the so-called 'magic angle', all the terms in the interaction Hamiltonians containing $(3\cos^2\theta - 1)$ tend to zero. In other words,

$$\mathscr{H}_{\mathrm{TOTAL}} = \mathscr{H}_Z + \mathscr{H}_\delta + \mathscr{H}_{\mathrm{II}}(\theta) + \mathscr{H}_{\mathrm{SS}}(\theta) + \mathscr{H}_{\mathrm{IS}}(\mathscr{J}) \tag{1.57}$$

In practical terms this means spinning the sample at an angle of $54°44'$ with respect to $\mathbf{B}_0$, which is known as magic-angle spinning (MAS), as shown in Fig. 1.42.

The effect of MAS on the spectrum of glycine (in conjunction with cross-polarization—see Section 1.2.3) is illustrated in Fig. 1.40(d–f). Also shown is the use of sideband suppression pulse sequences such as TOtal Sideband Suppression (TOSS),[29] which removes the rotational sidebands without enhancing the intensity of the centre band (as opposed to using higher spinning rate MAS, which does enhance the centre band intensity).

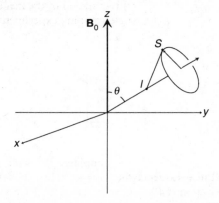

Fig. 1.41 A diagram showing the angle θ between the applied magnetic field $\mathbf{B}_0$ and the sample rotation axis, for an isolated pair of dipolar coupled spins, *IS*.

1.2.3 Cross-polarization

Another trick for obtaining narrow lines in solid-state NMR spectra is to use 'dilute spins'. An example of this is the use of natural abundance ^{13}C, where the odds of there being a ^{13}C within interaction distance of another ^{13}C are so low that the term $\mathcal{H}_{SS}(\theta)$ can be neglected:

$$\mathcal{H}_{\text{TOTAL}} = \mathcal{H}_Z + \mathcal{H}_\delta + \mathcal{H}_{II}(\theta) + \mathcal{H}_{IS}(\mathcal{J}) \tag{1.58}$$

If high-power decoupling is applied to the I spins, then the total Hamiltonian simplifies even further:

$$\mathcal{H}_{\text{TOTAL}} = \mathcal{H}_Z + \mathcal{H}_\delta \tag{1.59}$$

Enhancement of sensitivity for the dilute spins can be achieved through a technique called *cross-polarization*. This technique relies on polarization transfer from the abundant spins, for example ^{1}H, to the rare spins such as ^{13}C, through matching the $\mathbf{B}_1$ RF fields of the ^{1}Hs and the ^{13}Cs. It is analogous to the solution-state INEPT experiment (see Section 1.1.7). This is known as fulfilling the *Hartmann–Hahn condition*, which is as follows:

$$\gamma_H \mathbf{B}_{1H} = \gamma_C \mathbf{B}_{1C} \tag{1.60}$$

where $\mathbf{B}_{1H}$ and $\mathbf{B}_{1C}$ are known as the RF *spin locking* fields. Since $\gamma_H/\gamma_C = 4$, then $\mathbf{B}_{1H} = 4\mathbf{B}_{1C}$, and the sensitivity of the ^{13}C spins are enhanced by the ^{1}H spins by a factor of 4.

The pulse sequence for cross-polarization (CP) is shown in Fig. 1.43. In the first step,

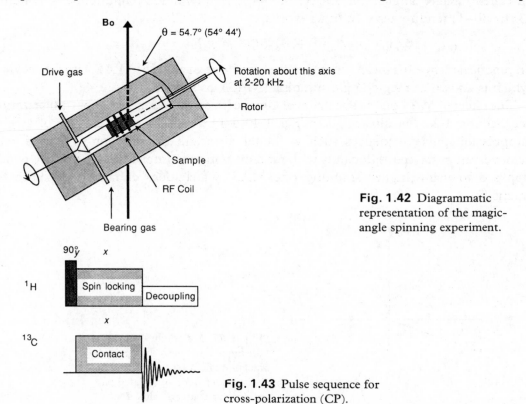

Fig. 1.42 Diagrammatic representation of the magic-angle spinning experiment.

Fig. 1.43 Pulse sequence for cross-polarization (CP).

the proton magnetization is rotated through $90°$ to the x axis and then held there by a 'spin locking' pulse (see Fig. 1.44). The proton spins are kept locked for a time period t, known as the *contact time*. During this time period, a strong on-resonance pulse is applied to the ^{13}C spins which also orient along the x axis. If the Hartmann–Hahn condition is met, and the two spin locking fields are matched, the net carbon magnetization will be enhanced by the proton reservoir. Since the ^{13}C spins are very dilute in the system, they will adopt the more favourable spin distribution of the proton system, while the total proton magnetization will be affected minimally. After the carbon magnetization has built up during the contact time, the carbon field is switched off, and the FID recorded. The proton field is kept on for high-power decoupling. One of the important consequences of this pulse sequence is that the carbon magnetization which yields an FID does not depend on the regrowth of the carbon magnetization in between scans, but arises entirely from contact with the proton spins. This means that the intensity of the carbon spectrum effectively depends on the relaxation of the proton spin system. This relaxation time, for protons spin locked in the rotating frame, is a rather special quantity denoted $T_{1\rho}$. The great advantage of this is that, generally, the ^{1}H relaxation times are much shorter than the ^{13}C relaxation times, so that successive scans may be recycled much faster than for normal ^{13}C acquisitions, yielding better signal-to-noise in a given time period.

When CP is combined MAS, high-resolution spectra, comparable to solution state spectra are obtained. For example, in the ^{13}C NMR spectrum of glycine shown in Fig. 1.40, CP–MAS leads to narrow lines, and when combined with sideband suppression pulse sequences, such as TOSS, give rise to high-resolution spectra.

1.2.4 Shielding anisotropy

By appropriate choice of a coordinate system, the chemical shift tensor σ, which is a three by three matrix, or second-rank tensor, may be converted to three principal elements, σ_{11}, σ_{22}, and σ_{33}, which serve to characterize the three-dimensional nature of shielding (see Section 1.1.10). From these elements we may define the *shielding anisotropy* $\Delta\sigma$ and the shielding *asymmetry parameter* η as follows:

$$\Delta\sigma = \sigma_{33} - \tfrac{1}{2}(\sigma_{11} + \sigma_{22}) \tag{1.61}$$

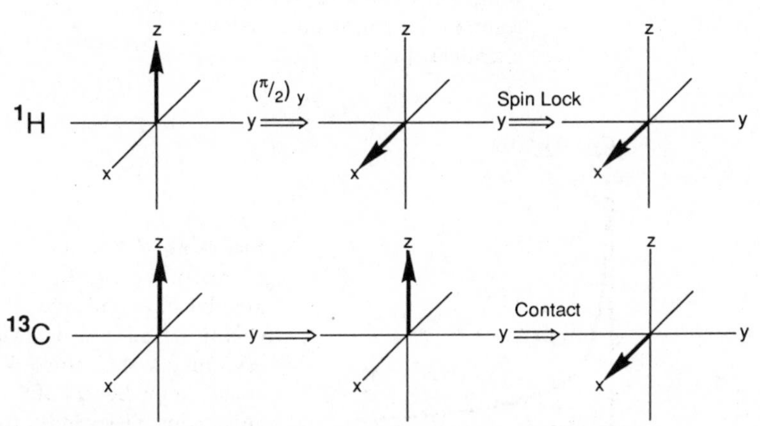

Fig. 1.44 Vector representation for cross-polarization (CP).

$$\eta = \frac{\sigma_{22} - \sigma_{11}}{\sigma_{33} - \sigma_{iso}} \tag{1.62}$$

The principal elements of the shift tensor can be obtained by rotational sideband intensity analysis.[30] The parameter $\Delta\sigma$ provides an estimate of the departure from spherical symmetry of the electrons around the nucleus.

Recall that in solution the isotropic molecular motion averages the shift anisotropy, $\sigma_{iso} = \frac{1}{3}(\sigma_{11} + \sigma_{22} + \sigma_{33})$. In the solid state, the chemical-shift anisotropy will be manifested according to the state of the sample:

(a) In a single crystal in a fixed orientation to the magnetic field, a single sharp line will be observed for each magnetically unique orientation of a particular nucleus with respect to the field direction, and the positions of these lines will change as the orientation of the crystal is changed;

(b) For a powdered sample, signals as in (a) will arise from each random crystallite orientation and a broad line will result, the shape of which will depend on the principal elements of the shielding tensor (see Fig. 1.45).

In case (a) in Fig. 1.46 given below, the principal tensor elements have the values indicated. Another common situation is illustrated in Fig. 1.46(b), in which two of the

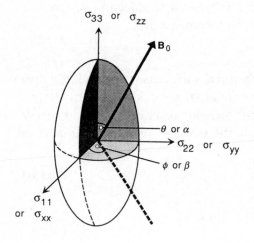

Fig. 1.45 The chemical shielding ellipsoid, which is used to indicate that different orientations of the magnetic field relative to the molecular framework result in different resonance positions for the same chemical species.

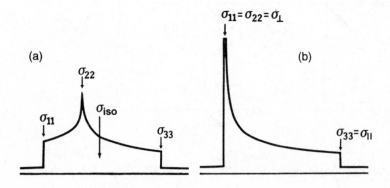

Fig. 1.46 Schematic representation of theoretical powder line shapes for the chemical-shift tensor, (a) asymmetric shift anisotropy, (b) axially symmetric shift anisotropy. (Reprinted from Ref. 28 with permission.)

elements are identical, and the shielding pattern is called *axially symmetric*. The shielding element of the unique axis is described as $\sigma_{\parallel}$ and the other two as $\sigma_{\perp}$. Such a situation can arise either by chance, or by the presence of molecular motion which partially averages the shift anisotropy.

1.2.5 Quadrupolar effects

For quadrupolar nuclei, such as ^{2}H, the interaction Hamiltonian is given by:

$$\mathscr{H}_Q = \frac{eQ}{2I(2I-1)}\, h\mathbf{I}\cdot\mathbf{V}\cdot\mathbf{I} \tag{1.63}$$

where $\mathbf{V}$ is the electric field gradient tensor at the nuclear site, eQ is the quadrupole moment for a single spin with nuclear spin quantum number, I and $\mathbf{I}$ is the angular momentum operator (I_x, I_y, I_z). If it is assumed that the Zeeman interaction is much larger than the quadrupolar interaction, the so-called high-field approximation, which is reasonable for a nucleus such as ^{2}H, then the energy shifts due to the quadrupolar interaction for a single nucleus may be obtained as follows:

$$\Delta E_Q = 3/8\!\left(\frac{e^2qQ}{h}\right)(3\cos^2\theta - 1) \tag{1.64}$$

where ΔE_Q is the energy shift due to the quadrupolar interaction, and θ is the angle between the principal component of the electric field gradient tensor (eq) and the applied magnetic field vector.

The energy level diagram for the combined Zeeman and quadrupolar interactions for a spin-1 nucleus is shown in Fig. 1.47. Note that $m = -1$ and $m = +1$ levels are affected equally and in the same direction by the quadrupolar interaction. The allowed transitions are shown, and the spectrum for a single spin-1 nucleus interacting with an axially symmetric field gradient consists of a doublet with peak separations given by:

$$\Delta v = 3/4\!\left(\frac{e^2qQ}{h}\right)(3\cos^2\theta - 1) \tag{1.65}$$

where (e^2qQ) is the *quadrupole coupling constant*, θ is the Euler angle between the principal axis system and the laboratory frame.

For a polycrystalline material, the energy is averaged over all possible crystal orientations, yielding the spectrum shown, the characteristic *Pake doublet* (see Fig. 1.48).

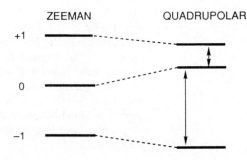

Fig. 1.47 Energy level diagram for the combined Zeeman and quadrupolar interactions for a spin-1 nucleus.

The 'steps' at the wings of the spectrum correspond to $\theta = 0°$, and the central pair of peaks to $\theta = 90°$ and the mid-point to $\theta = 54°44'$ where the two peaks from the quadrupolar splitting coalesce, and the quadrupolar interaction, in effect, vanishes. The total spectrum is the summation of two axially symmetric distributions reversed in sign and corresponding to the two lines in the doublet. The dashed curves indicate the envelopes from the separate contributions of each peak. For a rigid deuteron, the separation of the central doublet is:

$$\Delta v_1 = 3/4\left(\frac{e^2 qQ}{h}\right) \tag{1.66}$$

and that of the outermost steps is:

$$\Delta v_2 = 3/2\left(\frac{e^2 qQ}{h}\right) \tag{1.67}$$

An example of this is shown in Fig. 1.49.

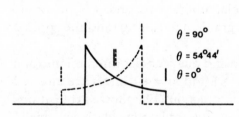

Fig. 1.48 Schematic representation of the powder pattern resulting from the sum of the contributions from the random orientations of a spin-1 nucleus with respect to the magnetic field in a polycrystalline sample. (Reprinted from Ref. 28 with permission.)

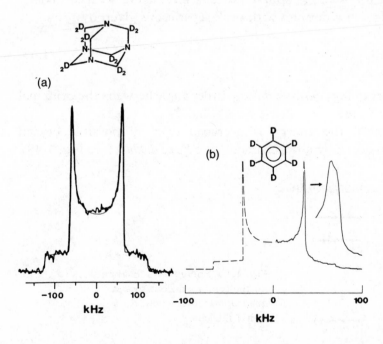

Fig. 1.49 (a) ^{1}H FT spectrum of hexamethylenetetramine-d^{12} fitted to a Pake diagram. (b) Experimental (solid line) and calculated (dashed line) deuterium spectra of polycrystalline benzene at 200 K. (Reprinted from Ref. 28 with permission.)

1.3 KINETICS

1.3.1 The NMR timescale

The NMR timescale is defined by the nuclear precession frequency, and kinetic processes or molecular motions that occur on this time scale are reflected directly in the chemical shift (δ), scalar coupling $(\mathcal{J})$, and relaxation parameters $(T_1$ and $T_2)$. For molecules in solution, there are many situations in which groups of nuclei exchange reversibly between two or more different environments. This exchange may reflect conformational flexibility, chemical reactions, formation of intermolecular complexes, or other events. The timescales in relation to the motional correlation time τ_c which are probed by NMR are summarized in Table 1.5.

Depending on the nature of the exchange process, the different environments undergoing exchange will be characterized by different values of the NMR parameters: chemical shift, coupling constant, and relaxation rates. If the measured variable is chemical shift, with exchange occurring between two environments characterized by shifts δ_A and δ_B, *slow exchange* occurs if the exchange rate $k \ll |\delta_A - \delta_B|$, *intermediate exchange* if $k \approx |\delta_A - \delta_B|$, and *fast exchange* if $k \gg |\delta_A - \delta_B|$. Similar expressions can be derived for the other NMR parameters, and is expressed in terms of lifetimes, $\tau (= 1/k)$ (see Table 1.6).

Note that the three parameters define quite different, although partially overlapping, timescales. For example, chemical-shift differences are of the order of 10–500 Hz (at ^{1}H frequencies of 200–400 MHz), so that fast exchange with respect to chemical shift will be observed for systems in which the lifetime is less than a few milliseconds. Since the chemical shift is field dependent, so is the chemical-shift difference between two environments, and populations of nuclei that appear to be in fast exchange on low-field spectrometers can sometimes be found in the intermediate or even slow exchange region on high-field spectrometers. For protons, typical differences in coupling constants are of the order of 10 Hz, and these coupling constants will be averaged even if the lifetime is as long as ≈ 100 ms. For the much larger coupling constants seen for other nuclei (e.g. ^{15}N—^{1}H) the timescale is comparable to the chemical-shift scale. Much shorter lifetimes, about 10^{-4} s, may be required for relaxation time averaging, since relaxation times of 10^{-3} s can often be observed, especially with macromolecular systems.

Whether a system is in the fast, intermediate, or slow exchange range with respect to chemical shift can often be judged from the appearance of the spectrum. For fast exchange a single resonance line is observed, with a chemical shift that is a weighted average of the individual species:

$$\delta_{obs} = \alpha \delta_A + (1 - \alpha)\delta_B \tag{1.68}$$

where α is the fractional population of species A, for the equilibrium:

$$A \underset{k_b}{\overset{k_a}{\rightleftharpoons}} B$$

In the intermediate exchange region, either single or multiple resonances occur, with

Table 1.5 NMR methods for determining motional correlation times, τ_c

$1 \text{ s} < \tau_c$	Real-time monitoring after initial perturbation
$10 \text{ ms} < \tau_c < 10 \text{ s}$	Two- or three-dimensional exchange spectroscopy (EXSY)
$100 \text{ }\mu\text{s} < \tau_c < 1 \text{ s}$	Lineshape analysis, exchange broadening, and exchange narrowing
$1 \text{ }\mu\text{s} < \tau_c < 10 \text{ ms}$	Measurements of relaxation time $T_{1\rho}$ in the rotating frame
$30 \text{ ps} < \tau_c < 1 \text{ }\mu\text{s}$	Measurements of relaxation time T_1 in the laboratory frame
$\tau_c < 100 \text{ ps}$	Averaged order parameter values

Table 1.6

Timescale	Exchange rate Slow	Intermediate	Fast
δ	$\tau \gg 1/\lvert\delta_A - \delta_B\rvert$	$\tau \approx 1/\lvert\delta_A - \delta_B\rvert$	$\tau \ll 1/\lvert\delta_A - \delta_B\rvert$
J	$\tau \gg 1/\lvert J_A - J_B\rvert$	$\tau \approx 1/\lvert J_A - J_B\rvert$	$\tau \ll 1/\lvert J_A - J_B\rvert$
$1/T_{(1 \text{ or } 2)}$	$\tau \gg 1/\lvert 1/T_A - 1/T_B\rvert$ $\gg \left\lvert \dfrac{T_A T_B}{T_A - T_B} \right\rvert$	$\tau \approx 1/\lvert 1/T_A - 1/T_B\rvert$ $\approx \left\lvert \dfrac{T_A T_B}{T_A - T_B} \right\rvert$	$\tau \ll 1/\lvert 1/T_A - 1/T_B\rvert$ $\ll \left\lvert \dfrac{T_A T_B}{T_A - T_B} \right\rvert$

characteristic lineshapes. Typical spectra[31] reflecting exchange at different rates relative to the chemical shift are shown in Fig. 1.50.

1.3.2 Slow and fast exchange

If a nucleus is changing between two or more environments, its resonance behaviour will be described by the Bloch equations only in the two extreme limits of slow and fast exchange. To take into account the intermediate region, in which exchange rates contribute to relaxation, the Bloch equations must be modified. For a simple two-site exchange, with $\delta_A \neq \delta_B$ but with no spin–spin coupling ($\mathcal{J}_A = \mathcal{J}_B = 0$), and with $T_{2A} = T_{2B} = T_2$, the modified Bloch equations, known as the McConnell equations, may be applied. Additional assumptions are that (i) the lifetime of the transition state is short, so that the exchange rate is determined by the lifetimes in either state, and that (ii) the rate of disappearance of nuclei from either environment follows first-order kinetics. The McConnell equations are:

$$\frac{d\mathbf{M}_{Ax}}{dt} = \gamma(\mathbf{M}_{Ax}\mathbf{B}_0 + \mathbf{M}_{Az}\mathbf{B}_1 \sin \omega t) - \frac{\mathbf{M}_{Ax}}{T_2} - \frac{\mathbf{M}_{Ax}}{\tau_A} + \frac{\mathbf{M}_{Bx}}{\tau_B} \tag{1.69}$$

$$\frac{d\mathbf{M}_{Ay}}{dt} = \gamma(\mathbf{M}_{Az}\mathbf{B}_1 \cos \omega t - \mathbf{M}_{Ax}\mathbf{B}_0) - \frac{\mathbf{M}_{Ay}}{T_2} - \frac{\mathbf{M}_{Ay}}{\tau_A} + \frac{\mathbf{M}_{By}}{\tau_B} \tag{1.70}$$

$$\frac{d\mathbf{M}_{Bx}}{dt} = \gamma(\mathbf{M}_{By}\mathbf{B}_0 + \mathbf{M}_{Bz}\mathbf{B}_1 \sin \omega t) - \frac{\mathbf{M}_{Bx}}{T_2} - \frac{\mathbf{M}_{Bx}}{\tau_B} + \frac{\mathbf{M}_{Ax}}{\tau_A} \tag{1.71}$$

$$\frac{dM_{By}}{dt} = \gamma(M_{Bz}B_1 \cos \omega t - M_{Bx}B_1) - \frac{M_{By}}{T_2} - \frac{M_{By}}{\tau_B} + \frac{M_{Ay}}{\tau_A} \qquad (1.72)$$

where $M_{A,B,x,y,z}$ is the expectation value for the bulk magnetization along a particular Cartesian axis in the rotating frame, B_0 is the static magnetic field, B_1 is the applied RF field, ω is the Larmor frequency, and τ_A and τ_B are the lifetimes of the nucleus in each site. Thus the total rate of change of the nuclear magnetization in state A will depend on the rate of change of the magnetization in state B. The general solution to the McConnell equations is rather complex:

$$f(\omega) = \tfrac{1}{2}\gamma M_0 \frac{(1 + \tau/T_2) + QR}{P^2 + R^2} \qquad (1.73)$$

where

$$\tau = \tau_A\tau_B/(\tau_A + \tau_B) \qquad (1.74)$$

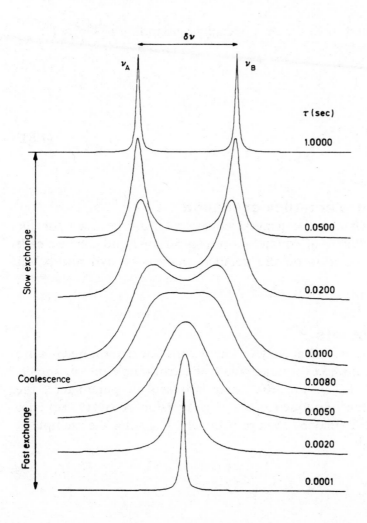

Fig. 1.50 Typical spectra reflecting exchange at different rates relative to the chemical shifts. (Reprinted from Ref. 31 with permission.)

is the mean lifetime, and

$$P = \tau(1/T_2)^2 - [\tfrac{1}{2}(\omega_A + \omega_B) - \omega]^2 + \tfrac{1}{4}(\omega_A - \omega_B)^2 + 1/T_2 \qquad (1.75)$$

$$Q = \tau[\tfrac{1}{2}(\omega_A + \omega_B) - \omega - \tfrac{1}{2}(p_A - p_B)(\omega_A - \omega_B)] \qquad (1.76)$$

$$R = [\tfrac{1}{2}(\omega_A + \omega_B) - \omega](1 + 2\tau/T_2) + \tfrac{1}{2}(p_A - p_B)(\omega_A - \omega_B) \qquad (1.77)$$

where p_A and p_B are the fractional populations of nuclei A and B.

This rather complicated equation can be simplified considerably if we consider effective transverse relaxation rates of equally populated sites (which are related to linewidth) for slow exchange:

$$\frac{1}{T_{2obs}} = \frac{1}{T_{2A}} + k \qquad (1.78)$$

and for fast exchange:

$$\frac{1}{T_{2obs}} = \frac{1}{T_2} + \frac{(\omega_A - \omega_B)^2}{8k_a} \qquad (1.79)$$

If the sites are unequally populated, then for slow exchange (assuming $p_B \ll p_A \approx 1$):

$$\frac{1}{T_{2obs}} = \frac{1}{T_{2A}} + k_a p_B \qquad (1.80)$$

and for fast exchange:

$$\frac{1}{T_{2obs}} = \frac{p_A}{T_{2A}} + \frac{p_B}{T_{2B}} + \frac{p_A p_B(\omega_A - \omega_B)^2}{k_a + k_b} \qquad (1.81)$$

1.3.3 Intermediate and more complex exchange

In the case of intermediate exchange the general equations must be used for the calculations (Equation (1.73)). This is of course, more complicated and the reader is referred to a number of excellent texts on the detailed analysis of dynamic NMR spectra.[32–34]

1.3.4 Measuring rate constants

The value of measuring rate constants is that enzyme kinetics may be determined *in vitro* and also *in vivo*.[35] This may be achieved experimentally by determining either δ_A and δ_B, $1/T_{2A}$ and $1/T_{2B}$, or $1/T_{1A}$ and $1/T_{1B}$. For the latter two sets of parameters, the Carr–Purcell–Meiboom–Gill spin echo sequence, and the inversion–recovery sequence (and saturation transfer or NOESY) may be used respectively. Consider the example of ligand binding to a macromolecule:

$$A + B \underset{k_{-1}}{\overset{k_1}{\rightleftharpoons}} AB$$

Let the parameters being measured be P, so that

$$P_{obs} \approx P_A p_A + P_{AB} p_{AB} \tag{1.82}$$

where $p_A = [A]/A_{total}$, $p_{AB} = [AB]/A_{total}$, and $A_{total} = [A] + [AB]$ and $B_{total} = [B] + [AB]$. Let $\Delta_0 = P_{AB} - P_A$ and $\Delta = P_{obs} - P_A$. Since $p_A + p_{AB} = 1$,

$$\Delta = \Delta_0 p_{AB} \tag{1.83}$$

Also at equilibrium,

$$K_d = \frac{[A][B]}{[AB]} = \frac{k_{-1}}{k_1} \tag{1.84}$$

So we can write p_{AB} as:

$$\frac{[B]}{[B] + K_d} \quad \text{or} \quad \frac{[A]B_{total}}{A_{total}([A] + K_d)} \tag{1.85}$$

If the experiment is carried out at constant A_{total}, and B_{total} is varied, then

$$\Delta = \frac{\Delta_0 [B]}{[B] + K_d} \tag{1.86}$$

which is of the same form as the Michaelis–Menten equation, so K_d can readily be determined. If A_{total} is varied, the equations are less simple, although if $[A] \gg [AB]$ then

$$\Delta = \frac{\Delta_0 B_{total}}{A_{total} + K_d} \tag{1.87}$$

Note that for the simple acid–base equilibrium:

$$A^- + H^+ \underset{k_{-1}}{\overset{k_1}{\rightleftharpoons}} AH$$

the equation has the same form:

$$\Delta = \frac{\Delta_0 [H^+]}{[H^+] + K_a} \tag{1.88}$$

so if Δ is plotted against pH, $pK_a = pH$ when $\Delta = \Delta_0/2$.

1.3.5 Saturation transfer

In the case of slow exchange, irradiation of resonance A will cause changes in intensity of the other resonances (B) because of *transfer of saturation* from one to the other as a result of exchange (provided that the exchange rate is comparable to the longitudinal relaxation rate $1/T_1$). This is exactly analogous to the NOE, and a one-dimensional saturation transfer experiment is analogous to the one-dimensional NOE difference experiment. The fractional change in intensity of the B resonance is given by the equation:

$$\frac{I_{B'}}{I_B} = \frac{1/T_{1B}}{1/T_{1B} + k_b} \tag{1.89}$$

where I_B and $I_{B'}$ are the intensities of the B resonance before and after irradiation of the A resonance respectively. Note that

$$\frac{1}{T_{1\text{obs}}} = \frac{1}{T_{1A}} + \frac{p_B k_b}{T_{1B}} \tag{1.90}$$

An example is shown in Fig. 1.51.

1.3.6 Two-dimensional exchange spectroscopy

The two-dimensional NOESY pulse sequence may be used to measure cross-relaxation due to chemical exchange as well as due to dipolar coupling. The pulse sequence is essentially identical to the NOESY although the name EXSY (for EXchange SpectroscopY) has been argued to be more appropriate for this usage[36]. An example is shown in Fig. 1.52. The cross-peak intensities may be used to calculate site-to-site rate constants, although the intensities need to be evaluated as a function of τ_m (mixing time):

$$I_{ij}(\tau_m) = [\delta_{ij} - \tau_m R_{ij} + \tfrac{1}{2}\tau_m^2 \sum_k R_{ik} R_{kj} - \cdots]M_j^0 \tag{1.91}$$

where M_j^0 refers to the initial magnetization of j, $R_{ij} = -k_{ji}$, and are off-diagonal elements of the exchange matrix (**R**), whose diagonal elements are $R_{ii} = T_{1,i}^{-1} + \Sigma_l k_{il}$. This matrix contains all the site-to-site rate constants. Furthermore $I_{ij} = I_{ji}$ even if $R_{ij} \neq R_{ji}$, which implies that the two-dimensional EXSY spectrum must be symmetric about the diagonal, and that unequal populations do *not* lead to asymmetry. At short mixing times, Equation (1.91) simplifies to:

$$I_{ij}(\tau_m) \approx -\tau_m R_{ij} M_j^0 = k_{ji} \tau_m M_j^0 \tag{1.92}$$

for $i \neq j$. Now the two-dimensional spectrum becomes a graphic display directly

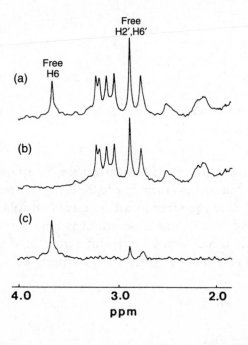

(a) Free H6 Free H2′,H6′

(b)

(c)

4.0   3.0   2.0
ppm

Fig. 1.51 One-dimensional saturation transfer ¹H NMR spectra of dihydrofolate reductase (a) without irradiation; (b) with irradiation at the resonance frequency of the 6-proton in the enzyme-free ligand trimethoprim; and (c) the difference spectrum (showing resonances due to the 6-proton of the enzyme-bound ligand trimethoprim, which is relatively broad, and within the free ligand). (Reprinted from Cayley et al., *Biochemistry*, **18**, 3886 (1979) with permission.)

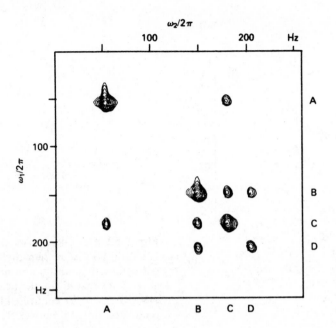

Fig. 1.52 Two-dimensional exchange spectrum of the protons in heptamethyl benzenonium ion in 9.4 M H_2SO_4. (Reprinted from Ref. 4 with permission.)

proportional to the exchange matrix **R**. The complexity of the exact equation (1.91) arises because cross-peaks between i and j can arise indirectly, by exchange from i to k and from k to j before the mixing period has expired. However, at very short τ_m these indirect cross-peaks vanish, so that the equations simplify. Equation (1.92) therefore permits solving for the rate constants, but it is inherently inaccurate, since at short values of τ_m the cross-peak intensities are small and therefore difficult to measure. Furthermore, at τ_m values that are practical, there is a risk of indirect cross-peaks which could be mistaken for evidence for a direct exchange process. A solution to this is to use several short τ_m values and determine k_{ij} as the slope of a plot of $I_{ij}(\tau_m)$ versus $M_j^0\tau_m$. This distinguishes direct cross-peaks, which have a non-zero slope, from the indirect ones, which show zero slope. However multiple τ_m experiments can consume large amounts of instrument time and using the general solution to Equation (1.91) is preferable. For simple two-site exchange between two uncoupled systems of spins A and B, assuming equal populations and $T_{1,A} = T_{1,B} = T_1$, and exchange rate constant, $k = k_{AB} + k_{BA}$, the intensities of the diagonal peaks and cross-peaks reduce to:

$$k = \frac{1}{\tau_m} \ln \frac{r+1}{r-1} \tag{1.93}$$

where $r = (I_{AA} + I_{BB})/(I_{AB} + I_{BA})$, or with unequal populations $r = 4p_A p_B(I_{AA} + I_{BB})/(I_{AB} + I_{BA}) - (p_A - p_B)^2$. By simply knowing the cross-peak intensity ratio, the rate constant for chemical exchange can be calculated. Of course, the great value of the two-dimensional experiment is its ability to measure multisite exchange rate constants, for which the two-site approximation is of no help, and the more complex matrix analysis is required (see Section 3.8).

An alternative approach to the EXSY experiment, which is effectively a three-dimensional NMR experiment (the three time variables being t_1, t_2, and τ_m), is the

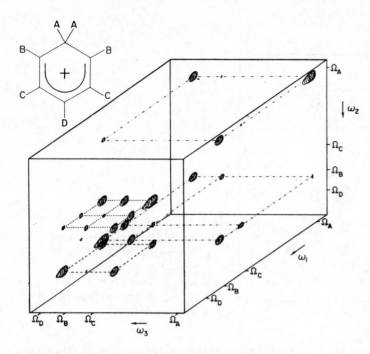

Fig. 1.53 Three-dimensional EXSY–EXSY of heptamethyl-benzenonium ion in sulphuric acid. Peaks that lie on identical planes (ω_1, ω_3) are connected by lines. (Reprinted from Ref. 38 with permission.)

so-called 'ACCORDION' experiment, in which t_1 and τ_m are varied together, thus reducing the experiment to a two-dimensional experiment.[37] The full three-dimensional EXSY–EXSY experiment can also be used as illustrated in Fig. 1.53 for heptamethylbenzenonium sulphate in sulphuric acid. The possible exchange processes are due to the 1,2 methyl group shifts: A⇌B, B⇌C, and C⇌D.[38]

1.4 MOLECULAR MOTION
1.4.1 Temperature dependence of chemical shifts
Note that *prima facie* it is not easy to tell if a particular line is affected by chemical exchange or not, particularly when approaching fast exchange. Thus it will not necessarily be obvious from examination of the resonances in a ^{1}H NMR spectrum of a protein whether a particular line is exchange broadened or shifted. Such exchange might be due to exchange with solvent, exchange from ionization, exchange from ligand binding, or exchange from conformational equilibria. Examination of the temperature dependence of the spectrum can help to reveal exchange, and furthermore may be used to examine molecular dynamics, for example the flipping of an aromatic ring (see Fig. 1.54).

1.4.2 Use of relaxation rates and NOEs
T_1s, T_2s, and NOEs may be used for measuring molecular motion in the form of local librational motion. For example, when ^{13}C T_1 measurements were made on sonicated dipalmitoyl lecithin suspensions, the results shown in Fig. 1.55 were obtained. The relaxation is dominated by the dipolar interaction from the hydrogens on the same bond. Apart from the carboxyls, which have no attached hydrogens, there is an increase in T_1, both towards the CH_3 group and towards the $(CH_3)_3N^+$ group. This means that there is increased mobility towards the ends of these chains.

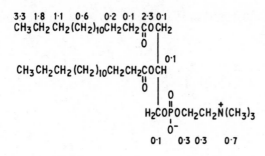

3·3 1·8 1·1   0·6   0·2 0·1 2·3 0·1
CH₃ CH₂ CH₂ (CH₂)₁₀CH₂CH₂ C̈O CH₂
                              ‖
                              O
                                        | 0·1
CH₃ CH₂ CH₂ (CH₂)₁₀CH₂CH₂COCH
                              ‖
                              O   O
                                     ‖
                        H₂COP̈OCH₂CH₂N̈(CH₃)₃
                            |
                            O⁻
                        0·1   0·3 0·3   0·7

Fig. 1.54 The effect of temperature on the spectrum of 8 mM hen lysozyme contaning 4.1 mM GlcNAc at pH 5.3. Temperatures are: (a) 55°C; (b) 45°C; (c) 37°C. (Reprinted R. A. Dwek (ed.), *NMR in Biology*, Academic Press, Orlando FL (1977) with permission.)

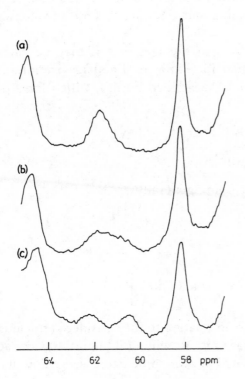

(a)

(b)

(c)

```
  6·4        6·2        6·0      5·8  ppm
```

Fig. 1.55 ¹³C T_1 values for sonicated dipalmitoyl lecithin suspensions.

1.4.3 Second moment analysis in solids

The quadrupolar interactions are sensitive to the presence of molecular motion. Indeed, this is perhaps the most important application of ²H NMR in biological studies. The parameters most often used to define the Pake doublet are the *moments* of the spectrum. The jth moment, M_j, is defined (in terms of angular frequency) for a lineshape $f(\omega)$ as:

$$M_j = \frac{\displaystyle\int_{-\infty}^{\infty} (\omega - \langle\omega\rangle)^j f(\omega)\, d\omega}{\displaystyle\int_{-\infty}^{\infty} f(\omega)\, d\omega} \tag{1.94}$$

where the denominator is a normalizing factor and $\langle\omega\rangle$ is the mean angular frequency of the line:

$$\langle\omega\rangle = \frac{\int_{-\infty}^{\infty} \omega f(\omega)\, d\omega}{\int_{-\infty}^{\infty} f(\omega)\, d\omega} \qquad (1.95)$$

If $f(\omega)$ is an even function of ω, $M_j = 0$ for all odd j, as is the case for dipolar interactions in high magnetic fields. Usually only the second moment, M_2 is used. Thus the moment of a curve is its mean squared width.

The importance of the second moment lies in the fact that it may be calculated theoretically if the lattice positions of the nuclei are known. The pairwise interactions summed over the lattice for a single crystal are described by M_2, which for a dipolar interaction is given by:

$$M_2(\text{heteronuclear}) = [\tfrac{1}{2}(2\pi D)(1 - 3\cos^2\theta)]^2 \qquad (1.96)$$

$$M_2(\text{homonuclear}) = [\tfrac{3}{4}(2\pi D)(1 - 3\cos^2\theta)]^2 \qquad (1.97)$$

where D is the dipolar coupling constant, or, more generally:

$$M_2(\text{heteronuclear}) = \left(\frac{1}{2}\gamma_A\gamma_B\hbar\,\frac{\mu_0}{4\pi}\right)^2 \sum_k \left[(1 - 3\cos^2\theta_{jk})^2/r_{jk}^6\right] \qquad (1.98)$$

$$M^2(\text{homonuclear}) = \left(\frac{3}{4}\gamma^2\hbar\,\frac{\mu_0}{4\pi}\right)^2 \sum_k \left[(1 - 3\cos^2\theta_{jk})^2/r_{jk}^6\right] \qquad (1.99)$$

where the sum runs over all relevant nuclei, k, in relation to the considered nucleus j. As we have discussed before, any rapid molecular motions will tend to partially average the dipolar interactions, so the second moment will appear to be decreased below its rigid

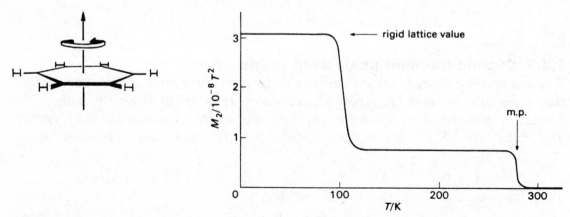

Fig. 1.56 Variation of the intramolecular homonuclear contribution to the second moment of the proton resonance of benzene as a function of temperature. (Reprinted from Ref. 39 with permission.)

lattice value. Consider a simple molecular motion such as rotation about an axis fixed in the molecule, for example the proton spectrum of solid benzene rotating about the plane of the ring is shown in Fig. 1.56.

The variation of M_2 with temperature is also shown in Fig. 1.56.[39] Some molecular motion at ≈ 100 K causes a reduction in the second moment (by a factor of 4). Incidentally, a moment analysis can be used to obtain the chemical-shift anisotropy parameters $\delta(=(\sigma_{33}-\sigma_1))$ and $\eta(=(\sigma_{11}-\sigma_{22})/(\sigma_{33}-\sigma_i))$:

$$M_2 = (\delta^2/15)(3+\eta^2) \tag{1.100}$$

and the third moment,

$$M_3 = (2\delta^3/35)(1-\eta^2) \tag{1.101}$$

1.4.4 Quadrupolar nuclei

One of the problems of solid-state ^{2}H NMR is the high RF powers that are involved, and the possibility of distortion to the spectra as a result of what is known as 'breakthrough' (which is the receiver detecting the end of the transmitter pulse, since at high power levels it takes some time for the RF coil to dissipate the power). One pulse sequence which helps in this area is related to the Hahn spin echo sequence, and is called the quadrupolar echo sequence:

$$(\pi/2)_x - \tau - (\pi/2)_y - 2\tau - t_2 - (\text{acquisition})$$

Unfortunately, because spin-1 nuclei have three available spin states, it is not possible to represent the behaviour of the spins in terms of simple magnetization vectors as is possible for spin-$\frac{1}{2}$ nuclei. However, the effect of the sequence is highly analogous to the Hahn spin echo sequence for spin-$\frac{1}{2}$ nuclei. The time-dependent behaviour of the spins is reversed and the spins are refocused to give an echo. The important feature of this application of echo formation is that the loss of critical data points from the beginning of the FID (caused by the receiver dead-time and pulse breakthrough) is avoided in the echo, which is well removed in time from the RF pulse. Since these initial data points are crucial in defining the outer limits of the spectrum, a dramatic improvement in spectral lineshape is observed (see Fig. 1.57).

When molecular motion is present, the situation can become more complex, and the detailed spectral shape may be dependent on the pulse interval used to produce the echo (2τ), and may be used to ascribe a mechanism to the motion.

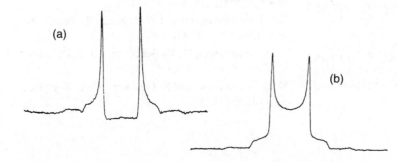

(a)

(b)

Fig. 1.57 Spectra of perdeuterated Perspex (plexiglass) illustrating the effect of the quadrupolar echo sequence. (a) Spectrum obtained using $\pi/2$ pulses. (b) Spectrum obtained with quadrupolar echo sequence. (Reprinted from Ref. 28 with permission.)

REFERENCES

1. B. Jacobson, W. A. Anderson, and J. T. Arnold, *Nature*, **17**, 772–3 (1954).

2. M. Saunders, A. Wishnia, and J. G. Kirkwood, *J. Am. Chem. Soc.*, **79**, 3289 (1957).

3. A. Abragam, *Principles of Nuclear Magnetism*, Oxford University Press (1961).

4. R. R. Ernst, G. Bodenhausen, and A. Wokaun, *Principles of Nuclear Magnetic Resonance in One and Two Dimensions*, Oxford University Press (1987).

5. D. M. S. Bagguley (ed.). *Pulsed Magnetic Resonance: NMR, ESR, and Optics (A Recognition of E. L. Hahn)* Oxford University Press (1992).

6. C. P. Slichter, *Principles of Magnetic Resonance*, 2nd edn, Springer-Verlag, Berlin (1978).

7. A. E. Derome, *Modern NMR Techniques for Chemistry Research*, Pergamon, Oxford (1987).

8. J. K. M. Sanders and B. K. Hunter, *Modern NMR Spectroscopy, A Guide for Chemists*, 2nd edn, Oxford University Press (1993).

9. R. A. Dwek, *NMR in Biochemistry*, Clarendon Press, Oxford (1973).

10. T. L. James, *Nuclear Magnetic Resonance in Biochemistry*, Academic Press, New York (1975).

11. K. Wüthrich, *NMR in Biological Research: Peptides and Proteins* American Elsevier, New York (1976).

12. O. Jardetsky and G. C. K. Roberts, *NMR in Molecular Biology*, Academic Press, New York (1981).

13. K. Wüthrich, *NMR of Proteins and Nucleic Acids* John Wiley and Sons, New York (1986).

14. P. G. Morris, *NMR Imaging in Medicine and Biology*, Oxford University Press (1986).

15. L. Kaufman, L. E. Crooks, and A. R. Margulis, *NMR Imaging in Medicine*, Igakun-shoin, Tokyo (1981).

16. J. W. Shriver, *J. Magn. Reson.*, **94**, 612 (1991).

17. P. Güntert, N. Schaefer, G. Otting, and K. Wüthrich, *J. Magn. Reson. Series A*, **101**, 103–5 (1993).

18. See, for example: G. W. Vuister, R. Boelens, R. Kaptein, R. E. Hurd, B. John, and P. C. M. Van Zijl, *J. Am. Chem. Soc.*, **113**, 9688–90 (1991) and references cited therein.

19. O. W. Sørensen, G. W. Eich, M. H. Levitt, G. Bodenhausen, and R. R. Ernst, *Prog. in NMR Spectros.*, **16**, 163–92 (1983).

20. R. G. Brewer and E. L. Hahn, *Sci Am.*, **251**, 50–7 (1984).

21. I. Solomon, *Phys. Rev.*, **99**, 559 (1955).

22. D. Neuhaus and M. Williamson, *The Nuclear Overhauser Effect in Structural and Conformational Analysis*, VCH Publishers, New York (1989).

23. M. Karplus, *J. Chem. Phys.*, **30**, 11 (1959).

24. M. Karplus, *J. Am. Chem. Soc.*, **85**, 2870 (1963).

25. M. Mehring, *High Resolution NMR in Solids*, 2nd edn, Springer-Verlag, New York (1983).

26. U. Haeberlin, *High Resolution NMR in Solids: Selective Averaging*, Academic Press, New York (1976).

27. B. C. Gerstein and C. R. Dybrowski, *Transient Techniques in NMR of Solids*, Academic Press, Orlando FL (1985).

28. C. A. Fyfe, *Solid State NMR for Chemists*, CFC Press, Guelph (1983).

29. W. T. Dixon, *J. Magn. Reson.*, **44**, 220 (1981).

30. J. Herzfeld and A. E. Berger, *J. Chem. Phys.*, **73**, 6021–30 (1980).

31. H. Günther, *NMR Sepctroscopy*, John Wiley and Sons, New York (1980).

32. L. M. Jackman and F. A. Cotton (eds.) *Dynamic NMR Spectroscopy*, Academic Press, New York (1975).

33. J. L. Kaplan and G. Fraenkel, *NMR of Chemically Exchanging Systems*, Academic Press, New York (1980).

34. J. Sandström, *Dynamic NMR Spectroscopy*, Academic Press, New York (1982).

35. K. M. Brindle, *Prog. in NMR Spectroscopy*, **20**, 257–93 (1988).

36. C. L. Perrin and T. J. Dwyer, *Chem. Rev.*, **90**, 935–67 (1990).

37. G. Bodenhausen and R. R. Ernst, *J. Am. Chem. Soc.*, **104**, 1304 (1982).

38. C. Griesinger, O. W. Sørensen, and R. R. Ernst, *J. Magn. Reson.*, **84**, 14–63 (1989).

39. E. R. Andrew and R. G. Eades, *Proc. Roy. Soc. A.*, **218**, 537 (1953).

2 · Methods for spectral assignment— multidimensional NMR

The first step in biological NMR spectroscopy of complex biomolecules, in particular by ^{1}H NMR, is the assignment of the spectrum. For this, a general strategy for identification and assignment of resonances is usually adopted, which employs a number of two-dimensional or higher (three-dimensional, four-dimensional) NMR experiments. For large biomolecules these approaches fail and we will consider alternative methods in Chapters 3 and 4 for such situations.

2.1 TWO-DIMENSIONAL NMR METHODS

2.1.1 Adding the second dimension

The essence of all two-dimensional NMR experiments is the modulation of a normal spectrum as a function of a variable time interval, t_1. Consider a sample containing a single line, with chemical shift v. After a $(\pi/2)_x$ pulse, ignoring longitudinal relaxation but not transverse relaxation, the line is allowed to precess at v Hz for time t_1, followed by a second $(\pi/2)_x$ pulse (see Fig. 2.1). The vector has precessed through an angle $2\pi v t_1$ during the interval, so if its length is $\mathbf{M}$, then the component along the y axis is $\mathbf{M} \cos 2\pi v t_1$ and that along the x axis is $\mathbf{M} \sin 2\pi v t_1$, by simple trigonometry. The value of $\mathbf{M}$ is related to the initial magnetization $\mathbf{M}_0$ by:

$$\mathbf{M} = \mathbf{M}_0 e^{-t_1/T_2^*} \tag{2.1}$$

by definition of T_2^*. The second $(\pi/2)_x$ pulse rotates the y axis component through a further 90° to place it along the z axis, while leaving the x axis component unchanged. So

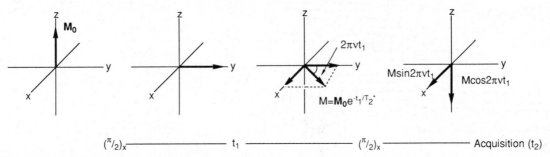

$(^\pi/_2)_x$ ——————— t_1 ——————— $(^\pi/_2)_x$ ——————— Acquisition (t_2)

Fig. 2.1 The simple COSY experiment: amplitude modulation of the NMR signal is induced by varying the interval, t_1, between the pulses.

the amount of magnetization left in the xy plane, which determines the size of the NMR signal, is $\mathbf{M} \sin 2\pi v t_1$. The spectrum obtained is normal apart from this change in amplitude. If we now perform a number of experiments with different values of t_1, starting from 0 up to several seconds, then this corresponds to discrete sampling of the t_1 time interval. If we transformed this into a spectrum, we would see a peak which oscillated sinusoidally (because the transverse magnetization is given by $\mathbf{M} \sin 2\pi v t_1$ with a frequency v, as a function of t_1 (see Fig. 2.2). If we select a point from each spectrum corresponding with the top of the peak and plot this as a function of t_1, we obtain an FID oscillating with frequency v (see Fig. 2.3). Fourier transformation of these data generates another spectrum just as it did for transformation of the first set of FIDs. This is a two-dimensional data set, and the two-dimensional FT converts the data to a two-dimensional frequency spectrum $f(v_1, v_2)$. Now both v_1 *and* v_2 correspond to the chemical shift v, and we obtain a square spectrum, with a peak on the diagonal at (v, v). Cross-sections through the peak along either axis are Lorentzian with linewidth $1/\pi T_2^*$ (see Fig. 2.4).

2.1.2 COSY

The basic experiment outlined in Section 2.1.1 was the COSY experiment introduced by Jeener in 1971. In a coupled system, the second pulse of the experiment causes magnetization which arose from one transition during t_1 to be redistributed amongst all the others with which it is associated.

We have seen how two-dimensional data sets are created, and this applies equally well

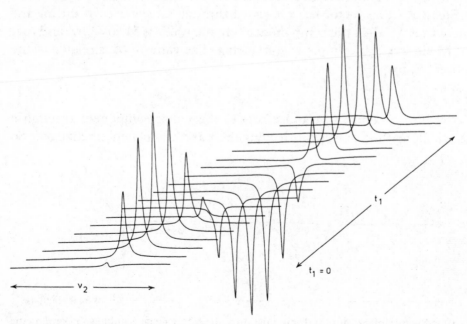

Fig. 2.2 The spectra obtained using the sequence shown in Fig. 2.1 with a variable t_1. (Reprinted from ref. 7, Ch. 1 with permission.)

to all two-dimensional NMR experiments. Now let us turn to the particular details of the sequence $90° - t_1 - 90°$, to see how, in the presence of homonuclear coupling, it gives us information about coupling networks. For the example of a single line free from coupling, this sequence generates a square spectrum with a peak on the diagonal which is not particularly helpful because it contains no more information than an ordinary spectrum. There was no difference between the situation during the intervals t_1 and t_2; the magnetization simply evolved with its chemical shift during each time period. For two-dimensional experiments to be informative, it is necessary for at least part of the magnetization to evolve with different frequencies during the two time intervals. With this sequence of pulses, this may arise in the presence of coupling, because magnetization

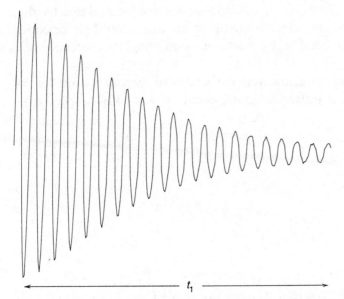

Fig. 2.3 A slice from the data of Fig. 2.2, taken through the tops of the peaks. Note that this is simply an FID in the second dimension. (Reprinted from ref. 7, Ch. 1 with permission.)

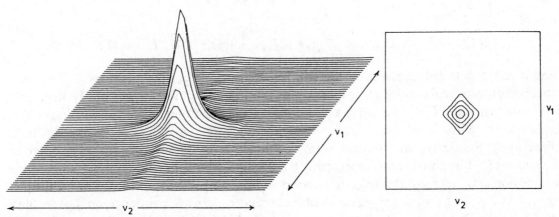

Fig. 2.4 The spectra from Fig. 2.2 after a two-dimensional Fourier transformation, presented as both a stacked plot (left) and a contour plot (right). (Reprinted from ref. 7, Ch. 1 with permission.)

(or, more properly, *coherence*—see Section 1.1.9) may be transferred between multiplets during the second pulse.

We considered the vector picture for the COSY experiment in Section 2.1.1. However, strictly speaking, if we had considered only the motion of the bulk magnetization vector $\mathbf{M}_0$, then following the second 90° pulse, the magnetization is aligned along the $-z$ axis, which is not observable. In order to understand this classical picture, we have to invoke the components of the bulk magnetization, whose precession under the influence of scalar coupling gives rise to observable magnetization. In my opinion this is harder to understand than the product operator formalism, because when is it appropriate to invoke such components, and when can they be neglected? Obviously the answer to this lies in the type of pulse sequence being analysed, and if it involves scalar coupling the approach outlined in Fig. 2.1 is appropriate. However, provided you are not intimidated by the, sometimes rather lengthy and tedious, algebra involved in using product operator formalism, no implicit assumptions need to be made in calculating the outcome of a particular pulse sequence.

Using product operator formalism, we show here the results of the calculation for the COSY experiment at each step of the pulse sequence, assuming a weakly coupled two-spin system:

$$\sigma_0 = I_{1z} + I_{2z}$$

$$\Big\downarrow \quad \pi/2(I_{1x} + I_{2x})$$

$$\sigma_1 = -I_{1y} - I_{2y}$$

$$\Big\downarrow \quad \Omega_1 t_1 I_{1z} + \Omega_2 t_1 I_{2z} + \pi \mathcal{J}_{12} t_1 2 I_{1z} I_{2z}$$

$$\sigma_2 = (-I_{1y} \cos \Omega_1 t_1 + I_{1x} \sin \Omega_1 t_1 - I_{2y} \cos \Omega_2 t_1 + I_{2x} \sin \Omega_2 t_1) \cos \pi \mathcal{J}_{12} t_1$$

$$+ (2 I_{1x} I_{2z} \cos \Omega_1 t_1 + 2 I_{1y} I_{2z} \sin \Omega_1 t_1 + 2 I_{1z} I_{2x} \cos \Omega_2 t_1 + 2 I_{1z} I_{2y} \sin \Omega_2 t_1) \sin \pi \mathcal{J}_{12} t_1$$

$$\Big\downarrow \quad \pi/2(I_{1x} + I_{2x})$$

$$\sigma_3^{obs} = (I_{1x} \sin \Omega_1 t_1 + I_{2x} \sin \Omega_2 t_1) \cos \pi \mathcal{J}_{12} t_1 - (2 I_{1y} I_{2z} \sin \Omega_1 t_1 + 2 I_{1z} I_{2y} \sin \Omega_2 t_1) \sin \pi \mathcal{J}_{12} t_1$$

For a more detailed explanation of this calculation, refer to Appendix 1. The first observable term will continue to precess at $\Omega_1 \pm \pi \mathcal{J}_{12}$ in the detection period. Hence, after two-dimensional FT, it will lead to a two-dimensional multiplet pattern on the diagonal at $\omega_1 = \omega_2 = \Omega_1$ with in-phase doublet structure (cosine dependence on $\mathcal{J}_{12}$) in both directions. Similarly, the second term leads to an in-phase diagonal multiplet at $\omega_1 = \omega_2 = \Omega_2$. The third term, representing antiphase spin-1 magnetization, will resume precession at $\Omega_1 \pm \pi \mathcal{J}_{12}$ in the detection period and therefore leads to a cross-peak multiplet at $\omega_1 = \Omega_2$, $\omega_2 = \Omega_1$, with antiphase doublet structure (sine dependence on $\mathcal{J}_{12}$) in both dimensions. The last term represents a cross-peak multiplet at $\omega_1 = \Omega_1$, $\omega_2 = \Omega_2$. The relationship between product operators and lineshape is shown in Fig. 2.5.

(a) I_{1y}

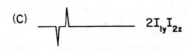

(b) I_{2y}

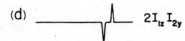

(c) $2I_{1y}I_{2z}$

(d) $2I_{1z}I_{2y}$

Fig. 2.5 The one-dimensional spectra representing the physical interpretation of Cartesian product operators involving transverse terms (I_{1y}, I_{2y}, $2I_{1y}I_{2z}$, and $2I_{1z}I_{2y}$). (Reprinted from M. H. Levitt, in *Pulse Methods in 1D and 2D Liquid-Phase NMR* (ed. W. S. Brey), Academic Press, Orlando FL (1988) with permission.)

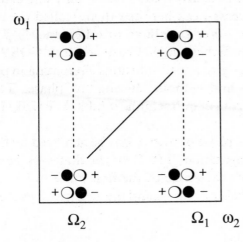

Fig. 2.6 Theoretical appearance of a two-dimensional phase sentisive DQF–COSY spectrum (with positive and negative peaks appropriately labelled).

In summary the two-dimensional map will have the appearance shown in Fig. 2.6. Thus cross-peaks are dispersed by two different chemical shifts in both dimensions, so that the likelihood of overlapping resonances is much smaller. Note that the experiment is relatively fast (compared with single-frequency decoupling experiments), and the experiment time remains constant regardless of the number of correlations in the spectrum.

In order to predict the appearance of the cross-peak multiplets from Cartesian product operators, note that the operators in $\sigma(t_1, t_2 = 0)$ (I_{kx}, $2I_{ky}I_{lz}$, etc.) are responsible for the phases in the ω_2-domain, while the trigonometric functions determines the phases in the ω_1-direction. In the ω_2-domain, one obtains:

I_{kx}: disperse in-phase doublet

I_{ky}: absorptive in-phase doublet

$I_{kx}I_{lz}$: dispersive antiphase doublet

$I_{ky}I_{lz}$: absorptive antiphase doublet

In the ω_1-domain one obtains:

$$\cos \Omega_k t_1 \cos \pi \mathcal{J}_{kl} t_1:\quad \text{absorptive in-phase doublet}$$

$$\sin \Omega_k t_1 \cos \pi \mathcal{J}_{kl} t_1:\quad \text{dispersive in-phase doublet}$$

$$\cos \Omega_k t_1 \sin \pi \mathcal{J}_{kl} t_1:\quad \text{dispersive antiphase doublet}$$

$$\sin \Omega_k t_1 \sin \pi \mathcal{J}_{kl} t_1:\quad \text{absorptive antiphase doublet}$$

In general these can be represented schematically as shown in Fig. 2.7.

Although when COSY was first introduced the phase information was thrown away (absolute-value mode COSY), the so-called *phase-sensitive* mode provides a wealth of useful information. For example, a proton may have several couplings to various groups of other protons, and in a complex spectrum the resulting multiplet pattern may be impossible to interpret. A basic COSY experiment tells us that two protons are coupled, but it does not tell us *which* of several couplings is involved. That is, only one coupling between a pair of protons gives rise to the cross-peak between them (called the *active* coupling), but each proton may have many other couplings to other nuclei (*passive* couplings), and there is no obvious way to tell which is which. Phase-sensitive COSY does indicate this, because only the active couplings give rise to splittings alternating in phase, whereas all the passive couplings split the line without altering the phase. This is illustrated for a cross-peak of the COSY spectrum of bull seminal inhibitor (BUSI IIA) in Fig. 2.8.

Acquisition of two-dimensional data that is phase-sensitive can be achieved by one of two methods: time-proportional phase incrementation (TPPI)[1] or the States–Habekorn–Ruben method.[2] The first method increments the phase of the first pulse in 90° steps, together with the incrementation of the t_1 variable. The second method involves shifting

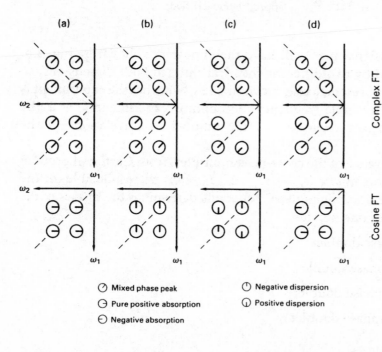

(a)   (b)   (c)   (d)

Complex FT

Cosine FT

⊘ Mixed phase peak   ◔ Negative dispersion

⊖ Pure positive absorption   ◑ Positive dispersion

⊖ Negative absorption

Fig. 2.7 Two-dimensional signal patterns associated with the typical Cartesian operators: (a) $I_{ky} \cos \Omega_k t_1 \cos \pi \mathcal{J}_{kl} t_1$; (b) $I_{kx} \sin \Omega_k t_1 \cos \pi \mathcal{J}_{kl} t_1$; (c) $2I_{kx} I_{lz} \cos \Omega_k t_1 \sin \pi \mathcal{J}_{kl} t_1$; (d) $2I_{ky} I_{lz} \sin \Omega_k t_1 \sin \pi \mathcal{J}_{kl} t_1$. If complex Fourier transformation is carried out (upper), all the peaks have composite lineshapes, whereas cosine Fourier transformation (lower) leads to pure two-dimensional absorption or dispersion lineshapes. (Reprinted from ref. 4, Ch. 1 with permission.)

the phases of both the first pulse and the receiver by 90° (or, alternatively, shifting the phase of the second pulse by −90°) for each t_1 value. This technique is also the means by which *quadrature detection* is achieved in the second dimension. Since both methods are largely equivalent (although some pulse sequences have phase cycles that favour one method over the other) we will not consider the details of how the methods work. The interested reader should refer to the original papers for a full description.

One of the problems that plague two-dimensional spectra is the so-called t_1 noise, which arises from a variety of reasons including instrumental artefacts,[3] which are sometimes removed using *symmetrization* (Fig. 2.9). This technique discards cross-peaks which are not symmetrical about the diagonal. As you can see, this can be a fatal trap for the unwary,

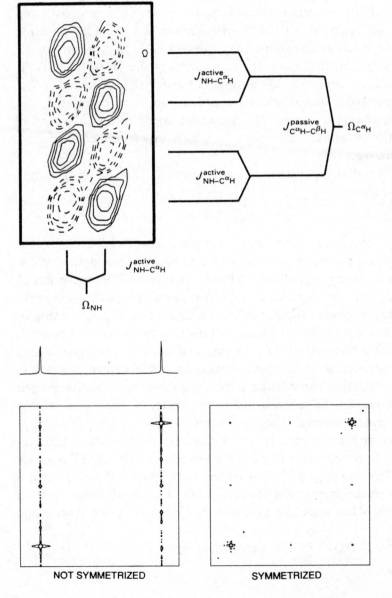

Fig. 2.8 Detail of a phase sensitive COSY for BUSI IIA, showing the NH—C$^\alpha$H cross-peak of cysteine-57 (with negative contours represented by dashed lines). The passive coupling between the C$^\alpha$H proton and the C$^\beta$H proton (which is not involved in coherence transfer) leads to duplication of the basic square of four signals with alternating signs. The coupling between C$^\alpha$H and C$^\beta$H′ is not resolved. (Reprinted from ref. 4, Ch. 1 with permission.)

Fig. 2.9 The problem of t_1-noise (the ridge along the vertical axis) and the dangers of symmetrization. (Reprinted from ref. 7, Ch. 1 with permission.)

leading to false cross-peaks. Furthermore, in phase-sensitive COSY spectra, in which it is common to have poorer digital resolution in F_1 vs. F_2, the cross-peak multiplet structure might be well resolved only in one quadrant of the spectrum. Symmetrization results in a spectrum in which the multiplet with the poorest resolution before symmetrization dominates on both sides of the diagonal. While popular for a while in the early 1980s, this technique is almost never used with biomolecules nowadays, and is better not used at all.

2.1.3 Multiple-quantum filtration

A number of forms of one- and two-dimensional spectroscopy employ multiple-quantum coherences either directly, by converting them into observable magnetization, or through filtration for the simplification of spectra. Multiple-quantum spectroscopy can be important in biomolecular NMR in certain situations, for example where two-step connectivities are required (as seen in RELAY—see Section 2.1.4). The multiple quantum correlation experiment has greater sensitivity than RELAY in cases where the linewidths are large. An example of this class of experiments is the INADEQUATE (Incredible Natural Abundance DoublE QUantum Transfer Experiment) pulse sequence.[4,5] This experiment enables homonuclear $\mathcal{J}$-coupled spins to be correlated, and is usually employed in a two-dimensional manner. However, more commonly used in biomolecular NMR is multiple-quantum filtration, in which unwanted resonances—singlets for example—are removed.

In the multiple-quantum filtration experiment, we add an extra $\pi/2$ pulse immediately after the end of the COSY sequence:

$$(\pi/2)_\phi - t_1 - (\pi/2)_\phi (\pi/2)_x - t_2$$

where the third pulse follows instantaneously, and multiple-quantum coherences that happened to exist before the third pulse are converted back to observable magnetization. With a suitable choice of phase cycling, signals which have arisen from different orders of multiple-quantum coherence can be separated out. For instance, double-quantum coherence is twice as sensitive to phase changes in its excitation sequence as is single-quantum coherence. Therefore, if we shift the phase ϕ of the first two pulses (in general, *all* pulses before the coherence is created) by 90°, the phase of the detected signal which came via double-quantum coherence is inverted. Inverting the receiver phase (i.e. subtracting the 90° from the 0° experiment) then selects the component which passed through double-quantum coherence. To select p-quantum coherence, the rule is to step the phases of the excitation pulses through the sequence $0, 180/p \ldots (2p-1) \times 180/p$, alternating the receiver phase on each scan. The advantages are considerable improvements in spectral simplification. Singlets are largely eliminated in DQF–COSY and AX and AB systems in TQF–COSY (see Fig. 2.10). Furthermore the diagonal *and* the cross-peaks are in pure absorption phase, improving the resolution. The disadvantage is that, for example, the DQF–COSY is less sensitive by a factor of two than is a conventional COSY.

The double-quantum filtered (DQF) COSY experiment is:

$$(\pi/2)_\phi - t_1 - (\pi/2)_\phi - \Delta - (\pi/2)_x - t_2 \quad \text{(acquire)}$$

where Δ is a short phase switching delay. For a weakly coupled two spin-$\frac{1}{2}$ system, the density operators (σ) for each step in the pulse sequence (assuming no evolution during Δ) are given below. For σ_4, we consider the sums and differences of product operators and indicate their relationship to coherence transfer. For the final step, assuming evolution of couplings and chemical shifts during t_2, the observables are expressed in terms of the differences of $\frac{1}{2}\sin$ or $\frac{1}{2}\cos$ functions, and the receiver phase is assumed to be along the $+x$ axis. Assuming a 90° linear phase correction of the spectrum in t_1, the calculation is as follows:

Let us begin the analysis with $\phi = 0$, then the state of the spin system after the second pulse is defined by σ_3:

$$\sigma_3 = (-I_{1z}\cos\Omega_1 t_1 + I_{1x}\sin\Omega_1 t_1 - I_{2z}\cos\Omega_2 t_1 + I_{2x}\cos\Omega_2 t_1)\cos\pi\mathcal{J}_{12}t_1$$

$$+ (2I_{1x}I_{2y}\cos\Omega_1 t_1 + 2I_{1y}I_{2x}\sin\Omega_1 t_1 + 2I_{1z}I_{2y}\cos\Omega_2 t_1 + 2I_{1y}I_{2z}\sin\Omega_2 t_1)\sin\pi\mathcal{J}_{12}t_1$$

Expanding in terms of zero-quantum and double-quantum coherence components:

$$\sigma_3 = (-I_{1z}\cos\Omega_1 t_1 + I_{1x}\sin\Omega_1 t_1 - I_{2z}\cos\Omega_2 t_1 + I_{2x}\cos\Omega_2 t_1)\cos\pi\mathcal{J}_{12}t_1$$

$$+ \{\tfrac{1}{2}[2I_{1x}I_{2y} + 2I_{1y}I_{2x}) - (2I_{1y}I_{2x} - 2I_{1x}I_{2y})]\cos\Omega_1 t_1$$

$$+ \tfrac{1}{2}[(2I_{1x}I_{2y} + 2I_{1y}I_{2x}) - (2I_{1y}I_{2x} - 2I_{1x}I_{2y})]\cos\Omega_2 t_1$$

$$+ 2I_{1z}I_{2y}\sin\Omega_1 t_1 + 2I_{1y}I_{2z}\sin\Omega_2 t_1\}\sin\pi\mathcal{J}_{12}t_1$$

During the phase cycle, all terms except those corresponding to double-quantum coherence cancel:

$$\sigma_3 = \{\tfrac{1}{2}(2I_{1x}I_{2y} + 2I_{1y}I_{2x})\cos\Omega_1 t_1 + 1/2(2I_{1x}I_{2y} + 2I_{1y}I_{2x})\cos\Omega_2 t_1\}\sin\pi\mathcal{J}_{12}t_1$$

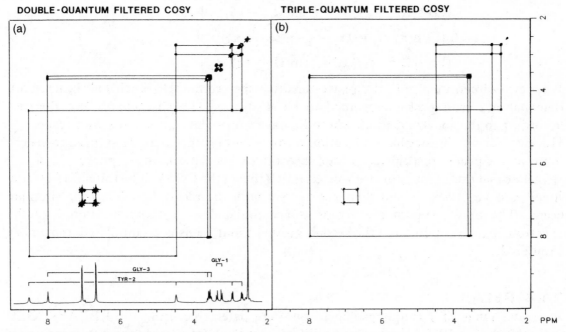

DOUBLE-QUANTUM FILTERED COSY   **TRIPLE-QUANTUM FILTERED COSY**

(a)   (b)

GLY-1
GLY-3
TYR-2

8   6   4   2   8   6   4   2 PPM

Fig. 2.10 Double- and triple-quantum filtered COSY spectra of the tripeptide Gly—Tyr—Gly. (Reprinted from ref. 7, Ch. 1 with permission.)

After the mixing pulse:

$$\sigma_4 = \{\tfrac{1}{2}(2I_{1x}I_{2z} + 2I_{1z}I_{2x})\cos\Omega_1 t_1 + 1/2(2I_{1x}I_{2z} + 2I_{1z}I_{2x})\cos\Omega_2 t_1\}\sin\pi\mathcal{J}_{12}t_1$$

Each term evolves in t_2:

$$\sigma_5 = \tfrac{1}{2}I_{1y}\sin\pi\mathcal{J}_{12}t_2\cos\Omega_1 t_1\sin\pi\mathcal{J}_{12}t_1\cos\Omega_1 t_2$$

$$-\tfrac{1}{2}I_{1x}\sin\pi\mathcal{J}_{12}t_2\cos\Omega_1 t_1\sin\pi\mathcal{J}_{12}t_1\sin\Omega_1 t_2$$

$$+\tfrac{1}{2}I_{2y}\sin\pi\mathcal{J}_{12}t_2\cos\Omega_1 t_1\sin\pi\mathcal{J}_{12}t_1\cos\Omega_2 t_2$$

$$-\tfrac{1}{2}I_{2x}\sin\pi\mathcal{J}_{12}t_2\cos\Omega_1 t_1\sin\pi\mathcal{J}_{12}t_1\sin\Omega_2 t_2$$

$$+\tfrac{1}{2}I_{1y}\sin\pi\mathcal{J}_{12}t_2\cos\Omega_1 t_1\sin\pi\mathcal{J}_{12}t_1\cos\Omega_1 t_2$$

$$-\tfrac{1}{2}I_{1x}\sin\pi\mathcal{J}_{12}t_2\cos\Omega_1 t_1\sin\pi\mathcal{J}_{12}t_1\sin\Omega_1 t_2$$

$$+\tfrac{1}{2}I_{2y}\sin\pi\mathcal{J}_{12}t_2\cos\Omega_1 t_1\sin\pi\mathcal{J}_{12}t_1\cos\Omega_2 t_2$$

$$-\tfrac{1}{2}I_{2x}\sin\pi\mathcal{J}_{12}t_2\cos\Omega_1 t_1\sin\pi\mathcal{J}_{12}t_1\sin\Omega_2 t_2$$

If the receiver is along the $+x$ axis, then the observable magnetization is:

$$\sigma_5^{\text{obs}} = -\tfrac{1}{2}I_{1x}[\tfrac{1}{2}\sin(\pi\mathcal{J}_{12}t_1 + \Omega_1 t_1) - \tfrac{1}{2}\sin(\Omega_1 t_1 - \pi\mathcal{J}_{12}t_1)]$$

$$\times [\tfrac{1}{2}\cos(\Omega_1 t_2 - \pi\mathcal{J}_{12}t_2) - \tfrac{1}{2}\sin(\Omega_1 t_2 - \pi\mathcal{J}_{12}t_2)]$$

$$-\tfrac{1}{2}I_{2x}[\tfrac{1}{2}\sin(\pi\mathcal{J}_{12}t_1 + \Omega_1 t_1) - \tfrac{1}{2}\sin(\Omega_1 t_1 - \pi\mathcal{J}_{12}t_1)]$$

$$\times [\tfrac{1}{2}\cos(\Omega_2 t_2 - \pi\mathcal{J}_{12}t_2) - \tfrac{1}{2}\sin(\Omega_2 t_2 - \pi\mathcal{J}_{12}t_2)]$$

$$-\tfrac{1}{2}I_{1x}[\tfrac{1}{2}\sin(\pi\mathcal{J}_{12}t_1 + \Omega_2 t_1) - \tfrac{1}{2}\sin(\Omega_1 t_1 - \pi\mathcal{J}_{12}t_1)]$$

$$\times [\tfrac{1}{2}\cos(\Omega_1 t_2 - \pi\mathcal{J}_{12}t_2) - \tfrac{1}{2}\sin(\Omega_1 t_2 - \pi\mathcal{J}_{12}t_2)]$$

$$-\tfrac{1}{2}I_{2x}[\tfrac{1}{2}\sin(\pi\mathcal{J}_{12}t_1 + \Omega_2 t_1) - \tfrac{1}{2}\sin(\Omega_2 t_1 - \pi\mathcal{J}_{12}t_1)]$$

$$\times [\tfrac{1}{2}\cos(\Omega_2 t_2 - \pi\mathcal{J}_{12}t_2) - \tfrac{1}{2}\sin(\Omega_2 t_2 - \pi\mathcal{J}_{12}t_2)]$$

After two-dimensional FT, the phase-sensitive spectrum appears with all peaks, both diagonal and cross-peaks, presumed to be absorptive. This is actually not the case according to the above equation, where we have sinusoidal components with respect t_1. However, the 90° linear phase correction restores the absorption mode in this dimension. Thus all the peaks are a cluster of four absorptive peaks alternating in phase ($-$, $+$, $-$, $+$). Note that the intensity of cross-peaks is half that of the COSY. The full phase-cycling involves $\phi = x, y, -x, -y$ and then $y, -x, -y, x$ with the receiver $= +, -$, for alternate scans. The first group of four scans is accumulated in memory location 1 of the spectrometer's computer and the second group of four scans is accumulated in memory location 2.

2.1.4 RELAY

A greater range of $\mathcal{J}$ couplings (up to two) are detectable using the RELAY–COSY experiment. The RELAY experiment differs from the COSY experiment in that there is an additional delay and RF pulses which serve to allow relayed coherence transfer between two spins which are not directly coupled, but instead are coupled through a common partner. The pulse sequence is as follows:

$$(\pi/2)_x - t_1 - (\pi/2)_\phi - \tau - (\pi)_{\phi + \pi/2} - \tau - (\pi/2)_\phi - t_2$$

with the π-pulse in the middle of the 2τ delay to refocus the chemical shifts. The type of spin system amendable to relayed coherence transfer would be $I - S - M$, where $\mathcal{J}_{IM} = 0$, and the efficiency of coherence transfer depends on the magnitudes of the coupling constants $\mathcal{J}_{IS}$ and $\mathcal{J}_{SM}$, and also on the value of τ. Using product operator formalism, we already know, from the calculation given in the previous section, the density matrix after the second $\pi/2$ pulse. The term corresponding to transverse S magnetization ($2I_z S_y$ $\sin \omega_I t_1 \sin \pi \mathcal{J}_{IS} t_1$) can be subject to relayed coherence transfer from I to M via S. During the 2τ interval the evolution of chemical shifts can be ignored, since they are refocused by the π-pulse (see Section 1.1.4), and so this term evolves only under scalar coupling:

$$2I_z S_y \sin \omega_I t_1 \sin \pi \mathcal{J}_{IS} t_1 \xrightarrow{\pi \mathcal{J}_{IS} 2\tau 2I_z S_z} \xrightarrow{\pi \mathcal{J}_{SM} 2\tau 2S_z M_z} [2I_z S_y \cos \pi \mathcal{J}_{IS} 2\tau \cos \pi \mathcal{J}_{SM} 2\tau$$

$$- 4I_z S_x M_z \cos \pi \mathcal{J}_{IS} 2\tau \sin \pi \mathcal{J}_{SM} 2\tau - S_x \sin \pi \mathcal{J}_{IS} 2\tau \sin \pi \mathcal{J}_{SM} 2\tau$$

$$- 2S_y M_z \sin \pi \mathcal{J}_{IS} 2\tau \sin \pi \mathcal{J}_{SM} 2\tau] \sin \omega_I t_1 \sin \pi \mathcal{J}_{IS} t_1 \qquad (2.2)$$

In order for coherence to be transferred from I to M, the transverse S magnetization evolves in phase with respect to $\mathcal{J}_{IS}$ and in antiphase with respect to $\mathcal{J}_{SM}$. The last term in the preceding expression is converted into observable M magnetization after the last 90° pulse:

$$- 2S_y M_z \sin \pi \mathcal{J}_{IS} 2\tau \sin \pi \mathcal{J}_{SM} 2\tau \sin \omega_I t_1 \sin \pi \mathcal{J}_{IS} t_1$$

$$\xrightarrow{\pi/2(I_x + S_x + M_x)} 2S_z M_y \sin \pi \mathcal{J}_{IS} 2\tau \sin \pi \mathcal{J}_{SM} 2\tau \sin \omega_I t_1 \sin \pi \mathcal{J}_{IS} t_1 \qquad (2.3)$$

This term defines a cross-peak at $\omega_1 = \omega_I$ and $\omega_2 = \omega_M$ with antiphase doublet structure in both dimensions. The value for τ has to be chosen carefully so that the coherence transfer function is not zero, taking relaxation into account. In practice a number of values need to be tried out before the best RELAY data are obtained.

2.1.5 Total correlation spectroscopy (TOCSY/HOHAHA)

TOtal Correlation SpectroscopY (TOCSY) or HOmonuclear HArtmann–HAhn (HOHAHA) spectroscopy are two variants of an experiment which has largely superseded the RELAY experiment. This method relies on cross-polarization rather than the coherence transfer methods of the RELAY–COSY experiment. The pulse sequence is shown in Fig. 2.11.

The pulse sequence is essentially the same as the ROESY (see Section 2.1.7), and the HOHAHA pulse sequence involves a modified spin lock which we will discuss shortly. The spin-locking field causes cross-polarization, or an oscillatory exchange of spin-locked magnetization between two spins, provided the effective local RF fields experienced by the two spins are identical. This causes the spins to become temporarily equivalent. The spin-lock sequences are termed 'isotropic mixing' (no effective field) or 'Hartmann–Hahn mixing' (with effective field, i.e. this is the HOHAHA experiment). Isotropic mixing leads

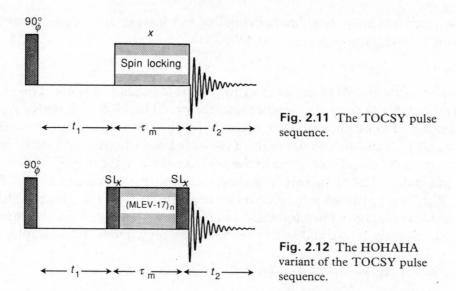

Fig. 2.11 The TOCSY pulse sequence.

Fig. 2.12 The HOHAHA variant of the TOCSY pulse sequence.

to spectra with mixed phases (phase modulated), whereas in the Hartmann–Hahn mixing pure phases are obtained. For a simple two-spin system, complete exchange of magnetization occurs for contact times (or spin-lock mixing times) equal to $1/2J$, where J is the two-spin scalar coupling constant. For biological macromolecules, which involve large spin systems, there is no simple analytical solution, and in practice spectra are usually obtained with a variety of spin-lock mixing times.

The pulse sequence depicted in Fig. 2.11 represents the simplest form of the TOCSY pulse sequence. The HOHAHA variant simply involves alternative decoupling schemes during the spin-lock period. The schemes which are commonly employed include composite pulses such as MLEV-17 and DIPSI-X, and the pulse sequence usually employed in biomolecular NMR is depicted in Fig. 2.12.

As we shall see in Section 2.1.7, the basic TOCSY pulse sequence is identical to the ROESY pulse sequence, except that the length of the mixing time is different. This means that ROESY cross-peaks can occur in a TOCSY spectrum and TOCSY cross-peaks can occur in a ROESY spectrum. The two cross-peaks have opposite phase and are relatively easily distinguished, although pulse sequences have been developed for use in crowded biomolecule spectra for 'cleaning up' the spectrum. One example is the 'Clean TOCSY',[6] which involves introducing delays between the composite pulses in the MLEV-17 spin-locking field with the delay $=2.6 \times \tau_{90}$, where $\tau_{90}=$ time for 90° pulse in MLEV-17 sequence.

2.1.6 Heteronuclear correlation spectroscopy—HETCOR, HMQC, and HSQC

The conventional method for obtaining $^1\mathrm{H}-^{13}\mathrm{C}$ correlations until recently made use of the heteronuclear shift correlation experiment,[7-9] which in its simplest form is as follows:

$I:$ $\quad (\pi/2)-t_1-(\pi/2)$

$S:$ $\quad\quad\quad\quad (\pi/2)-\text{Acquire }(t_2)$

where $I = {}^1H$ and $S =$ heteronucleus. This experiment is analogous to the COSY experiment, where the only difference is that the coherence transfer processes of the second pulse have been to another coupled nucleus, by providing a simultaneous pulse at another frequency.

In terms of product operators:

$$\sigma_0 = I_z + S_z \xrightarrow{\pi/2(I_x)} \sigma_1 = -I_y + S_z \xrightarrow{\Omega_I t_1 I_z + \Omega_S t_1 S_z + \pi \mathcal{J}_{IS} 2 I_z S_z}$$

$$\sigma_2 = (-I_y \cos \Omega_I t_1 + I_x \sin \Omega_I t_1) \cos \pi \mathcal{J}_{IS} t_1 + (2 I_x S_z \cos \Omega_I t_1 + 2 I_x S_z \sin \Omega_I t_1) \sin \pi \mathcal{J}_{IS} t_1$$

where the S_z term is dropped for convenience

$$\xrightarrow{\pi/2(I_x + S_x)} \sigma_3^{\text{obs}} = (I_x \sin \Omega_I t_1 - I_z \cos \Omega_I t_1) \cos \pi \mathcal{J}_{IS} t_1 - (2 I_x S_y \cos \Omega_I t_1$$

$$+ 2 I_z S_y \sin \Omega_I t_1) \sin \pi \mathcal{J}_{IS} t_1$$

The $-2 I_z S_y$ term represents antiphase magnetization which precesses at the Larmor frequency of spin I during t_1, and precesses at the Larmor frequency of spin S during t_2. It will evolve during t_2 to give observable magnetization. It is thus analogous to a cross-peak in homonuclear COSY experiments, and demonstrates that the correlated nuclei are scalar coupled. However, since a $(\pi/2)$ pulse is not applied to the S spins before the t_1 period, there are no diagonal peaks in the heteronuclear correlation spectrum. The resulting spectrum is shown in Fig. 2.13.

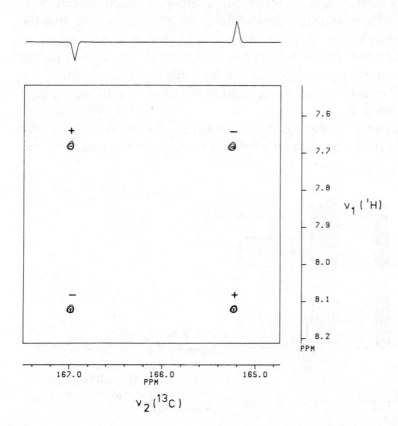

Fig. 2.13 The basic ${}^1H—{}^{13}C$ heteronuclear shift correlation experiment for an AX system. Both the ${}^1H$ and ${}^{13}C$ dimensions contain antiphase doublets, and a slice taken parallel with ν_2 is shown above the contour plot. (Reprinted from ref. 7, Ch. 1 with permission.)

Clearly, the experiment as it stands does not give us quite what we expect. Normally we would apply broadband proton decoupling during the acquisition of the heteronuclei, so that there are no multiplet structures due to proton coupling. Also, ω_1-dimension contains the frequencies of proton lines coupled to the heteronucleus (i.e. the ^{13}C satellites of the proton lines, if carbon is being observed). It would be simpler if we could eliminate the heteronuclear coupling in *both* dimensions without eliminating the experiment, which relies on that coupling for its operation.

The necessary modification is relatively simple—we need to allow the antiphase S magnetization to refocus before detection, which in practice may be achieved by waiting a delay before and after the mixing pulse (see Fig. 2.14). The delays allow the coherence to dephase before and to rephase after the coherence transfer. If the delays are chosen as $\tau = \tau' = 1/(2\mathcal{J}_{IS})$, then the effects of the delays are equivalent to $(\pi/2)$ rotations about the z axis. The π pulses shown remove offset-dependent phase errors (see Fig. 2.14). This is sometimes referred to as the HETCOR (for HETeronuclear CORrelation) experiment. In this case the I_x term is relevant, and it is transformed by the $\tau - (\pi/2)(I_x + S_x) - \tau'$ sequence as follows:

$$I_x \xrightarrow{(\pi/2)2I_zS_z} 2I_yS_z \xrightarrow{\pi/2(I_x + S_x)} -2I_zS_y \xrightarrow{(\pi/2)2I_zS_z} S_x \qquad (2.4)$$

giving in-phase S spin magnetization. Thus it is now possible (but optional) to decouple the I spins during the detection period. This will collapse all multiplets to singlets in t_2, with an effective increase in resolution and sensitivity. In addition, inspection of the $+I_x$ term given for σ_3^{obs} above shows a cosinusoidal modulation $\cos \pi \mathcal{J}_{IS} t_1$. It is thus possible (but again optional) to decouple the S spins during t_1, since this will result in $\cos \pi \mathcal{J}_{IS} t_1 = 1$, and the $+I_x$ term will not be eliminated. It should also be noted that the experiment can be tuned to the magnitude of the scalar coupling by exploiting the condition $\tau = \tau' = 1/(2\mathcal{J})$. This condition arises simply from the sinusoidal dependence of the antiphase magnetization upon $\mathcal{J}_{IS}$.

The problems for these schemes for biological studies is that their sensitivity is very low, due to the detection of rare-spin magnetization. More recently the heteronuclear

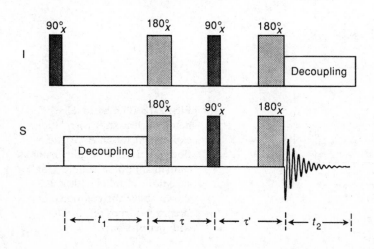

Fig. 2.14 Modified heteronuclear shift correlation pulse sequence, with heteronuclear decoupling. The π pulses remove offset-dependent phase errors.

multiple-quantum coherence (HMQC) experiment (sometimes referred to as the 'forbidden echo' experiment) has been introduced.[10-12] Optimum sensitivity would be obtained starting with proton polarization and ending up with proton detection. The basic pulse sequence is outlined in Fig. 2.15.

For a heteronuclear spin systems IS, in terms of product operators:

$$I_z \xrightarrow{A} -2I_xS_y \xrightarrow{t_1} -2I_xS_y \cos \omega_S t_1 \xrightarrow{B} -I_y \cos \omega_S t_1 \qquad (2.5)$$

where A and B are operators representing the effect of the pulse sequence before and after the t_1 period, respectively, and the pulse phases are assumed to be $+x$. Equation (2.6) indicates that the proton signal (I_y) which is detected will be in-phase and amplitude modulated by ω_S, giving rise to a single absorption line in the two-dimensional spectrum. If we consider a second coupled proton M, then the effect of the small homonuclear coupling $\mathscr{J}_{IM}$ during the short delays Δ can be neglected. During t_1, dephasing caused by $\mathscr{J}_{SM}$ coupling is refocused by the 180° pulse. However, the effect of any $\mathscr{J}_{IM}$ coupling is not refocused, since both coupled nuclei are influenced by the 180° pulse. Therefore, the effect of the pulse sequence is as follows:

$$I_z \xrightarrow{At_1B} -I_y \cos \pi \mathscr{J}_{IM}t_1 \cos \omega_S t_1 + 2I_xM_z \sin \pi \mathscr{J}_{IM}t_1 \cos \omega_S t_1 \qquad (2.6)$$

The signal which is detected will consist of four-multiplet components at $\omega_S \pm \pi \mathscr{J}_{IM}$ and $\omega_I \pm \pi \mathscr{J}_{IM}$, and will contain both in-phase absorptive (I_y) and antiphase dispersive $(2I_xM_z)$ components. An example of an HMQC spectrum for an enzyme-inhibitor complex is shown in Fig. 2.16.

A related pulse sequence is the heteronuclear single quantum coherence (HSQC) experiment, as outlined in Fig. 2.17. This is a variation on an experiment introduced by Bodenhausen and Ruben[13] and employs two INEPT-type transfers (see Section 1.1.7) to transfer magnetization from the protons to the low-γ nucleus and back to the protons. The enhancement in sensitivity is much greater than that obtainable by exploiting the nuclear Overhauser effect in simple heteronuclear correlated experiments, so it has been described as the Overbodenhausen experiment. The INEPT sequence transfers I_{spin} (e.g. ^{1}H) I_z magnetization into antiphase S-spin magnetization. Decoupling the I-spin during t_1 is accomplished by application of a 180° pulse at the midpoint of t_1. A subsequent

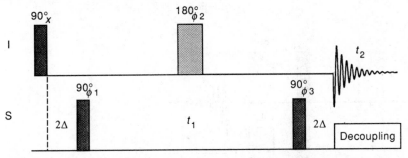

Fig. 2.15 The basic HMQC pulse sequence. The delay Δ is usually set to $\sim 20\%$ less than $1/4(\mathscr{J}_{IS})$. The phase cycling is as follows: $\phi_1 = x, -x$; $\phi_2 = x, x, y, y, -x, -x, -y, -y$; receiver $= 2(x, -x, -x, x)$. To suppress incomplete steady-state effects, ϕ_3 is incremented from x to $-x$ after 8 scans, together with the receiver phase.

INEPT transfer reconverts the transverse S-magnetization into observable I magnetization. In terms of product operators:

$$I_z \xrightarrow{\text{INEPT}} -2I_zS_y \xrightarrow{t_1} -2I_zS_y \cos \omega_S t_1 \xrightarrow{\text{reverse INEPT}} -I_x \cos \omega_S t_1 \tag{2.7}$$

where for clarity only those terms that contribute to the final spectrum are retained, and the phases of the pulses are given that correspond to the first step in the phase cycle. Pure absorptive lineshapes are obtained. Furthermore, the linewidth is narrower than with the

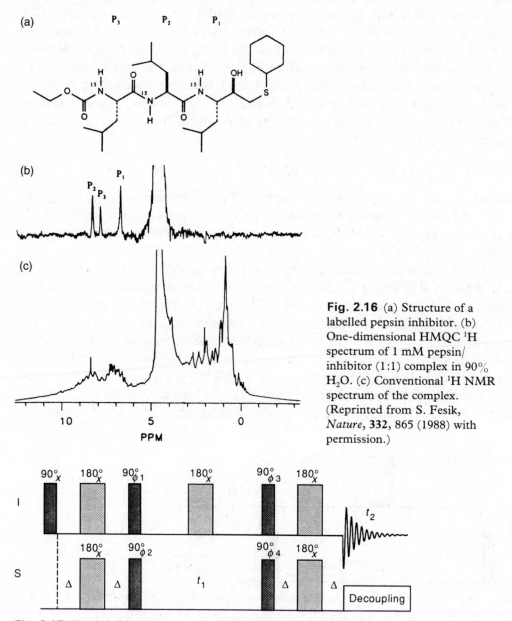

Fig. 2.16 (a) Structure of a labelled pepsin inhibitor. (b) One-dimensional HMQC ^{1}H spectrum of 1 mM pepsin/inhibitor (1:1) complex in 90% H$_2$O. (c) Conventional ^{1}H NMR spectrum of the complex. (Reprinted from S. Fesik, *Nature*, **332**, 865 (1988) with permission.)

Fig. 2.17 The HSQC pulse sequence. The phase cycling is as follows: $\phi_1 = y, -y$; $\phi_2 = x, x, -x, -x$; $\phi_3 = x, x, x, x, -x, -x, -x, -x$; $\phi_4 = 8(x), 8(-x)$; receiver $= x, -x, -x, x, 2(-x, x, x, -x)$.

HMQC experiment, since it is not affected by homonuclear $\mathcal{J}$ couplings and determined by the average relaxation rates of S_x and I_zS_y. There are a number of advantages that the HSQC experiment has over the HMQC, principally in terms of sensitivity.[14,15]

One last variation worth mentioning is based on either the HMQC or the HSQC experiments and is called the heteronuclear multiple-bond correlation (HMBC) experiment.[15,16] This is a sensitive technique for the determination of long-range (two- and three-bond) heteronuclear connectivities. For example, with proteins, it correlates the amide nitrogen with the $C^\alpha H$ of the next residue.

2.1.7 Nuclear Overhauser effect spectroscopy (NOESY) and rotating frame Overhauser effect spectroscopy (ROESY)

The two-dimensional experiment which permits the detection of NOEs is the NOESY pulse sequence:

$$(\pi/2)_\phi - t_1 - (\pi/2)_\phi - \tau_m - (\pi/2)_\phi - \text{Acquire } (t_2)$$

Unlike the majority of two-dimensional experiments, this can be understood classically using the vector formalism (see Fig. 2.18).

The first 90°_x pulse creates transverse xy magnetization, and the spins now precess during t_1 as in the COSY sequence. The second 90°_x pulse rotates components of the magnetization along the $-z$ axis. During the subsequent mixing period τ_m, z-magnetization components exchange under the influence of cross-relaxation. The transverse components are not required and are removed by phase cycling. The magnetization vectors therefore have a component in the z direction only after these procedures. Finally, a third 90°_x pulse regenerates observable magnetization. If this sequence is repeated for a larger value of t_1, the magnetization vectors would dephase further, and a smaller $-z$ component is thus created, which would pass through zero and become positive for increasing values of t_1. The changes are 'read' by the final 90° pulse. We observe diagonal peaks which fail to migrate during τ_m, and cross-peaks which are generated from the magnetization transfer (cross-relaxation) between two spins.

Note that in any practical application, a variety of cross-peaks can be generated by mechanisms other than cross-relaxation, and these cause confusion in interpreting the data. For example, chemical exchange cross-peaks occur (which can be valuable if we already know chemical exchange is taking place), but can be a source of undesirable

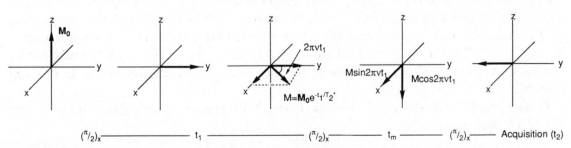

Fig. 2.18 The vector model for the NOESY pulse sequence.

confusion. Furthermore, in a weakly coupled system we have to consider the cross-relaxation pathways in systems of $\mathcal{J}$ coupled spins. This generates $\mathcal{J}$ cross-peaks which arise from coherence transfer. For this we have to use the product operator formalism to describe the system.

By analogy with the COSY pulse sequence, the system remains the same until after the second $90°$ pulse, when the density operator takes the form:

$$\sigma_3 = [-I_z \cos \omega_I t_1 + I_x \sin \omega_I t_1 - S_z \cos \omega_S t_1 + S_x \sin \omega_S t_1] \cos \pi \mathcal{J}_{IS} t_1$$

$$-[2I_x S_y \cos \omega_I t_1 + 2I_y S_x \cos \omega_S t_1 + 2I_z S_y \sin \omega_I t_1 + 2I_y S_z \sin \omega_S t_1] \sin \pi \mathcal{J}_{IS} t_1 \quad (2.8)$$

The terms $2I_x S_y$ and $2I_y S_x$ consist of a superposition of double- and zero-quantum coherence, which can be shown by expanding σ_3:

$$\sigma_3 = [-I_z \cos \omega_I t_1 + I_x \sin \omega_I t_1 - S_z \cos \omega_S t_1 + S_x \sin \omega_S t_1] \cos \pi \mathcal{J}_{IS} t_1$$

$$+ \{\tfrac{1}{2}[(2I_x S_y + 2I_y S_x)^{(1)} - (2I_y S_x - 2I_x S_y)^{(2)}] \cos \omega_I t_1$$

$$+ \tfrac{1}{2}[(2I_x S_y + 2I_y S_x)^{(3)} - (2I_y S_x - 2I_x S_y)^{(4)}] \cos \omega_I t_1$$

$$+ 2I_z S_y \sin \omega_I t_1 + 2I_y S_z \sin \omega_S t_1\} \sin \pi \mathcal{J}_{IS} t_1 \quad (2.9)$$

where the terms (1) and (3) represent double-quantum coherence, and (2) and (4) represent zero-quantum coherence. Since we are only interested in longitudinal magnetization components during the mixing period τ_m, it is necessary to remove all the components except $-I_z \cos \omega_I t_1$ and $-S_z \cos \omega_S t_1$. In order to do this, we can phase cycle the first two pulses by $90°$ each: $x, x, y, y, -x, -x, -y, -y$, and add each separate response, then this will cancel coherences of order $1, 2, 3, 5 \ldots$ and leave the zero-quantum coherences. Thus at the beginning of the mixing period we find that:

$$\sigma_3 = (-I_z \cos \omega_I t_1 - S_z \cos \omega_S t_1) \cos \pi \mathcal{J}_{IS} t_1 + (I_y S_x - I_x S_y)$$

$$\times [\cos \omega_S t_1 - \cos \omega_I t_1] \sin \pi \mathcal{J}_{IS} t_1 \quad (2.10)$$

The zero-quantum term $(I_y S_x - I_x S_y)$ evolves in a way that depends on the difference in the chemical shifts, and the longitudinal $(-I_z)$ terms mix under the effects of cross-relaxation. After the final $90°_x$ pulse, the observable magnetization is:

$$\sigma_4^{\text{obs}} = [I_y a_{II} \cos \omega_I t_1 + S_y a_{SS} \cos \omega_S t_1 + I_y a_{SI} \cos \omega_S t_1 + S_y a_{IS} \cos \omega_I t_1]$$

$$\times \cos \pi \mathcal{J}_{IS} t_1 + (I_z S_x - I_x S_z) \cos(\omega_I - \omega_S) \times \tau_m (\cos \omega_S t_1 - \cos \omega_I t_1) \sin \pi \mathcal{J}_{IS} t_1 \quad (2.11)$$

The terms in σ_4 represent diagonal peaks proportional to the mixing coefficients a_{II} and a_{SS}, and NOE cross-peaks proportional to a_{SI} and a_{IS}. These are all in phase with respect to $\mathcal{J}_{IS}$. Also, we find diagonal and cross-peaks which are antiphase with respect to $\mathcal{J}_{IS}$. These are the $\mathcal{J}$ cross-peaks derived from zero-quantum coherence which could not be removed by phase cycling, and have to be eliminated by other methods (e.g. random incremental variation of τ_m).

If we consider a three-spin system as shown, with typical internuclear distances (see Fig. 2.19) and define

$$\sigma_{IS} = W_2 - W_0 \quad (2.12)$$

and

$$\rho_I = 2W_1^I + W_2 + W_0 \tag{2.13}$$

then

$$\eta = \frac{\sigma_{IS}}{\rho_I + \rho^*} \tag{2.14}$$

where ρ^* takes into account additional relaxation mechanisms other than cross-relaxation. The ρ's represent the direct relaxation rates of the spins, and the σ's represent the cross-relaxation rates between them. However, equation (2.14) describes the *steady-state* NOE. In the case of the NOESY experiment, it is the *kinetic* NOE which is measured. The equations for the development of cross-peak intensities with time are therefore considerably more complex, and are given in Section 3.8. However, the development of NOESY cross-peak intensity with time is shown in Fig. 2.20.

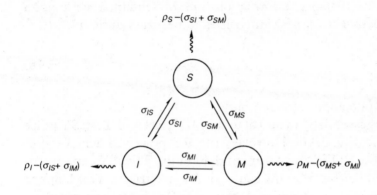

Fig. 2.19 A typical three-spin system. The ρs represent the direct relaxation rates of the spins, and the σs represent the cross-relaxation rates. (Redrawn from S. W. Homans, *A Dictionary of NMR Concepts*, 2nd edn, Oxford University Press (1992) with permission.)

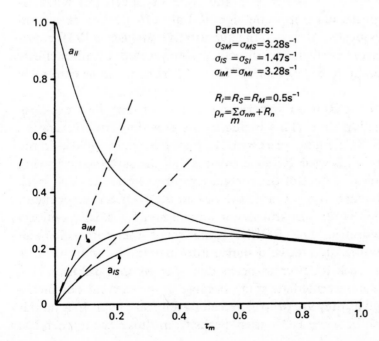

Parameters:
$\sigma_{SM} = \sigma_{MS} = 3.28s^{-1}$
$\sigma_{IS} = \sigma_{SI} = 1.47s^{-1}$
$\sigma_{IM} = \sigma_{MI} = 3.28s^{-1}$

$R_I = R_S = R_M = 0.5s^{-1}$
$\rho_n = \sum_m \sigma_{nm} + R_n$

Fig. 2.20 Time development of the intensities of the I spin diagonal peak (a_{II}) and the S and M spin cross-peaks (a_{IS} and a_{IM}) as a function of mixing time (τ_m) of the NOESY sequence. The Rs are the total direct relaxation rates of the spins. (Reprinted from S. W. Homans, *A Dictionary of NMR Concepts*, 2nd edn, Oxford University Press (1992) with permission.)

The I spin diagonal peak intensity is governed by a_{II} decays with increasing τ_m, whereas the S and M spin cross-peak intensities generated by a_{IS} and a_{IM} increase before decaying with increasing τ_m. The importance of the time development of cross-peak intensities is that the initial build-up rates of the S and M spin cross-peaks (i.e. at $\tau_m = 0$) are proportional to σ_{IS} and σ_{IM}. If, therefore, a_{IM} corresponds to a known fixed internuclear distance r_{IM}, r_{IS} can be calculated from the r^{-6} dependence of σ.

$$\frac{r_{IM}}{r_{IS}} = \left[\frac{\sigma_{IS}}{\sigma_{IM}} \right]^{1/6} \tag{2.15}$$

This is the principal method by which quantitative distance information is obtained in biomolecular NMR. A series of NOESY spectra is recorded as a function of mixing time in order to generate experimental plots similar to those obtained experimentally, as shown above. Alternatively, if one can be certain that the measured cross-peak intensities in a single NOESY experiment for a given mixing time were developed within the initial rate approximation, then quantitative distance information can be obtained with good accuracy by measurement of relative cross-peak intensities rather than initial rates:

$$\frac{r_{IM}}{r_{IS}} = \left[\frac{a_{IS}}{a_{IM}} \right]^{1/6} \tag{2.16}$$

Note that this is true for initial rate approximation only.

The ROESY pulse sequence is essentially identical to the HOHAHA or TOCSY pulse sequence (see Section 2.1.5, and Fig. 2.11). However, the RF power to achieve spin locking is significantly lower in the ROESY experiment than in the TOCSY experiment (and also the transmitter is usually offset from the centre of the spectrum). This helps to prevent the occurrence of TOCSY cross-peaks in the ROESY spectrum, which is important since the TOCSY peaks have opposite sign and are often much greater in intensity than the ROESY cross-peaks. Although the quantitative analysis of ROE cross-peaks for the calculation of internuclear distances directly is complicated (by offset effects and $\mathbf{B}_1$ field inhomogeneity), qualitative use of the ROESY experiment can be extremely useful.

The most valuable feature of the ROESY experiment is the fact that the cross-peak intensity increases with correlation time. This is due to the fact that cross relaxation occurs in the presence of a weak RF field rather than a large static magnetic field as in the case of the NOESY experiment. This weak RF field ensures that the extreme narrowing limit ($\omega_0 \tau_c \ll 1$) is maintained for all values of the correlation time. Thus the cross-peak intensities will not tend to zero when $\omega_0 \tau_c = 1$, as is the case in the NOESY experiment. For ^{1}H NMR at high field, this latter scenario is not uncommon for small peptides, oligonucleotides, and carbohydrates. The ROESY experiment therefore offers an important alternative for obtaining qualitative distance information. In addition, it is possible to determine whether weak ROE cross-peaks derive from spin diffusion (i.e. magnetization transfer through an intermediate spin), because in the extreme narrowing limit such peaks alternate in sign as they are propagated along a linear chain. In a similar way, it is possible to distinguish between ROE cross-peaks from those due to chemical exchange, again because they have opposite sign, whereas in the NOESY experiment, all spin diffusion and EXSY cross-peaks have the same sign. Thus the ROESY experiment is useful for intermediate sized molecules that give little or no conventional NOE (see Fig. 1.28), and it minimizes spin-diffusion for large molecules, although at the expense of a reduction in sensitivity.

2.1.8 Solvent suppression

Since the molarity of protons in water is 110 M, for ^{1}H NMR of 1–5 mM protein in 90% H_2O, there is a real problem in detecting the protein resonances in the presence of such a large H_2O resonance. There are two types of approach to water (or solvent) suppression:

 (i) pre-irradiation with low-power frequency-selective continuous-wave RF;

(ii) selective excitation of the water resonance with a series of either long, weak ('soft') pulses or short, strong ('hard') pulses separated by delays.

Method (i) is generally employed in protein NMR,[17] although there are disadvantages:

(a) requires considerable adjustment of magnetic field homogeneity ('shimming' the magnet), in order to include solvent outside the volume enclosed by the receiver coils;

(b) Bloch–Siegert effects at high decoupler power (slight changes in chemical shift of resonances near the resonance being irradiated);

(c) weak cross-peaks near the solvent are eliminated ('spill-over saturation').

Three variations of method (ii) are commonly employed with biological macromolecules. The Redfield 2–1–4 pulse[18] employs a sequence with 'soft' pulses:

$$0.2\tau - (\alpha)_0 - 0.4\tau - (\alpha)_{180} - 0.2\tau$$

where $\alpha = 0.1\tau$ in length with the power set such that $\alpha \approx 44°$. The disadvantages of this sequence are:

(a) exchange with solvent causes appearance of large magnetization transfer cross-peaks between exchanging protons and solvent peak;

(b) not possible to obtain all cross-peaks for whole spectral width.

The Guéron pulse sequence[19] uses 'hard' pulses:

$$(\pi/2)_y - \tau - (\pi/2)_{-y}$$

This so-called 'jump-and-return' sequence suffers from the same disadvantage (b) as the Redfield pulse sequence. Another pulse sequence, introduced by Hore,[20] recognized that the Guéron pulse sequence is a member of a complete family of sequences, of which the $1\bar{3}3\bar{1}$ is the best:

$$(\alpha)_x - \tau - 3(\alpha)_{-x} - \tau - 3(\alpha)_x - \tau - (\alpha)_{-x}$$

where $\alpha \approx 11°$. This is a good method, although not employed as often as it should be, principally because of the difficulty in setting up the experiment. Finally, there is a selective excitation-relaxation approach, in which a soft π pulse (such as the DANTE pulse sequence[21,22]) is used to invert the water resonance. When the magnetization has relaxed back to an exact null, the non-selective excitation (for example a COSY or NOESY pulse sequence) is applied.[23] The disadvantage of this method is that sequences such as DANTE have excitation sidebands that can suppress resonances of interest.

Since the introduction of these schemes, a number of alternative strategies for water suppression in multidimensional NMR have been proposed.[24,25] The effectiveness of each of these methods is very much dependent on the individual hardware, such as the quality of the RF, the positioning of the downleads of the probe coil, the quality of the magnet shimming, and so forth. The best advice for approaching this is to try out as many different water suppression techniques as possible, and find which works best for a given spectrometer configuration and user ability. A good review of solvent suppression methods has appeared recently.[26]

2.2 THREE-DIMENSIONAL NMR METHODS
2.2.1 Adding the third dimension

One of the problems with ^{1}H NOESY NMR of proteins is that all the NOEs must be resolved. This becomes increasingly difficult with higher molecular weight proteins, such as T$_4$ lysozyme ($M_r = 18.7$ kDa, 164 amino acids) (see Fig. 2.21). To solve this problem, a

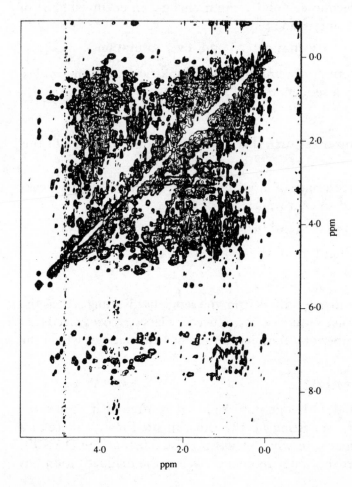

Fig. 2.21 The aliphatic region of a two-dimensional NOE spectrum of 4 mM T$_4$ lysozyme in D$_2$O. (Reprinted from S. Fesik and E. Zuiderweg, *Quat. Rev. Biophys.*, **23**, 97–131 (1990) with permission.)

whole family of pulse sequences has been introduced that involve three time variables rather than two, so that three-dimensional Fourier transformation leads to a cube rather than a contour map. They involve using an isotopic label for indirect detection via HMQC methods in combination with a type of COSY or NOESY sequence. For example, a simple ^{15}N-labelled tripeptide, whose one-dimensional HMQC was shown in Fig. 2.16 has the spectrum shown in Fig. 2.22. The ^{1}H NOESY—^{15}N HMQC three-dimensional experiment of T$_4$ lysozyme yields a spectrum in which all the cross-peaks are now resolved (see Fig. 2.23).

Almost all of these new three-dimensional (and four-dimensional) pulse sequences rely on heteronuclear and homonuclear couplings for correlating resonances from one residue to the next, in, for example, an oligopeptide, oligonucleotide, or oligosaccharide, isotopically labelled with ^{15}N and/or ^{13}C. We will discuss the most important of these pulse sequences in the next few sections. The $\mathcal{J}$ couplings that are involved, and the

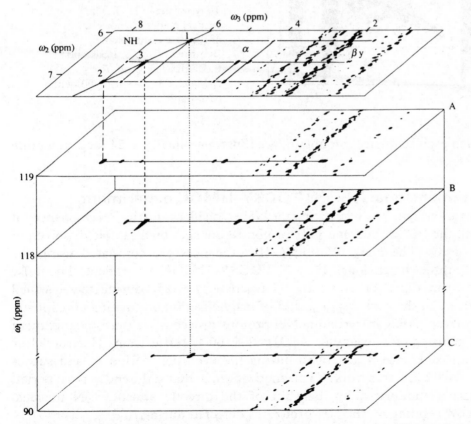

Fig. 2.22 (Top) ^{15}N-filtered two-dimensional NOESY spectrum of the ^{15}N-labelled tripeptide shown in Fig. 2.16(a). (Bottom) Three-dimensional HMQC–NOESY spectrum of the tripeptide. The vertical dashed lines connect amide diagonal peaks of the two-dimensional NOESY to the corresponding peaks in the individual ^{15}N planes of the three-dimensional data set. The horizontal lines indicate the NOE cross-peaks connected to each of the amides. (Reprinted from S. Fesik and E. Zuiderweg, *Quat. Rev. Biophys.*, **23**, 97–131 (1990) with permission.)

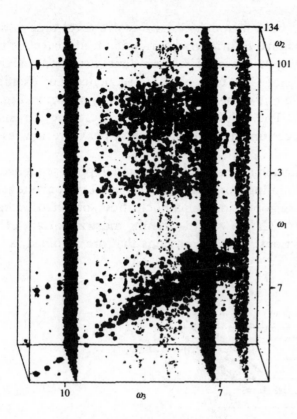

Fig. 2.23 Three-dimensional contour map of a ^{1}H—^{15}N NOESY–HMQC data set acquired on a 4 mM sample of T$_4$ lysozyme in H$_2$O. The vertical planes are noise artefacts of the data acquisition. (Reprinted from S. Fesik and E. Zuiderweg, *Quat, Rev. Biophys.*, **23**, 97–131 (1990) with permission.)

correlations which each sequence can obtain, are illustrated in Fig. 2.24 for a dipeptide fragment.

2.2.2 Three-dimensional ^{1}H—^{15}N TOCSY–HMQC experiment

The spin systems are assigned by correlating the NH resonances and the ^{15}N resonances of each residue with the ^{13}C$^\alpha$ carbon and C$^\alpha$H proton resonances on a sample dissolved in 90% H$_2$O/10% D$_2$O. The NH, ^{15}N and C$^\alpha$H resonances are correlated using the double-resonance three-dimensional ^{1}H—^{15}N TOCSY–HMQC experiment. The pulse sequence is shown in Fig. 2.25. Aliphatic ^{1}H resonances evolve during the so-called evolution period, t_1. In the next step, transfer of magnetization originating on aliphatic protons to the corresponding intraresidue NH protons proceeds via isotropic mixing of ^{1}H magnetization using, for example, a [DIPSI-2–60°]$_n$ pulse train. Heteronuclear multiple-quantum coherence is generated during the evolution period t_2, and subsequently converted back into transverse ^{1}H magnetization, so that at the end of the t_2 period ^{1}H magnetization is modulated by the shift of the directly attached ^{15}N nucleus. Magnetization now residing on the NH protons is detected during t_3.

In addition to providing intraresidue correlations between aliphatic and NH protons, the ^{15}N-separated TOCSY–HMQC experiment recorded with a short mixing time (<35 ms) can be used to obtain semi-quantitative estimates of $^3J_{\alpha\beta}$ coupling constants from the intensities of the well-resolved HN—C$^\beta$H cross-peaks. There is a dependency of the pertinent J coupling constants on the mixing time and a detailed account of this can be

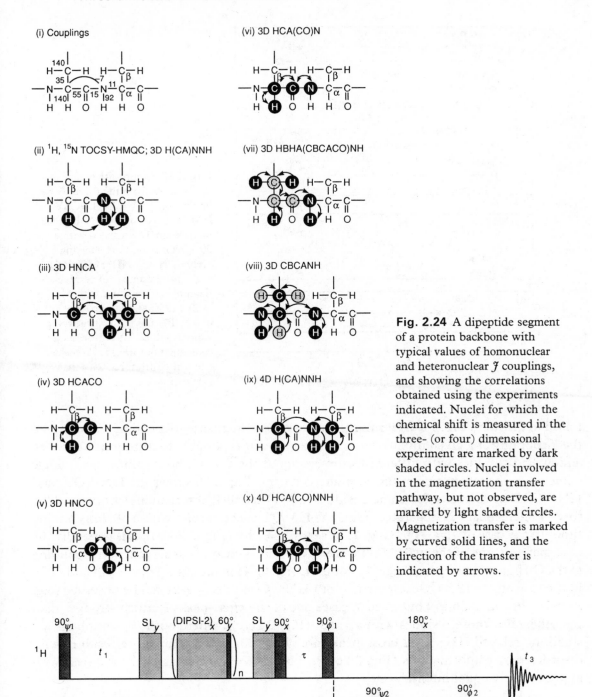

Fig. 2.24 A dipeptide segment of a protein backbone with typical values of homonuclear and heteronuclear $\mathcal{J}$ couplings, and showing the correlations obtained using the experiments indicated. Nuclei for which the chemical shift is measured in the three- (or four) dimensional experiment are marked by dark shaded circles. Nuclei involved in the magnetization transfer pathway, but not observed, are marked by light shaded circles. Magnetization transfer is marked by curved solid lines, and the direction of the transfer is indicated by arrows.

Fig. 2.25 Pulse sequence for the three-dimensional ^{1}H—^{15}N TOCSY–HMQC experiment. The phase cycling is as follows: $\psi_i = x, -x$; $\phi_2 = 2(x), 2(-x)$; $\phi_1 = 4(x), 4(y), 4(-x), 4(-y)$; receiver: $x, 2(-x), x, -y, -x, 2(y), -y, -x, 2(x), -x, y, 2(-y), y$. The delay τ is equal to half the DIPSI-2 mixing time to minimize ROE effects, and the delay Δ is set to slightly less than $1/(2\mathcal{J}_{\text{NH}})$.

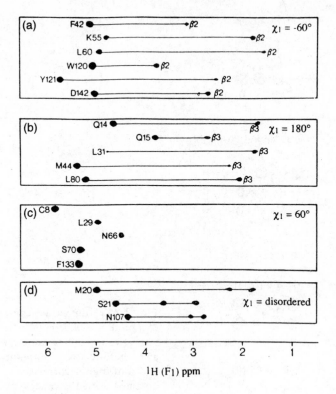

Fig. 2.26 Composite of amide strips taken from the three-dimensional ^1H—^{15}N TOCSY–HMQC spectrum of ^{15}N-labelled interleukin-1β recorded with a 30.7 ms mixing time showing NH—$C^\alpha H$ and NH—$C^\beta H$ cross-peaks for a number of residues. Examples of residues with $\chi_1 = -60°$, $180°$, and $60°$ are shown in (a), (b), and (c), respectively, while residues with disordered χ_1 conformations are shown in (d). (Reprinted from ref. 27 with permission.)

found in the literature.[27] Furthermore, the cross-peak intensities depend upon the size of the J coupling at a given mixing time. For example, it is possible to distinguish between amino acid residues with different rotamers about the C^α—C^β bond, since the χ_1 angle determines the magnitude of the coupling constant. Fig. 2.26 shows the HN—$C^\alpha H$ and HN—$C^\beta H$ cross-peaks for some residues in ^{15}N-labelled interleukin-1β obtained by three-dimensional ^{15}N-separated TOCSY–HMQC spectroscopy with a 30.7 ms mixing time. In the case of $\chi_1 = -60°$ (Fig. 2.26(a)) or $\chi_1 = 180°$ (Fig. 2.26(b)), it is generally the case that only one of the two possible HN—$C^\beta H$ cross-peaks is observed, corresponding to the $C^\beta H$ proton with the larger $^3J_{\alpha\beta}$ coupling (i.e. $C^{\beta2}H$ in the case of $\chi_1 = -60°$ and $C^{\beta3}H$ in the case of $\chi_1 = 180°$). Occasionally, both HN—$C^\beta H$ cross-peaks can be observed (e.g. Asp-142 in Fig. 2.26(a)), but in such cases one of the cross-peaks is much stronger than the other. For those residues with $\chi_1 = 60°$ (Fig. 2.26(c)), where both $^3J_{\alpha\beta}$ couplings are small, no relayed HN—$C^\beta H$ cross-peaks are observed. Finally, for those residues with disordered χ_1 conformations (Fig. 2.26(d)), both HN—$C^\beta H$ cross-peaks are seen with approximately equal intensity.

2.2.3 Three-dimensional H(CA)NNH experiment

The NH, ^{15}N and $C^\alpha H$ resonances can also be correlated using the triple resonance $H^\alpha(C^\alpha)$—^{15}N—NH (also known as the H(CA)NNH) experiment.[28,29] The pulse sequence is shown in Fig. 2.27. The H^α chemical shifts evolve during the evolution period t_1 and ^1H magnetization is subsequently transferred by an INEPT sequence to the directly coupled C^α spin. Immediately prior to the transfer, the water and antiphase H^α spins are

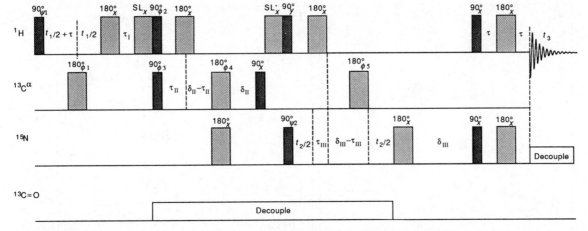

Fig. 2.27 Pulse sequence for the three-dimensional H(CA)NNH experiment. The phase cycling is as follows: $\psi_1 = x$; $\psi_2 = 16(x), 16(-x)$; $\phi_1 = \phi_2 = x, -x$; $\phi_3 = 2(x), 2(-x)$; $\phi_4 = 4(x), 4(y), 4(-x), 4(-y)$; $\phi_5 = 8(x)$, $8(-x)$. Typical durations for the delays are $\tau_1 = 1.5$ ms, $\tau_{II} = 1.7$ ms, $\delta_{II} = 12.5$ ms, $\tau_{III} = 2.75$ ms, $\delta_{III} = 11.5$ ms, and $\tau = 2.25$ ms, slightly less than $1/(4J_{NH})$. Water suppression is achieved using two purge pulses SL_x and SL'_x, typically applied for 1.5 and 9 ms respectively. GARP modulation is used for carbonyl decoupling, while ^{15}N decoupling during acquisition can be accomplished with either WALTZ or GARP modulation.

orthogonal so that a ^{1}H trim pulse can be applied to suppress the water signal. After application of the 90° ^{1}H and ^{13}C$^\alpha$ pulse, the antiphase C$^\alpha$ polarization is refocused with respect to the H$^\alpha$ spins during the period $2\tau_{II}$, after which ^{13}C$^\alpha$—^{15}N dephasing occurs during the interval $2\delta_{II}$. By concatenating the 180° pulses, the ^{13}C$^\alpha$—^{1}H$^\alpha$ refocusing and ^{13}C$^\alpha$—^{15}N defocusing periods overlap, thereby shortening the total length of the delays required. The next ^{13}C 90° pulse creates zz carbon–nitrogen magnetization ($S_z N_z$, where S and N are carbon and nitrogen respectively), and the following ^{1}H trim and $90°_y$ pulses serve to suppress the water resonance further. Magnetization is then transferred to ^{15}N by the application of a ^{15}N 90° pulse and ^{15}N chemical shifts evolve during the period t_2. The effects of ^{1}H—^{15}N and ^{13}C$^\alpha$—^{15}N J coupling are removed during this time by application of ^{1}H and ^{13}C 180° pulses. ^{15}N magnetization is refocused with respect to the C$^\alpha$ spin during the delay $2\delta_{III}$ and defocused with respect to the directly coupled NH proton during $2\tau_{III}$ to allow transfer back to NH protons via another INEPT sequence. Finally, amide protons are detected during t_3 with ^{15}N decoupling. The effects of one-bond C$^\alpha$—C′ (C′ = carbonyl) and ^{1}H—^{15}N couplings are removed by composite pulse decoupling (the sequence GARP is often used[30]) during the time that C$^\alpha$ transverse magnetization evolves and during the acquisition period.

Two sets of correlations are observed in the three-dimensional H(CA)NNH experiment: the intraresidue connectivity between C$^\alpha$H(i), ^{15}N(i) and NH(i) arising from transfer of magnetization via one bond ^{13}C$^\alpha$—^{15}N coupling (≈ 11 Hz), and the sequential interresidue connectivity between C$^\alpha$H($i-1$), ^{15}N(i) and NH(i) via the two-bond ^{13}C$^\alpha$—^{15}N coupling. This is illustrated in Fig. 2.28 for two NH(F_3)—C$^\alpha$H(F_1) planes of the three-dimensional H(CA)NNH spectrum of [^{15}N, ^{13}C]calmodulin.

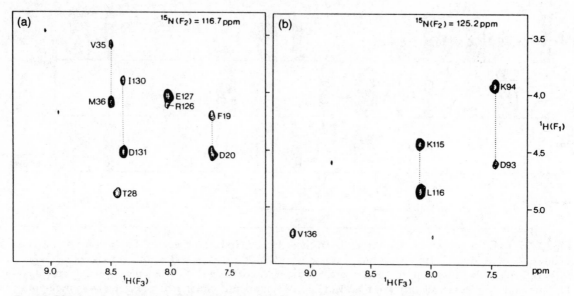

Fig. 2.28 Examples of two NH(F_3)—C$^{\alpha}$H(F_1) planes at different ^{15}N (F_2) frequencies of the three-dimensional H(CA)NNH spectrum of 1.5 mM ^{15}N/^{13}C labelled calmodulin complexed with a 26-residue fragment of myosin light chain kinase in 90% H$_2$O/10% D$_2$O. Strong intraresidue C$^{\alpha}$H(i)—NH(i)—^{15}N(i) and weaker interresidue C$^{\alpha}$H(i-1)—NH(i)—^{15}N(i) correlations are observed. (Reprinted from ref. 29 with permission.)

2.2.4 Three-dimensional HNCA experiment

The HNCA experiment correlates ^{15}N and NH chemical shifts with the intraresidue C$^{\alpha}$ shift. The HNCA and TOCSY–HMQC experiments together establish the intraresidue correlations between pairs of ^{15}N—NH and C$^{\alpha}$—H$^{\alpha}$ backbone resonances. These pairs of resonances are linked through two separate and independent pathways: for TOCSY–HMQC via ^{1}H—^{1}H $^{3}\mathcal{J}$ coupling (5–10 Hz) and for HNCA via $^{1}\mathcal{J}_{NC\alpha}$ coupling (8–12 Hz). For almost all residues this provides unambiguous correlation of the NH, H$^{\alpha}$, ^{15}N, and C$^{\alpha}$ shifts, despite overlap in the ^{15}N—^{1}H and ^{13}C$^{\alpha}$—^{1}H two-dimensional correlation spectra. The two-bond ^{13}C$^{\alpha}$—^{15}N connectivity, which can be as large as 7 Hz, observed in the HNCA spectrum, provides additional and independent sequential connectivity information. The pulse sequence for the HNCA experiment is shown in Fig. 2.29. An INEPT sequence is used to transfer magnetization originally residing on the NH protons to the directly attached ^{15}N spin. The ^{15}N chemical shifts evolve during the period t_1, with ^{13}C$^{\alpha}$ and ^{1}H decoupling achieved by the application of 180° ^{1}H and ^{13}C pulses in the middle of t_1. During the delay δ, ^{15}N magnetization becomes antiphase with respect to the coupled ^{13}C$^{\alpha}$ spins. The delay δ is set to an integral multiple of $1/(\mathcal{J}_{NH})$, so that ^{15}N magnetization remains antiphase with respect to the coupled proton. To minimize relaxation losses while still optimizing transfer via two-bond ^{15}N—^{13}C$^{\alpha}$ couplings, δ is best set to $1/(3\mathcal{J}_{NC\alpha})$ or 33 ms in practice. Subsequent application of 90° pulses simultaneously to both ^{1}H and C$^{\alpha}$ spins establishes three-spin NH—^{15}N—^{13}C$^{\alpha}$ coherence (i.e. the term, $I_y N_x A_y$ is generated where I, N, and A refer to ^{1}H, ^{15}N, and ^{13}C spins). Evolution of just the ^{13}C$^{\alpha}$ chemical shifts during the period t_2 is ensured by refocusing of the ^{1}H and ^{15}N chemical shifts through the

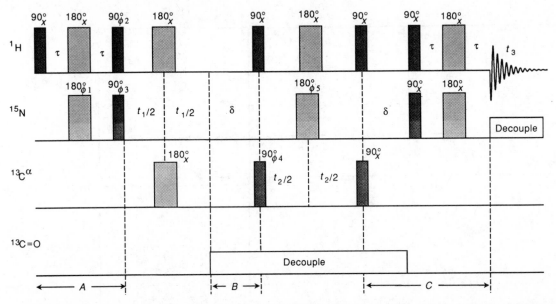

Fig. 2.29 Pulse sequence for the three-dimensional HNCA experiment. The delay δ is adjusted to be an integral multiple of $1/\mathcal{J}_{NH}$ and to allow maximal magnetization transfer between the ^{15}N and C^{α} spins. The phase cycling employed is as follows: $\phi_1 = x, -x$; $\phi_2 = y; -y$; $\phi_3 = x$; $\phi_4 = 2(x), 2(-x)$; $\phi_5 = 4(x), 4(y), 4(-x), 4(-y)$; receiver $= x, 2(-x), x, -x, 2(x), -x$.

application of ^{1}H and ^{15}N $180°$ pulses at the midpoint of the t_2 period. Magnetization is then transferred back to NH protons by simply reversing the above procedures.

In terms of product operator formalism, let us consider only those terms that contribute to the final spectrum, and the effects of certain groups of pulses can be combined into operators **A**, **B**, and **C**. Assuming that the two-bond coupling between ^{15}N(i) and ^{13}C$^{\alpha}(i-1)$ is zero and that the delay $\delta = 1/(2\mathcal{J}_{NA})$ (where $\mathcal{J}_{NA}$ is the one-bond ^{15}N—^{13}C$^{\alpha}$ coupling), the evolution of magnetization during the course of the experiment can be described as:

$$I_z \xrightarrow{\ \mathbf{A}\ } -2I_zN_y \xrightarrow{\ t_1\ } 2I_zN_y \cos \Omega_N t_1 \cos \pi\mathcal{J}_{NC'}t_1$$

$$\xrightarrow{\ \mathbf{B}\ } -4I_yN_xA_y \cos \Omega_N t_1 \cos \pi\mathcal{J}_{NC'}t_1$$

$$\xrightarrow{\ t_1\ } 4I_yN_xA_y \cos \Omega_N t_1 \cos \pi\mathcal{J}_{NC'}t_1 \cos \Omega_A t_2$$

$$\xrightarrow{\ \mathbf{C}\ } I_x \cos \Omega_N t_1 \cos \pi\mathcal{J}_{NC'}t_1 \cos \Omega_A t_2 \qquad (2.17)$$

where I, N, and A denote the intraresidue NH, ^{15}N, and ^{13}C$^{\alpha}$ spins, and the carbonyl spin of the preceding residue is denoted by C'. The chemical shifts of the intraresidue ^{15}N and ^{13}C$^{\alpha}$ spins are denoted as Ω_N and Ω_A, and $\mathcal{J}_{NC'}$ is the interresidue ^{15}N—^{13}CO coupling constant. If a significant two-bond coupling, $^2\mathcal{J}_{NC\alpha(i-1)}$, exists between ^{15}N($i$) and the preceding ^{13}C$^{\alpha}(i-1)$ spin, A', the signal immediately prior to detection is given by

$$I_x \cos \Omega_N t_1 \cos \pi\mathcal{J}_{NC'}t_1 \{\cos \Omega_A t_2 [\sin \pi\mathcal{J}_{NA}\delta \cos \pi\mathcal{J}_{NA'}\ \delta]^2$$

$$+ \cos \Omega_{A'} t_2 [\sin \pi\mathcal{J}_{NA'}\ \delta \cos \pi\mathcal{J}_{NA}\delta]^2\} \qquad (2.18)$$

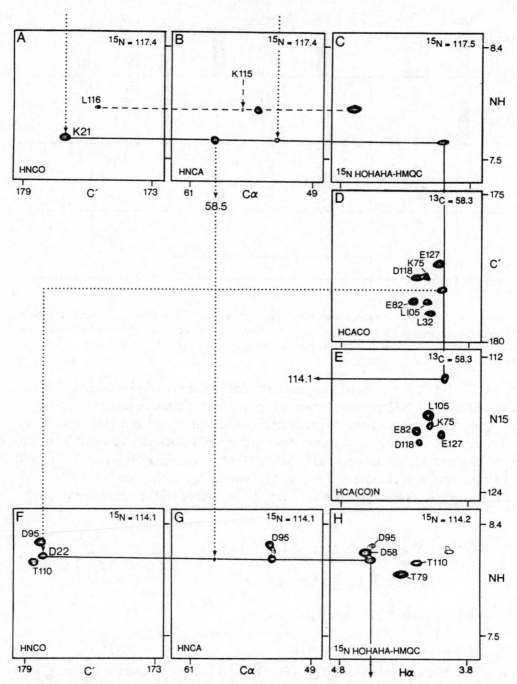

Fig. 2.30 Illustration of the three-dimensional triple resonance correlation experiments to obtain sequential assignments via one-bond and two-bond heternocuclear J couplings between Lys-21 and Asp-22 of $^{15}N/^{13}C$ labelled calmodulin. Solid and dotted lines trace the connectivity patterns for these two residues. Broken lines correspond to parts of the connectivity patterns observed for other residues. Slices A–C are taken at the Lys-21 ^{15}N chemical shift. Slices D and E are taken at the Lys-21 C^x shift, observed in B. Slices F–H are taken at the frequency of Asp-22, as measured in E. The three-dimensional ^1H—^{15}N TOCSY–HMQC, HNCA, and HNCO experiments are recorded in H_2O, while the HCACO, and HCA(CO)N experiments are recorded in D_2O. (Reprinted from M. Ikura, L. E. Kay, and A. Bax, *Biochemistry*, **29**, 4659–67 (1990) with permission.)

In both cases, Equations (2.17) and (2.18) indicate that the detected signal is amplitude modulated in both the t_1 and t_2 dimensions, leading to pure absorption lineshapes after three-dimensional Fourier transformation. Note also that since the acquisition time usually employed in the ^{15}N dimension is shorter than $1/(2\mathcal{J}_{NC'})$, then the $\mathcal{J}_{NC'}$ coupling present during t_1 has little effect on the signal.

Two sets of correlations are obtained for the HNCA experiment: the intraresidue $C^\alpha(i)$—^{15}N(i)—NH(i) connectivity via one bond $^1\mathcal{J}_{NC\alpha}$ coupling (8–12 Hz) and the interresidue $C^\alpha(i-1)$—^{15}N(i)—NH(i) connectivity via two-bond $^2\mathcal{J}_{NC\alpha}$ coupling (5–7 Hz). Examples of NH(F_3)—^{13}C(F_2) planes at different ^{15}N chemical shifts of the HNCA spectrum of ^{15}N/^{13}C labelled calmodulin are shown in Fig. 2.30.

2.2.5 Three-dimensional HCCH–COSY and HCCH–TOCSY experiments

The side-chain spin systems are identified by correlating aliphatic ^{1}H resonances with their attached ^{13}C resonances via HCCH–COSY and HCCH–TOCSY experiments on samples dissolved in D_2O. Both experiments are based on analogous principles and make use of the well-resolved one-bond ^{1}H—^{13}C (140 Hz) and ^{13}C—^{13}C (30–40 Hz) couplings to transfer magnetization along the side chain via the pathway:

$$^1H \xrightarrow{\ ^1\mathcal{J}_{CH}\ } {}^{13}C \xrightarrow{\ ^1\mathcal{J}_{CC}\ } {}^{13}C \xrightarrow{\ ^1\mathcal{J}_{CH}\ } {}^1H$$

This approach circumvents the problems of poorly resolved three bond ^{1}H—^{1}H couplings (3–12 Hz).

The pulse sequences are shown in Fig. 2.31. After the evolution period t_1, ^{1}H magnetization (H_A) is transferred to its directly bonded ^{13}C nucleus (C_A) via the $^1\mathcal{J}_{CH}$ coupling in an INEPT-type sequence. The ^{13}C magnetization which is antiphase with respect to the polarization of H_A becomes in-phase during the delay $2\delta_1$, and ^{13}C magnetization evolves during the evolution period t_2 under the influence of the ^{13}C chemical shift. The 180° ^{1}H pulse during t_2 removes the effects of ^{1}H—^{13}C coupling and the 180° ^{13}CO pulse eliminates $\mathcal{J}_{C\alpha C'}$ dephasing of the ^{13}C$^\alpha$ magnetization. During t_2 and the delay δ_1, C_A magnetization becomes antiphase with respect to its ^{13}C coupling partner C_B at a rate that is determined by $^1\mathcal{J}_{CC}$. In the case of the HCCH–COSY experiment, a 90° ^{13}C pulse then transfers this antiphase C_A magnetization into antiphase C_B magnetization in a COSY-like fashion. During the interval $2(\Delta+\delta_2)$, the antiphase C_B magnetization becomes in-phase with respect to C_A, and the 180° ^{1}H pulse (applied at time δ_2 before the final set of simultaneous 90° pulses) ensures that the refocused C_B magnetization is antiphase with respect to its attached proton H_B. Finally, C_B magnetization is transferred to H_B by a reverse refocused INEPT which is detected during t_3. In the case of the HCCH–TOCSY, a short ^{13}C trim pulse is applied along the x axis at the end of the t_2 period, which defocuses all in-phase ^{13}C magnetization that is not parallel to the effective field axis, and the subsequent isotropic mixing (using the DIPSI-3 composite pulse) transfers the net ^{13}C magnetization to its neighbours. Finally, the ^{13}C magnetization is transferred back to the attached protons by a reverse INEPT sequence and detected in t_3.

The final result in both cases is a three-dimensional spectrum in which each

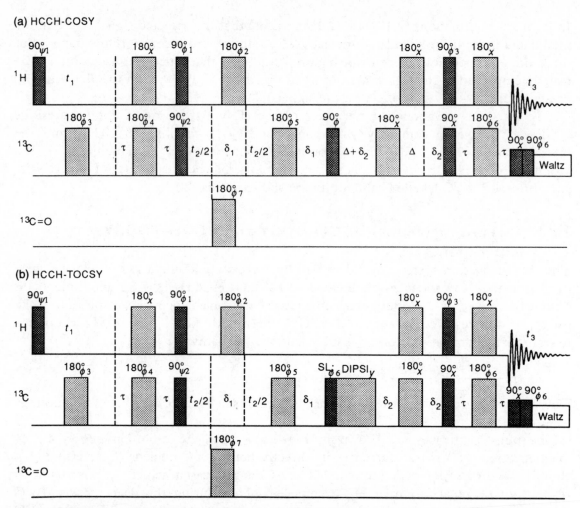

Fig. 2.31 Pulse sequences for the three-dimensional HCCH–COSY and HCCH–TOCSY experiments. The phase cycling is as follows: $\phi_1 = y, -y$; $\phi_2 = 4(x), 4(y), 4(-x), 4(-y)$; $\phi_3 = 8(x), 8(-x)$; $\phi_4 = 2(x), 2(-x)$; $\phi_5 = 2(x), 2(y), 2(-x), 2(-y)$; $\phi_6 = 4(x), 4(-x)$; $\phi_7 = 8(x)$; $\psi_1 = x$; $\psi_2 = x$; receiver $= 2(x, -x, -x, x), 2(-x, x, x, -x)$. The delay τ is set to 1.5 ms, slightly less than $1/(4J_{CH})$, the delays δ_1 and δ_2 are set to $\sim 1/(6J_{CH}) \sim 1.1$ ms, and the delay $\Delta + \delta_1$ is set to $\sim 1/(8J_{CC}) \sim 3.25$ ms. The 180_{ϕ_7} carbonyl pulse can be an off-resonance DANTE pulse. ^{13}C decoupling during the acquisition can be achieved with WALTZ or GARP modulation. The 90_x–90_{ϕ_6} pulse pair immediately prior to the start of the WALTZ decoupling serves to reduce the intensity of modulation sidebands originating from incompletely refocused ^{1}H magnetization ($I_y S_z$ type terms). This pulse pair has the effect of inverting S_x on alternate scans, cancelling signals originating from $I_y S_z$.

^{1}H(F_1)—^{1}H(F_3) has an appearance similar to that of a two-dimensional ^{1}H—^{1}H COSY or TOCSY, but is edited by the ^{13}C chemical shifts for ^{13}C nuclei directly bonded to protons, at the diagonal where the magnetization originates. Also, in contrast to two-dimensional correlation experiments, the cross-peaks are not symmetric about the diagonal. This is illustrated in Fig. 2.32, where magnetization transferred from proton A to proton B shows a correlation between the diagonal at $(F_1, F_3) = (\delta_A, \delta_A)$ and the cross-peak at $(F_1, F_3) = (\delta_A, \delta_B)$ in one half of the spectrum. The symmetric correlation between the diagonal

$(F_1, F_3) = (\delta_B, \delta_B)$ and the cross-peak at $(F_1, F_3) = (\delta_B, \delta_A)$ occurs in the plane corresponding to the chemical shift of ^{13}C nucleus directly bonded to proton B.

The HCCH–COSY experiment is particularly useful for identifying Gly, Ala, Thr, and Val spin systems, as well as for amino acids with the AMX spin system (e.g. Ser, Cys, Asn, Asp, His, Thr, Phe, and Trp). Fig. 2.33 illustrates the use of the HCCH–COSY

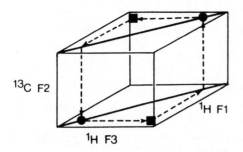

Fig. 2.32 Schematic diagram of two slices of an HCCH spectrum at different ^{13}C(F_2) chemical shifts showing the diagonal and cross-peaks expected for a simple two-spin system. The diagonal peaks are represented by circles and the cross-peaks by squares. Note that each slice is asymmetric about the diagonal, and cross-peaks only appear in the F_3 dimension. This is due to the fact that magnetization originating from a proton attached to a given ^{13}C nucleus is only visible in the F_1–F_3 plane at the ^{13}C(F_2) chemical shift of this particular ^{13}C nucleus. The corresponding cross-peak at the other side of the $F_1 = F_3$ diagonal is found in the plane taken at the ^{13}C frequency of the destination carbon. (Reprinted from G. M. Clore, A. Bax, P. C. Driscoll, P. T. Wingfield, and A. M. Gronenborn, *Biochemistry*, **29**, 8172–84 (1990) with permission.)

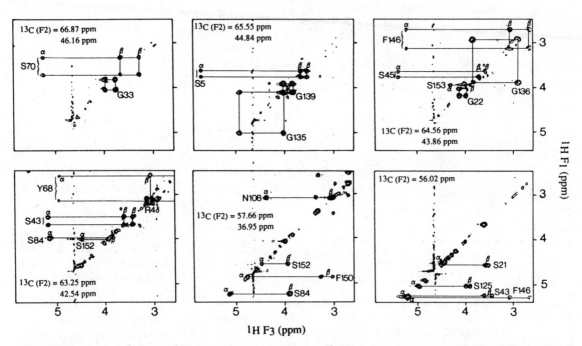

Fig. 2.33 Selected ^{1}H(F_3)—^{1}H(F_1) planes at different ^{13}C(F_2) frequencies of the three-dimensional HCCH–COSY spectrum of 1.7 mM ^{13}C/^{15}N-labelled interleukin-1β in D$_2$O, principally illustrating connectivities involving Gly and Ser residues. Because extensive folding is employed (i.e. resonances outside the sweep width used appear in the spectrum with opposite phase, folded about the axis of the edge of the sweep width), the ^{13}C chemical shifts are given by $x \pm nSW$, where x is the ppm value listed in the diagram, n is an integer, and SW is the sweep width (20.71 ppm). (Reprinted from G. M. Clore, A. Bax, P. C. Driscoll, P. T. Wingfield, and A. M. Gronenborn, *Biochemistry*, **29**, 8172–84 (1990) with permission.)

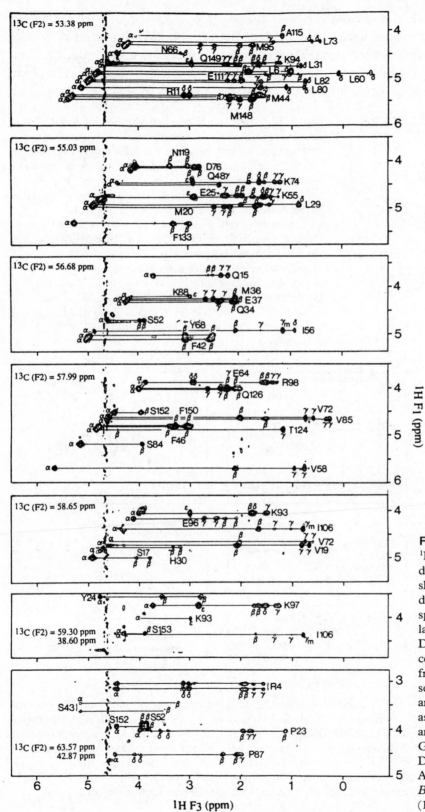

Fig. 2.34 Selected $^1H(F_3)$—$^1H(F_1)$ planes at different $^{13}C(F_2)$ chemical shifts of the three-dimensional HCCH–TOCSY spectrum of 1.7 mM $^{15}N/^{13}C$ labelled interleukin-1β in D$_2$O illustrating relayed connectivities originating from the C^{x}H proton of several longer side-chain amino acid spin systems such as Glu, Gln, Met, Pro, Arg, and Lys. (Reprinted from G. M. Clore, A. Bax, P. C. Driscoll, P. T. Wingfield, and A. M. Gronenborn, *Biochemistry*, **29**, 8172–84 (1990) with permission.)

experiment in the assignment of Gly and Ser spin systems in ^{15}N/^{13}C labelled interleukin-1β. Both these residues tend to pose problems in conventional ^{1}H—^{1}H correlation spectra as their proton resonances are very close to each other. Each slice corresponds to several ^{13}C shifts separated by 20.71 ppm. Gly ^{13}C$^\alpha$ shifts occur at 42–46 ppm, whereas Ser ^{13}C$^\alpha$ shifts occur at 55–60 ppm and Ser ^{13}C$^\alpha$ shifts occur at 61–67 ppm. The two Gly C$^\alpha$H protons are attached to the same C$^\alpha$ carbon, so that a symmetric pattern appears about the diagonal. This is also true for the β-methylene protons of Ser, but these are also associated with cross-peaks to the C$^\alpha$H protons. Therefore there is no difficulty in deciding which peaks originate from C$^\alpha$H protons of Gly and C$^\beta$H protons of either Ser or another AMX spin system such as Phe, whose C$^\beta$ has a similar chemical shift to Gly C$^\alpha$. The assignment of the Ser spin system is easily checked by examining slices corresponding to the C$^\alpha$ and C$^\beta$ shifts of Ser.

In the case of Gln, Glu, Met, and Leu, where the chemical shifts of the C$^\beta$H and C$^\gamma$H protons are often degenerate, and also the longer spin systems such as Arg, Pro, and Lys, then relayed through-bond connectivities are required. The ambiguities in an HCCH–COSY can usually be resolved using the HCCH–TOCSY. Examples of different sorts of relayed connectivities are shown in Fig. 2.34, which shows HCCH–TOCSY spectra of ^{15}N/^{13}C labelled interleukin-1β. The relayed connectivities originating from the C$^\alpha$H proton of different amino acid spin systems are illustrated. Complete spin systems are clearly delineated not only for medium length side chains such as those of Glu, Gln, Met, and Val, but also for the long side chains of Leu, Ile, Pro, Arg, and Lys.

2.2.6 Three-dimensional HCACO experiment

The HCACO three-dimensional spectra link the carbonyl resonances with the intraresidue C$^\alpha$ and H$^\alpha$ resonances. The pulse sequence is shown in Fig. 2.35. Magnetization is transferred from the H$^\alpha$ protons to the directly coupled C$^\alpha$ spins via an INEPT sequence, and C$^\alpha$ magnetization evolves during the period t_1 under the influence of ^{13}C$^\alpha$ chemical shifts, as well as the ^{13}C$^\alpha$—^{13}CO and ^{13}C$^\alpha$—^{13}C$^\beta$ J couplings, while the H$^\alpha$—^{13}C$^\alpha$ J coupling is removed by the 180° ^{1}H pulse applied at the midpoint of t_1. Transfer of magnetization occurs next in a COSY-like fashion from C$^\alpha$ spins to C′ spins via the $^1J_{C\alpha C'}$ coupling, by application of simultaneous 90° C$^\alpha$ and C′ pulses at the end of the t_1 period. Carbonyl evolution proceeds during t_2, while the net effects of ^{1}H—C′ and C$^\alpha$—C′ J couplings are removed by 180° ^{1}H and ^{13}C$^\alpha$ pulses at the midpoint of t_2. Decoupling of ^{15}N during t_2 is accomplished by irradiation of ^{15}N with the WALTZ composite decoupling scheme.[31] At the end of the t_2 period, C′ magnetization is transferred back to the C$^\alpha$ spin, at which point the C$^\alpha$ magnetization is antiphase with respect to the C′ spins. This antiphase magnetization refocuses during the subsequent interval 2Δ, and to minimize the loss of magnetization due to relaxation and dephasing by coupling of the ^{13}C$^\alpha$ spin to the ^{13}C$^\beta$ spin, 2Δ should be set to $\approx 1/(3J_{C\alpha C'})$, 6 ms in practice. The dephasing arising from the C$^\alpha$—C$^\beta$ coupling is also reduced by application of a semiselective 180° C$^\alpha$ pulse, which results in good inversion of the C$^\alpha$ spins, but poor inversion of the C$^\beta$ resonances. At the end of the 2Δ delay, magnetization is transferred back to the C$^\alpha$H spins by a reverse INEPT sequence and detected during t_3. Since the

HCACO experiment detects H^α resonances during t_3 and does involve the amide protons, it can be performed in D_2O.

Using product operator formalism, the HCACO experiment may be described as follows:

$$I_z \xrightarrow{\quad \mathbf{A} \quad} -2I_zA_y \xrightarrow{\quad t_1 \quad} -4I_zA_xS_z \cos \Omega_A t_1 \sin \pi \mathcal{J}_{AS}t_1 \cos \pi \mathcal{J}_{AB}t_1$$

$$\xrightarrow{\quad \mathbf{B} \quad} -4I_zA_zS_y \cos \Omega_A t_1 \sin \pi \mathcal{J}_{AS}t_1 \cos \pi \mathcal{J}_{AB}t_1$$

$$\xrightarrow{\quad t_2 \quad} -4I_zA_xS_z \cos \Omega_A t_1 \sin \pi \mathcal{J}_{AS}t_1 \cos \pi \mathcal{J}_{AB}t_1 \cos \Omega_S t_2$$

$$\xrightarrow{\quad \mathbf{C} \quad} I_x \cos \Omega_A t_1 \sin \pi \mathcal{J}_{AS}t_1 \cos \pi \mathcal{J}_{AB}t_1 \cos \Omega_S t_2 \qquad (2.19)$$

where I, A, and S refer to the H^α, C^α, and C' spins, Ω_A and Ω_S are the C^α and C' chemical shifts, and $\mathcal{J}_{AS}$ is the one-bond C^α—C' coupling (≈ 55 Hz) and $\mathcal{J}_{AB}$ is the C^α—C^β coupling (≈ 40 Hz). Equation (2.19) indicates that both the active $\mathcal{J}_{AS}$ and passive $\mathcal{J}_{AB}$ couplings are present during the evolution period t_1. Phasing the F_1 dimension in absorption mode would result in the superposition of two antiphase F_1 doublet components. Consequently, it is wise to phase F_1 in dispersive mode. As equation (2.19) indicates, the lineshapes in both the F_2 and F_3 dimensions are purely absorptive. An example is shown in Figure 2.30.

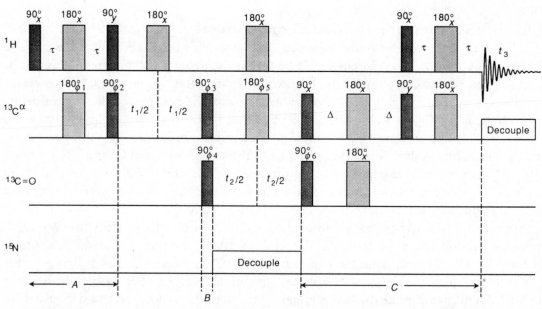

Fig. 2.35 Pulse sequence for the three-dimensional HCACO experiment. Typical delay durations are $\tau = 1.5$ ms and $\Delta = 3$ ms. The power levels of the C^α and C' pulses are adjusted such that during application of a C^α pulse minimal excitation of C' occurs, and vice versa. In order to minimize the effects of homonuclear $\mathcal{J}$ modulation by the passive C^α—C^β couplings during the intervals Δ, a semiselective 180°_x (C^α) pulse may be used. The phase cycling employed is as follows: $\phi_1 = x, -x$; $\phi_2 = x$; $\phi_3 = y, -y$; $\phi_4 = 2(x), 2(-x)$; $\phi_5 = 4(x)$, $4(-x)$; $\phi_6 = 8(x), 8(-x)$; receiver $= 2(x, -x, -x, x), 2(-x, x, x, -x)$.

2.2.7 Three-dimensional HNCO experiment

The HNCO three-dimensional experiment provides crucial sequential connectivities linking the ^{15}N—^{1}H pair of one residue with the carbonyl resonance of the preceding residue. Because the amide protons are involved, this experiment must be recorded in H_2O. When the absolute chemical shifts of the carbonyl resonances in H_2O and D_2O are compared, a small isotope shift (see Chapter 6 for a general discussion of this effect) of ≈ 0.08 ppm has to be taken into account. Limited digital resolution and minor pH differences can lead to difficulties in assigining carbonyl resonances in overlapping regions of the spectrum.

The pulse sequence is shown in Fig. 2.36. Magnetization originating from NH protons is transferred to the directly bonded ^{15}N spins using an INEPT sequence, following which ^{15}N magnetization evolves exclusively under the influence of the ^{15}N chemical shifts as a result of ^{1}H, C′, and C$^{\alpha}$ decoupling through the application of 180° pulses at the midpoint of t_1. During the delay δ, ^{15}N magnetization becomes antiphase with respect to the polarization of the carbonyl spin of the preceding residue, via the one-bond coupling $^{1}J_{NC'}$, which is of the order of 15 Hz. The first 90° C′ pulse converts the antiphase ^{15}N magnetization into ^{15}N—C′ zero- and double-quantum coherence. Evolution of C′ chemical shifts then occurs during the period t_2, and the contribution of ^{15}N chemical shifts and ^{1}H—^{15}N J coupling are removed by the application of a 180° ^{15}N pulse at the midpoint of t_2, and the C$^{\alpha}$—C′ J coupling removed with a 180° C$^{\alpha}$ pulse at the same point. Magnetization is then transferred back to the NH proton by reversing the transfer steps, and is detected in t_3.

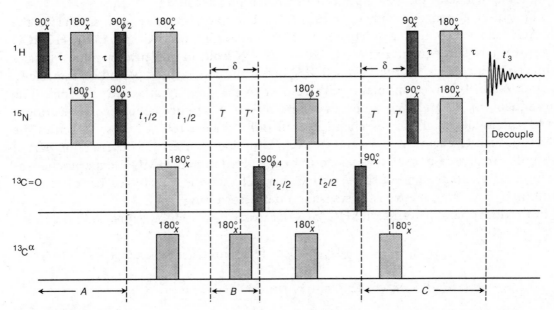

Fig. 2.36 Pulse sequence for the three-dimensional HNCO experiment. Pre-saturation of the water resonance is required during the recovery period and during both δ periods. The phase cycling employed is as follows: $\phi_1 = x, -x$; $\phi_2 = 2(y), 2(-y)$; $\phi_3 = x$; $\phi_4 = 4(x), 4(-x)$, $\phi_5 = 8(x), 8(y), 8(-x), 8(-y)$; receiver = $2(x)$, $4(-x), 2(x), 2(-x), 4(x), 2(-x)$.

Using product operator formalism, the HNCO experiment is as follows:

$$I_z \xrightarrow{\text{A}} -2I_zN_y \xrightarrow{t_1} 2I_zN_y \cos \Omega_N t_1 \xrightarrow{\text{B}} 4N_xS_yI_z \cos \Omega_N t_1$$

$$\xrightarrow{t_2} 4N_xS_yI_z \cos \Omega_N t_1 \cos \Omega_S t_2 \cos\{\pi \mathcal{J}_{NC\alpha}\{t_2-2(T-T')]\} \left\{ \prod_k \cos \pi \mathcal{J}_{kS} t_2 \right\}$$

$$\xrightarrow{\text{C}} I_x \cos \Omega_N t_1 \cos \Omega_S t_2 \cos\{\pi \mathcal{J}_{NC\alpha}[t_2-2(T-T')]\} \left\{ \prod_k \cos \pi \mathcal{J}_{kS} t_2 \right\} \qquad (2.20)$$

where the NH, ^{15}N, and C$'$ spins are denoted by I, N, and S respectively, and Ω_N and Ω_S are the ^{15}N and C$'$ chemical shifts, $\mathcal{J}_{NC\alpha}$ is the one-bond intraresidue ^{15}N—^{13}C$^\alpha$ $\mathcal{J}$ coupling, and $\mathcal{J}_{kS}$ is the long-range coupling between the carbonyl carbon and other protons, k. The detected signal, I_x, is amplitude modulated by Ω_N in the t_1 dimension and by Ω_S in the t_2 dimension, so that the cross-peaks in the fully processed three-dimensional spectrum have pure absorption lineshape. By setting $T > T'$, the magnetization envelope in t_2 reaches a maximum value at $2(T - T')$, which is equivalent to resolution enhancement by a shifted sine bell function. An example is shown in Figure 2.30.

2.2.8 Three-dimensional HCA(CO)N experiment

The HCA(CO)N experiment correlates the H$^\alpha$ and C$^\alpha$ resonances of one amino acid with the ^{15}N resonance of the next amino acid. This experiment is conducted in D$_2$O since NH protons are not involved, and in comparing the ^{15}N chemical shifts with those obtained in D$_2$O, an upfield shift of ≈ 0.7 ppm must be taken into account.

The pulse sequence is shown in Fig. 2.37. The experiment is very similar to the HCACO one, except that magnetization which transferred to the C$'$ spin in the HCACO experiment is subsequently transferred to the ^{15}N spin of the next residue. This is achieved by including an interval $\delta \approx 0.3/\mathcal{J}_{NC'}$ after the end of the t_1 period, so that the C$'$ magnetization becomes antiphase with respect to the directly attached ^{15}N spin. The subsequent ^{15}N 90° pulse generates two-spin ^{15}N—C$'$ coherence which evolves during t_2 under the influence of the ^{15}N chemical shift only, since a 180° C$'$ pulse refocuses the effects of both C$'$ chemical shifts and C$'$—C$^\alpha$ $\mathcal{J}$ coupling. The C$'$—^{15}N correlation part of the pulse sequence is exactly analogous to the ^{1}H-detected HMQC experiment (see Section 2.1.6). At the end of the t_2 period, magnetization is transferred back to the H$^\alpha$ protons by reversing the transfer steps, and detected during t_3.

In product operator terms, the HCA(CO)N experiment can be expressed as:

$$I_z \xrightarrow{\text{A}} -2I_zA_y \xrightarrow{t_1} -4I_zA_xS_z \cos \Omega_A t_1 \sin \pi \mathcal{J}_{AS} t_1 \cos \pi \mathcal{J}_{AB} t_1$$

$$\xrightarrow{\text{B}} -8I_zA_zS_xN_y \cos \Omega_A t_1 \sin \pi \mathcal{J}_{AS} t_1 \cos \pi \mathcal{J}_{AB} t_1$$

$$\xrightarrow{t_2} -8I_zA_zS_xN_y \cos \Omega_A t_1 \sin \pi \mathcal{J}_{AS} t_1 \cos \pi \mathcal{J}_{AB} t_1 \cos \Omega_N t_2$$

$$\xrightarrow{\text{C}} I_x \cos \Omega_A t_1 \sin \pi \mathcal{J}_{AS} t_1 \cos \pi \mathcal{J}_{AB} t_1 \cos \Omega_N t_2 \qquad (2.21)$$

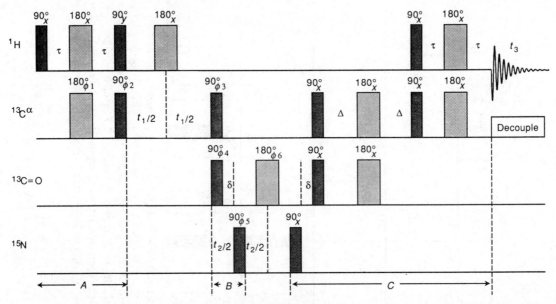

Fig. 2.37 Pulse sequence for the three-dimensional HCA(CO)N experiment. Typical delay durations are $\tau = 1.5$ ms, $\Delta = 3$ ms, and $\delta = 18$–20 ms. The power levels of the C^α and C' pulses are adjusted such that during application of a C^α pulse minimal excitation of C' occurs, and vice versa. In order to minimize the effects of homonuclear $\mathcal{J}$ modulation by the passive C^α—C^β couplings during the intervals Δ, a semiselective 180°_x (C^α) pulse may be used. The phase cycling employed is as follows: $\phi_1 = x, -x$; $\phi_2 = x$; $\phi_3 = y, -y$; $\phi_4 = 2(x), 2(-x)$; $\phi_5 = 4(x), 4(-x)$; $\phi_6 = 8(x), 8(y), 8(-x), 8(-y)$; receiver $= x, 2(-x), x, -x, 2(x), 2(-x), 2(x), x, -x, 2(-x), x$.

where the notation is the same as before, and N refers to the ^{15}N spin of the next residue. The effect of the ^{15}N—C^α coupling during t_1 has ben neglected, and because $1/(2\mathcal{J}_{NC\alpha})$ is smaller than either t_1 or t_2, the presence of this coupling leads only to a small unresolvable broadening in the F_1 and F_2 dimensions. Equation (2.21) indicates that both the active coupling, $\mathcal{J}_{AS}$ and the passive coupling, $\mathcal{J}_{AB}$ are present during t_1. Just as in the HCACO experiment, the pure phase signal in the F_1 dimension comprises a pair of antiphase doublets, so that the F_1 is best phased in dispersive mode. If the t_1 acquisition time is set to about $(1/(2\mathcal{J}_{AB})$ the $\sin \pi \mathcal{J}_{AS} t_1 \cos \pi \mathcal{J}_{AS} t_1$ time dependence results in an envelope similar to that of an FID resolution enhanced with a sine-bell function. After Fourier transformation and phasing to dispersive mode, the spectrum yields a non-Lorentzian lineshape that resembles a sine-bell line narrowed singlet. The lineshapes in both the F_2 and F_3 dimensions are purely absorptive. An example is given in Figure 2.30.

2.2.9 Three-dimensional HBHA(CBCACO)NH experiment

The HBHA(CBCACO)NH experiment[32] correlates the amide ^1H and ^{15}N resonances of one residue with the H^α and H^β resonances of its preceding residue. Thus for each amide, two or three (in the case of non-equivalent H^β protons) resonances are observed, with the chemical shifts corresponding to $^1H^\alpha$ and $^1H^\beta$ in F_1, ^{15}N in F_2 and $^1H^N$ in F_3. The pulse sequence is outlined in Fig. 2.38.

The H^α and H^β transverse magnetization is created by the initial ^1H 90° pulse, and evolves during the time period t_1. At the end of this period the H^α and H^β are antiphase

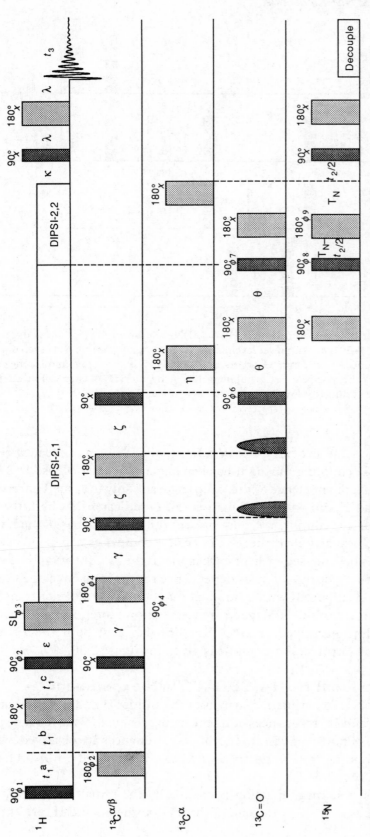

Fig. 2.38 Pulse sequence for the three-dimensional HBHA(CBCACO)NH experiment. The ^{1}H carrier is set to the H_2O frequency for the first part of the pulse sequence up to the $90^\circ_{\phi_7}$ pulse, and switched to the centre of the amide region (8.4 ppm) thereafter. Similarly, the broadband DIPSI-2 is switched between the frequencies (designated DIPSI-2, 1 and DIPSI-2, 2). ^{15}N-decoupling is achieved with WALTZ-16 modulation. Carbonyl pulses have a shaped amplitude profile, corresponding to the centre lobe of a sin x/x function, and a duration of 202 μs. Typical delays are: $\varepsilon = 2.1$ ms; SL $= 1.0$ ms; $\gamma = 3.1$ ms; $\eta = 4.7$ ms; $\zeta = 3.7$ ms; $\theta = 11.4$ ms; $T_N = 11.2$ ms; $\lambda = 2.25$ ms; initial delays $t_1^a = 1.5$ ms, $t_1^b = 0$, and $t_1^c = 1.5$ ms. The phase cycling is as follows: $\phi_1 = y$; $\phi_2 = x$, $-x$; $\phi_3 = y$; $\phi_4 = 8(x)$, $8(y)$, $8(-x)$, $8(-y)$; $\phi_5 = 4(x)$, $4(-x)$; $\phi_6 = 2(x)$, $2(-x)$; $\phi_7 = 48.5^\circ$; $\phi_8 = 4(x)$, $4(-x)$; $\phi_9 = 8(x)$, $8(-x)$; receiver $= x$, $2(-x)$, x, $-x$, $2(x)$, $2(-x)$, $2(x)$, $-x$, $-x$, x, $2(-x)$, x.

with respect to their directly attached C^α and C^β spins, and then a pair of 90° ($^1H/^{13}C$) pulses is applied, transferring ^1H magnetization perpendicular to the ϕ_2 axis into antiphase magnetization. Refocusing of this antiphase magnetization occurs during a short delay ε, and at the end of this a water purge pulse, labelled SL, is applied along with ^1H decoupling. The transverse carbon magnetization dephases due to homonuclear J coupling with adjacent carbons, although the effect between aliphatic and carbonyl carbons is eliminated by ensuring that the $180°_{\phi4}$ pulse does not excite the carbonyls. After the $C^{\alpha/\beta}$ $90°_x$ pulse, the transverse carbon magnetization dephases due to J_{CC} coupling to their adjacent carbons, and this time the coupling evolves since a pair of 180° pulses are applied simultaneously to the C′ and $C^{\alpha/\beta}$ spins. After a time 2ζ, a pair of 90° pulses is

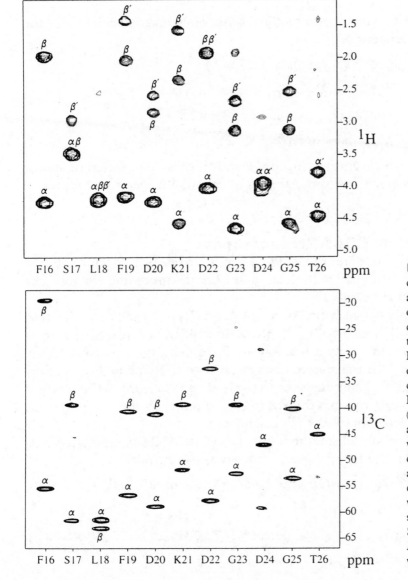

Fig. 2.39 Strip plot of the correlations observed for the amides of residues Phe[16]—Thr[26] of the calmodulin-peptide complex obtained using the three-dimensional HBHA(CBCACO)NH experiment. Each amide correlates (A) with the H^α and H^β of the preceding residue or (B) with the corresponding C^α and C^β frequencies. Resonances which are not marked by α or β correspond to correlations to amide ^1H—^{15}N pairs that are close in frequency to the one for which the strip has been selected. (Reprinted from S. Grzesiek and A. Bax, *J. Biomol. NMR*, **3**, 185–204 (1993) with permission.)

applied to the C′ and C$^{\alpha/\beta}$ spins, and during the next transfer interval, 2θ, antiphase C′ magnetization refocuses for a time period 2η, and the transverse carbonyl ^{13}C dephases with respect to the neighbouring nitrogen for a total time 2θ. After this time period, simultaneous 90° ^{13}C′ and ^{15}N pulses transfer the antiphase C′ magnetization into antiphase N magnetization. During the subsequent ^{15}N period $2T_N$, the transverse ^{15}N magnetization rephases with respect to its carbonyl coupling partner. The effect of the ^{13}C$^{\alpha}$—^{15}N $\mathcal{J}$ coupling is eliminated during the first $2T_N$–t_2 fraction of the ^{15}N evolution period by the 180° ^{15}N pulse and by the fact that the 180° C′ pulse is adjusted so it does not excite the C$^{\alpha}$ spins. During the last fraction, t_2, of the ^{15}N evolution time, the 180° ^{13}C$^{\alpha}$ pulse decouples the ^{15}N/^{13}C$^{\alpha}$ interaction. During the delay κ, ^{1}H decoupling is switched off and ^{15}N magnetization becomes antiphase with respect to its attached proton spin. The final reverse INEPT converts the ^{15}N magnetization back into observable magnetization detected during time t_3.

In terms of product operator formalism, the final observable magnetization after the HBHA(CBCACO)NH experiment is:

$$H_z^{\alpha} \rightarrow H_x^N \cos^m(2\pi\gamma\mathcal{J}_{C\alpha C\beta})\cos^m(2\pi\zeta\mathcal{J}_{C\alpha C\beta})\sin(2\pi\zeta\mathcal{J}_{C\alpha C'})$$

$$\times \sin(2\pi\eta\mathcal{J}_{C\alpha C'})\sin(2\pi\theta\mathcal{J}_{C'N})\sin(2\pi T_N\mathcal{J}_{C'N})\cos(2\pi\,\delta_{H\alpha}t_1)\cos(2\pi\,\delta_N t_2)$$

$$H_z^{\beta} \rightarrow H_x^N \cos^n(2\pi\gamma\mathcal{J}_{C\beta C\gamma})\sin(2\pi\gamma\mathcal{J}_{C\alpha C\beta})\sin(2\pi\zeta\mathcal{J}_{C\alpha C\beta})\sin(2\pi\zeta\mathcal{J}_{C\alpha C'})$$

$$\times \sin(2\pi\eta\mathcal{J}_{C\alpha C'})\sin(2\pi\theta\mathcal{J}_{C'N})\sin(2\pi T_N\mathcal{J}_{C'N})\cos(2\pi\,\delta_{H\beta}t_1)\cos(2\pi\,\delta_N t_2) \qquad (2.22)$$

where H^{α} and H^{β} are the operators representing the H$^{\alpha}$ and H$^{\beta}$ spins. An example is shown in Fig. 2.39, which shows correlation for the amides of calmodulin-peptide complex.

2.2.10 Three-dimensional CBCANH experiment

The CBCANH experiment[33] correlates the amide ^{1}H and ^{15}N resonances with both the intraresidue C$^{\alpha}$ and C$^{\beta}$ resonances, and with the C$^{\alpha}$ and C$^{\beta}$ of the preceding residue. The pulse sequence is outlined in Fig. 2.40.

An INEPT transfer enhances polarization of both C$^{\alpha}$ and C$^{\beta}$ spins, and their chemical shifts evolve during $2T_{AB}$. When the 90° ^{13}C pulse is applied, C$^{\beta}$ magnetization is transferred to C$^{\alpha}$, but at the same time a fraction of C$^{\alpha}$ magnetization remains on C$^{\alpha}$. Subsequently, C$^{\alpha}$ magnetization is transferred to its intraresidue ^{15}N and to the ^{15}N of the next residue, via the $^1\mathcal{J}_{C\alpha N}$ and $^2\mathcal{J}_{C\alpha N}$ couplings respectively. The ^{15}N chemical shifts evolve during $2T_N$, and then ^{15}N magnetization is transferred to the amide protons with a reverse INEPT sequence, prior to detection of the H^N resonances.

In terms of product operator formalism, the result of the CBCANH experiment can be expressed in terms of the amplitudes of the four cross-peaks as follows:

$$A \rightarrow H_{1x} \cos^m(2\pi\mathcal{J}_{AB}T_{AB})\sin(2\pi\mathcal{J}_{N1A}\zeta)\cos(2\pi\mathcal{J}_{N2A}\zeta)\cos^m(2\pi\mathcal{J}_{AB}\zeta)\sin(2\pi\mathcal{J}_{N1A}T_N)$$

$$\times \cos(2\pi\mathcal{J}_{N1C\alpha}T_N)\cos(2\pi\Omega_A t_1)\cos(2\pi\Omega_{N1}t_2)$$

$$A \rightarrow H_{2x} \cos^m(2\pi\mathcal{J}_{AB}T_{AB})\sin(2\pi\mathcal{J}_{N2A}\zeta)\cos(2\pi\mathcal{J}_{N1A}\zeta)\cos^m(2\pi\mathcal{J}_{AB}\zeta)\sin(2\pi\mathcal{J}_{N2A}T_N)$$

$$\times \cos(2\pi\mathcal{J}_{N2C\alpha}T_N)\cos(2\pi\omega_A t_1)\cos(2\pi\Omega_{N2}t_2)$$

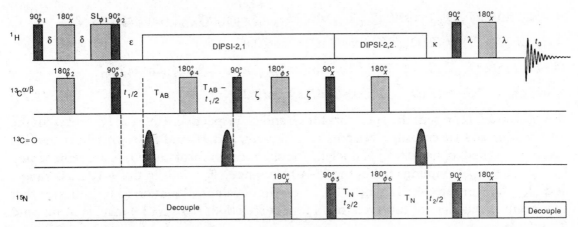

Fig. 2.40 Pulse sequence for the three-dimensional CBCANH experiment. The ^{1}H carrier is set to the H_2O frequency for the first part of the pulse sequence up to the $90^\circ_{\phi 7}$ pulse, and switched to the centre of the amide region (8.4 ppm) thereafter. Similarly the broadband DIPSI-2 is switched between the frequencies (designated DIPSI-2,1 and DIPSI-2,2). ^{15}N-decoupling is achieved with WALTZ-16 modulation. Carbonyl pulses have a shaped amplitude profile, corresponding to the centre lobe of a sin x/x function, and a duration of 202 μs. Typical delays are: $\delta = 1.5$ ms; $\varepsilon = 2.1$ ms; $T_{AB} = 3.3$ ms; $\zeta = 11$ ms; $T_N = 11.2$ ms; $\kappa = 5.4$ ms; $\lambda = 2.25$ ms. The phase cycling is as follows: $\phi_1 = y$; $\phi_2 = x, -x$; $\phi_3 = x$; $\phi_4 = 8(x), 8(y), 8(-x), 8(-y)$; $\phi_5 = 2(x), 2(-x)$; $\phi_6 = 4(x), 4(-x)$; receiver $= x, 2(-x), 2(x), 2(-x), x, -x, 2(x), 2(-x), 2(x) - x$.

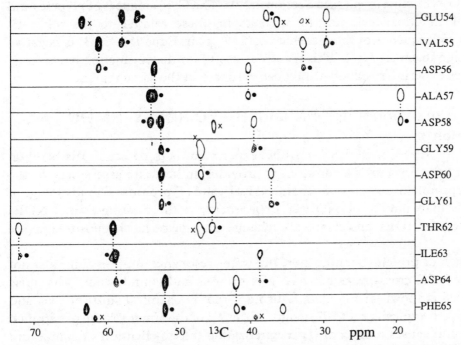

Fig. 2.41 Strip plot of the correlations observed for the amides of residues Glu54—Phe65 of a calmodulin-peptide complex obtained using the three-dimensional CBCANH experiment. Each amide correlates with its C^α and C^β nuclei and with the C^α and C^β of the preceding residue. Interresidue connectivities are marked with dots. Negative peaks, corresponding to C^β resonances and to C^α of glycine residues, are displayed with fewer contour levels. Peaks marked with an 'X' correspond to correlations to amide ^{1}H—^{15}N pairs that are close in frequency to that of the strip selected. (Reprinted from ref. 33 with permission.)

$$B \rightarrow H_{1x} \sin(2\pi \mathcal{J}_{AB}T_{AB})\cos^n(2\pi \mathcal{J}_{BC}T_{AB})\sin(2\pi \mathcal{J}_{N1A}\zeta)\cos(2\pi \mathcal{J}_{N2A}\zeta)\sin(2\pi \mathcal{J}_{AB}\zeta)$$

$$\times \sin(2\pi \mathcal{J}_{N1A}T_N)\cos(2\pi \Omega_{N1C\alpha}T_N)\cos(2\pi \Omega_B t_1)\cos(2\pi \Omega_{N1}t_2)$$

$$B \rightarrow H_{2x} \sin(2\pi \mathcal{J}_{AB}T_{AB})\cos^n(2\pi \mathcal{J}_{BC}T_{AB})\sin(2\pi \mathcal{J}_{N2A}\zeta)\cos(2\pi \mathcal{J}_{N1A}\zeta)\sin(2\pi \mathcal{J}_{AB}\zeta)$$

$$\times \sin(2\pi \mathcal{J}_{N2A}T_N)\cos(2\pi \Omega_{N2C\alpha}T_N)\cos(2\pi \Omega_B t_1)\cos(2\pi \Omega_{N2}t_2) \tag{2.23}$$

where A and B represent the operators for C^α and C^β respectively, N_1 and N_2 represent the intraresidue and succeeding ^{15}N spins respectively, and H_1 and H_2 represent the amide protons attached to these two ^{15}N nuclei. The chemical shifts of nucleus X are represented by Ω_X. The $\mathcal{J}_{C\alpha C\beta}$ couplings fall in the 35–40 Hz range, $^1\mathcal{J}_{NC_\alpha}$ falls in the 9–12.5 Hz range, and $^2\mathcal{J}_{NC_\alpha}$ falls in the 6–9 Hz range. Based on these values, Equation (4.6) indicates that significant intensities are expected for all four correlations if $T_{AB} = 3.3$ ms, $\zeta = 11$ ms, and $T_N = 11$ ms.

An example of this type of experiment is shown in Fig. 2.41, which shows strips from the three-dimensional spectrum of a calmodulin-peptide complex for residues Glu54—Phe65. Each strip represents a narrow F_3 region of an F_1/F_3 cross-section of the three-dimensional spectrum at the $^1H/^{15}N$ frequency of an amino acid. Thus each strip displays the C^α and C^β correlations observed for the amide of a given residue. These latter correlations are marked by solid dots and can tentatively be differentiated from the intraresidue ones by their weaker intensities or unambiguously identified from the HBHA(CBCACO)NH experiment (see Section 2.2.9). The C^α resonances of glycines and C^β resonances of all other residues are opposite in phase relative to the other C^α correlations, as a consequence of the fact that $\cos(2\pi \mathcal{J}_{AB}\zeta)$ in Equation (2.23) is negative and opposite in sign to $\sin(2\pi \mathcal{J}_{AB}\zeta)$. This feature is useful for discriminating between C^α and C^β resonances of serine residues, which can resonate in the same region.

2.2.11 Three-dimensional ^1H—^{15}N and ^1H—^{13}C ROESY–HMQC experiments

The main use of the ^1H—^{15}N ROESY–HMQC experiment is to detect NOEs between NH protons and bound water. In addition, it provides an accurate assessment of the intraresidue interproton distances between the NH, H^α, and H^β protons. The advantage of the experiment is that, unlike the NOESY experiment, in the rotating frame, NOEs (hereafter referred to as ROEs) and chemical exchange cross-peaks have opposite sign (see Section 1.3.6).

For the detection of protein-bound water, the water resonance suppression must not interfere with the ROE cross-peaks between NH protons and bound water. The pulse sequence employed is shown in Fig. 2.42. During t_1 the ^1H chemical shifts evolve, and then a spin-lock pulse, typically set to the length of the $T_{1\rho}$ of the amide protons, is applied along the y axis, and causes mixing of ^1H magnetization due to chemical exchange and ROEs. At this stage the ^1H magnetization is aligned along the effective field, in the yz plane and close to the y axis. After the dephasing caused by the ^{15}N—^1H $\mathcal{J}$ coupling during the time period 2τ (which is set to $\approx 1/(2\mathcal{J}_{NH})$), all magnetization residing on the ^{15}N-attached protons is subsequently converted to ^1H—^{15}N multiple quantum coherence by the application of a ^{15}N $90°$ pulse. Magnetization from protons not attached to ^{15}N

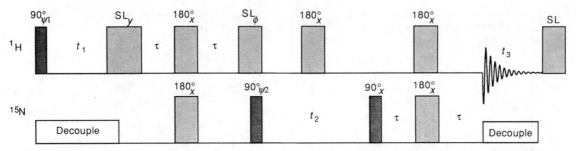

Fig. 2.42 Pulse sequence for the ^{1}H—^{15}N three-dimensional ROESY–HMQC experiment. The phase cycling is as follows: $\psi_1 = x$; $\psi_2 = 2(x), 2(-x)$; $\phi = 4(x), 4(-x)$. ROESY mixing is achieved by a spin lock along the y axis (SL$_y$). The two other spin lock pulses (SL$_\phi$ and SL), related to water suppression, are applied only for 2 ms. The delay τ is set to 2.25 ms, slightly less than $1/(4J_{NH})$.

remain aligned along the y axis at the end of the interval 2τ. This includes the large water magnetization which needs to be suppressed. By applying a short (≈ 2 ms) ^{1}H trim pulse along the x axis, this magnetization is effectively removed by randomization. As a result, the necessity for pre-saturation of the water resonance, with the consequent perturbation of ROEs between the water and ^{15}N-attached protons, is avoided. During the evolution period t_2, ^{1}H offset effects are suppressed by the 180° ^{1}H pulse, and multiple-quantum coherence evolves with the ^{15}N chemical shifts. At the end of the t_2 period, multiple-quantum coherence is converted back into observable magnetization. To ensure identical amounts of z magnetization at the start of every sequence, a trim pulse is applied along the x axis immediately after acquisition.

Negative ROE cross-peaks between NH protons and bound water can arise through two different mechanisms. The first involves an ROE only and is due to the close proximity of the NH proton and the bound water molecule. The second involves an indirect pathway whereby magnetization is transferred by an ROE from an NH proton to a rapidly exchanging side-chain proton (i.e. hydroxyl group of Ser, Thr, and Tyr, the side-chain amide group of Asn, Gln, and Lys, and the guanidinium group of Arg), followed by chemical exchange between the latter proton and water. As long as the three-dimensional structure of the protein being studied is known, it is easy to assess the possible contribution of this indirect magnetization transfer pathway.

The application of the three-dimensional ^{1}H—^{15}N ROESY–HMQC experiment is illustrated in Fig. 2.43, which shows several NH(F_3)—F_1(^{1}H) planes at different ^{15}N frequencies of the spectrum of ^{15}N-labelled IL-1β. Seven ROE cross-peaks between NH protons in the F_3 dimension and the water frequency in the F_1 dimension are clearly seen in Fig. 2.43, and, in total, 17 such cross-peaks were identified in the complete data set, of which 15 could be attributed unambiguously to ROEs between NH protons and bound water.

The main use of the three-dimensional ^{1}H—^{13}C ROESY–HMQC experiment is to determine the relative intensities of intraresidue ROEs between H$^\alpha$ and H$^\beta$ protons for the purposes of stereospecific assignment of β-methylene protons. The pulse sequence is shown in Fig. 2.44. It is essential to place the ^{1}H carrier frequency downfield of the H$^\alpha$

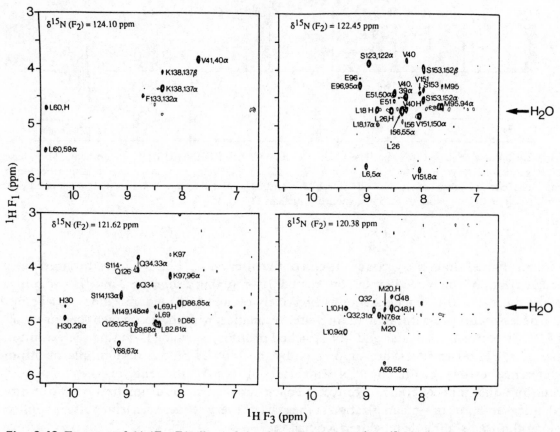

Fig. 2.43 Four out of 64 (F_1, F_3) slices of the three-dimensional ^{1}H—^{15}N three-dimensional ROESY–HMQC spectrum of 1.7 mM uniformly ^{15}N-labelled interleukin-1β in 90% H$_2$O/10% D$_2$O. Peaks are labelled i, j, where i refers to the NH proton (with the residue name and number indicated) and j refers either to water (indicated by the letter H) or to an aliphatic proton (indicated by residue number and proton type). The water frequency is at 4.67 ppm. (Reprinted from G. M. Clore, A. Bax, P. T. Wingfield, and A. M. Gronenborn, *Biochemistry*, **29**, 5671–6 (1990) with permission.)

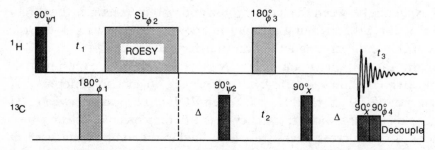

Fig. 2.44 Pulse sequence for the ^{1}H—^{13}C three-dimensional ROESY–HMQC experiment. The phase cycling is as follows: $\psi_1 = 4(x), 4(y)$; $\psi_2 = x, -x$; $\phi_1 = 4(x), 4(-x)$; $\phi_2 = 4(y), 4(-x), 4(-y), 4(x)$; $\phi_3 = 2(x), 2(y), 2(-y), 2(x), 2(-x), 2(-y), 2(y), 2(-x)$; $\phi_4 = 8(x), 8(-x)$; receiver $= x, 2(-x), x, y, 2(-y), y$. The ^{13}C 180$^\circ_{\phi 1}$ pulse is a composite 180° pulse of the type 90$^\circ_x$180$^\circ_y$90$^\circ_x$. SL$_{\phi 2}$ is the ROESY spin lock. The delay Δ should be set to ~ 3 ms, slightly less than $1/(2\mathcal{J}_{CH})$. ^{13}C decoupling during the acquisition period is achieved using coherent decoupling, and the ^{13}C 90$^\circ_x$90$^\circ_{\phi 4}$ pulse pair at the beginning of the acquisition period (and prior to the start of the decoupling) reduces the intensity of modulation side bands.

resonances to avoid artifacts from Hartmann-Hahn transfer between H^α and H^β protons in cases where the matching for this process is near perfect.

In principle, the H^α—H^β NOE cross-peak intensities could also be obtained from a short mixing time three-dimensional ^1H—^{13}C NOESY–HMQC spectrum. However, it is difficult in practice to obtain such a spectrum because the diagonal resonances at short mixing times are very intense, resulting in substantial amounts of t_1 noise which obscure many of the weaker cross-peaks. Since the ROE is positive for all values of the molecular correlation time τ_c, indirect ROE contributions are of opposite sign relative to direct ROE effects in the case of one intervening spin, and positive indirect effects involving an even number of intermediate spins are generally not observable because positive and negative contributions tend to cancel one another out. Therefore, ROESY spectra recorded with reasonably long mixing times still provide a faithful representation of internuclear distances not only for small proteins and oligonucleotides, but also for larger proteins as well.

The optimal value of the ROESY mixing time is equal to the approximate average value of the spin-locked relaxation time $T_{1\rho}$, and when the mixing time is set correctly, the diagonal resonances are significantly attenuated and reasonably strong cross-peaks can be observed. This is illustrated by several $^1H(F_3)$—$^1H(F_1)$ planes of the ^{13}C-separated ROESY–HMQC spectrum of $^{13}C/^{15}N$ labelled IL-1β shown in Fig. 2.45. As a consequence of the short mixing time (<25 ms for a protein of ~18 kDa), interproton distances larger than about 3.5 Å do not give rise to observable ROE effects.

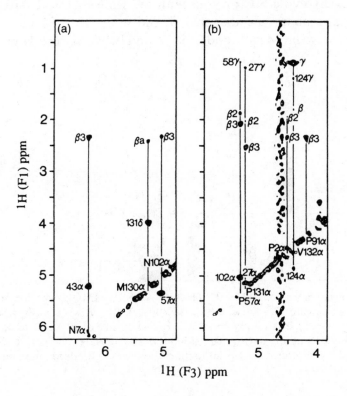

Fig. 2.45 Two $^1H(F_1)$—$^1H(F_3)$ planes at different ^{13}C frequencies of the three-dimensional ^1H—^{13}C three-dimensional ROESY–HMQC spectrum of 1.7 mM $^{13}C/^{15}N$ labelled interleukin-1β in D_2O recorded with a 22 ms mixing time. (a) $\delta_C = 51.4$ ($\pm nSW$) ppm; (b) $\delta_C = 61.9$ ($\pm nSW$) ppm. (The sweep width is 20.71 ppm.) Only positive levels are shown and diagonal resonances are therefore not observed. Note that in addition to intraresidue ROEs, a number of both short- and long-range interresidue ROEs are seen in both slices. (Reprinted from ref. 27 with permission.)

2.2.12 Three-dimensional ^{1}H—^{15}N HMQC–NOESY–HMQC experiment

The ^{1}H—^{15}N HMQC–NOESY–HMQC experiment is useful in cases in which protons have degenerate chemical shifts, since their corresponding directly attached ^{15}N often have non-degenerate chemical shifts. NOEs can be detected in the three-dimensional spectrum in which the heteronuclear chemical shifts are labelled in the F_1 and F_2 dimensions, and the ^{1}H chemical shift in the F_3 dimension. The pulse sequence is shown in Fig. 2.46.

Heteronuclear multiple-quantum coherence is generated during the t_1 period which is subsequently converted back into transverse magnetization. At the end of the t_1 period, ^{1}H magnetization is modulated by the directly attached ^{15}N spins. During the subsequent NOESY mixing period τ_m, ^{1}H magnetization is transferred back to its immediate spatial neighbours. At the end of τ_m, NH magnetization is converted back into heteronuclear multiple-quantum coherence during t_2, before being detected during t_3. Clearly, the same pulse sequence can be used to detect NOEs between aliphatic or aromatic protons with degenerate chemical shifts by replacement of the ^{15}N pulses with ^{13}C pulses.

Two ^{15}N(F_2)—^{15}N(F_1) planes and one NH(F_3)—^{15}N(F_1) plane of the three-dimensional HMQC–NOESY–HMQC spectrum of ^{15}N-labelled calmodulin are showin in Fig. 2.47. In the slices taken perpendicular to the F_3 axis, the F_1 and F_2 coordinates are those of the ^{15}N atoms directly attached to the originating and destination protons respectively. Thus in the case of NH protons with different chemical shifts, the NOE cross-peaks appear in only one half of the spectrum in a given (F_1, F_2) slice (i.e. for an NOE interaction between amide protons A and B, one NOE appears in the plane $F_3 = \delta$A, and the other one in the slice with $F_3 = \delta$B). For NOEs involving NH protons with the same chemical shifts, the cross-peaks occur symmetrically about the diagonal in the same (F_1, F_2) plane. An example is shown in Fig. 2.47, between Met-76 and Lys-77, and between the NH protons of Met-144 and Met-145.

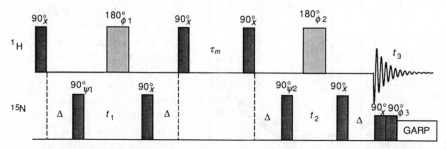

Fig. 2.46 Pulse sequence for the three-dimensional ^{1}H—^{15}N HMQC–NOESY–HMQC experiment. The phase cycling is as follows: $\psi_1 = x, -x$; $\psi_2 = 2(x), 2(-x)$; $\phi_1 = 4(x), 4(y), 4(-x), 4(-y)$; $\phi_2 = 8(x), 8(y), 8(-x),$ $8(-y)$; $\phi_3 = x, -x$; receiver $= (x, -x, -x, x), 2(-x, x, x, -x), (x, -x, -x, x)$. The delay Δ should be set to 4.5 ms, slightly less than $1/(2\mathcal{J}_{NH})$. ^{15}N decoupling during acquisition is achieved using coherent ^{15}N GARP modulation and the $90°_x$–$90°_{\phi 3}$ ^{15}N pulse pair immediately prior to the start of the GARP decoupling serves to reduce the intensity of modulation sidebands.

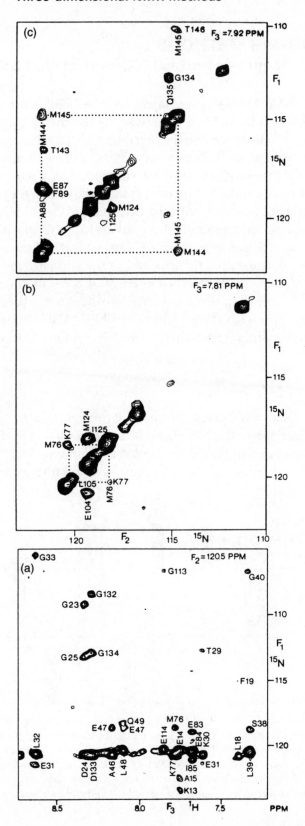

Fig. 2.47 One selected $^{15}N(F_1)$—$^1H(F_3)$ slice (a), and two selected $^{15}N(F_1)$—$^{15}N(F_2)$ slices at different $^1H(F_3)$ chemical shifts (b and c) of the three-dimensional ^1H—^{15}N HMQC–NOESY–HMQC spectrum of 1.5 mM ^{15}N-labelled calmodulin in 90% H_2O/10% D_2O. The NOE mixing time τ_m is 140 ms. (Reprinted from M. Ikura, A. Bax, G. M. Clore, and A. M. Gronenborn, *J. Am. Chem. Soc.*, **112**, 9020 (1990) with permission.)

2.3 FOUR-DIMENSIONAL NMR METHODS

2.3.1 Four-dimensional $^{13}C/^{15}N$-edited and $^{13}C/^{13}C$-edited NOESY experiments

The three-dimensional ^{1}H—^{15}N NOESY–HMQC experiment spreads NOE cross-peaks between NH protons and aliphatic protons into a third dimension by the chemical shift of the directly attached ^{15}N atoms. This three-dimensional experiment effectively removes, except in exceptional cases, the chemical shift degeneracy associated with the NH protons; it still leaves ambiguities associated with severely overlapping aliphatic protons. Even if a cross-peak connecting an aliphatic and amide proton is well resolved in the three-dimensional spectrum, it is frequently not possible, with the exception of cases involving the H^{α} resonances, to identify conclusively the aliphatic proton involved, on the basis of its chemical shift. This problem can be overcome by introducing a fourth dimension[34] involving the chemical shifts of the ^{13}C spins directly attached to the aliphatic protons.

The pulse sequence of the four-dimensional $^{13}C/^{15}N$-edited NOESY experiment is shown in Fig. 2.48. The experiment comprises three separate two-dimensional experiments, namely the ^{1}H—^{13}C HMQC, ^{1}H—^{1}H NOESY and ^{1}H—^{15}N HMQC sequences. The transfer of magnetization from the aliphatic protons to the NH protons follows the following pathway:

$$^{1}H \xrightarrow{^{1}J_{CH}} {^{13}C} \xrightarrow{^{1}J_{CH}} {^{1}H} \xrightarrow{NOE} NH \xrightarrow{^{1}J_{NH}} {^{15}N} \xrightarrow{^{1}J_{NH}} NH$$

The chemical shifts of ^{13}C, ^{1}H, and ^{15}N evolve during the variable time periods t_1, t_2, and t_3, which are incremented independently, and the NH signal is detected during t_4. There are three aspects of practical importance in this four-dimensional experiment. The first is that the number of peaks in the four-dimensional spectrum is the same as in the

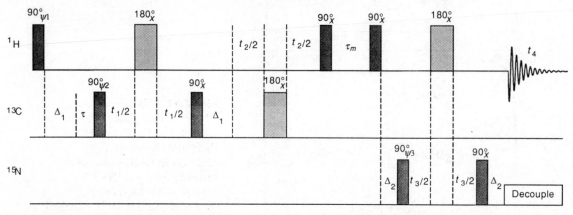

Fig. 2.48 Pulse sequence of the four-dimensional $^{13}C/^{15}N$-edited NOESY expriment. The phase cycling is as follows: $\psi_1 = 4x$; $\psi_2 = 2(x, -x)$; $\psi_3 = 2(x), 2(-x)$; receiver $= x, 2(-x), x$. To minimize relaxation losses, the delays Δ_1 and Δ_2 (that allow for efficient creation of heteronuclear multiple-quantum coherence) should be set to 3.0 ms and 4.5 ms, slightly less than $1/(2J_{HC})$ and $1/(2J_{HN})$ respectively. The delay, τ, immediately prior to application of the first ^{13}C 90° pulse is included to compensate for the ^{13}C 180° pulse, so that no first-order phase correction is needed in F_2. ^{15}N decoupling during acquisition can be achieved using incoherent GARP or WALTZ decoupling.

corresponding $^{15}N/^{13}C$-edited three-dimensional and two-dimensional spectra, so that extension to the fourth dimension increases resolution without a corresponding increase in complexity. The second is that the through-bond transfer steps are highly efficient as they involve one-bond heteronuclear couplings (90–130 Hz) which are much larger than the linewidths. As a result, the sensitivity of the experiment is high and can easily be performed on 1–2 mM samples of uniformly labelled $^{15}N/^{13}C$ protein. Third, extensive spectral folding can be employed to maximize resolution in the $^{13}C(F_1)$ dimension, so that each ^{13}C coordinate corresponds to a series of ^{13}C chemical shifts separated by intervals of typically ≈ 20 ppm. This does not complicate the interpretation of the four-dimensional spectrum, since all the ^{13}C resonances would have been assigned previously using the three-dimensional HNCA, HCCH–COSY, and HCCH–TOCSY experiments (described in Sections 2.2.4 and 2.2.5), and the appropriate ^{13}C chemical shift is easily ascertained on the basis of the ^1H chemical shift of the aliphatic proton from which the magnetization originates.

Fig 2.49 shows selected $F_4(NH)$—$F_2(^1H)$ slices of the four-dimensional $^{13}C/^{15}N$-edited NOESY experiment on $^{13}C/^{15}N$ labelled interleukin-1β at two $^{15}N(F_3)$ and several $^{13}C(F_1)$ frequencies, together with the corresponding $F_3(NH)$—$F_1(^1H)$ slices of the three-dimensional $^{13}C/^{15}N$-edited ^1H—^{15}N NOESY–HMQC spectrum at the same $^{15}N(F_2)$ chemical shifts.[8] The improved resolution relative to the three-dimensional spectrum is clearly apparent. For example, consider the NOEs between an aliphatic resonance at 0.89 ppm and the NH protons of Ser-21 Lys-27, Gln-39, and Val-58. Between 1.2 and 0.8 ppm alone there are some 57 separate ^1H resonances, so that the assignments of these NOEs from the three-dimensional spectrum is virtually impossible. The two NOEs involving Ser-21(NH) and Lys-27(NH) are seen in the slice at $\delta_C = 65.9/45.2/21.9$ ppm, while those involving Gln-39(NH) and Val-58(NH) are observed in the slice at $\delta_C = 63.3/42.5/21.9$ ppm. From ^1H and ^{13}C assignments already available, it is a simple matter to assign the NOEs to Ser-21(NH) and Lys-27(NH) as originating from one of the methyl groups of Leu-26. The NOEs to Gln-39(NH) and Val-58(NH) could involve the same or a different methyl group of either Val-19 or Val-100. However, the NOE to Val-58(NH) is maximal in this slice, whereas the NOE to Gln-39(NH) is maximal in an adjacent slice downfield in ^{13}C chemical shift. This suggests that the NOEs arise from methyl groups with different ^{13}C chemical shifts, with the NOE to Val-58(NH) originating from Val-100($C^{\gamma b}H_3$) and to Gln-39(NH) arising from Val-19($C^{\gamma a}H_3$).

The pulse sequence for the four-dimensional $^{13}C/^{13}C$-edited NOESY experiment[35] is shown in Fig. 2.50. As in the four-dimensional $^{13}C/^{15}N$-edited NOESY experiment, it comprises a central NOESY sequence sandwiched between two HMQC sequences so that the transfer of magnetization follows the pathway:

$$^1H \xrightarrow{^1J_{CH}} {}^{13}C \xrightarrow{^1J_{CH}} {}^1H \xrightarrow{NOE} {}^1H \xrightarrow{^1J_{CH}} {}^{13}C \xrightarrow{^1J_{CH}} {}^1H$$

The magnetization originating from and destined to reside on the protons evolves during t_2 and t_4, while the corresponding ^{13}C chemical shifts of the attached carbon atoms evolve during t_1 and t_3. To prevent magnetization originating on ^{13}C spins being transferred to coupled ^1H spins via a DEPT type mechanism, a 1 ms ^{13}C saturation pulse is applied immediately prior to the first ^{1}H 90° pulse. Similarly, the ^{13}C 90$_x$ pulse applied at the beginning of the NOESY mixing period converts all residual $I_z S_z$ magnetization to $I_z S_x$, and any residual ^{13}C magnetization is removed by the 1 ms ^{13}C saturation pulse at the end of the NOESY mixing period. The spin-lock pulse along the $-y$ axis is applied prior to the second ^{1}H 90° pulse and ensures that all ^{1}H magnetization except that along $-y$ is dephased. The delay Δ is set to $1/(2\mathcal{J}_{CH})$ to allow efficient creation of ^{1}H—^{13}C

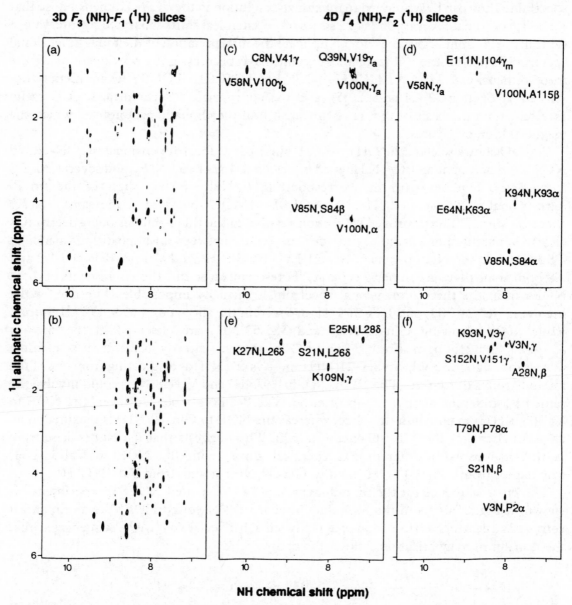

Fig. 2.49 Representative F_4(NH)—F_2(^{1}H) planes of the four-dimensional ^{15}N/^{13}C edited NOESY spectrum of ^{13}C/^{15}N labelled interleukin-1β together with the F_3(NH)—F_1(^{1}H) slices of the three-dimensional ^{15}N/^{13}C edited NOESY at the corresponding ^{15}N frequencies. The acquired four-dimensional data matrix comprised 16 complex (t_1) × 64 complex (t_2) × 16 complex (t_3) × 512 real (t_4) data points. (Reprinted from ref. 27 with permission.)

multiple-quantum coherence. The delay τ, immediately prior to the application of the first ^{13}C $90°$ pulse, is set to the duration of the ^{13}C $180°$ composite pulse. This compensates for the ^{13}C $180°$ pulse in the middle of t_2 so that no first-order phase correction is necessary in F_2. The length of the initial t_1 and t_3 delays should be adjusted to exactly half a dwell time such that the zero- and first-order phase corrections in F_1 and F_3 are $90°$ and $-180°$ respectively. This is achieved by setting $[t_{1,3}(0) + \tau_{180}(^1\text{H}) + 4\tau_{90}(^{13}\text{C})/\pi] = \Delta t_{1,3}/2$, where $\Delta t_{1,3}$ is the value of the t_1 and t_3 increment, and $t_{1,3}(0)$ is the initial value of the t_1 and t_3 delays. This ensures that the folded peaks are all absorptive and have positive or negative amplitudes depending on whether the number of times they are folded is even or odd respectively. ^{13}C decoupling during the acquisition is achieved with GARP modulation.

The relationship between three-dimensional ^{13}C-edited NOESY and four-dimensional ^{13}C/^{13}C-edited NOESY spectra is illustrated in Fig. 2.51. The three-dimensional spectrum consists of a single cube in which NOEs appear in different ^{1}H—^{1}H planes determined by the ^{13}C chemical shift of the carbon atom attached to the destination proton. Each ^{1}H—^{1}H slice of the three-dimensional cube constitutes a cube in the four-dimensional spectrum in which the NOEs are further edited by the ^{13}C chemical shift of the carbon atom bonded to the originating proton. Thus, the four-dimensional spectrum consists of a series of three-dimensional cubes at different ^{13}C chemical shifts in the F_3 dimension, corresponding to the resonance frequency of the carbon atom directly bonded to the destination proton(s), and the axes of each cube are the ^{1}H chemical shifts of the originating and destination protons in F_2 and F_4, respectively, and the ^{13}C chemical shift of the carbon atom attached to the originating proton in the F_1 dimension.

In principle, a three-dimensional ^{13}C-edited NOESY spectrum contains approximately the same information as the four-dimensional ^{13}C/^{13}C-edited spectrum. However, in any given ^{13}C slice of the three-dimensional spectrum, the NOE peak between the originating proton H_k and the destination proton H_i is only labelled by the two ^{1}H chemical shifts and the chemical shift of the carbon atom directly bonded to the destination proton C_i. It is often true that there are many protons which resonate at the same chemical shift as H_k, and therefore in the three-dimensional spectrum it is essential to be able to locate the symmetry-related NOE cross-peak at the chemical shift of the carbon atom C_k directly bonded to H_k in order to assign the identity of H_k with any degree of confidence. This can

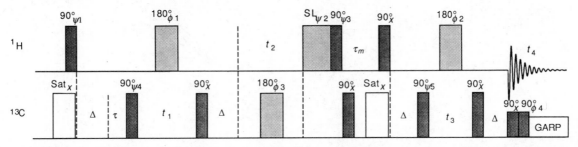

Fig. 2.50 Pulse sequence of the four-dimensional ^{13}C/^{13}C-edited NOESY experiment. The phase cycling is as follows: $\psi_1 = x$; $\psi_2 = -y$; $\psi_3 = x$; $\psi_4 = x, -x$; $\psi_5 = 2(x), 2(-x)$; $\phi_1 = \phi_2 = 4(x), 4(y)$; $\phi_3 = 2(x), 2(-x)$; $\phi_4 = x$, $-x$; receiver $= x, 2(-x), x$; The ^{13}C $180°$ pulse is a composite pulse $(90°_x - 180°_y - 90°_x)$; Sat$_x$ is a 1 ms ^{13}C saturation pulse; and SL$_{\psi 2}$ is a 500 μs ^{1}H spin-lock pulse. The delay $\Delta = 1/(2\mathcal{J}_{\text{CH}})$ (~ 3.4 ms).

be difficult due to the presence of t_1 noise and extensive spectral overlap in the $^1H(F_1)$ dimension, as well as the fact that many of the NOE peaks may be of very weak intensity, and may not appear in both locations in the three-dimensional spectrum. In the case of the four-dimensional spectrum, all this information is contained in a single peak. Also, each NOE appears twice in the four-dimensional spectrum (since the ^{13}C shifts of the destination and originating carbon atoms are swapped), providing a means to improve the quality of the spectrum by symmetrization. A further advantage of the four-dimensional over the three-dimensional spectrum is that in the four-dimensional spectrum, genuine 'diagonal' peaks corresponding to magnetization that has not been transferred from one proton to another, as well as the intense NOE peaks involving protons attached to the same carbon atom (i.e. methylene protons), appear in only a single $F_2(^1H)$, $F_4(^1H)$ plane of each $F_1(^{13}C)$, $F_2(^1H)$, $F_4(^1H)$ cube at the carbon frequency where the originating and destination carbon atoms coincide (i.e. at $F_1 = F_3$). Therefore, these intense resonances do not obscure NOEs between protons with similar or even degenerate chemical shifts, thus providing additional valuable interproton distance restraints for the three-dimensional structure determination.

In the schematic example illustrated in Fig. 2.51, there are three protons close in space: the methylene protons i and i' attached to the same carbon atom C_i, and a methyl proton k attached to the carbon atom C_k. In the three-dimensional spectrum, the diagonal peaks for the methylene protons i and i', and NOE cross-peaks between the methylene protons i and i' from the methyl proton k to the methylene protons i and i' are seen at the $^{13}C(F_2)$

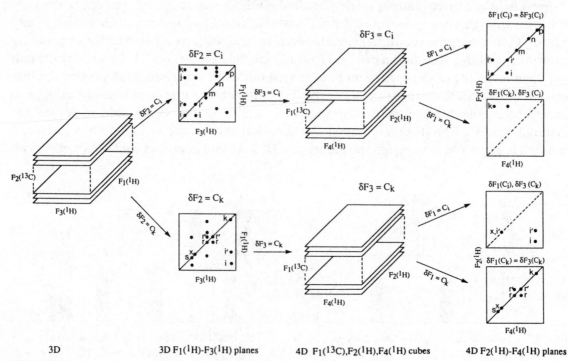

Fig. 2.51 Schematic representation of the relationship between three-dimensional ^{13}C-edited NOESY and four-dimensional $^{13}C/^{13}C$-edited NOESY spectra. (Reprinted from ref. 35 with permission.)

frequency of C_i; the diagonal peak for k and the NOEs from i and i' to k are seen at the $^{13}C(F_2)$ frequency of C_k. In the four-dimensional spectrum, the diagonal peaks for the methylene protons i and i', and the methyl proton k are seen in the $^1H(F_2)$—$^1H(F_4)$ planes at $\delta F_1(^{13}C_i) = \delta F_3(^{13}C_i)$ and $\delta F_1(^{13}C_k) = \delta F_3(^{13}C_k)$ respectively. The NOEs from k to i are seen in the $^1H(F_2)$—$^1H(F_4)$ plane at $\delta F_1(C_k), \delta F_3(C_i)$, while the NOEs from i and i' to k are seen in the $^1H(F_2)$—$^1H(F_4)$ planes at $\delta F_1(C_1), \delta F_3(C_k)$. Also shown in the latter plane is an NOE between two protons x and i' with identical chemical shifts, where the chemical shift of the carbon atom attached to x is the same as that of C_k.

Examples of several different slices of the four-dimensional $^{13}C/^{13}C$-edited NOESY spectrum of $^{13}C/^{15}N$-labelled IL-1β are shown in Figs 2.52 and 2.53, demonstrating their simplicity and resolution. As in the four-dimensional $^{13}C/^{15}N$-edited case, interpretation of the complete four-dimensional $^{13}C/^{13}C$-edited NOESY spectrum is straightforward using a simple search algorithm to match up the four chemical-shift coordinates of each peak with the complete list of ^1H and ^{13}C assignments already available from three-dimensional methods.

Fig. 2.52 shows the $F_2(^1H)$—$F_4(^1H)$ plane at $\delta F_1(^{13}C), \delta F_3(^{13}C) = 44.3, 34.6$ ppm. As extensive folding is employed, each ^{13}C frequency corresponds to several ^{13}C chemical shifts given by $x \pm nSW$, where x is the shift specified, n is an integer, and SW is the spectral width (in this case 20.71 ppm). A number of long-range NOE cross-peaks are clearly discerned involving both aromatic and aliphatic protons. In addition, there are a number of intraresidue cross-peaks. The intraresidue NOE between the $C^\beta H$ and $C^\alpha H$ protons of Lys-77 occurs in the region between 1 and 2 ppm. A plot of this region of the two-dimensional spectrum is shown in Fig. 2.52(b) and the position of the peak between $C^\gamma H$ and $C^{\beta a}H$ protons of Lys-77 which is indicated by the arrow if Fig. 2.52(a), is marked by the letter X in Fig. 2.52(b). It is clear from Fig. 2.52(b) that no individual cross-peaks can be resolved in this region of the two-dimensional spectrum. Figs. 2.52(c) and (d) show the positive and negative contours, respectively, of the $F_1(^{13}C)$—$F_3(^{13}C)$ plane at $F_2(^1H)$, $F_4(^1H) = 1.39, 1.67$ ppm corresponding to the ^1H coordinates of the cross-peak between the $C^\gamma H$ and $C^{\beta \alpha}H$ protons of Lys-77. There are four positive cross-peaks (in which the number of times the peaks are folded is even), and three negative ones (in which the number of times the peaks are folded is odd) clearly resolved in this single $F_1(^{13}C)$—$F_3(^{13}C)$ plane of the spectrum. Thus, seven NOE cross-peaks are superimposed at the ^1H coordinates of 1.39, 1.67 ppm, which could not have been assigned in either a two- or a three-dimensional ^{13}C-edited NOESY spectrum.

Figures 2.53(a) and (b) show two $F_2(^1H)$—$F_4(^1H)$ planes at $\delta F_1(^{13}C), \delta F_3(^{13}C) = 42.4$, 35.9 ppm and 35.9, 42.4 ppm respectively. Thus these two planes are related by symmetry, and the same NOEs are observed with comparable intensities in both planes with the $F_2(^1H)$ and $F_4(^1H)$ frequencies of the cross-peaks interchanged. A number of intraresidue and long-range interresidue NOEs are observed. In addition, there is a sequential NOE between the $C^\alpha H$ protons of Tyr-90 and Pro-91 (indicative of a *cis* proline). The resonances of these protons have near identical chemicals shifts (4.19 and 4.25 ppm, respectively) so that this particular NOE would have been very difficult to observe in either a two-dimensional spectrum or a three-dimensional ^{13}C-edited NOESY spectrum, as the cross-peak is located so close to the diagonal. Another example of an

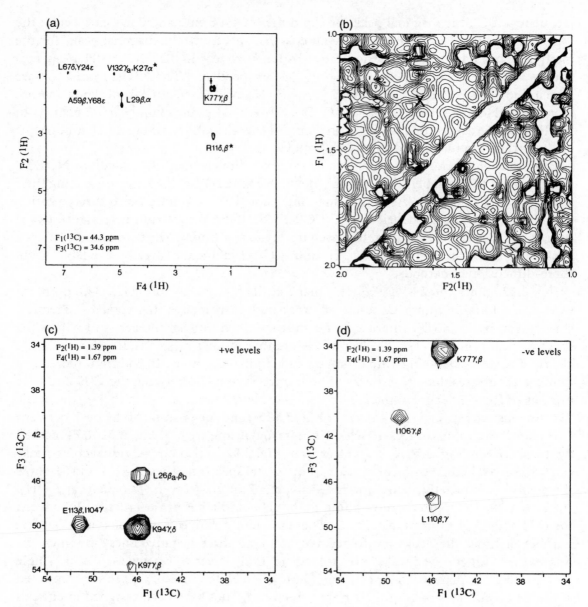

Fig. 2.52 Selected $F_2(^1H)$—$F_4(^1H)$ and $F_1(^{13}C)$—$F_3(^{13}C)$ planes of the four-dimensional $^{13}C/^{13}C$-edited NOESY spectrum of 1.7 mM $^{13}C/^{15}N$ labelled interleukin-1β in D_2O. (a) $F_2(^1H)$—$F_4(^1H)$ slice at $\delta F_1(^{13}C) = 44.3$ ppm and $\delta F_3(^{13}C) = 34.6$ ppm; (c) and (d) positive and negative contours of the $F_1(^1H)$—$F_3(^{13}C)$ plane at $\delta F_2(^1H) = 1.39$ ppm, $\delta F_4(^1H) = 1.67$ ppm corresponding to the ^1H chemical shifts of the cross-peak between the $C^\gamma H$ and $C^{\beta a}H$ protons of Lys-77 shown by an arrow in (a); (b) region between 1 and 2 ppm of the 110 ms two-dimensional NOESY spectrum of interleukin-1β corresponding to the boxed region in (a). In (a) there are two positive cross-peaks indicated by an asterisk, while the remaining cross-peaks are negative. (Reprinted from ref. 35 with permission.)

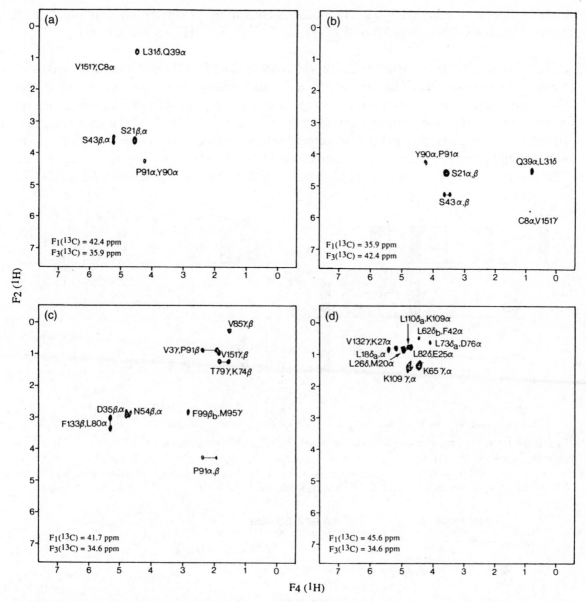

Fig. 2.53 Selected $F_2(^1H)$—$F_4(^1H)$ slices at several $F_3(^{13}C)$ and $F_1(^{13}C)$ chemical shifts of the four-dimensional $^{13}C/^{13}C$-edited NOESY spectrum of IL-1β. (a) and (b) constitute 'mirror-image' planes displaying NOE interactions between the same set of protons. (c) and (d) show two ^1H—^1H planes taken at the same $F_3(^{13}C)$ frequency, but different $F_1(^{13}C)$ frequencies. A single slice of a regular three-dimensional ^{13}C-edited NOESY spectrum in the present case would consist of 32 such planes superimposed on one another. Only positive contours are displayed in (a), (b), and (c), while only negative ones are plotted in (d). (Reprinted from ref. 35 with permission.)

NOE between protons with near identical chemical shifts can be seen in Fig. 2.53(c) between Phe-99($C^{\beta b}$H) and Met-95(C^γH) at $\delta F_2(^1H)$, $\delta F_4(^1H) = 2.87$, 2.75 ppm.

2.3.2 Four-dimensional HCANNH and HCA(CO)NNH experiments

An alternative to the use of the three-dimensional experiments outlined in Section 2.2 for the main chain directed sequential assignments is the use of two four-dimensional experiments, the four-dimensional HCANNH and HCA(CO)NNH experiments.[36] Intraresidue (and many sequential interresidue) connections are obtained from the four-dimensional version of the three-dimensional H(CA)NNH experiment, except that the $^{13}C^\alpha$ chemical shift is also measured, so it is called the four-dimensional HCANNH.

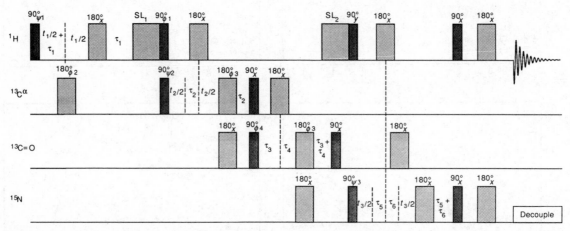

Fig. 2.54 Pulse sequence of the four-dimensional HCA(CO)NNH experiment. The phase cycling is as follows: $\phi_1 = 4(y)$, $4(-y)$; $\phi_2 = x$, $-x$; $\phi_3 = 2(x)$, $2(y)$, $2(-x)$, $2(-y)$; $\phi_4 = 2(x)$, $2(-x)$; $\psi_1 = x$; $\psi_2 = 8(x)$, $8(-x)$; $\psi_3 = x$, $-x$; $\psi_4 = (x, -x, -x, x)$, $2(-x, x, x, -x)$, $(x, -x, -x, x)$. Typical values are: $SL_1 = 1$ ms; $SL_2 = 9$ ms; $\tau_1 = 1.5$ ms; $\tau_2 = 1.7$ ms; $\tau_3 = 4.5$ ms; $\tau_4 = 9.5$ ms; $\tau_5 = 2.75$ ms; $\tau_6 = 11.0$ ms; and $\tau_7 = 2.25$ ms.

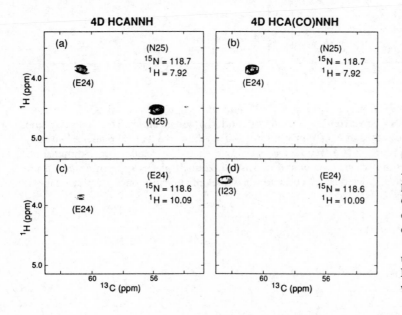

Fig. 2.55 Representative $F_1(^1H)$—$F_2(^{13}C)$ planes at particular ^{15}N (F_3) and ^1H (F_4) chemical shifts of the four-dimensional HCA(CO)NNH experiment on 1.1 mM $^{13}C/^{15}N$-labelled human ubiquitin in 90% H_2O/10% D_2O. (Reprinted from ref. 36 with permission.)

Sequential assignments are based on the complementary four-dimensional HCA-(CO)NNH experiment.

The four-dimensional HCANNH pulse sequence is a simple extension of the three-dimensional H(CA)NNH experiment. By allowing the magnetization to evolve according to the $^{13}C^{\alpha}$ chemical shifts, $^1H^{\alpha}$, $^{13}C^{\alpha}$, ^{15}N, and $^1H^N$ chemical shifts within each residue can be correlated (see Fig. 2.24). Interresidue cross-peaks are also sometimes observed as a result of transfer of magnetization from the $^{13}C^{\alpha}$ spin of the preceding residue to the ^{15}N spin via an interresidue $^2J_{C\alpha N}$ (≈ 7 Hz) coupling (see Fig. 2.24). Although this cross-peak is often too weak to be observed, when both are present there is ambiguity as to their correct assignments.

In the four-dimensional HCA(CO)NNH experiment, this interresidue correlation is detected specifically (see Fig. 2.24). The pulse sequence is shown in Fig. 2.54. Because most of the magnetization transfer steps rely on relatively large one-bond couplings ($J_{HC} \approx 147$ Hz, $J_{C\alpha C'} \approx 55$ Hz, $J_{CN} \approx 15$ Hz, and $J_{NH} \approx 94$ Hz), the sensitivity of this experiment is very good, even though it involves four magnetization transfer steps. Taken together, the two four-dimensional experiments provide the correlated chemical shifts of $^1H^N$ and ^{15}N with $^{13}C^{\alpha}(i)$ and $^1H^{\alpha}(i)$, as well as $^1H^N$ and ^{15}N with $^{13}C^{\alpha}(i\text{-}1)$ and $^1H^{\alpha}(i\text{-}1)$. As an illustration of the assignment strategy, Fig. 2.55 shows a comparison of identical $^{13}C^{\alpha}(F_2)$—$^1H^{\alpha}(F_1)$ slices at fixed $^{15}N(F_3)$—$^1H^N(F_4)$ chemical shifts from the two four-dimensional spectra uniformly $^{13}C/^{15}N$ labelled ubiquitin. In the HCANNH spectrum, two correlations are observed to the amide protons of Asn-25, one intraresidue correlation from the $^1H^{\alpha}$ proton of Asn-25 and the other from the interresidue correlation of the $^1H^{\alpha}(i-1)$ proton of Glu-24. In the corresponding F_2F_1 slice of the HCA(CO)NNH spectrum, only the interresidue correlation is observed. Although the interresidue cross-peak is often weaker in the HCANNH spectrum, as expected from the relative magnitudes of the $^1J_{C\alpha N}$ and $^2J_{C\alpha N}$ couplings, the specific detection of this cross-peak in the HCA(CO)NNH spectrum allows its unambiguous identification. In the next step, the cross-peak due to the intraresidue correlation for Glu-24 is located in the HCANNH spectrum by searching at the $^1H^{\alpha}$, and $^{13}C^{\alpha}$ chemical shifts are determined for the previous interresidue correlation; the cross-peak is found at the F_2F_1 slice shown in Fig. 2.55. In this case, the next interresidue correlation to Ile-23 is missing from the HCANNH spectrum. However, this correlation is easily detected in the corresponding F_2F_1 slice of the HCA(CO)NNH spectrum, allowing the sequential assignment to be carried out.

REFERENCES

1. D. Marion and K. Wüthrich, *Biochem. Biosphy. Res. Commun.*, **113**, 967–74 (1983).

2. D. J. States, R. A. Habekorn, and D. J. Ruben, *J. Magn. Res.*, **48**, 286–92 (1982).

3. A. F. Mehlkopf, D. Korbee, T. A. Tiggelman, and R. Freeman, *J. Magn. Reson.*, **58**, 315–23 (1984).

4. A. Bax, R. Freeman, and S. P. Kempsell, *J. Am. Chem. Soc.*, **102**, 4849 (1980).

5. A. Bax, R. Freeman, and S. P. Kempsell, *J. Magn. Reson.*, **41**, 349 (1980).

6. C. Griesinger, G. Otting, K. Wüthrich, and R. R. Ernst, *J. Am. Chem. Soc.*, **110**, 7870–2 (1988).

7. A. Bax and G. A. Morris, *J. Magn. Reson.*, **42**, 501–5 (1981).

8. A. Bax, *J. Magn. Reson.*, **53**, 517–20 (1983).

9. V. Rutar, *J. Magn. Reson.*, **58**, 306–10 (1984).

10. L. Müller, *J. Am. Chem. Soc.*, **101**, 4481 (1979).

11. A. Bax, R. H. Griffey, and B. L. Hawkins, *J. Magn. Reson.*, **55**, 301 (1983).

12. A. G. Redfield, *Chem. Phys. Lett.*, **96**, 537 (1987).

13. G. Bodenhausen and D. J. Ruben, *Chem. Phys. Lett.*, **69**, 185 (1980).

14. A. Bax, M. Ikura, L. E. Kay, D. A. Torchia, and R. Tschudin, *J. Magn. Reson.*, **86**, 304–18 (1990).

15. T. J. Norwood, J. Boyd, J. E. Heritage, N. Soffe, and I. D. Campbell, *J. Magn. Reson.*, **87**, 488–501 (1990).

16. A. Bax and M. F. Summers, *J. Am. Chem. Soc.*, **108**, 2093–4 (1986).

17. G. Wider, R. V. Hosur, and K. Wüthrich, *J. Magn. Reson.*, **52**, 130–5 (1983).

18. A. G. Redfield, S. D. Kunz, and E. K. Ralph, *J. Magn. Reson.*, **19**, 114–17 (1975).

19. P. Plateau and M. Guéron, *J. Am. Chem. Soc.*, **104**, 7310–11 (1982).

20. P. J. Hore, *J. Magn. Reson.*, 55, 283 (1983).

21. G. A. Morris and R. Freeman, *J. Am. Chem. Soc.*, **101**, 760–2 (1979).

22. R. Freeman, *Chem. Rev.*, **91**, 1397–412 (1991).

23. G. A. G. Haasnoot, *J. Magn. Reson.*, **53**, 153–8 (1983).

24. A. Bax, V. Sklenar, G. M. Clore, and A. M. Gronenborn, *J. Am. Chem. Soc.*, **109**, 6511–15 (1987).

25. D. Marion, M. Ikura, and A. Bax, *J. Magn. Reson.*, **84**, 425–30 (1989).

26. M. Guéron, P. Plateau, and M. Decorps, *Prog. in NMR Spectr.*, **23**, 135–209 (1991).

27. G. M. Clore and A. M. Gronenborn, *Prog. in NMR Spectr.*, **23**, 32–92 (1991).

28. G. T. Montelione and G. Wagner, *J. Magn. Reson.*, **87**, 183 (1990).

29. L. E. Kay, M. Ikura, and A. Bax, *J. Magn. Reson.*, **91**, 84–92 (1991).

30. A. J. Shaka, P. Barker, and R. Freeman, *J. Magn. Reson.*, **64**, 547 (1985).

31. A. J. Shaka, J. Keeler, and R. Freeman, *J. Magn. Reson.*, **53**, 313 (1983).

32. S. Grzesiek and A. Bax, *J. Am. Chem. Soc.*, **114**, 6291–93 (1992).

33. S. Grzesiek and A. Bax, *J. Magn. Reson.*, **99**, 201–7 (1992).

34. L. E. Kay, G. M. Clore, A. Bax, and A. M. Gronenborn, *Science*, **249**, 411–14 (1990).

35. G. M. Clore, L. E. Kay, A. Bax, and A. M. Gronenborn, *Biochemistry*, **30**, 12–18 (1991).

36. W. Boucher, E. D. Laue, S. Campbell-Burk, and P. J. Domaille, *J. Amer. Chem. Soc.*, **114**, 2262–4 (1992).

3 · Obtaining NMR structures

NMR spectroscopy both in solution and in the solid state can generate a variety of distance restraints which can be used to compute the three-dimensional structure of a biomolecule. The accuracy of these distance measurements depends largely on the technique, with those methods that yield the lowest number of restraints tending to give more accurate data, although in the case of solid-state NMR data this usually defines local rather than global structure. However, accuracy is not always required in order to determine a structure with high precision. As we shall see in Chapter 4, protein structures can be obtained with high precision with relatively poorly defined distance restraints such as are obtained from NOESY experiments. However, as will be considered in Chapter 9, the same does not hold for DNA, for which accurate distances are required. Furthermore, the use of the NOEs is confined to those molecules whose molecular weight (M_r) is <30 kDa (or whose overall rotational correlation time, τ_c, is long). For larger molecules alternative strategies have to be employed, which include solid-state NMR techniques and paramagnetic probes.

In Sections 3.1–3.4, we shall consider the ways in which NMR can be used to obtain distance and torsion-angle information, and in Sections 3.5–3.10 we shall consider how that NMR data can be used to generate and refine three-dimensional structures. The ways in which structures (or structural information) are obtained are illustrated in the examples given in Chapters 4, 6, 7, 8, 9, and 10.

3.1 NOEs

Nuclear Overhauser enhancements can be used to obtain distances between protons up to about 5 Å apart. In general, NOESY spectra are obtained at a variety of mixing times, τ_m. This enables the region where the NOE build-up rates are linear to be determined. Note that at long mixing times or for large biomolecules ($M_r > 30$ kDa), *spin diffusion* takes place, in which cross-relaxation is propagated across the molecule from one nucleus to another. This can be avoided by ensuring that the initial rate condition is fulfilled. The cross-peak intensities in NOESY spectra can be measured to an accuracy of about 10%. The NOE build-up rate depends on r^{-6}, where r is the distance between the two protons, and therefore it might be hoped that distances can be obtained very accurately. However, as has already been pointed out in Section 1.1.6, local internal motions, together with other contributory forms of relaxation or exchange, are the reasons that high accuracy is not obtainable in practice.

Thus, in general, NOESY cross-peak intensities are classified into three different categories, with distances of 1.8–2.7 Å, 1.8–3.3 Å, and 1.8–5.0 Å for strong, medium, and weak NOEs. Although this classification is somewhat arbitrary, it works well in practice. Usually, the NOE cross-peak intensities are calibrated against some internal standard fixed length, and the choice of such a standard is important, since local motion can lead to erroneous results. For example, calibration using the δ and ε protons of an aromatic ring is quite common, and yet it is not uncommon for aromatic side chains (at least, in proteins) to give NOEs which vary too much to be reliable. Care has to be taken to ensure that there is relatively little molecular motion in the functional group being used for distance calibration. Strong NOEs usually appear as intense cross-peaks in NOESY spectra even at short mixing times (e.g. $\tau_m = 50$ ms for a protein whose $M_r < 20$ kDa). In the same spectra, medium NOEs appear as weak cross-peaks. The weak NOEs appear only at longer mixing times (e.g. $\tau_m = 200$ ms for a protein whose $M_r < 20$ kDa) and may be due to indirect effects. The lower bound is kept the same for each category because local motion could severely attenuate the NOE, even for short distances. These distance bounds or *restraints* are used as input into computer algorithms for the calculation of the three-dimensional structure (see Section 3.4). Even though only ranges of distances for a given pair of atoms can be found, these restrict the number of possible structures drastically, and refinement protocols (see Sections 3.5–3.10) ensure that a low energy structure can be determined.

3.2 COUPLING CONSTANTS AND STEREOSPECIFIC ASSIGNMENTS

We have already indicated that judicious use of the Karplus relation can yield information on the dihedral angles. The constants A, B and C in Equation (1.39) have been determined empirically. In proteins, the couplings of interest are the three-bond couplings; in particular $^3\mathcal{J}_{H\alpha,NH}$ is related to the angle ϕ, and $^3\mathcal{J}_{H\alpha,H\beta}$ and $^3\mathcal{J}_{N,H\beta}$ are related to χ_1. These relations have been calibrated on known structures[1,2] and are given below (in Hertz):

$$
\left.
\begin{aligned}
^3\mathcal{J}_{H\alpha,HN}(\phi) &= 6.4\ \cos^2(\phi - 60°) - 1.4\ \cos(\phi - 60°) + 1.9 \\
^3\mathcal{J}_{H\alpha,H\beta2}(\chi_1) &= 9.5\ \cos^2(\chi_1 - 120°) - 1.6\ \cos(\chi_1 - 120°) + 1.8 \\
^3\mathcal{J}_{H\alpha,H\beta3}(\chi_1) &= 9.5\ \cos^2\chi_1 - 1.6\ \cos\chi_1 + 1.8 \\
^3\mathcal{J}_{N,H\beta2}(\chi_1) &= -4.5\ \cos^2(\chi_1 + 120°) + 1.2\ \cos(\chi_1 + 120°) + 0.1 \\
^3\mathcal{J}_{N,H\beta3}(\chi_1) &= 4.5\ \cos^2(\chi_1 - 120°) + 1.2\ \cos(\chi_1 - 120°) + 0.1
\end{aligned}
\right\} \quad (3.1)
$$

Using these equations, the dihedral angles ϕ and χ_1 may be calculated from both homonuclear and heteronuclear couplings.[3] It is well known that the dihedral angles ϕ and ψ are extremely important in defining the secondary structure of a protein as shown in Fig. 3.1. There are corresponding important dihedral angles in other types of biomolecules (see Chapters 9 and 10). The typical Ramachandran plot depicting the regions of dihedral angle space for the protein structural elements, α-helix and β-sheet, is shown in Fig. 3.2.

Accurate measurement of $\mathcal{J}$ couplings in an overlapping one-dimensional spectrum of a

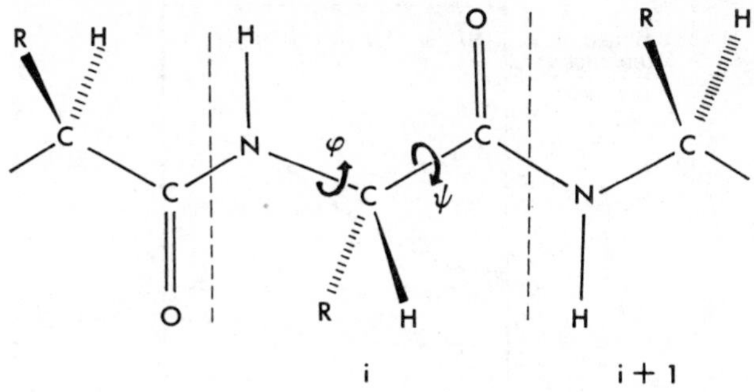

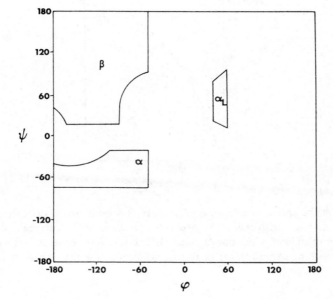

Fig. 3.1 Definition of dihedral angles ϕ and ψ. (Reprinted from ref. 22, Ch. 1 with permission.)

Fig. 3.2 Ramachandran plot showing the low-energy conformations of the peptide chain as a function of ϕ and ψ. Regions corresponding to α-helix, β-sheet, and left-handed α-helix are indicated. (Reprinted from ref. 22, Ch. 1 with permission.)

biomolecule can be very difficult. Two-bond and three-bond coupling constants may be extracted with care from the antiphase multiplet structure of the cross-peaks in phase-sensitive COSY or DQF–COSY spectra (see Fig. 3.3(a)). There can be a serious overestimate of the actual coupling constant if the linewidth is greater than the coupling. This is illustrated in the plot shown in Fig. 3.3(b). Therefore it is essential to obtain the two-dimensional spectra with very high digital resolution in the F_2 dimension, in which the coupling is usually measured. Careful use of this procedure, together with lineshape analysis of the antiphase multiplet structure can yield accurate J values which correlate well with X-ray backbone dihedral angles.

For the measurement of protein χ_1 angles in residues that contain a single β proton (Thr, Val, Ile), the same method as has already been outlined may be adopted (provided the NH protons have been exchanged with deuterium), as shown in Fig. 3.4. Couplings between $C^\alpha H$—$C^\beta H$ may be measured in a similar fashion, but these are usually extremely difficult to measure. A better approach is to use one of a number of modified COSY experiments which simplify the cross-peak multiplet structure, such as the β-COSY,[4] the E.COSY[5] (Exclusive COSY), the z-COSY[6] or the PE.COSY[7] (Primitive Exclusive COSY). The

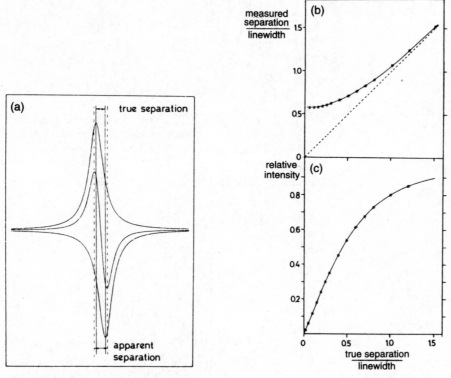

Fig. 3.3 Effects of line broadening on the apparent line separation and the overall cross-peak intensity in antiphase multiplets. (a) Definition of *true separation* and *apparent separation* in an antiphase doublet simulated by the addition of two Lorentzian lines with opposite sign. (b) and (c) Plots versus the ratio of true separation over linewidth for the apparent experimental peak separation and for the observed relative intensity.

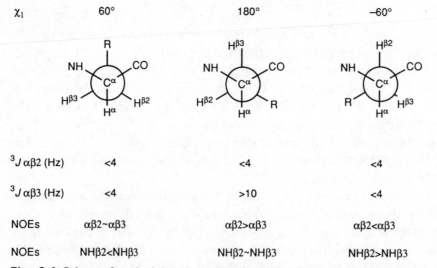

Fig. 3.4 Scheme for obtaining stereospecific assignments of β-methylene protons on the basis of $^3J_{\alpha\beta}$ coupling constants and intraresidue NOEs involving the NH, C^αH, and C^βH protons. (Redrawn from G. M. Clore and A. M. Gronenborn, *Crit. Rev. Biochem. Mol. Biol.*, **24**, 479–557 (1989) with permission.)

β-COSY resembles the COSY experiment except that a mixing pulse with a small flip angle is used. The disadvantage of this experiment is that the diagonal is largely dispersive, which can make measuring couplings for resonances close to the diagonal difficult. The idea of the E.COSY pulse sequences is to coadd the DQF–COSY and TQF–COSY spectra, as depicted in Fig. 3.5. The disadvantage of the E.COSY experiment is that it has very low sensitivity. The z-COSY fails to suppress zero-quantum transitions, which can obscure the correlation peaks. The PE.COSY experiment is a variant of the β-COSY in which the dispersive character of the diagonal is removed by subtracting a $0°$-mixing pulse COSY spectrum from a $35°$-mixing pulse COSY spectrum. It is as least four times more sensitive than the E.COSY experiment.

For larger biomolecules, measurement of $\mathcal{J}$ couplings from the PE.COSY spectrum is difficult because of extensive resonance overlap. In these cases, an estimate of the values for $\mathcal{J}$ couplings may be obtained with ^{15}N-labelled material, from the cross-peaks in a three-dimensional ^{1}H–^{15}N TOCSY–HMQC spectrum. The extent of magnetization transfer during the Hartmann–Hahn spin lock depends on the magnitude of the scalar $\mathcal{J}$ coupling.

The measured coupling constant represents the time average (on the 100 ms time scale) of the actual $\mathcal{J}$ coupling. Thus the interpretation of a measured $\mathcal{J}$ coupling can be hampered by uncertainty as to whether it represents a $\mathcal{J}$ value that corresponds to a single dihedral angle, or whether it represents a rapid averaging of $\mathcal{J}$ values from quite different dihedral angles. When a biological molecule does not exist as a single conformation, but as an equilibrium of rapidly interconverting conformers, the NOE and $\mathcal{J}$ contain time-averaged distance and angle information. Both averaging processes are strongly non-linear in their dependence on geometry, and the presence of multiple conformations can lead to NOEs and $\mathcal{J}$ couplings that cannot simultaneously be consistent with a single conformation.

The stereospecific assignment of resonances in a biomolecule, for example the β-methylene protons or methyl resonances of leucine or valine, can be extremely important in defining local conformations. Early NMR structures were obtained by treating NOEs involving chiral protons with indistinguishable chemical shifts (<0.01 ppm) as a single distance to the geometric centre between the two methylene protons, with a 0.3 Å correction added to the upper bound of the measured distance to correct for the difference in position of the so-called 'pseudo-atom' relative to the true proton. Similarly, the distance to the geometric centre of a methyl group was used with a

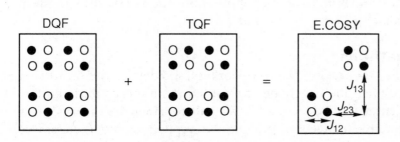

Fig. 3.5 E. COSY experiment to simplify the multiplet structure of cross-peaks. The DQF–COSY and TQF–COSY cross-peaks of a three-spin system are combined to produce an E.COSY pattern. Positive and negative multiplet components are distinguished by empty and filled circles.

corresponding increase of 1 Å to the upper bound for the measured distance. Instead of using pseudo-atoms, a more rigorous approach is to match the experimental NOEs and $\mathcal{J}$ coupling constants with those calculated for conformations present in a database. Both possible stereospecific assignments are considered, and if the database contains only conformations consistent with one of the two possibilities, then the correct assignment, together with the allowed ranges for ϕ, ψ, and χ_1, can be determined. At the time of writing, in practice only about 50% of the possible stereospecific assignments can be made in an unambiguous manner.

The $^3\mathcal{J}_{HN\alpha}$ can be used in crude calculations to provide ϕ backbone torsion-angle restraints, which for proteins can be classified into two ranges: $^3\mathcal{J}_{HN\alpha} < 6$ Hz corresponding to $-35° < \phi < -85°$; and $^3\mathcal{J}_{HN\alpha} > 8$ Hz corresponding to $-80° < \phi < -175°$. Refinement of these numbers using the methods already outlined, including more accurate ϕ, ψ, and χ_1 torsion angles and stereospecific assignments, can almost double the number of experimental restraints that are used as input to distance geometry and related algorithms (see Section 3.5).

3.3 DIPOLAR COUPLED SPINS IN SOLIDS

It may seem curious to some to have a section on solid-state NMR placed among solution-state NMR distance and angle constraint methods. However, another source of distance and/or torsion-angle information from NMR is the determination of the dipolar coupling interaction in the solid state. These new techniques are just beginning to gain popularity, with a number of variants being developed at the time of writing, and will undoubtedly become important in the future. Torsion angles can be determined very accurately from the angular dependence of the dipolar coupling, which can be used to determine the structure of a small biomolecule. Distances calculated from the measured dipolar coupling constant in the solid state can also be used, and are significantly more accurate than distances from NOEs in solution. The main disadvantage is that currently only pairwise distances or angles between isolated spins can be measured, so that there is a serious paucity of constraints compared with solution-state methods. On the other hand, the accuracy of the measurements is of the order of $\pm 3°$ for torsion angles or $\pm 0.1 - 0.5$ Å for distances. Therefore, these approaches involve generating a very limited set of accurate torsion-angle or distance measurements which can be used to define parts of a molecular structure. Furthermore, in the case of the distance measurement methods, far longer distances (up to 12 Å) can be measured. For instance, functionally crucial distances may be measured, such as between a substrate and an enzyme, or a protein and DNA, or an antibody and its antigen (see Fig. 3.6). Examples of the application of these methods to structures are given in Sections 4.6, 7.1.5, 8.4.3, and 10.3.

3.3.1 Torsion-angle methods

This method was developed by Opella, Cross, and co-workers, and determines the torsion angles, for example in the polypeptide backbone of a protein, as a means to determine structure based solely on orientational restraints.[8-10] We will consider the theory of this technique as applied to proteins. However, it could be applied to other classes of biomolecules such as DNA, although to date it has been limited primarily to membrane proteins, examples of which are presented in Sections 10.3.2 and 10.3.4.

The α-carbons of adjacent amino acids in a polypeptide chain are joined by the amide C—N bonds. The six atoms that form this peptide linkage lie roughly in a plane, and the relative orientation of adjacent planes is defined by the ϕ and ψ torsion angles (see Fig. 3.7). The secondary structure of the peptide backbone can be determined by establishing the sequential orientation of each peptide plane relative to a common axis. Measurements of both the dipolar and chemical-shift interactions are necessary to limit the number of possible orientations and define the peptide structure.[8,10–12] The orientation dependence of the dipolar interaction has the form:

$$\Delta v = D(3 \cos^2 \theta - 1) \tag{3.2}$$

where Δv is the observed dipolar splitting, D is the dipolar coupling constant, and θ is the angle between the internuclear vector connecting the spins and the external magnetic field. The dipolar interactions point along the bond axes and are axially symmetric. The observed dipolar splitting ranges from zero for bonds aligned at the magic angle, to $2D$ for

1. Studies of Protein–Ligand Interactions (Enzymatic Mechanisms)

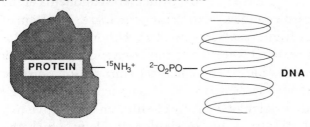

2. Studies of Protein–DNA Interactions

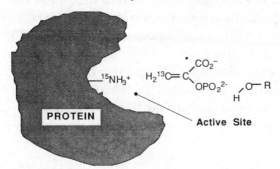

3. Studies of Protein–Protein Interactions

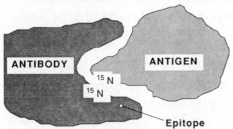

Fig. 3.6 Illustration of the cases in which distance measurements which can be made by solid-state NMR techniques would be useful in biological systems.

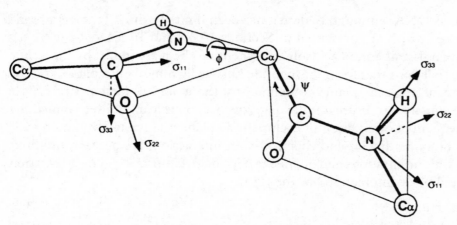

Fig. 3.7 Illustration of the peptide planes of a dipeptide and the appropriate orientations of the amide ^{15}N and carbonyl ^{13}C chemical-shift tensors with respect to the molecular frame. The ϕ and ψ angles define the orientation of the planar peptide linkages and are determined by measuring the N—C and N—H bond orientations, as well as the orientation of the ^{15}N and ^{13}C shift tensors. (Reprinted from S. Smith and O. Peersen, *Ann Rev. Biophys. Biomol. Struct.*, **21**, 25–47 (1992) with permission.)

bonds parallel to the z axis of the magnetic field. One of the problems in determining peptide plane orientations is that the experimental Δv specifies two or four different values for θ depending on the size of D relative to Δv. Consequently, several bond orientations need to be measured to arrive at a unique solution for the peptide-plane orientation. The ^{15}N—^{1}H and ^{15}N—^{13}C bond orientations are the most readily measured dipolar interactions because they involve spin-$\frac{1}{2}$ nuclei, although ^{14}N quadrupole and dipole splittings can also be measured and used as orientation restraints.[13,14] Additional restraints for peptide-plane orientations can be derived from the amide ^{15}N and carbonyl ^{13}C chemical-shift tensors. The orientation dependence of the chemical-shift interaction is given by:

$$\sigma = \sigma_{11} \cos^2 \alpha \sin^2 \beta + \sigma_{22} \sin^2 \alpha \cos^2 \beta + \sigma_{33} \cos^2 \beta \tag{3.3}$$

where σ is the observed chemical shift, σ_{11}, σ_{22}, and σ_{33} are the principal components of the chemical-shift tensor, and α and β are the Euler angles relating the principal axis system of the chemical-shift tensor to the laboratory frame (see Section 1.2.4). Although the dipolar interaction points along the internuclear axis and is related to the magnetic field axis by a single rotation about an angle θ, the orientation of the chemical-shift tensor must be determined relative to a molecular axis and relative to the magnetic field axis. The standard approach for establishing the orientation of a chemical-shift tensor relative to a molecular axis system has been through NMR studies of single crystals in which the molecular axis orientation can be determined independently using X-ray methods.[15–17] More recent methods have been developed that make use of the axial orientation of dipolar interactions as a convenient frame of reference.[18–21]

3.3.2 Homonuclear distance-dependent methods

A number of methods have been introduced recently whose aim is to reintroduce the dipolar coupling for strongly coupled homonuclear spins and to use this to measure selected distances in biomolecules, particularly those with $M_r > 30$ kDa.

The first technique, called rotational resonance (or R^2), introduced by Griffin and coworkers,[22,23] relies on the fact that magnetization transfer between two homonuclear spins is most efficient when the shift difference equals an integral of the magic-angle spinning frequency, which can place the spinning sideband of one resonance overlapping with the isotropic resonance of another.[24] This phenomenon is called the rotational resonance condition, and is given by

$$\Delta v = n\omega_r \tag{3.4}$$

where Δv is the chemical-shift difference of the cross-relaxing resonances, n is an integer, and ω_r is the spinning speed. Clearly, the requirement for isotropic resonances to be well separated places some restrictions on the applicability of this method. Griffin and co-workers made use of this phenomenon using isotopic labels, although others have used it at natural abundance.[25,26] The pulse sequence is shown in Fig. 3.8. It consists of a standard cross-polarization sequence (see Section 1.2.3), followed immediately by a flip-back pulse and selective inversion of one of the two resonances being studied. After a variable mixing time for magnetization exchange (τ_m), the NMR signal is acquired. If the two rotationally coupled spins are also close in space, and therefore coupled by a dipolar interaction, their resonance intensities will vary as a function of the mixing time due to a transfer of magnetization. When spinning at the $n = 1$ rotational resonance condition, the initial rate of magnetization transfer is dominated by the strength of the dipolar coupling. As the value of n increases, the rate of transfer decreases and contributions from the orientations of the chemical-shift tensors become more significant, making interpretation of the transfer rate more difficult. Therefore the $n = 1$ experiments are the most useful, although it is still necessary to know the isotropic shifts of the two resonances, the zero-quantum transverse decay T_2^{ZQ}, the principal values of the shift tensors (which are known for many functional groups[27]), and, less important, the mutual orientation of the two tensors. The zero-quantum T_2^{ZQ} decay influences magnetization transfer, since the difference polarization is continuously exchanged with zero-quantum coherence under the influence of dipolar coupling, and may be crudely estimated as the sum of the single-quantum transverse relaxation rates (T_2, obtained from linewidths) of the two spins. Theoretical simulations of strongly coupled sites also predict T_2^{ZQ} and oscillations in the transfer curves that result from the relative orientations of the exchanging sites.[28]

Since the magnetization transfer is proportional to dipolar coupling (D), measurement

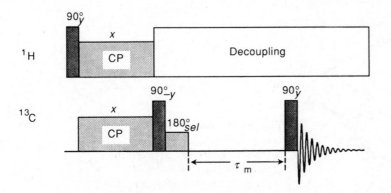

Fig. 3.8 Pulse sequence for rotational resonance experiment. After cross-polarization (CP), the $90°_{-y}$ pulse returns the ^{13}C magnetization to the z-axis (a 'flipback pulse'). The weak $180°_{sel}$ inverts selectively one of the two resonances separated by Δv, and, after a variable mixing time τ_m, the spectrum is recorded.

of D (in Hz) can be used to calculate internuclear distance (r) according to the following relationship:

$$D = \left(\frac{\mu_0}{4\pi}\right)\frac{\gamma_I\gamma_S\hbar}{r^3} \tag{3.5}$$

where γ_I and γ_S are the gyromagnetic ratios for the two spins involved (in the case of the R^2 experiment, two homonuclear spins, but obviously this applies to two heteronuclear spins), $\hbar$ is Planck's constant divided by 2π, and μ_0 is the permeability of a vacuum. The distance for an isolated spin pair (e.g. ^{13}C—^{13}C or ^{31}P—^{31}P) can be determined to an accuracy of about ± 0.5 Å, and can be used as a restraint in a distance geometry calculation. An example of such a theoretical curve for a simple model system is outlined for doubly labelled tyrosine ethyl ester ([2-^{13}C]-ethyl-[4'-^{13}C]tyrosine ester) in Fig. 3.9. The structure is depicted in Fig. 3.10. The experimental data fit with the simulated curves to an accuracy of ± 0.5 Å. The limits of distances which may be detected are around 7 Å.

Other homonuclear distance-dependent methods have been introduced recently, but, at the time of writing, have not been applied to biomolecules. These include the 'Simple Excitation for the Dephasing of Rotational Echo Amplitudes' (SEDRA) experiment[29] and the 'Dipolar Recovery At the Magic Angle' (DRAMA) experiment[30] and its variants.[31,32]

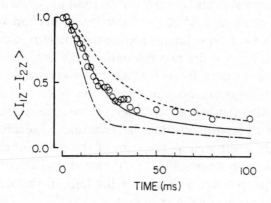

Fig. 3.9 Calculated and experimental evolution of the difference polarization for the $n=1$ rotational resonance in tyrosine ethyl ester. Experimental points are represented by open circles, and calculated curves for distances of 0.555 nm ($-\cdot-\cdot-$), 0.505 nm ($\ldots$), and 0.455 nm ($------$). (Reprinted from ref. 23 with permission.)

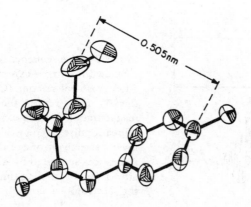

Fig. 3.10 The structure of tyrosine ethyl ester, showing the 0.505 nm separation between the labelled carbons. (Reprinted from ref. 23 with permission.)

The SEDRA experiment works outside the rotational resonance condition and under the following inequality:

$$\frac{\omega_r}{2} < \Delta v < 2\omega_r \tag{3.6}$$

whereas DRAMA only works when $\Delta v \approx 0$.

Experimentally, R^2, SEDRA, and DRAMA require high proton $\mathbf{B}_1$ field strengths (80–100 kHz, or 90° ^{1}H pulse widths (PW) $< 3.1\ \mu s$, since $\mathbf{B}_1$ RF field strength $= 1/(4 \times \text{PW})$), and, in the case of SEDRA and DRAMA, require carefully calibrated pulse widths. Furthermore, all three experiments require specialized software (available from the authors) for the analysis of the data.

3.3.3 Heteronuclear distance-dependent methods

The principal heteronuclear technique introduced recently by Schaefer and co-workers[33–35] is called rotational echo double resonance (or REDOR), and relies on the fact that the effect of the dipolar interation between two spins on the rotational echo can be manipulated by π pulses. The dephasing of magnetization of one spin involved in dipolar coupling to another heteronucleus in the presence and absence of these π pulses, and subsequent refocusing as a function of the magic angle spinning frequency, leads to a variation of resonance intensities. This intensity variation is related to the dipolar coupling constant. The pulse sequence is outlined in Fig. 3.11. The first spectrum is obtained using a standard cross-polarization pulse sequence with a π pulse on the observed nucleus (e.g. ^{13}C) in the middle of the evolution period. During this period, the

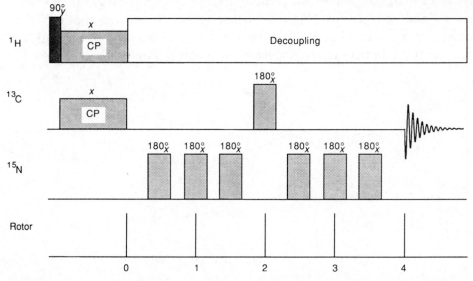

Fig. 3.11 Pulse sequence for a version of rotational echo double resonance (REDOR) ^{13}C NMR. Two equally spaced 180° ^{15}N pulses per rotor period result in dephasing of transverse carbon magnetization produced by cross-polarization (CP) transfer from dipolar-coupled protons. The 180° ^{13}C pulse replaces the ^{15}N pulse in the middle of the dephasing period and refocuses isotropic ^{13}C chemical-shift differences at the beginning of data acquisition. High-power ^{1}H-decoupling is maintained throughout the ^{15}N dephasing sequence and the acquisition.

observable magnetization evolves under the influence of the chemical shifts and the heteronuclear dipolar interaction. The 180° pulse refocuses both interactions, leading to a signal S during the acquisition period. The second spectrum is obtained with an additional train of π pulses on the dipolar-coupled spin (e.g. ^{15}N). These pulses affect the observed signal by preventing rotational refocusing of the dipolar interaction. The magnetization is therefore not completely refocused, and the signal intensity drops by an amount ΔS. For weak dipolar coupling, the change in signal intensity is related to the distance between the coupled spins by

$$\Delta S/S = KD^2 N_c^2 v_r^{-2} \tag{3.7}$$

where N_c is the number of rotor cycles during the evolution period, v_r is the spinning speed, D is the dipolar coupling, and K is a dimensionless constant ($=1.066$). If performed at relatively slow spinning speeds and over several rotor cycles, the size of the difference signal can be increased. This experiment therefore yields the internuclear distance, which, for an isolated spin pair (e.g. ^{13}C—^{15}N or ^{31}P—^{13}C), can be determined to an accuracy around ± 0.1 Å. An example is shown in Fig. 3.12, which illustrates the relationship between $\Delta S/S$ and $\lambda_D (= N_c T_r D$, where T_r is the rotor period (i.e. the time taken to complete one rotation of the rotor)).

One advantage this experiment has over the current homonuclear methods is that the curve shown in Fig. 3.12 is universal. There is no dependence on the values or orientations of the chemical-shift tensors, or on the zero-quantum T_2^{ZQ}. Therefore once a particular instrument has been calibrated with a known sample, such a curve can be used to read off dipolar couplings directly. Furthermore, the distances which can be measured are significantly larger than the homonuclear methods, for example <5 Å for ^{13}C—^{15}N, <8 Å for ^{13}C—^{31}P, and <12 Å for ^{13}C—^{19}F. Also, REDOR can be applied[36-38] to quadrupolar nuclei, for example ^{2}H and ^{14}N. A more detailed description of the REDOR experiment, from the product operator perspective, is given in Appendix 1.

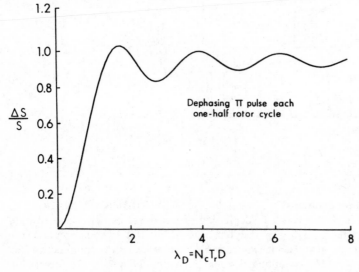

Fig. 3.12 Calculated plot of $\Delta S/S$ for a powder sample as a function of λ_D.

Other variants of REDOR include two-dimensional versions and experiments involving three or more S spins. The two-dimensional experiments yield ΔS directly as a function of the evolution time in the second dimension, although the sensitivity is reduced compared with the one-dimensional version, and therefore has limited applicability to biomolecules. Rotational echo triple-resonance (RETRO) experiments[39] and transferred-echo double resonance (TEDOR) experiments[40] are promising methods for editing out the natural abundance background signals by selecting unique spin triplets. A third rare spin, such as ^2H, ^{19}F, or ^{31}P is used to distinguish the signal arising from a specific $^{13}C—^{15}N$ pair from that of the natural abundance background, although they require specialized quadruple resonance instrumental capabilities (see Section 4.6.1 for an example).

Experimentally, like the homonuclear experiments, the heteronuclear experiments all require high proton field strengths (preferably 100 kHz or greater) and accurately calibrated π pulse widths. An additional requirement is that the instrument have three RF channels (or four channels for RETRO and TEDOR), and a three- or four-channel high-power probe. Although software is not strictly necessary for a simple two-spin REDOR experiment, software is also available from the authors.

3.4 USE OF PARAMAGNETIC IONS

The magnetic moment of an unpaired electron is 658 times greater than that of a nucleus. Thus the presence of a paramagnetic centre can have a drastic effect on the NMR spectrum.[41] For a protein, the binding of a paramagnetic ion can give rise to observed shifts arising from a scalar interaction (called the 'contact shift') or a dipolar interaction (called the 'pseudo-contact shift') or changes in the relaxation parameters. These changes are generally distant-dependent, and through the use of extrinsic paramagnetic probes,[42] or the use of intrinsic paramagnetic centres in metalloproteins[43] (for example the haem group in cytochrome c), distances can be mapped out from the centre. At one point in the 1970s, it was suggested that the approach with extrinsic paramagnetic probes could be used to determine the structure of a biomolecule in solution. We know now of course that that hope was never really fulfilled, and was largely superseded by the now widespread and more useful NOE measurements. The advantage of this approach, however, is that it may be used with larger proteins than NOE measurements. The disadvantage is that the paramagnetically shifted resonances need to be assigned rigorously prior to their pseudo-contact shift, since paramagnetic ions can bind to multiple sites and lead to erroneous interpretations of the data (see Section 7.2.1 for an example). However, when used with appropriate caution, the technique can be very useful for probing metal ion binding sites, spin-label binding sites, or metal centres in biomolecules.

The contact shift results from the delocalization of unpaired electron spin density at the resonating nucleus, which is usually transmitted through chemical bonds. The contact shift for first row transition metals is given by the expression:

$$\frac{(\omega_p - \omega_D)}{\omega_I} = -\left(\frac{A}{\hbar}\right)\frac{g\mu_B S(S+1)}{3k_B T \gamma_I} \tag{3.8}$$

where $A/\hbar$ is the hyperfine coupling constant (in rad s^{-1}), g is the Lande factor, assumed to be isotropic, T is the absolute temperature, k_B is the Boltzmann constant, γ_I is the nuclear gyromagnetic ratio (in rad s^{-1} T^{-1}), μ_B is the Bohr magneton, S is the electron spin quantum number, ω_p is the frequency shift in the paramagnetic complex, ω_D is the frequency shift in the diamagnetic complex, ω_I is the spectrometer irradiation frequency.

The pseudo-contact shift is dominated by a dipolar interaction between the nucleus being studied and an unpaired electron:

$$\frac{(\omega_p - \omega_D)}{\omega_I} = \frac{\mu_0}{4\pi} \frac{\mu_B^2 S(S+1)}{9 k_B T r^3} (3 \cos^2 \theta - 1)(g_\parallel^2 - g_\perp^2) \tag{3.9}$$

where μ_0 is the permeability of a vacuum, r is the length of the vector joining the metal ion and the resonating nucleus, θ is the angle between the electron–nuclear vector and the z axis of the magnetic susceptibility tensor, and the other symbols have already been defined. This equation assumes that the magnetic susceptibility tensor is axially symmetric.

In general, contact shifts and pseudo-contact shifts are used to analyse residues in a biomolecule that are in proximity to a covalent metal ion centre, such as in metalloproteins. The so-called extrinsic paramagnetic probe method involves the measurement of the effect on relaxation times of the paramagnetic ion. The relaxation of a spin close to a paramagnetic centre is called the 'proton relaxation enhancement' (PRE), and is given by the Solomon–Bloembergen equation:

$$\frac{1}{T_{1M}} = \frac{2}{15} \left(\frac{\mu_0}{4\pi} \right) \frac{\gamma_I^2 g^2 \mu_B^2 S(S+1)}{r^6} \left(\frac{3\tau_c}{1 + \omega_I^2 \tau_c^2} + \frac{7\tau_c}{1 + \omega_S^2 \tau_c^2} \right) \tag{3.10}$$

where $g^2 \mu_B^2 S(S+1)$ is the square of the electronic magnetic moment, μ_I^2 is proportional to the square of the nuclear magnetic moment, r is the electron-nucleus distance, τ_c is the correlation time, ω_S is the electronic and ω_I the nulear Lamor frequencies. The correlation time τ_c depends on the overall rotational correlation time τ_R, and also the electron-spin relaxation time of S, τ_S, and the chemical exchange lifetime τ_M. Thus τ_c is defined as:

$$\tau_c^{-1} = \tau_R^{-1} + \tau_M^{-1} + \tau_S^{-1} \tag{3.11}$$

For paramagnetic ions in solution such as Mn^{2+} and Gd^{3+}, $\tau_S \approx 10^{-9}$ sec, $\tau_M \approx 10^{-9}$ sec, and $\tau_R \approx 10^{-11}$ sec. When these ions are bound to a macromolecule, $\tau_R \approx 10^{-7}$ sec for molecules with $M_r \approx 100$ kDa and therefore τ_c^{-1} will change in magnitude, having contributions from both τ_S and τ_M. These ions result in changes in the proton relaxation rates of the macromolecules, and the longer the τ_S value the larger the change in proton relaxation rate. Once corrections have been made for diamagnetic contributions to $1/T_1$, the lifetime (τ_M) of the ligand in, say, an enzyme complex, can be evaluated. The relationship between the experimentally obtained paramagnetic relaxation rate ($1/T_{1p}$), the paramagnetic dipolar electron–nuclear relaxation ($1/T_{1M}$) in the enzyme complex, and the lifetime of this complex (τ_M):

$$1/T_{1p} = fq/(T_{1M} + \tau_M) \tag{3.12}$$

where $f = $ [paramagnetic ion]/[enzyme], and q is the coordination number. Thus in the extrinsic paramagnetic probe method, measurement of the biomolecule relaxation time in the presence of the paramagnetic species can be used to calculate selected distances for structure determination. The dangers and delights of this method are illustrated in Sections 7.2.1 and 7.2.2.

3.5 DISTANCE GEOMETRY

The first step in generating an NMR structure is the use of distance and angle restraints in an algorithm known as distance geometry.[44-46] Distance geometry involves taking the restraints together with the covalent structure, and creating a matrix of interatomic distances between all atoms so as to be consistent with the input. This approach is called the metric matrix distance geometry and the set of distances from n-dimensional space is projected into three-dimensional Cartesian coordinate space. This method operates in distance space, whereas there are also real space methods which operate in either Cartesian coordinate space directly, or in torsion-angle space.[47] The algorithm involves no initial priming with any kind of initial approximate structure. Indeed, most versions start from a set of randomly generated structures with the restraints 'embedded'. It is therefore a very unbiased way of determining structure. There are essentially three steps in the process:

(i) The initial *bounds* on the distances between all the atoms of the molecule are determined by triangulation from the experimental restraints and from the distance and planarity restraints derived from the primary structure. Since the set of distance restraints or bounds obtained from NOE data is usually incomplete and imprecise (in the case of other distance restraints such as those obtained from solid-state NMR, the number of restraints is very limited, although usually very much more accurate than NOE restraints), a more precise set of *bounds* is calculated using *bound smoothing*. This involves the selection of the smallest possible intervals between lower and upper bounds consistent with the triangle inequalities. The *triangle inequality* refers to the geometric constraints placed on the distances between three points, i.e. the sides of a triangle. There are two different triangle inequalities, one for the lower bounds and another for the upper bounds. For three points (i, j, k) the triangle relationships are:

$$u_{ij} \leq u_{ik} + u_{jk}; \; l_{ij} \geq l_{ik} - l_{jk} \tag{3.13}$$

where u_{ij} represents the upper bound on points i, j, and l_{ij} the lower bound for the same points. Thus if the upper bounds between atom pairs (i, j) and (j, k) are known, the upper bound between atoms (i, k) can be reduced until the triangle inequality is satisfied. This is carried out successively for all atom triplets until no further changes are made to the constraints. Then all the lower bounds are raised where appropriate, for all atom triplets. After the application of these triangle inequalities, the information present in the defined constraints (for example from NOEs) will have been passed on to some extent to those unknown bounds.

(ii) The values of the distances from within the bounds obtained by bound smoothing are guessed at random, and the atomic coordinates are generated which represent the best-fit to this guess. This is called *embedding*.

(iii) The deviations of the coordinates from the distance bounds, as well as the stereospecific assignments (if known), are minimized (by a number of methods). This is known as *optimization*.

Mathematically, the metric matrix distance geometry approach calculates a selection of trial distances to replace the upper and lower bounds in the distance matrix **D**. The trial distances d_{ij} are usually selected randomly, although they must still satisfy the bounds:

$$l_{ij} \le d_{ij} \le u_{ij} \tag{3.14}$$

Next the distance is calculated from each point i to the centre of mass, denoted as point 0, using Lagrange's theorem:

$$d_{i0}^2 = \frac{1}{N} \sum_{j=1}^{N} d_{ij}^2 - \frac{1}{N^2} \sum_{j=2}^{N} \sum_{k=1}^{j-1} d_{jk}^2 \tag{3.15}$$

The metric matrix g_{ij} is defined as the scalar product of the two vectors $\vec{u}_i$, $\vec{u}_j$, and can easily be calculated from the distances using the law of cosines:

$$g_{ij} \equiv \vec{u}_i \cdot \vec{u}_j = d_{i0}^2 \tfrac{1}{2}(d_{i0}^2 + d_{j0}^2 - d_{ij}^2) \tag{3.16}$$

Calculating the metric with respect to the centre of mass means that no coordinates are emphasized, and all the coordinates are comparable in magnitude. Repeating for all atom pairs yields the metric matrix **G**. The metric matrix can also be thought of as a matrix containing the projections of all atom pairs. The eigenvalues $\lambda_1, \lambda_2, \lambda_3, \ldots$ are determined and ranked according to magnitude. The three largest eigenvalues are selected and are by nature mutually orthogonal. The corresponding eigenvectors of the real symmetric metric matrix **G** comprise an N column matrix **W**, where N is called the *embedding dimensionality* and is usually three. The largest eigenvalues are chosen as the principal axes to minimize scatter into higher dimensions. Thus the embedded coordinates, v_{ij}, for point i in N dimensions, where $j = 1, 2 \ldots, N$ are given by:

$$v_{ij} = \lambda_N^{1/2} \, w_{ij} \tag{3.17}$$

One difficulty with the metric matrix distance geometry method is that distances alone cannot define the chirality of the structure, so that mirror images (local or global) of the correct structure can occur. In general, these can easily be rejected at the substructure stage as the chirality of single amino acids (L) and helices (right-handed) is known. A further problem associated with the method is that it does not sample efficiently the conformational space consistent with the stereochemical and experimental restraints. As a result, the atomic RMS distribution of a series of calculated structures tends to be underestimated, particularly in regions that are poorly defined by the experimental data. Subjecting these structures to restrained molecular dynamics, for example, results in an increase in the atomic RMS distribution, while improving the agreement with the experimental data and reducing the van der Waals energy.

In methods involving minimization in torsion-angle space,[48] bond lengths and angles are kept fixed during the minimization and only the torsion angles are varied. To ensure that correct folding occurs, the restraints have to be introduced gradually during the calculation. This is achieved in a variety of ways, so that the distances between residues further and further apart in the sequence are incorporated in successive cycles of minimization. In general, the structures found by these methods tend to have inordinately high energies and have to be subjected to extensive energy minimization with only minimal changes in atomic RMS deviations.

3.6 ENERGY MINIMIZATION

The next step in biomolecular structure determination is the refinement of the distance geometry-derived Cartesian coordinates. Energy minimization is generally used to find local minima in the potential energy for the structure. Commonly employed energy minimization algorithms[48,49] offer the choice of one of the following methods: steepest descent, conjugate gradient (also known as Powell[50]), and Newton–Raphson. All these assume that the energy surface is approximately harmonic, and physically adjust the atomic positions in Cartesian space over the energy surface, deeper into the local potential well. Although the rate of descent for the steepest descent method is initially fast, it is non-convergent and often results in large oscillations about the minimum. This may lead to structures which have climbed out of the local well to a higher energy point. The conjugate gradient method approaches the minimum more slowly, and results in a converged minimum point energy. However, for very poor starting structures, it is more likely to converge.

Mathematically, the steepest descent method starts by specifying the molecular coordinates prior to the kth iteration by a $3N$-dimensional vector $\mathbf{x}_{k-1}$. Each iteration then involves three steps: (i) a descent direction is chosen, represented by a $3N$-dimensional vector of unit length, $\mathbf{s}_k$; (ii) a descent step size, specified by the scalar λ_k, is determined; and (iii) the descent step is taken according to the relationship:

$$\mathbf{x}_k = \mathbf{x}_{k-1} + \lambda_k \mathbf{s}_k \tag{3.18}$$

The direction of displacement is parallel to the net force,

$$\mathbf{s}_k = \frac{\mathbf{F}}{|\mathbf{F}|} \tag{3.19}$$

where $\mathbf{F}$ is the force. Since $\mathbf{F}$ is the negative gradient of the potential,

$$\mathbf{F} = -\nabla V \tag{3.20}$$

where ∇ is the gradient and V the potential, then vector $\mathbf{s}_k$ is parallel to the negative gradient of the energy, and it points straight downhill.

The conjugate gradient method (or Powell method) is shown in Fig. 3.13 for a two-dimensional quadratic energy function. Starting from the initial point A, where the gradient is $\mathbf{g}_1$, the first search direction is taken to be along the negative gradient,

$$\mathbf{s}_1 = -\mathbf{g}_1 \tag{3.21}$$

which is along the line ABD. The minimum value of the energy along this search direction occurs at point B, where the negative gradient $\mathbf{g}_2$ is along the line BC. Whereas steepest descent would take the next step in that direction, conjugate gradient takes the second search direction to be a linear combination of the current gradient and the previous search direction,

$$\mathbf{s}_k = -\mathbf{g}_k + b_k \mathbf{s}_{k-1} \tag{3.22}$$

where the parameter b_k is a weighting factor equal to the ratio of the squares of the magnitudes of the current and previous gradient,

$$b_k = |\mathbf{g}_k|^2 / |\mathbf{g}_{k-1}|^2 \tag{3.23}$$

For a quadratic surface the search direction specified by these last two equations will pass through the minimum on the nth step for an n-dimensional step, as long as the minimum along the search direction of each successive step is found. Thus, for the two-dimensional case in Fig. 3.13, if the first step goes from A to B, the second search direction will pass through the minimum 0. Even if the step size is not optimum, the conjugate gradient method will still yield a search direction that is superior to that of steepest descent. If, for example, the first step in Fig. 3.13 were larger than the optimum, arriving at point D, steepest descent would follow the gradient in the direction DE, while the last two equations would produce a search direction $\mathbf{s}_2$ in the direction DF, which passes much nearer the minimum. As a consequence, this method is much more efficient than steepest descent.

The Newton–Raphson method is based on the assumption that, in the region of the minimum, the energy depends approximately quadratically on the independent variables. For the one-dimensional case, $V(x)$ close to the minimum is of the form:

$$V(x) = a + bx + cx^2 \tag{3.24}$$

where a, b, and c are constants. The first derivatives are:

$$V'(x) = b + 2cx \tag{3.25}$$

$$V''(x) = 2c \tag{3.26}$$

At the minimum, $V'(x^*) = 0$, and x^* can be calculated from (3.25):

$$x^* = -b/2c \tag{3.27}$$

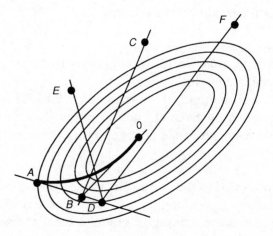

Fig. 3.13 Representation of paths followed by different energy minimization algorithms on a hypothetical two-dimensional potential surface. The minimum is at 0, and the ellipses represent contours of potential energy, spaced at equal intervals. Starting from point A, the path that would be taken by steepest descent with an infinitesimal step size is indicated by a heavy line. For larger step sizes, different paths would be followed. (Redrawn from ref. 51 with permission.)

Substituting (3.25) and (3.26) into (3.27) gives

$$x^* = x - V'(x)/V''(x) \tag{3.28}$$

If the system is at the point x, and if the first two derivatives are known at this point, (3.28) predicts the location of the minimum. This relationship can be generalized to the multidimensional case quite easily, and it serves as the basis for the Newton–Raphson method. For quadratic energy surfaces, no iteration is necessary. For non-quadratic but monotonic surfaces, this method and its various modifications have widespread utility. There are two reasons why it is rarely used as the sole energy minimization method in biomolecules. First the highly non-quadratic character of the energy surface and the presence of local minima render the surface unsuitable for this approach. Most biomolecular applications of the method have been to complete a minimization that was initiated by another method, which avoids the first problem. Second, the potential energy function is expressed in terms of bond lengths, angles, and Cartesian coordinates, and the calculation of the second derivative matrix is time-consuming for a large molecule. There are modifications, such as the adopted-basis set Newton–Raphson, or truncated Newton–Raphson, that attempt to address the second problem.

In general, for biomolecules, 100 or so steps of steepest descent minimization is followed by 500–1000 steps of either conjugate gradient or modified Newton–Raphson methods. These energy minimization steps can be carried out at a variety of the steps involved in generating NMR structures of biomolecules. Restraints can be included in some of the implementations of these algorithms. In fact, too many minimization steps without restraints (and without including solvent molecules) may lead to structures which are far from either X-ray crystal or NMR structures. Therefore, although these methods are used to generate theoretical model structures, without restraints, their indiscriminate application to biological problems is likely to lead to meaningless structures which are not at all helpful in understanding their function.

3.7 RESTRAINED MOLECULAR DYNAMICS AND SIMULATED ANNEALING

When restrained energy minimization methods are used, inevitable local energy minima are encountered which can lead to inaccurate structures. To circumvent this, restrained molecular dynamics (RMD) are usually employed. This involves including NMR restraints in one of the many molecular dynamics simulation programs. Molecular dynamics[51,52] solves Newton's equations of motion,

$$\mathbf{F}_i = m_i \mathbf{a}_i \tag{3.29}$$

where $\mathbf{F}_i$ is the force, m_i is the mass and $\mathbf{a}_i$ is the acceleration of atom i. The force on atom i can be computed directly from the derivative of the potential energy V with respect to the coordinates $\mathbf{r}_i$. The energy can be expressed in an explicitly differentiable form:

$$\frac{\mathrm{d}V}{\mathrm{d}\mathbf{r}_i} = m_i \frac{\mathrm{d}^2\mathbf{r}_i}{\mathrm{d}t_i^2} \tag{3.30}$$

Therefore, with an adequate expression for the potential energy and known masses, this differential equation can be solved for future positions in time, t_i. In general, this can be solved only approximately, since V is usually a complex function of the coordinates of all (or many) of the atoms (i.e. $V = V(r_1, r_2, r_3, \ldots, r_N)$). The temperature can be calculated from the atomic velocities

$$\frac{3N}{2} k_B T = \sum_{i=1}^{N} \tfrac{1}{2} m_i \mathbf{v}_i^2 \tag{3.31}$$

where k_B is Boltzmann's constant, m_i and $\mathbf{v}_i$ are the mass and velocity of atom i, and N is the number of atoms (and $3N$ is the number of degrees of freedom). For a simulation at constant energy, the temperature fluctuates due to the interconversion of kinetic and potential energy. If the temperature is held constant, then the atomic velocities can be adjusted accordingly. If the pressure is held constant, the volume must be allowed to fluctuate by rescaling the interatomic distances.

The total potential energy V_{total} is usually defined as the sum of a number of terms:

$$V_{\text{total}} = V_{\text{bond}} + V_{\text{angle}} + V_{\text{dihedr}} + V_{\text{vdW}} + V_{\text{coulomb}} + V_{\text{NMR}} \tag{3.32}$$

where V_{bond}, V_{angle}, and V_{dihedr} keep bond lengths, angles, and dihedral angles at their equilibrium values. The first five terms are empirical energy terms describing the physical interactions between the atoms, whereas the last term is a means of including the NMR information, but does not correspond to any real physical force. They can be summarized as follows:

$$V_{\text{bond}} = \sum_{\text{bonds}} \tfrac{1}{2} K_b (b - b_0)^2 \tag{3.33}$$

$$V_{\text{angles}} = \sum_{\text{angles}} \tfrac{1}{2} K_\theta (\theta - \theta_0)^2 \tag{3.34}$$

$$V_{\text{dihedr}} = \sum_{\text{dihedr}} K_\phi [1 + \cos(n\phi - \delta)] \tag{3.35}$$

These are pseudo-harmonic potentials that constrain bond lengths (b), bond angles (θ), and the rotamer angles (ϕ, δ) for staggered and eclipsed conformations, and K is a constant. The van der Waals and electrostatic interactions are described by V_{vdW} and V_{coulomb},

$$V_{\text{vdW}} = \sum_{\text{pairs } (ij)} [C_{12}/r_{ij}^{12} - C_6/r_{ij}^6] \tag{3.36}$$

$$V_{\text{coulomb}} = \sum_{\text{pairs } (ij)} q_i q_j / 4\pi\varepsilon_0 \varepsilon_r r_{ij} \tag{3.37}$$

where Equation (3.36) is the Lennard–Jones potential, containing repulsive and attractive terms (C is a constant), and Equation (3.37) describes the coulombic interactions between two charged particles (i, j) with partial charges q that are a distance r_{ij} apart in a dielectric medium described by $\varepsilon_0 \varepsilon_r$ term. The potential V_{NMR} contains the NMR restraints, and has the effect of pulling the protons that show an NOE interaction closer to the measured distance r_{ij}. Similarly, V_{NMR} may also contain J coupling information by including a torsion term. These potentials are also pseudo-harmonic functions of similar form to Equations (3.33)–(3.35). Distance constraints which can be reasonably accurately determined may therefore be defined as follows:

$$V_{NOE} = \begin{cases} K_1(r_{ij}-r_{ij}^0)^2 & \text{if} \quad r_{ij}>r_{ij}^0 \\ \\ K_2(r_{ij}-r_{ij}^0)^2 & \text{if} \quad r_{ij}<r_{ij}^0 \end{cases} \tag{3.38}$$

where r_{ij} and r_{ij}^0 are the calculated and experimental interproton distances, respectively, and K_1 and K_2 are force constants given by:

$$K_1 = k_B TS/[2(\Delta_{ij}^+)^2] \quad \text{and} \quad K_2 = k_B TS/[2(\Delta_{ij}^-)^2] \tag{3.39}$$

where k_B is Boltzmann's constant, T, the absolute temperature of the simulation (*not the* experiment), S a scale factor, and Δ_{ij}^+ and Δ_{ij}^- are the positive and negative error estimates, respectively, of r_{ij}. If, however, only ranges of distances can be specified, then the distance restraints are incorporated into a pseudo-square-well potential of the form:

$$V_{NOE} = \begin{cases} K_{NOE}(r_{ij}-r_{ij}^u)^2 & \text{if} \quad r_{ij}>r_{ij}^u \\ \\ 0 & \text{if} \quad r_{ij}^l \leq r_{ij} \leq r_{ij}^u \\ \\ K_{NOE}(r_{ij}-r_{ij}^l)^2 & \text{if} \quad r_{ij}<r_{ij}^l \end{cases} \tag{3.40}$$

where r_{ij}^u and r_{ij}^l are the upper and lower limits, respectively, of the target distances obtained from experiment, and K_{NOE} is the NOE force constant, which is typically chosen to be of the order of 1000 kJ mol^{-1} nm^{-2}. Similarly, torsion-angle restraints can be incorporated into pseudo-square-well potentials of the form:

$$V_{tor} = \begin{cases} K_{tor}(\phi_i-\phi_i^u)^2 & \text{if} \quad \phi_i>\phi_i^u \\ \\ 0 & \text{if} \quad \phi_i^l \leq \phi_i \leq \phi_i^u \\ \\ K_{tor}(\phi_i-\phi_i^l)^2 & \text{if} \quad \phi_i<\phi_i^l \end{cases} \tag{3.41}$$

where ϕ_i^u and ϕ_i^l are the upper and lower limits of the target range of a particular torsion angle obtained from experiment, ϕ_i is its calculated value, and K_{tor} is the torsion force constant, which is typically chosen to be of the order of 4000 kJ mol^{-1} rad^{-2}. To ensure that the experimental restraints are the dominating factor in determining the conformation of the molecule, it is very important that the force constants for the restraints are set sufficiently high that the experimental data are satisfied within the precision of the measurements. At the same time, the contribution from the empirical energy function should be such that for any individual RMD structure, the deviations from ideal geometry are small, and the non-bonded interactions are good (i.e. the Lennard–Jones potential is negative). This can be determined from a few trial calculations. Thus convergence on the structure is guided by the requirement to minimize NOE or other restraint violations. The number of distance restraint violations N_{viol} is counted when, for example, $r_{ij} \geq r_{ij}^0 + 1$, which would allow for 1 Å fluctuations. Another parameter which can be minimized in addition to N_{viol} is the sum of the distances in excess of the constraints $\sum \Delta r_{\text{viol}}$, which is defined as:

$$\sum \Delta r_{\text{viol}} = \sum_{k=1}^{N_{\text{viol}}} (r_{ij})_k - [(r_{ij}^0)_k + 1] \tag{3.42}$$

where the sum runs over all those interproton (or pseudo-atom) distances for which N_{viol} is defined. Although an arbitrary structure may be used for restrained molecular dynamics calculation, in practice a starting structure obtained from distance geometry and energy minimization is often used. Because of the kinetic energy present in the protein during the dynamics simulation, the problem of local minima can be overcome relatively easily.

The RMD approach requires a relatively large amount of computation time compared to distance geometry methods. This problem can be overcome by using a simplified potential energy function, where all non-bonded contact interactions are described by a single van der Waals repulsion term. Also by using a cutoff distance, in which non-bonded interactions for pairs of atoms that are separated by a distance greater than some reasonable value (e.g. 5–10 Å) are excluded, the number of non-bonded interactions is decreased significantly.

Simulated annealing[53] involves raising the temperature of the system followed by slow cooling in order to overcome local minima and locate the global minimum region of the target function. It is computationally more efficient than RMD and yields structures of similar quality. The potentials are very similar to RMD and again Newton's laws of motion are solved as a function of time. However, in implementations found in commercial programs, the non-bonded interaction potential is modified so that there is a simple van der Waals repulsion term with a variable force constant K_{rep}:

$$V_{\text{repel}} = \begin{cases} 0 & \text{if } r \geq s \cdot r_{\text{min}} \\ K_{\text{rep}}(s^2 r_{\text{min}}^2 - r^2)^2 & \text{if } r < s \cdot r_{\text{min}} \end{cases} \tag{3.43}$$

The values of r_{min} are given by the sum of the standard values of the van der Waals radii between two atoms as represented by the Lennard–Jones potential used in, for example, the empirical energy function CHARMM (see Section 3.9). A van de Waals scale factor, s, typically should be set to 0.8 to account for the fact that interatomic separations slightly smaller than the sum of the hard-sphere van der Waals radii can easily occur due to the attractive component of the van der Waals interaction.

3.8 ITERATIVE RELAXATION MATRIX APPROACHES

A further refinement worth mentioning is the iterative relaxation matrix approach. Relaxation matrix approaches (RMA) involve the calculation of the rates of cross-relaxation of protons, taking into account spin diffusion. This can be used in one of three ways. First, RMA can be used to calculate distance restraints from an initial model structure with the correct fold and experimental NMR data. Iterative cycles of the relaxation matrix approach and restrained molecular dynamics yield refined structures that eventually converge and result in a best-fit with the experimentally observed NOE intensities. Second, RMA can be used to calculate initial NOE distance restraints from experimental NOE intensities and an arbitrary starting structure. These restraints can be used in distance geometry or simulated annealing calculations, the output from which may be further refined by iterative RMA. Third, RMA can be used to calculate theoretical NOE intensities from a model structure using either the full-spin analysis or a two-spin approximation, and the theoretical NOESY spectra compared with the experimental data for help in cross-peak assignments.

Just as we saw for the chemical exchange relaxation matrix (see Section 1.3.6), the rate matrix $\mathbf{R}$ representing the cross-relaxation rates can be solved by solving a set of coupled differential equations describing the time development of the matrix of normalized cross-peak intensities $\mathbf{A}$ from a NOESY experiment:

$$\frac{d\mathbf{A}}{dt} = -\mathbf{R}\mathbf{A} \tag{3.44}$$

which can be formally solved for a mixing time τ_m as:

$$\mathbf{A} = \exp[-\mathbf{R}\tau_m] \tag{3.45}$$

where $\mathbf{R}$ is the cross-relaxation matrix with:

$$\mathbf{R} = \begin{bmatrix} \rho_1 & \sigma_{12} & \cdots & \sigma_{1n} \\ \sigma_{21} & \rho_2 & & \sigma_{2n} \\ \vdots & & & \\ \sigma_{n1} & \sigma_{n2} & & \rho_n \end{bmatrix} \tag{3.46}$$

The relaxation constants in the dipolar cross-relaxation matrix for a rigid molecule with N protons are defined by:

$$\rho_i = K \sum_{j=1, j \neq i}^{N} \left(\frac{1}{r_{ij}^6}\right) [6\mathcal{J}_2(2\omega) + 3\mathcal{J}_1(\omega) + \mathcal{J}_0(0)] \tag{3.47}$$

and

$$\sigma_{ij} = K \left(\frac{1}{r_{ij}^6}\right) [6\mathcal{J}_2(2\omega) - \mathcal{J}_0(0)] \tag{3.48}$$

where r_{ij} is the distance between protons i and j and $K = 0.1\gamma^4\hbar^2$. For isotropic tumbling of the molecule with a single correlation time τ_c, the spectral densities[54] (see Section 1.1.6) $\mathcal{J}_n(\omega)$ take the form

$$\mathcal{J}_n(\omega) = \frac{\tau_c}{1 + n^2\omega^2\tau_c^2} \tag{3.49}$$

where ω is the Larmor frequency of the protons. A step beyond this simple model is the 'model-free' approach by Lipari and Szabo,[55] where the internal motion is described by two parameters, an effective correlation time τ_e and an order parameter S^2:

$$\mathcal{J}(\omega) = S^2 \frac{\tau_c}{1 + \omega^2\tau_c^2} + (1 - S^2) \frac{\tau_e}{1 + \omega^2\tau_e^2} \tag{3.50}$$

which is used in some algorithms of RMA. The exponential matrix $\exp[-\mathbf{R}\tau_m]$ can be expanded in a power series:

$$\exp[-\mathbf{R}\tau_m] = 1 - \mathbf{R}\tau_m + 0.5\mathbf{R}^2\tau_m^2 \ldots \tag{3.51}$$

For sufficiently short mixing times only the first two terms of Equation (3.51) contribute appreciably and the NOE intensity a_{ij} becomes $\sigma_{ij}\tau_m$ and builds up linearly with the mixing time. Alternatively, the matrix equations can be solved as

$$\mathbf{A} = \mathbf{X} \exp[-\Lambda\tau_m]\mathbf{X}^{-1} \tag{3.52}$$

where Λ is the eigenvalue matrix obtained after diagonalization of $\mathbf{R}$,

$$\mathbf{X}^{-1}\mathbf{R}\mathbf{X} = \Lambda \tag{3.53}$$

Therefore, given a molecular model, the NOE matrix can be calculated from the relaxation matrix for each mixing time τ_m of a two-dimensional NOE experiment, and the build-up of NOE intensities in a time series can be obtained.

In the same fashion, the reverse procedure is also possible. When the complete NOE matrix $\mathbf{A}$ is known, the relaxation matrix $\mathbf{R}$ can be obtained after diagonalization of the NOE matrix:

$$\mathbf{X}^{-1}\mathbf{A}\mathbf{X} = \mathbf{D} = \exp[-\Lambda\tau_m] \tag{3.54}$$

$$\mathbf{R} = -\mathbf{X}\left[\frac{\ln\mathbf{D}}{\tau_m}\right]\mathbf{X}^{-1} \tag{3.55}$$

Thus, theoretically, the matrix $\mathbf{R}$ can be obtained from one two-dimensional NOE experiment taken at a suitably chosen mixing time, although in practice averaging over a series of τ_m values is more accurate.

3.9 GENERATING NMR STRUCTURES—WHICH SOFTWARE?

There are a number of software packages available for the calculation of NMR structures from a set of distance restraints, some commercial and some available for a nominal fee from the authors. Since the software field is developing at an incredible pace, I have confined this section to those programs that at the time of writing are being used most widely, with due apologies to those who feel that their software is the software to beat all and should have been mentioned.

For distance geometry there are a number of programs available, such as: DISGEO[56] and its newer commercial implementation, DGII (implemented in NMR*chitect* from Biosym Inc.); X-PLOR; DSPACE (which is similar to DISGEO); and DISMAN.[57] The program DISMAN is unlike the others in that it operates in torsion-angle space. The root-mean-square deviation (RMSD) of a set of NMR structures, calculated from the same set of data but using randomly selected starting conditions, is often used to assess the quality of an NMR structure (see Section 3.10). It was found that the RMSDs for the DISGEO tended to be smaller than for DISMAN. However, comparison of the DISMAN and DISGEO structures with the original crystal structure from which the artificial NMR restraints were derived showed that the difference in RMSD was largely caused by the fact that the DISGEO program tends to produce slightly expanded structures, reflecting the lack of NMR restraints insufficiently. Better structures could be obtained by subjecting the DISGEO structures to restrained molecular dynamics. A newer program DIANA,[58] also takes the torsion-angle approach.

For molecular dynamics and restrained molecular dynamics, the most commonly applied programs are the commercially available programs, CHARMM[59] (Polygen/Molecular Simulations Ltd.), DISCOVER (Biosym Inc.), and GROMOS[60] (Biomos f.v.). The program X-PLOR[61,62] includes the CHARMM algorithms and also includes simulated annealing, or there are standalone programs, such as SA (as implemented in NMR*chitect*, Biosym Inc.).

For iterative relaxation matrix approaches, there is IRMA[63] (as implemented in NMR*chitect*, Biosym Inc.), CORMA,[64,65] MARDIGRAS,[66] and X-PLOR has been modified to carry out such calculations. In some commercial implementations, such as from Biosym Inc., data from NMR experiments processed via FELIX can be passed directly to distance geometry, simulated annealing and/or restrained molecular dynamics modules, and back to IRMA for backcalculating spectra for further analysis or comparison in FELIX. The final structures can be visualized with INSIGHT II. A similar protocol is available from Molecular Simulations Ltd, with QUANTA and X-PLOR (with the equivalent to FELIX called NMR Pipe).

3.10 QUALITY OF NMR STRUCTURES—PRECISION VERSUS ACCURACY

The typical procedure in obtaining an NMR structure (and we will consider this in more detail with examples in Chapters 4 and 9) involves using a large number of different starting structures for the distance geometry and/or restrained molecular dynamics. If significant fractions of the calculated structures converge, satisfying all the NMR restraints, and showing a root-mean-square deviation (RMSD) from one another of <2 Å, then this indicates that the calculated structures must be close to the actual structure. The RMSD is generated by superimposing the centroids of the structures and calculating a rotation that minimizes the overall value:

$$RMSD = \sqrt{\frac{1}{N} \sum_{i=1}^{N} (r_i - r'_i)^2} \tag{3.56}$$

where N is the number of atoms being compared, r_i and r'_i are the atomic coordinates for the standard structure and the rotated structure respectively. For comparisons of families of NMR structures, each of the structures is rotated so as to provide the best-fit with a predetermined member of the family. RMSD values are reported for all pairwise structure comparisons. If serious violations of the NMR restraints remain, or if the RMSD between the various structures is too large, then reanalysis of the NMR spectra is necessary. A further check can be made using the iterative relaxation matrix approach and backcalculating the NOESY spectrum from the structure. An outline of the overall procedure is given in Fig. 3.14.

The precision of the measurements, for example how narrowly the bounds for NOE-derived distances are defined, will improve the RMSD between the set of calculated structures and therefore the accuracy of the structure. NMR protein structures which show a relatively low level of detail typically have RMSDs of at least 1.5–2 Å. Higher resolution structures show RMSDs of the order of 0.8 Å for the backbone atoms, 1 Å for backbone atoms plus some interior side-chain atoms, and around 1.5 Å for all the atoms in the molecule. Such improvements in resolution can be attributed in part to the availability of stereospecific assignments, which have the effect of narrowing the distance bounds. As we shall see in Chapter 4, relatively few restraints are sufficient to define the global fold in proteins, whereas the same is not true for nucleic acids, where there are no possible long-range restraints (see Chapter 9). Here iterative relaxation matrix approaches to refinement of the precision of the distance measurements from NOESY cross-peak volumes are very important.

Another method that is beginning to be used in assessing the accuracy of a structure is

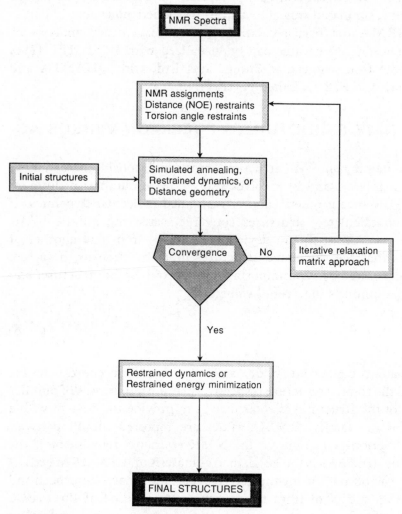

Fig. 3.14 The strategy for solving three-dimensional structure of biological macromolecules on the basis of NMR data.

the R-factor, or structure factor, which, by analogy with the term in X-ray crystallography, is the normalized mean deviation between the structure factors derived from the model and the experimental data. For NMR data, this usually involves[67] a direct measure of the fit between the experimental and theoretical NOE intensities when the final model is subjected to back-calculation of the NOESY spectrum. In its simplest, most general form the NMR R-factor is defined as:

$$R = \frac{\sum\limits_{i,j} W_{ij}(\tau_\mathrm{m})|A_{ij}^\mathrm{calc}(\tau_\mathrm{m}) - A_{ij}^\mathrm{exp}(\tau_\mathrm{m})|}{\sum\limits_{i,j} W_{ij}(\tau_\mathrm{m}) A_{ij}^\mathrm{exp}(\tau_\mathrm{m})} \tag{3.57}$$

where $A_{ij}^\mathrm{calc}(\tau_\mathrm{m})$ and $A_{ij}^\mathrm{exp}(\tau_\mathrm{m})$ are the elements connecting protons i and j of the theoretical and experimental NOE intensity matrices for a given mixing time τ_m. Weight factors $W_{ij}(\tau_\mathrm{m})$ are included to account for measurement errors. A number of other definitions of the R-factor have appeared, and some of the available algorithms (e.g. IRMA from Biosym Ltd) calculate several different types of R-factor simultaneously. The convergence limit on as low an R-factor as is achievable in an iterative relaxation matrix structure calculation is usually used as the stopping point for the refinement. However, as with the term in X-ray crystallography, the R-factor only measures the agreement of the calculated structure with the experimental data, so that a low R-factor does not necessarily mean an accurate structure if the precision of the measurements is poor. One type of R-factor recently introduced[68] that attempts to circumvent this problem relies on cross-validation, which is a statistical method that estimates the quality of the fit to the observed data without making any assumptions about the distribution of errors in the NMR data.

Despite all our best efforts at obtaining precise distance measurements for use in calculating accurate structures, there are often still regions of a molecular structure for which NMR restraints are absent. These regions are therefore disordered during all the refinement procedures, and it is tempting to suggest that this region has considerable conformational flexibility. In other words, a common argument is that absence of NMR restraints means an absence of structure. But the negative observation of a less accurately determined structure in a certain region (resulting from the lack of NOEs, for example) is not sufficient to establish that a structure is disordered and conformationally labile. Recently this problem has been addressed[69-71] in the case of ^{15}N-labelled protein by using the heteronuclear ^{15}N T_1 and T_2 relaxation rates and ^{15}N NOEs as a measure of backbone dynamics. This is a more rigorous way for determining if molecular motion is the real reason for the absence of NOEs, although it can only detect motion that is faster than the overall tumbling of the molecule. The method involves measuring the ^{15}N relaxation rates and ^{15}N NOEs by indirect detection using variants[69] on the HSQC pulse sequence (discussed in Section 2.1.6), and determining the spectral densities from the following equations:

$$1/T_1 = (\tfrac{1}{4})D^2\{\mathcal{J}(\omega_\mathrm{H} - \omega_\mathrm{N}) + 3\mathcal{J}(\omega_\mathrm{N}) + 6\mathcal{J}(\omega_\mathrm{H} + \omega_\mathrm{N})\} + (\tfrac{1}{3})c^2(\omega_\mathrm{N}) \tag{3.58}$$

$$1/T_2 = (\tfrac{1}{8})D^2\{4\mathcal{J}(0) + \mathcal{J}(\omega_\mathrm{H} - \omega_\mathrm{N}) + 3\mathcal{J}(\omega_\mathrm{N}) + 6\mathcal{J}(\omega_\mathrm{H}) + 6\mathcal{J}(\omega_\mathrm{H} + \omega_\mathrm{N})\}$$
$$+ (\tfrac{1}{18})c^2\{3\mathcal{J}(\omega_\mathrm{N}) + 4\mathcal{J}(0)\} \tag{3.59}$$

$$f_I\{S\} = (I - I_0)/I_0 = (\tfrac{1}{4}) \left(\frac{\gamma_H}{\gamma_N}\right) D^2 \{6\mathcal{J}(\omega_H + \omega_N) - \mathcal{J}(\omega_H - \omega_N)\} T_1 \tag{3.60}$$

where $D = \gamma_H \gamma_N \hbar (\mu_0/4\pi)(1/r_{NH}^3)$ and $c = \omega_N(\sigma_\parallel - \sigma_\perp)$, ω_N and ω_H are the Larmor frequencies of ^{1}H and ^{15}N (3142.8 and -318.5 rad s^{-1} respectively), r_{NH} is the N—H bond length (0.103 nm), $\mathcal{J}(\omega_i)$ is the spectral density function, and $\sigma_\parallel$ and $\sigma_\perp$ are the parallel and perpendicular components of the axially symmetric ^{15}N chemical shift tensor (which for a peptide bond, $\sigma_\parallel - \sigma_\perp = -160$ ppm[72]). The NOE enhancement, $f_I\{S\}$ is defined as the fractional enhancement in Equation (3.60), where I is the cross-peak intensity in the presaturated spectrum, and I_0 is the cross-peak intensity in the control spectrum. Using the model-free approach of Lipari and Szabo given in Equation (3.50), and substituting this for the spectral densities in Equations (3.58)–(3.60), fits of S^2 and τ_e can be obtained for a given value of τ_c. This then allows a determination of the internal motion at each particular residue in a protein as defined by the ^{15}N spectral parameters, assuming the motion is isotropic. For anisotropic motion, corrections have to be made[73] to the equations of Lipari and Szabo. In a similar fashion, side-chain dynamics can be probed through the use of ^{13}C relaxation and NOE data. Examples of the analysis of backbone dynamics from ^{15}N relaxation and NOE data are given in Section 4.4.

3.11 OUTLOOK

The methods by which NMR data are obtained, either solution-state NOE and $\mathcal{J}$-coupling data, or solid-state torsional and distance restraints, are continually being improved. As greater accuracy is attained, greater confidence can be placed in the precision of the structures which are solved. Obviously refinement and model-building is an integral part of biomolecular NMR structure determination, and faster, more effective algorithms for such calculations will also be important in the future. Intrinsic disorder, either through the absence of distance restraints, or through local dynamics, will always result in structures which are underdetermined in certain regions. However, the study of dynamics from relaxation measurements, at least in the MHz to kHz motion regimes, provided these are faster than the overall tumbling of the molecule, sets NMR apart from other structural methods in being able to quantitate molecular dynamics.

REFERENCES

1. A. DeMarco, M. Llina, and K. Wüthrich, *Biopolymers*, **17**, 2727–42 (1978).

2. A. Pardi, M. Billeter, and K. Wüthrich, *J. Mol. Biol.*, **180**, 741–51 (1984).

3. G. T. Montelione, M. E. Winkler, P. Rauenbühler, and G. Wagner, *J. Magn. Reson.*, **82**, 198 (1989).

4. A. Bax and R. Freeman, *J. Magn. Reson.*, **44**, 542–61 (1981).

5. C. Griesinger, O. W. Sørensen, and R. R. Ernst, *J. Am. Chem. Soc.*, **107**, 6394–6 (1985).

6. H. Oschkinat, A. Pastore, P. Pfändler, and G. Bodenhausen, *J. Magn. Reson.*, **69**, 559–66 (1986).

7. L. Müller, *J. Magn. Reson.*, **72**, 191 (1987).

8. T. A. Cross, *Biophys. J.*, **49**, 124–6 (1986).

9. T. A. Cross and S. J. Opella, *J. Am. Chem. Soc.*, **105**, 306–8 (1983).

10. S. J. Opella, P. L. Stewart, and K. G. Valentine, *Q. Rev. Biophys.*, **19**, 7–49 (1987).

11. L. K. Nicholson and T. A. Cross, *J. Mol. Biol.*, **218**, 621–37 (1991).

12. Q. Teng and T. A. Cross, *J. Magn. Reson.*, **85**, 439–47 (1989).

13. S. J. Opella, *Annu. Rev. Phys. Chem.*, **33**, 553–62 (1982).

14. S. J. Opella and P. L. Stewart, *Methods in Enzymol.*, **176**, 242–75 (1989).

15. P. Lauterbur, *Phys. Rev. Lett.*, **1**, 343–4 (1958).

16. R. G. Griffin, J. D. Ellet Jr, M. Mehring, J. G. Bullitt, and J. S. Waugh, *J. Chem. Phys.*, **57**, 2147–55 (1972).

17. S. Pausak, A. Pines, and J. S. Waugh, *J. Chem. Phys.*, **59**, 591–5 (1973).

18. C. J. Hartzell, T. K. Pratum, and G. P. Drobny, *J. Chem. Phys.*, **87**, 4324–31 (1987).

19. C. J. Hartzell, M. Whitfield, T. G. Oas, and G. P. Drobny, *J. Am. Chem. Soc.*, **109**, 5966–9 (1987).

20. T. G. Oas, C. J. Hartzell, F. W. Dahlquist, and G. P. Drobny, *J. Am. Chem. Soc.*, **109**, 5962–6 (1987).

21. T. G. Oas, C. J. Hartzell, T. J. McMahon, G. P. Drobny, and F. W. Dahlquist, *J. Am. Chem. Soc.*, **109**, 5956–62 (1987).

22. D. P. Raleigh, M. H. Levitt, and R. G. Griffin, *Chem. Phys. Lett.*, **146**, 71–6 (1988).

23. D. P. Raleigh, F. Creuzet, S. K. Das Gupta, M. H. Levitt, and R. G. Griffin, *J. Am. Chem. Soc.*, **111**, 4502–3 (1989).

24. E. R. Andrew, A. Bradbury, R. G. Eades, and V. T. Wynn, *Phys. Lett.*, **4**, 99 (1963).

25. M. G. Columbo, B. H. Meier, and R. R. Ernst, *Chem. Phys. Lett.*, **146**, 189 (1988).

26. W. E. J. R. Maas and W. S. Veeman, *Chem. Phys. Lett.*, **149**, 170 (1988).

27. W. S. Veeman, *Prog. NMR Spectroscop.*, **16**, 193 (1984).

28. M. H. Levitt, D. P. Raleigh, F. Creuzet, and R. G. Griffin, *J. Chem. Phys.*, **92**, 6347–64 (1990).

29. T. Gullion and S. Vega, *Chem. Phys. Lett.*, **194**, 423–8 (1992).

30. R. Tycko and G. Dabbagh, *Chem. Phys. Lett.*, **173**, 461–5 (1990).

31. R. Tycko and G. Dabbagh, *J. Am. Chem. Soc.*, **113**, 9444–8 (1991).

32. R. Tycko and S. O. Smith, *J. Chem. Phys.*, **98**, 932 (1993).

33. T. Gullion and J. Schaefer, *J. Magn. Reson.*, **81**, 196–200 (1989).

34. T. Gullion and J. Schaefer, *Adv. Magn. Reson.*, **13**, 57–83 (1989).

35. G. R. Marshall, D. D. Beusen, K. Kociolek, A. S. Redlinski, M. T. Leplawy, Y. Pan, and J. Schaefer, *J. Am. Chem. Soc.*, **112**, 963–6 (1990).

36. A. Schmidt, R. A. McKay, and J. Schaefer, *J. Magn. Reson.*, **96**, 644–50 (1992).

37. C. P. Grey and W. S. Veeman, *Chem. Phys. Lett.*, **192**, 379 (1992).

38. C. P. Grey, W. S. Veeman, and A. J. Vega, *J. Chem. Phys.*, **98**, 7711–24 (1993).

39. S. M. Holl, R. A. McKay, T. Gullion, and J. Schaefer, *J. Magn. Reson.*, **89**, 620–6 (1990).

40. S. M. Holl, G. R. Marshall, D. D. Beusen, K. Kociolek, A. S. Redlinski, M. T. Leplawy, R. A. McKay, S. Vega, and J. Schaefer, *J. Am. Chem. Soc.*, **114**, 4830–3 (1992).

41. R. A. Dwek, *NMR and Biochemistry*, Oxford University Press, Oxford (1973).

42. J. J. Villafranca, *Methods in Enzymol.*, **177**, 403–13 (1989).

43. I. Bertini, L. Banci, and C. Luchinat, *Methods in Enzymol.*, **177**, 247–63 (1989).

44. G. M. Crippen, *J. Comput. Phys.*, **24**, 96 (1977).

45. I. D. Kuntz, J. F. Thompson, and C. M. Oshiro, *Methods in Enzymol.*, **177**, 159–205 (1989).

46. T. F. Havel, *Prog. Biophys. Mol. Biol.*, **56**, 43–78 (1991).

47. W. Braun, *Quart. Rev. Biophys.*, **19**, 115–57 (1987).

48. R. Fletcher, *Practical Methods of Optimization*, Volume 1, John Wiley and Sons, New York (1980).

49. W. H. Press, B. P. Flannery, S. A. Teukolsky, and W. T. Vetterling, In *Numerical Recipes, The Art of Scientific Computing*, Cambridge University Press (1986).

50. M. J. D. Powell, *Mathematical Programming*, **12**, 241–54 (1977).

51. J. A. McCammon and S. C. Harvey, *Dynamics of Proteins and Nucleic Acids*, Cambridge University Press (1987).

52. R. M. Scheek, W. F. van Gunsteren, and R. Kaptein, *Methods in Enzymol.*, **177**, 204–18 (1989).

53. M. Nilges, G. M. Clore, and A. M. Gronenborn, *FEBS Lett.*, **239**, 129–36 (1988).

54. I. Solomon, *Phys. Rev.*, **99**, 559–65 (1955).

55. G. Lipari and A. Szabo, *J. Am. Chem. Soc.*, **104**, 4546–59 (1982).

56. T. F. Havel and K. Wüthrich, *Bull. Math. Biol.*, **46**, 673–98 (1984).

57. W. Braun and N. Gō, *J. Mol. Biol.*, **186**, 611–26 (1985).

58. M. Billeter, Y. Qian, G. Otting, M. Müller, W. J. Gehring, and K. Wüthrich, *J. Mol. Biol*, **214**, 183–97 (1990).

59. B. R. Brooks, R. E. Bruccoleri, B. O. Olafson, D. J. States, S. Swaninathan, and M. Karplus, *J. Comput. Chem.*, **4**, 187–217 (1983).

60. W. F. van Gunsteren, R. Boelens, R. Kaptein, and E. R. P. Zuidereg In Nucleic acid conformation and dynamics, *Proceedings of NATO/CECAM Workshop*, ed. W. K. Olson, pp. 79–82 (1983).

61. A. T. Brünger, G. M. Clore, A. M. Gronenborn, and M. Karplus, *Proc. Natl. Acad. Sci. USA*, **83**, 3801–5 (1986).

62. A. T. Brünger, *J. Mol. Biol.*, **203**, 803–16 (1988).

63. R. Boelens, T. M. G. Konig, and R. Kaptein, *J. Mol. Struct.*, **173**, 299–311 (1988).

64. J. W. Keepers and T. L. James, *J. Magn. Reson.*, **57**, 404–26 (1984).

65. B. A. Borgias and T. L. James, *J. Magn. Reson.*, **79**, 493–512 (1988).

66. B. A. Borgias and T. L. James, *J. Magn. Reson.*, **79**, 475–87 (1990).

67. C. Gonzalez, J. A. C. Rullman, A. M. J. J. Bonvin, R. Boelens, and R. Kaptein, *J. Magn. Reson.*, **91**, 659–64 (1991).

68. A. T. Brünger, G. M. Clore, A. M. Gronenborn, R. Saffrich, and M. Nilges, *Science*, **261**, 328–31 (1993).

69. L. E. Kay, D. A. Torchia, and A. Bax, *Biochemistry*, **28**, 8972–9 (1989).

70. G. Barbato, M. Ikura, L. E. Kay, R. W. Pastor, and A. Bax, *Biochemistry*, , **31**, 5269–78 (1992).

71. B. L. Grasberger, A. M. Gronenborn, and G. M. Clore, *J. Mol. Biol.*, **230**, 364–72 (1993).

72. Y. Hiyama, C. Niu, J. V. Silverton, A. Bavoso, and D. Torchia, *J. Am. Chem. Soc.*, **110**, 2378–83 (1988).

73. M. J. Dellwo and A. J. Wand, *J. Am. Chem. Soc.*, **111**, 4571 (1989).

Part II · *Proteins*

4 · Protein structure

The application of NMR, particularly solution-state NMR, to the determination of the three-dimensional structure of proteins is now well-established.[1] Currently there are now around 100 NMR structures of proteins with up to 157 residues, and a number of proteins with up to 300 residues whose structure determinations are in progress. While not yet rivalling X-ray crystallography in terms of the number of structures deposited with the Brookhaven Protein Data Bank, the growth in NMR structures deposited has been very rapid in the last few years, around 21% of the structures in 1991. The application of two-dimensional NMR methods to proteins was pioneered in the early 1980s by Kurt Wüthrich and extended to three-dimensional methods by Richard Ernst, Marius Clore and Angela Gronenborn. Wüthrich put forward a strategy for carrying out the sequential assignments of protein spectra that are now the basis for all high-resolution ^{1}H NMR of proteins. Since then a large number of laboratories have been active in determining protein structures. The laboratories of Ad Bax, and of Marius Clore and Angela Gronenborn have been particularly instrumental in the development and application of three-dimensional and four-dimensional NMR methods to isotopically enriched proteins, and have developed strategies based on Wüthrich's approach with appropriate modifications. Although reliable automated methods for sequential assignments are not yet available,[2] the methods for determining the three-dimensional structures of globular proteins up to $M_r = 30$ kDa can now be achieved relatively readily, providing high quality spectra can be obtained.

In the following sections we will consider the methods by which protein structures may be obtained, for $M_r < 15$ kDa by homonuclear methods, $M_r < 30$ kDa by heteronuclear methods, and $M_r > 30$ kDa by a variety of other methods. Such molecular weights represent rough guidelines for well-behaved soluble globular proteins. However, it should be stressed that molecular weight is not the only determinant for a successful NMR structure determination. Initially studies have to be carried out to establish whether the peptide or protein aggregates (by monitoring the concentration dependence of the NMR linewidths or the circular dichroism spectra, or by ultracentrifugation studies), whether the solution viscosity is too great (which also affects linewidth), whether the folded state of the protein is stable to temperature and pH, the degree of internal flexibility (also reflected by the linewidths), and the extent of chemical shift dispersion for the protein resonances. If the lines are too broad, and cannot be narrowed significantly by varying the conditions, or the chemical shift dispersion is not large enough (which would often arise in unstructured proteins, or proteins containing a large number of degenerate residue types), then a protein whose M_r is 5 kDa might be harder to study than one which is 15 kDa. Thus there is no substitute for preliminary 1D NMR experiments on a new protein, irrespective of molecular weight, which under a variety of conditions quickly reveal the feasibility of an NMR study.

4.1 PEPTIDE AND PROTEIN STRUCTURES WITH $M_r < 15$ kDa

4.1.1 Sequential assignments

There is a standard notation used for distances in proteins, which is that the distance between the hydrogen atoms A and B located in the amino acid residues in the sequence positions i and j, respectively, is denoted by $d_{AB}(i, j)$. This notation can be applied for all hydrogen atoms, but in practice the distances between different backbone hydrogen atoms and between backbone hydrogens and βCH_n groups are of particular interest, as shown in Fig. 4.1.

The standard approach for the sequential assignments of a small ($M_r < 15$ kDa) protein is to carry out the following four steps:

(i) The spin systems of non-labile protons in individual amino acids are identified for the native protein in D_2O by DQF–COSY and TOCSY experiments.

(ii) From studies in H_2O, the identity of amino acid spin systems of the individual residues are completed through identification of the J connectivities with the labile protons.

(iii) Sequentially neighbouring amino acid ^1H spin systems are identified from observation of the sequential NOE connectivities $d_{\alpha N}$ or d_{NN}, or possibly $d_{\beta N}$.

(iv) The aim of steps (i)–(iii) is to identify groups of ^1H NMR lines corresponding to peptide segments that are sufficiently long to be unique in the primary sequence, or a single residue which is unique. Sequential assignments are then obtained by matching the peptide segments thus identified by NMR with the corresponding segments in the independently determined (not by NMR) amino acid sequence.

An alternative to the strategy outlined in (i)–(iv) has been called the main-chain-directed (MCD) assignment strategy.[3] This places less emphasis on identifying the spin systems at the start of the process, and, instead, sequential NOEs and NOEs typical for

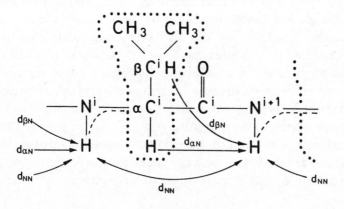

Fig. 4.1 Polypeptide segment with indication of the spin systems of non-labile protons in the individual residues (inside dotted lines), the H^α—NH COSY connectivities (broken lines), and the sequential NOE connectivities (arrows). (Reprinted from ref. 1 with permission.)

regular secondary structures are analysed simultaneously to obtain sequential assignments. This approach has advantages when there is already information about secondary structure.

As an example of the sequential assignment procedure, consider the DNA-binding domain 1–51 of the *E.coli lac* repressor headpiece, which contains three unique amino acids, Gly-14, His-29 and Ile-48, and numerous dipeptide segments with unique NMR properties (see Fig. 4.2). We will discuss the structure of this protein in more detail in Section 4.3.1. For unique residues, the sequential assignment follows directly from the identification of the spin system. For the resonances of unique dipeptide segments, sequential assignments result if a sequential connectivity between the ^{1}H spin systems of the two residues can be established. The unique residues and dipeptide segments can then be used as reference locations in the primary structure, so that further sequential identification steps directly provide the sequential assignments for the adjacent residues.

The types of spin systems that are present in amino acids are shown in Table 4.1. Their characteristic chemical shifts for random extended chain structures (sometimes inappropriately called 'random coil') are given in Appendix 2. They have characteristic cross-peak fine structures in their DQF–COSY spectra. In Appendix 2 we illustrate this by showing the DQF–COSY and TOCSY spectra for 19 of the 20 amino acids obtained at 11.7 T (500 MHz ^{1}H) and pD 7.0. Using this type of information, generally all but the side chains of Lys and Arg beyond the βCH$_2$ can be readily identified for proteins with $M_r \leq 15$ kDa. Other techniques (TOCSY and RELAY) are needed for the complete assignment of all the resonances for Lys and Arg.

The short distances found in the major types of secondary structure in peptides and proteins are summarized[4] in Table 4.2, and displayed on the common protein structural motifs in Fig. 4.3. The type III β-turn is essentially identical to a type I turn, or to part of a 3_{10} helix. The intraresidue vicinal distances must always be short, and constitute the majority of the strong NOEs seen. These enhancements can be identified by comparing NOESY and COSY spectra. The next most common NOEs correspond to the sequential distances for the i to the $i+1$ residues. The importance of observing the NH signals in H$_2$O is clear. Finally, comparison of the distances for α-helix and β-sheet shows that regular secondary structure can be spotted and categorized by its NOE pattern. α-helices have particularly short sequential d_{NN}, and β-sheets have short $d_{\alpha N}$. Even more characteristic are the short $d_{\alpha N}(i, i+3)$ and $d_{\alpha\beta}(i, i+3)$ found in helices, and the short

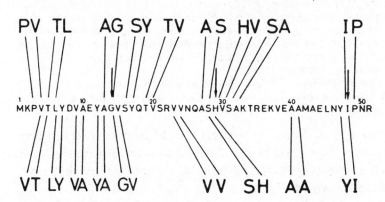

Fig. 4.2 Unique amino acid residues (arrows) and selected unique dipeptide segments (large letters) in the *lac* repressor headpiece, which are suitable starting points for making sequential assignments. (Reprinted from ref. 1 with permission.)

Table 4.1 Side chains R and three-letter and one-letter symbols for the 20 common amino acids, and spin systems of the non-labile hydrogen atoms in the molecular fragments H—αC—R[a,b,c]

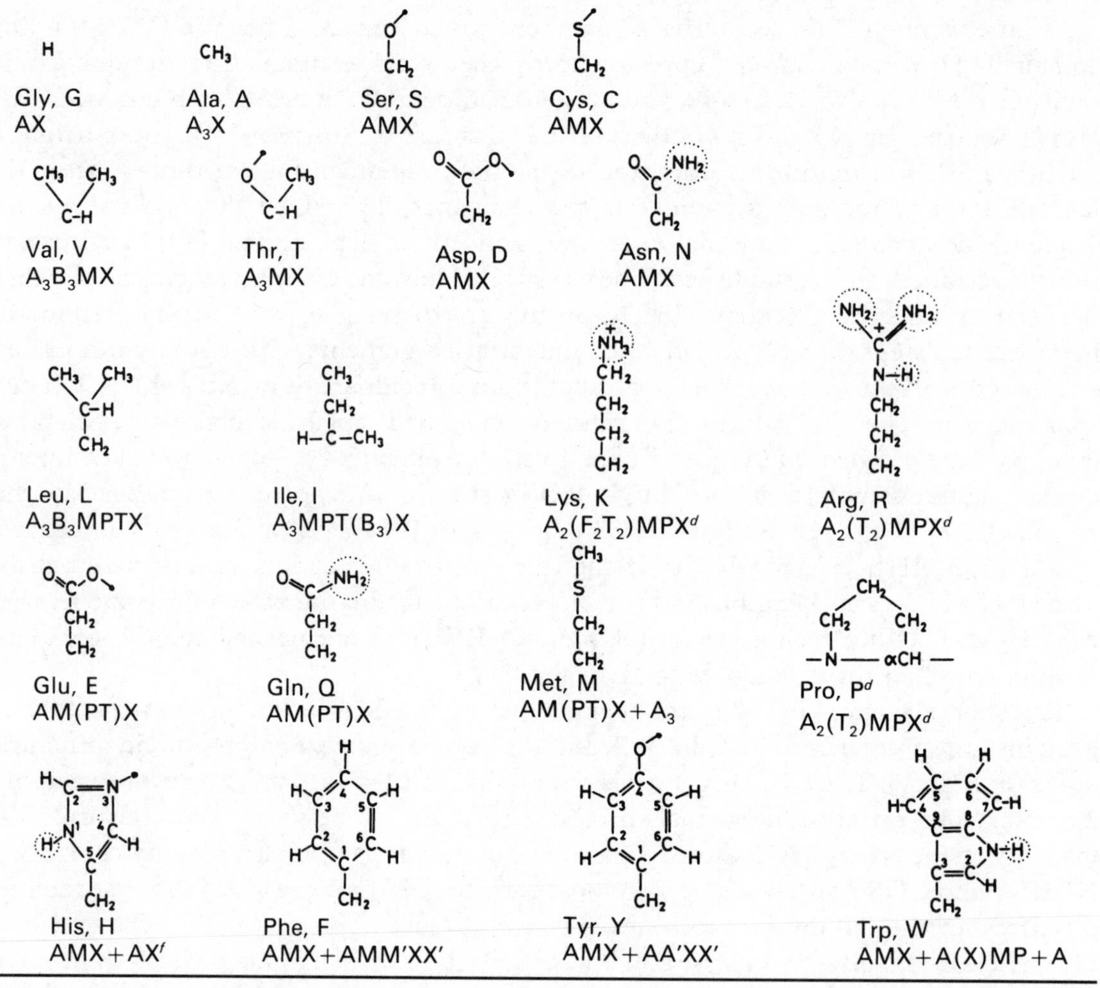

Gly, G
AX

Ala, A
A_3X

Ser, S
AMX

Cys, C
AMX

Val, V
A_3B_3MX

Thr, T
A_3MX

Asp, D
AMX

Asn, N
AMX

Leu, L
A_3B_3MPTX

Ile, I
A_3MPT(B_3)X

Lys, K
$A_2(F_2T_2)$MPX[d]

Arg, R
$A_2(T_2)$MPX[d]

Glu, E
AM(PT)X

Gln, Q
AM(PT)X

Met, M
AM(PT)X+A_3

Pro, P[d]
$A_2(T_2)$MPX[d]

His, H
AMX+AX[f]

Phe, F
AMX+AMM'XX'

Tyr, Y
AMX+AA'XX'

Trp, W
AMX+A(X)MP+A

[a]Reprinted from Ref. 1 with permission.

[b]Also indicated is the numeration used for the aromatic ring atoms. Not indicated is the standard identification of the other side-chain heavy atoms by lower case Greek letters, with βC being next to αC (IUPAC–IUB Commission on Biochemical Nomenclature, 1970).

[c]Labile protons that can under certain conditions be observed by NMR in aqueous solution are shown in dotted circles. Those labile protons, which are not usually observed, are indicated by small filled circles.

[d]For simplicity the spin systems for Arg, Lys, and Pro were written with the assumption that with the exception of the β position, each methylene group gives rise to a single two-proton resonance.

[e]The structure for Pro includes the backbone atoms αCH and N.

[f]In His, 2H and 4H appear often as two singlet lines, but the connectivity through the small four-bond coupling of approximately 1 Hz was observed in several proteins. 2H can be exchanged with deuterium of the solvent, D_2O, within a period of several hours to several months.

interstrand $d_{\alpha N}(i, j)$, $d_{NN}(i, j)$, and $d_{\alpha\alpha}(i, j)$ found in β-sheets. Indeed, even without any sequential assignments, the presence of $d_{\alpha\alpha}$ NOEs can be used as a crude diagnosis for antiparallel β-sheets. These characteristic NOE patterns are depicted in the, now standard, graphical representation shown in Fig. 4.4.

4.1.2 J coupling constants

We have already considered (in Section 3.2) how J coupling constants can be used to generate torsion-angle constraints in structure determination. Clearly there is a correlation between given values of ^3J coupling constants and regular secondary structure. Using Equation (3.1), the following values for $^3J_{HN\alpha}$ in regular secondary structures are obtained:

$$\alpha \text{ helix } (\phi = -57°), \, ^3J_{HN\alpha} = 3.9 \text{ Hz}$$
$$3_{10} \text{ helix } (\phi = -60°), \, ^3J_{HN\alpha} = 4.2 \text{ Hz}$$
$$\text{Antiparallel } \beta \text{ sheet } (\phi = -139°), \, ^3J_{HN\alpha} = 8.9 \text{ Hz}$$
$$\text{Parallel } \beta \text{ sheet } (\phi = -119°), \, ^3J_{HN\alpha} = 9.7 \text{ Hz}$$

These numbers provide a basis for the identification of regular secondary structure from measurements of $^3J_{NH\alpha}$. As a general rule of thumb, helices exhibit $^3J_{NH\alpha} < 6$ Hz, β-sheets exhibit $^3J_{NH\alpha} > 8$ Hz, and random extended chains exhibit $^3J_{NH\alpha} = 6$–8 Hz. Note, however, that deviations from this rule occur when proline is present.

4.1.3 Amide exchange rates

The formation of hydrogen bonds between backbone amide protons and backbone carbonyl oxygens is a characteristic of regular secondary structures. In the α-helix and 3_{10}-helix, hydrogen bonds occur between CO_i—NH_{i+4} and CO_i—NH_{i+3}, respectively, and in β-sheets a dense network of hydrogen bonds occurs between adjacent strands (see Fig. 4.3). In regular secondary structures all amide protons are involved in hydrogen bonds with the following exceptions: the first four residues in an α-helix; the first three

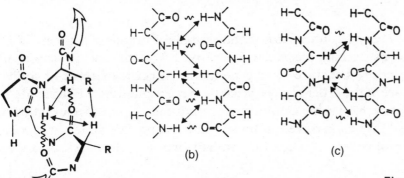

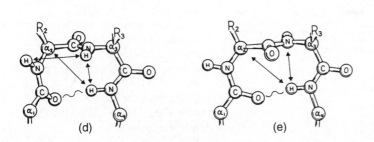

Fig. 4.3 Regular secondary structure found in proteins. Some short interproton distances are indicated by double-headed arrows, and hydrogen bonds by wavy lines: (a) α-helix, with the direction of the helix indicated by large arrows; (b) antiparallel β-sheet; (c) parallel β-sheet; (d) β-turn type I; (e) β-turn type II. (Reprinted from ref. 4 with permission.)

Table 4.2 Short distances found in protein structures[a]

Distance[b]	α-Helix[c]	3₁₀ helix	β-antiparallel sheet	β-parallel sheet	Type I turn	Type II turn
$d_{\alpha N}(i,i)$	2.7	2.7	2.8	2.8	2.8/2.8	2.7/22
$d_{\alpha\beta}(i,i)$[d]	2.2–2.9	2.2–2.9	2.2–2.9	2.2–2.9	2.2–2.9	2.2–2.9
$d_{\beta N}(i,i)$[d]	2.0–3.4	2.0–3.4	2.4–3.7	2.6–3.8	2.0–3.5	2.0–3.4/3.2–4.0
$d_{\alpha N}(i,i+1)$	3.5	3.4	2.2	2.2	3.4/3.2	2.2/3.2
$d_{NN}(i,i+1)$	2.8	2.6	4.3	4.2	2.6/2.4	4.5/2.4
$d_{\beta N}(i,i+1)$[d]	2.5–3.8	2.9–3.0	3.2–4.2	3.7–4.4	2.9–4.1/3.6–4.4	3.6–4.4
$d_{\alpha N}(i,i+2)$	4.4	3.8			3.6	3.3
$d_{NN}(i,i+2)$	4.2	4.1			3.8	4.3
$d_{\alpha N}(i,i+3)$	3.4	3.3			3.1–4.2	3.8–4.7
$d_{\alpha\beta}(i,i+3)$[d]	2.5–4.4	3.1–5.1				
$d_{\alpha N}(i,i+4)$	4.2					
$d_{\alpha N}(i,j)$[e]			3.2	3.0		
$d_{NN}(i,j)$[e]			3.3	4.0		
$d_{\alpha\alpha N}(i,j)$[e]			2.3	4.8		

[a]Data from Ref. 4. Only distances up to 4.5 Å are shown.
[b]$d_{AB(i,j)}$ means the distance between proton attached to heavy atom A on residue number i and proton attached to heavy atom B on residue number j.
[c]Secondary structure constructed as follows: α-helix $\phi = -57°$, $\psi = -47°$; 3₁₀-helix $\phi = -60°$, $\psi = -30°$; antiparallel-β $\phi = -139°$, $\psi = +135°$; parallel-β $\phi = -119°$; $\psi = +113°$; type I turn $\phi_2 = -60°$, $\psi_2 = -30°$, $\phi_3 = -90°$, $\psi = 0°$; type II turn $\phi_2 = -60°$, $\psi_2 = 120°$, $\phi_3 = 90°$, $\psi_3 = 0°$. For turns, treat residue i as residue 2/residue 3, except for $d_{\alpha N}(i,i+3)$, where i is residue 1.
[d]Distances involving β protons are the lower and upper limits to the range of distances possible. A CH₂ fragment is assumed, and the nearer hydrogen atom of the two is always considered.
[e]Interstrand distances. The indices (i,j) indicate the residues that give the shortest distance of a given type (see Fig. 4.3).

residues in a 3₁₀-helix; and every second residue in the peripheral strands of β-sheets. The rate at which amide protons exchange correlates very well with the manifestation of hydrogen bonding in secondary structure.[5–9] In general, the intrinsic lifetimes of amide protons in small peptides are of the order of a few seconds to a few minutes.[10] Even at pH 3 and 277 K, the lifetimes of exchangeable amides in unstructured peptides are usually no greater than 100 minutes.[11] When involved in hydrogen bonding, the amide exchange rates can increase by several orders of magnitude to half-lives ($t_{1/2}$) from five hours to

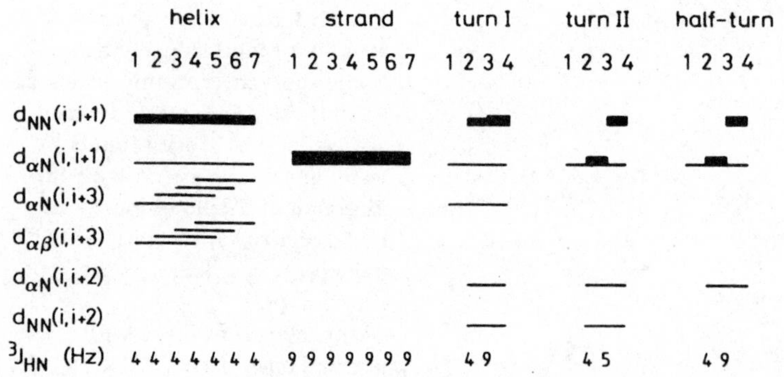

Fig. 4.4 Characteristic patterns of short-range NOEs involving the NH, C$^\alpha$H, and C$^\beta$H protons seen in various secondary structure elements. The NOEs are classified as strong, medium, and weak, as reflected by the thickness of the line. (Reprinted from G. M. Clore and A. M. Gronenborn, *Crit. Rev. Biochem. Mol. Biol.*, **24**, 479–557 (1989) with permission.)

several days/weeks. The observation of slowly exchanging amides is certainly indicative of secondary structure although it does not allow identification of the hydrogen bond acceptor, which can under certain circumstances be determined from a pH titration.[12] Amide exchange rates can also be used to measure intrinsic protein dynamics (see Section 4.1.6) and protein folding/unfolding (see Section 5.2).

4.1.4 Chemical shifts

The chemical shifts of NH and C^xH protons can be indicative of regular secondary structure.[13–18] In particular, in helices NH and C^xH have chemical shifts upfield (≈ 0.1 ppm for NH; ≈ 0.39 ppm for C^xH) from the average for extended chain conformations, and in β-sheets resonate at lower field (≈ 0.5 ppm for NH; ≈ 0.37 ppm for C^xH). Similar changes are also found for carbonyl ^{13}C, α-^{13}C and amide ^{15}N chemical shifts.

Another empirically useful indicator of secondary structure somewhat analogous to the amide exchange rate is the temperature dependence of the amide chemical shifts. For small peptides, these would also appear to correlate well with regular secondary structure and provide independent corroborating evidence for the existence of hydrogen bonds. Plots of the amide chemical shift against temperature are usually linear with a slope which is referred to as the *temperature coefficient*. In general, the temperature coefficients of amide proton resonances for extended chain structures are on the order of -6 to -10 ppb/K, whereas small temperature coefficients (~ -4.0 ppb/K) are correlated with solvent inaccessible or hydrogen-bonded environments.[19] Therefore low-temperature coefficients should correlate with the slowly exchanging amides, and both data can be used to identify the presence of regular secondary structure, particularly for small linear peptides. However, the temperature dependence of amide shifts has been found to be rarely useful for globular proteins.

4.1.5 Effect of solvent on secondary structure

The solvent in which small proteins and linear peptides are examined by NMR can have a profound influence on the conformation in solution. This is particularly true for small peptides in which the populations of the folded conformations can be small and measurements of structure are strongly influenced by populations of unstructured species. Many peptides have been studied in deuterated organic solvents owing to their excellent solubility properties and lack of interfering resonances. For example, Ni and co-workers[20] observed medium- and long-range NOEs characteristic of helical structure for the C-terminal thrombin fragment, hirudin 55–65, in 66% d_6-DMSO mixtures. Similarly, Kessler and co-workers[21] found that the structure of cyclic somatostatin analogues in d_6-DMSO corresponded well with crystallographic structures. In contrast, in another study Kessler and co-workers[22] found that the structures of cyclosporin A in hydrophobic solvents such as $CDCl_3$ and C_6D_6 were close to the X-ray crystallographic structure, whereas in d_6-DMSO they found a complex mixture of different conformations. Thus, stable structures for cyclic and linear peptides have been identified in organic solvents, but it is clear that the solvent plays a considerable rôle. It is generally believed that d_6-DMSO destabilizes all but highly preferred conformations, by strong polar

Table 4.3 Dynamic characteristics of protein structure

Motion	Timescale (s^{-1})	Method of study
Opening of secondary structure	10^{-4}–10^4	^{1}H—^{2}H exchange
Aromatic side-chain rotation	10^0–10^5	NMR line-shape analysis and selective saturation
Aromatic and aliphatic side-chain rotation	10^6–10^{10}	NMR relaxation time and fluorescence techniques
Segmental motion of the main chain	10^6–10^{10}	NMR relaxation times and fluorescence techniques
Overall tumbling of protein	10^6–10^{10}	NMR relaxation times, fluorescence and hydrodynamic techniques

interactions with the protein.[23] However, longer correlation times are often observed in d$_6$-DMSO, giving rise to NOESY cross-peaks that would otherwise be undetectable in aqueous solution and room temperature.

There are numerous examples of helical structures with varying degrees of stability in water and trifluoroethanol (TFE)–water solutions.[24–29] TFE solutions have proved useful for the characterization of helical regions in peptides because, while this solvent will not induce a helix in a peptide lacking an intrinsic helical tendency, it does act to stabilize a nascent helix. This may be due to the weaker basicity of TFE compared with water. With lessened hydrogen bonding of amide protons to the solvent, intramolecular hydrogen bonds are favoured, and secondary structures such as α-helices are stabilized. At the same time, the greater hydrophobicity of TFE may act to disrupt hydrophobic interactions important for stabilization of β-sheet structures.[30] It seems unlikely that TFE mixtures are a good mimic of a peptide's native environment.[31]

4.1.6 Protein dynamics

Throughout this Chapter we will encounter examples where the dynamics of a protein are discussed. In order to place this in the context of the NMR timescale, we need to consider the timescales of motion in a protein structure. These are illustrated in Table 4.3. The ways in which protein dynamics may be measured by NMR include the measurement of T_1, T_2 (MHz motions) and $T_{1\rho}$ (kHz motions) ^{15}N and ^{13}C relaxation rates and NOEs (see Sections 1.1.4, 1.1.5, 1.2.3, and 3.10) which probe fast motion, and amide exchange rates (see Sections 4.1.3 and 5.2), which measure much slower motions.

4.2 PROTEIN STRUCTURES WITH $M_r \approx$ 15–30 kDa
4.2.1 Sequential assignments

For larger proteins the problems of overlapping resonances and increased linewidths can become considerable, as we saw in Section 1.2.3, and the solution is to turn to isotopic labelling and three-dimensional heteronuclear NMR. This is because the heteronuclear couplings, $^1\mathcal{J}_{CH}$ (125–160 Hz) and $^1\mathcal{J}_{NH}$ ($\approx$92 Hz), are much larger than $^3\mathcal{J}_{HH}$, and frequenctly as much as 50–90% of the magnetization can be transferred from protons to their directly coupled heteronuclei.[32] Consequently, such heteronuclear shift correlation experiments are highly sensitive. The strategy for making three-dimensional heteronuclear NMR assignments relies exclusively on homonuclear and heteronuclear $\mathcal{J}$ coupling, avoids the use of NOE connectivities, and makes use of seven different experiments illustrated in Fig. 2.24. We have already considered these experiments in some detail in Chapter 2. The strategy was introduced originally by Markley and co-workers.[33,34] The advantage of avoiding NOE-based strategies for the assignment of larger protein spectra is that it is largely independent of protein conformation, and relies only on one-bond coupling. On the other hand, each sequential connectivity depends upon a single correlation. Degeneracy of carbon and nitrogen resonances can be more hampering than degeneracy of proton resonances in the case of the NOE-based assignments, where observation of more than one of the $d_{\alpha N}$, d_{NN}, and $d_{\beta N}$ connectivities can resolve ambiguities. Therefore, these heteronuclear strategies can also be combined with homonuclear NOESY experiments, although the presence of isotopic labelling, for example with ^{13}C, can exacerbate the linewidth problem for large proteins, because the ^{1}H—^{13}C dipolar interaction causes additional proton line broadening.

Typical values for the relevant coupling constants are indicated in Fig. 2.24. Equally important are the magnitudes of the T_2 relaxation times, which determine linewidth and depend approximately linearly on the rotational correlation time, and on the degree of internal mobility and local conformation. Generally, for residues that do not have a high degree of internal mobility in a globular protein of $M_r \approx 20$ kDa at 35°C, the linewidths ≈ 12 Hz for the NH proton, for ^{15}NH nitrogen ≈ 7 Hz (^{1}H-coupled) and ≈ 4 Hz (^{1}H-decoupled), ≈ 15 Hz for ^{13}C$^{\alpha}$, and ≈ 25 Hz for a ^{13}C-attached ^{1}H$^{\alpha}$. The linewidth of the carbonyl is dominated by chemical-shift anisotropy and is therefore proportional to the square of $\mathbf{B}_0$, and ≈ 6 Hz at 11.7 tesla. Comparison of these values with those shown in Fig. 2.24 indicates that for a 20 kDa protein most one-bond $\mathcal{J}$ couplings are significantly larger than the linewidths, which means that magnetization can be transferred efficiently from one spin to its directly attached coupling partner. Use of a combination of these experiments permits the complete sequential assignments for proteins up to a maximum $M_r \approx 30$ kDa. For proteins with favourable resonance dispersion and sensitivity, the HNCA, HCACO, and HN(CA)CO experiments may not be necessary to complete the assignments.[35]

4.2.2 Isotopic labelling of proteins

The key to carrying out multidimensional NMR studies on proteins with $M_r < 30$ kDa, and also structural investigations on selected regions of proteins with $M_r > 30$ kDa, involves the use of isotopic labelling. This can take two forms: universal labelling or selective labelling. The isotopes commonly employed under these circumstances are: ^{2}H (which will be considered in Section 4.2.3), ^{13}C, and ^{15}N. Less common isotopic labels which have been employed include ^{19}F, ^{3}H, ^{17}O, and ^{77}Se. In order for isotopic labelling to be practical, it is essential that the protein of interest is overexpressed at a sufficiently high level in an organism that is easy to handle. Generally this means that the gene encoding a particular protein should be inserted into a plasmid with a good promoter, and, most conveniently, in a suitable strain of *Escherichia coli*. A reasonable target for levels of overexpression for selected amino acid isotope labelling should be in the region of 50 mg of purified protein from 1 L of cells. Lower levels of overexpression can be tolerated for universal labelling with ^{15}N since ^{15}NH$_4$Cl is relatively cheap.

In recent years the methods of molecular biology have been developing at an extraordinary pace. There are an enormous number of techniques available for overexpression of proteins, and it is beyond the scope of this book to consider them in any detail. One system which has gained much in popularity in the last few years is the bacteriophage T7 expression system developed by Studier and co-workers,[36] with which it is not uncommon to achieve levels of expression of the order of 200 mg pure protein from 1 L of *E.coli*. The T7 RNA polymerase is very selective for specific promoters that are rarely encountered in DNA unrelated to T7 DNA.[37,38] Since efficient termination signals are also rare, T7 RNA polymerase is able to make complete transcripts of almost any DNA that is placed under the control of a T7 promoter. Furthermore, the T7 RNA polymerase elongates chains about five times faster than *E.coli* RNA polymerase.[39] The inclusion of the gene for T7 lysozyme, which binds T7 RNA polymerase, on plasmids pLysS and pLysE provides a means of controlling the basal levels of expressed protein prior to induction. pLysE provides higher levels of lysozyme and thus lower basal levels of expression of the target gene.

The target gene is inserted into a pET plasmid containing the T7 promoter and terminator, which is commercially available (Novagen Inc., Madison, WI). This plasmid is transformed into an *E.coli* strain providing T7 RNA polymerase, for example BL21 or HMS174, along with a pLys plasmid to suppress basal expression. On occasions, overexpression of proteins with a low solubility, or whose folding pathways are slow, can lead to *inclusion bodies*, in which the protein forms into insoluble clumps. This can be a mixed blessing. Sometimes the purification of such inclusion bodies is much easier, provided the protein can be refolded to the native form after purification. Sometimes the protein cannot be refolded and inclusion body formation must be avoided at all costs, for example by altering the growth conditions.[40] There are some ways to control inclusion body formation.[41] If the engineered strain is grown on minimal media (M9 salts[42]) in the presence of either ^{15}NH$_4$Cl or [^{13}C$_6$]glucose, or both together, then high incorporations can be achieved. The only disadvantages of this approach are that the cells usually grow to lower cell densities than in rich media (such as LB media), and the purification profile of the protein may not be identical to the unlabelled form.

Selective labelling of amino acid types is generally a considerably more expensive endeavour than universal labelling, primarily because of the cost of specifically labelled amino acids. Two approaches can be adopted: the first being growth of the strain into mid-logarithmic phase in a defined media containing unlabelled amino acids, and then administering labelled amino acids at the same time that the gene product is induced (for example by addition of IPTG, or isopropyl-1-thio-β-D-galactopyranoside);[43] and the second being the growth of an auxotrophic strain containing the overexpression plasmid in a defined media containing the labelled amino acid throughout growth and induction.[44] The first approach gives reasonable levels of incorporation for good overexpression systems, such as the T7 system, but suffers from the disadvantage that 100% labelling is

Table 4.4 *E.coli* amino acid metabolism[a]

Amino acid	Sole nitrogen source[b]	Key precursor	N source α-NH	Side chain	α-N product	Useful host genotype[c]
Glu	+	α-ketoglutarate; Gln	NH_3, Gln		All	gdhA, gltB (*aspC, avtA, ilvE, tyrB)
Gln	+	Glu	Glu	NH_3	All	glnA
Arg	+	Glu	Glu	Glu, Asp	·	argH
Pro	+	Glu	Glu	·	Glu, Asn, Lys, Met, Thr, NH_3	*aspC, tyrB
Asn	+	Asp	Asp	NH_3, Gln	·	asnA, asnB
Lys	−	Asp	Asp, Glu	Glu, Asp	·	lysA
Met	+	Asp	Asp	·	·	metC
Thr	−	Asp	Asp	·	Gly, NH_3	thrC
Ile	−	Thr	Glu	·	Glu	*ilvE
Leu	±	Pyruvate	Glu	·	Glu	*ilvE, tyrB
Val	−	Pyruvate	Glu, Ala	·	Glu, Ala	avtA, *ilvE, tyrB
Ser	+	3-phosphoglycerate (Gly)	Glu, Gly	·	Gly, Cys, Trp, NH_3	serA
Gly	+	Ser (Thr)	Ser	·	Ser, NH_3	glyA
Cys	−	Ser	Ser	·	·	cysE
Trp	±	Chorismate, Ser	Ser	Gln	Ser	trpA,B
Phe	−	Chorismate	Glu	·	Glu	*aspC, ilvE, tyrB
Tyr	+	Chorismate	Glu	·	Glu	*aspC, tyrB
His	±	Adenine	Glu	Gln (adenine) ·		hisD
Ala	+	Pyruvate	Glu, Val	·	Glu, Val, NH_3	*avtA

[a]Data from Ref. 46.
[b]Growth on the amino acid as the sole nitrogen source (+ indicates growth, ± indicates possible growth).
[c]Asterisk (*) indicates a transaminase deficiency.

usually not achievable because of isotopic dilution from other unlabelled amino acids via catabolic pathways. The second approach gives high levels of incorporation, but requires engineering of the auxotrophic strain. This latter point is no longer a particular problem, since, for example, commercially available kits can be obtained (Novagen Inc., Madison, WI) to insert the T7 RNA polymerase into auxotrophic strains of *E.coli* (which are available from the *E.coli* Genetics Centre, Yale University, New Haven, CT[45]). Both approaches can suffer from lowered incorporations when labelling with ^{15}N, due to exchange with ^{14}N catalysed by transaminases. This is summarized in Table 4.4, which gives information concerning amino acid metabolism in *E.coli*, with emphasis on the ^{15}N-labelling of proteins. Also included are the key precursors to each amino acid, which are useful for ^{13}C labelling, and a list of useful host genotypes which block various metabolic steps and lead to auxotrophy. Usually it is desirable to use auxotrophs which are blocked both in a particular amino acid biosynthetic pathway (the amino acid which will be labelled), and also in transaminase activity.[46]

When an amino acid occurs only once in the protein, then labelling a specific amino acid leads to a residue-specific assignment of the NMR spectrum. When an amino acid occurs more than once, then specific amino acid labelling provides a residue-specific but not a sequence-specific assignment. There are, however, a number of elegant solutions to the latter problem for making sequence-specific assignments. One approach, used by Knowles and co-workers[47] when studying triosephosphate isomerase (TIM) by ^{15}N NMR (see also Section 8.3), involved carrying out site-directed mutagenesis on two of the three histidines in the protein, leaving only the active-site histidine, which was then specifically labelled by placing the mutated gene into an auxotrophic strain grown in the presence of [^{15}N]histidine. Clearly, the natural extension to this approach is universal labelling of a wild-type protein and of a protein that is mutated site-specifically, since then the *absence* of particular resonances in the mutant protein spectra indicate sequence-specific assignments in the wild-type protein.[48] Another approach is the site-directed mutagenesis of the codon for a specific cysteine residue for a selenocysteine codon,[49] and subsequent growth of the organism containing the mutated gene in the presence of ^{77}Se salts. Currently, this represents the only facile way for sequence-specific labelling of one amino acid out of a number in a given primary sequence. There are, however, good prospects that alternative strategies will be available in the near future. Schultz, Wemmer, and co-workers[50] have used *in vitro* suppression of an Ala→TAG nonsense mutation with a chemically aminoacylated suppresser tRNA to introduce [^{13}C]Ala at position 82 of T4 lysozyme. This method is not yet very practical on the large scales necessary for protein NMR, but no doubt refinement of the method will solve such difficulties. One last method worth mentioning which works well for small peptides is the solid-phase synthesis of the peptide with the labelled residue precursor used at the appropriate point in the cycle. This is only satisfactory for short (less than 30-residue) peptides.

4.2.3 Fractional deuterium labelling of proteins

The use of deuterium labelling to simplify ^{1}H NMR spectra of proteins has a long history, dating back to the work of Jardetsky, Markley, and co-workers,[51] in which selected residues of staphylococcal nuclease were protonated in a perdeuterated background, and Katz and Crespi on a series of proteins from photosynthetic bacteria.[52] Since then the approach has largely been focused on reducing the ^{1}H—^{1}H dipolar interaction through fractional deuterium labelling.[53–55] An example of this is shown in Fig. 4.5, where the elimination of passive coupling in the COSY spectra of random fractionally deuterated thioredoxin improves the resolution considerably. This type of deuteration is achieved in an exactly analogous manner to the general methods for isotopic labelling outlined in Section 4.2.2. Universal labelling is typically carried out by growth of the overexpressing organism in a defined media dissolved in various ratios of D_2O/H_2O according to the required fraction of random deuteration. Specific amino acid types can also be labelled by incorporation of either specifically or randomly deuterated amino acids into a defined media during growth of the overexpressing organism, as we have already discussed in Section 4.2.2.

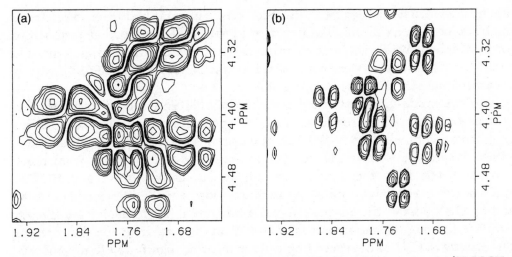

Fig. 4.5 Elimination of passive spin coupling by random fractional deuteration. (a) ^{1}H COSY spectrum of natural abundance *E. coli* thioredoxin; (b) 75% uniformly ^{2}H-labelled thioredoxin. The improved resolution observed in the fractionally deuterated sample is partly due to the longer T_2 relaxation time and hence narrower linewidth, but mainly it is the result of the elimination of other protons involved in coherence transfer. (Reprinted from ref. 53 with permission.)

There are a number of advantages to the use of deuteration for NMR studies of larger proteins. Random deuteration helps to reduce spin diffusion, and incorporation of stereospecifically deuterated amino acids helps to resolve degenerate proton resonances and enable quantitative NOE analysis. Furthermore, chiral deuteration can aid in stereospecific assignments, although since it is experimentally difficult, it has been rarely employed.

4.3 EXAMPLES OF LOW-RESOLUTION PROTEIN STRUCTURES

The earliest NMR structures were based solely on two-dimensional sequential assignments, in which often some resonances were not assigned, leading to ambiguities in all the NOEs in the same spectral regions as the unassigned resonances. Such structures were also of relatively low precision, and yet, even so, compared favourably with the corresponding X-ray structures (see Section 4.5). The pioneering protein structure determinations were carried out in the laboratories of Kurt Wüthrich and Robert Kaptein. In this section we will consider two of the more famous examples of these early structures, starting with the assignments and leading up to the three-dimensional structure. In both examples, further refinements place their more recent structures in the high-resolution category, although we shall concentrate on the early work as an example of the sequential assignment strategy and as testbeds for the refinement strategies that are now commonplace.

4.3.1 The *lac* repressor headpiece

The *lac* headpiece[56–58] is a separate domain, 51 residues long ($M_r = 5500$), of the *lac* repressor that is responsible for recognition of a specific sequence of base pairs of the *lac* operator. No X-ray crystal structure is known for either repressor or headpiece. However, on the basis of sequence homology (with cro and λ repressors) it had been suggested to contain the helix–turn–helix motif. The first step in the NMR structure determination carried out by Kaptein, Wüthrich, and co-workers was the sequential assignment. The connectivities for $d_{\alpha N}$ distances were established in combined COSY–NOESY experiments, an example of which is shown in Fig. 4.6.

The region of the amide-$C^{\alpha}H$ proton cross-peaks of the NOESY spectrum is combined with the corresponding spectral region of a COSY spectrum recorded under identical conditions. Such a combined COSY–NOESY plot contains both the $d_{\alpha N}(i, i+1)$ and the J connectivities. When one follows the polypeptide chain in the direction from the C terminus to the N terminus, the lines which indicate successive COSY and NOESY connectivities describe a clockwise spiral. As an illustration, consider residues 27–32. We start from the COSY peak of the unique His-29. A horizontal line to the left leads to the chemical-shift position of the amide proton of His-29 on the virtual diagonal. If present, the $d_{\alpha N}$ connectivity to $C^{\alpha}H$ of the preceding residue must be manifested in the NOESY spectrum by a cross-peak located on the vertical line through the His-29 amide proton position. Only two NOESY cross-peaks fall on this line; one accounts for the intraresidue

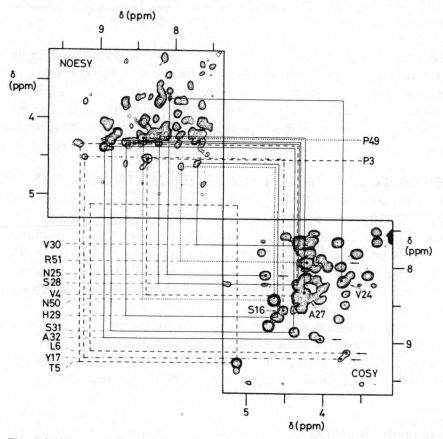

Fig. 4.6 Combined COSY–NOESY connectivity diagram of the *lac* repressor headpiece showing sequential assignments via NOEs between amide protons and the C^{α} protons of the preceding residues. Connectivities can be traced between residues 51–49 (dots), 32–27 (line), 25–24 (line), 17–16 (dot-dash), and 6–3 (dash). (Reprinted from ref. 56 with permission.)

NOE between amide proton and $C^\alpha H$ of His-29, and the other represents the $d_{\alpha N}$ connectivity to residue 28. The corresponding $C^\alpha H$ chemical shift corresponds to that of a serine spin system, which confirms that the $d_{\alpha N}$ connectivity is between His-29 and its nearest neighbour. The connectivity pattern continues with a horizontal line from the $d_{\alpha N}$ cross-peak to the virtual diagonal and from there with a vertical line into the COSY spectrum. There, this vertical line cuts through various cross-peaks with nearly identical $C^\alpha H$ shifts. The Ser-28 cross-peaks between amide proton and $C^\alpha H$ is identified as indicated in Fig. 4.6 from determination of the amide proton chemical shift by an intraresidue NOE from a serine $C^\beta H_2$ to the amide proton. That this assignment is correct is confirmed by the continuation of the sequential assignments via a horizontal line to the Ser-28 amide proton position on the diagonal and a vertical line into the NOESY spectrum which leads to a $d_{\alpha N}$ connectivity with an alanine (Ala-27—Ser-28 is a unique dipeptide sequence in the *lac* repressor headpiece—see Section 4.1.1). The additional $d_{\alpha N}$ connectivity from Ala-27 to Glu-26 is not resolved, but can be established from a NOESY spectrum recorded at higher pH. Starting again from the unique residue His-29 in the direction towards the C terminus, the $d_{\alpha N}$ connectivity leading to the amide proton of Val-30 can be found: a vertical line through the COSY cross-peak of His-29 leads to the $C^\alpha H$ proton position on the diagonal in the upper right. From there the horizontal line to the left into the NOESY spectrum leads to a well-resolved $d_{\alpha N}$ connectivity. A vertical line though this NOESY cross-peak leads to the position of the amide proton of position 30 on the virtual diagonal in the lower left of Fig. 4.6., from where a horizontal line connects to the amide–$C^\alpha H$ proton cross-peak of residue 30. This assignment was confirmed independently from relayed coherence transfer spectroscopy. Continuing with the anticlockwise assignment path, the connectivities to Ser-31 and Ala-32 are apparent.

In a similar manner d_{NN} connectivities were established, as shown in Fig. 4.7. Also, $d_{\beta N}$

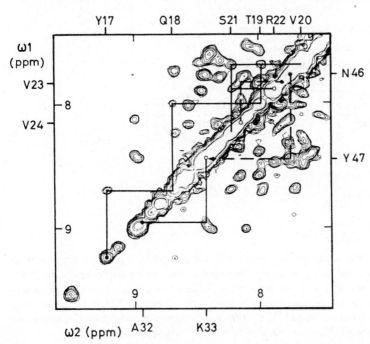

Fig. 4.7 NOESY data for the *lac* repressor headpiece showing sequential assignments via NOEs between amide protons (d_{NN}). Connectivities can be traced between residues 17–22 (line), 23–24 (dash), 32–33 (line), and 46–47 (dash). For each segment the start and end is indicated by closed and open circles respectively. (Reprinted from ref. 56 with permission.)

connectivities were obtained in a similar manner, giving the map of connectivities shown in Fig. 4.8 (where $d_{\alpha N}=d_1$; $d_{NN}=d_2$ and $d_{\beta N}=d_3$). Note that the pattern of $d_{\alpha N}$ and d_{NN} connectivities is characteristic of α-helix. These approaches lead to the complete ^{1}H assignments for all the backbone C$^\alpha$H and amide protons (except Ile 48, due to overlap problems). Most of the side-chain protons were also assigned, although for some of the long side chains of Lys, Arg, and Gln the assignments do not extend beyond the C$^\beta$ protons. By making use of a combination of $^3J_{C\alpha C\beta}$ couplings and NOEs, the prochiral groups of Val-9, -20, -23, -38 could be stereospecifically assigned. Longer-range NOEs (particularly the $(i,i+3)$) are summarized in Fig. 4.9. These permitted the identification

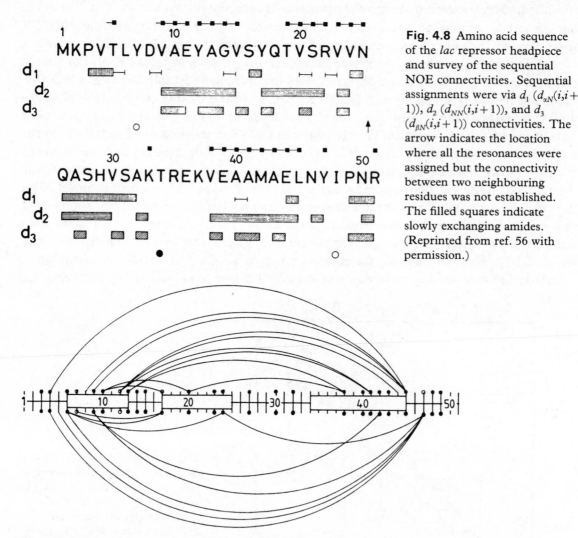

Fig. 4.8 Amino acid sequence of the *lac* repressor headpiece and survey of the sequential NOE connectivities. Sequential assignments were via d_1 ($d_{\alpha N}(i,i+1)$), d_2 ($d_{NN}(i,i+1)$), and d_3 ($d_{\beta N}(i,i+1)$) connectivities. The arrow indicates the location where all the resonances were assigned but the connectivity between two neighbouring residues was not established. The filled squares indicate slowly exchanging amides. (Reprinted from ref. 56 with permission.)

Fig. 4.9 Schematic representation of the primary and secondary structure of the *lac repressor* headpiece with long-range NOEs indicated. Helices are indicated by boxes. Non-polar side chains are indicated by filled circles and tyrosyl residues by open circles. The division of the lines indicating the long-range enhancements above and below the sequence was chosen for presentation purposes only. (Reprinted from E. Zuiderweg, M. Billeter, R. Kaptein, R. Boelens, R. M. Scheek, and K. Wüthrich, in *Progress in Bioorganic Chemistry and Molecular Biology* (ed. Yu and Ovchinikov), Elsevier, Amsterdam (1984) with permission.)

of three α-helices at 6–13, 17–25, and 34–45. Consistent with this is the observation of slowly exchanging amides in the helical hydrogen bonds, and 17 hydrogen-bonded restraints were added for which the acceptor was known, i.e. for the α-helical regions. Very few interhelix NOEs were found (see Fig. 4.9), and this small number of contacts meant that an approximate structure could be built by a computer model-building method. The three helices were constructed as regular α-helices and kept as rigid structures. All NOEs to side-chain protons were referred to the C^β carbons, with appropriate distance corrections. The helices were then moved manually, to try to satisfy all NOE constraints. This resulted in a high energy structure (total potential energy $V_{total} \approx 10^{10}$ kJ mol^{-1}), primarily due to errors made during the coordinate measurement, with about 20% of the 169 NOE constraints still outside the 3.5 Å bound imposed (see Fig. 4.10 and Fig. 4.11(a)). Finally, this high-energy structure was refined, first by restrained energy minimization (450 iterations), which reduced the energy very significantly (-1210 kJ mol^{-1}) without improving the violations of NOE contacts, and second by a 25 ps molecular dynamics calculation. In this calculation the pseudo-potential V_{NMR} was used as defined in Equation (3.40) with K_{NOE} being set arbitrarily to 250 kJ mol^{-1} nm^{-2}, which is such that $V_{NMR} = k_B T/2$ at 300 K when $r_{ij} - r_{ij}^0 = 1$ Å. Larger values for K_{NOE} up to 4000 kJ mol^{-1} nm^{-2} were found to speed up convergence and were used in some calculations. At the end of the RMD run the energy was reduced slightly and the number of NOE violations was reduced to only 7.5% of the total number of NOE constraints. Note that when the NOE restraints were moved altogether the structure did not change significantly during subsequent molecular dynamics calculations. Thus the NOE constraints do not act to 'hold the structure together'. Indeed, this should not be necessary for a good protein structure.

In subsequent refinements of the structure,[59,60] RMD for 60 ps (without solvent) resulted in a structure that satisfied the experimental restraints very well and also had a low energy (the energy dropped from $+4074$ kJ mol^{-1} for the starting structure to -3091 kJ mol^{-1} after RMD and restrained energy minimization). The remaining volations of the constraints were not greater than 0.5 Å, while the sum of all violations was

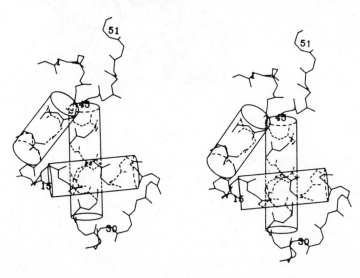

Fig. 4.10 Stereo diagram showing the backbone atoms of the *lac* repressor headpiece taken as a snapshot during a restrained molecular dynamics calculation. The cylinders represent the three α-helices at positions 6–13, 17–25, and 34–45. (Reprinted from ref. 60 with permission.)

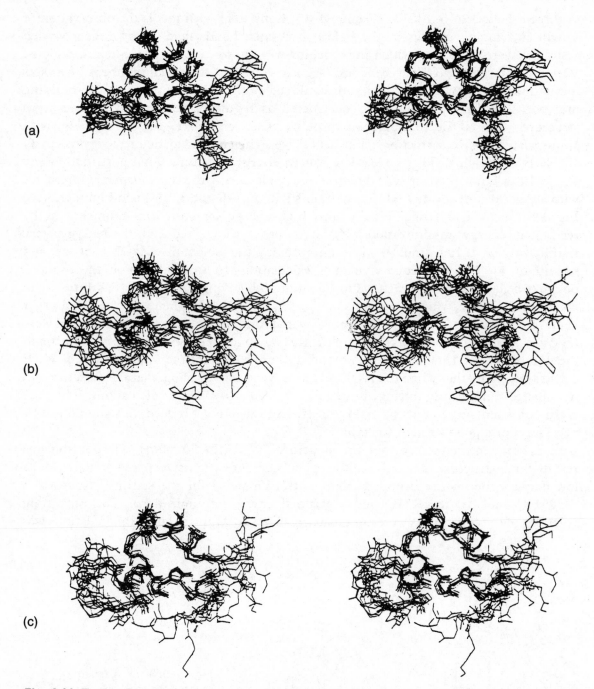

Fig. 4.11 Family of ten conformations of the *lac* repressor headpiece obtained after distance geometry (a), distance bounds driven dynamics (b), and restrained molecular dynamics (c). Stereo diagrams of the backbone atoms are shown. (Reprinted from ref. 60 with permission.)

5.8 Å. When distance geometry, followed by a more elaborate version of restrained molecular dynamics calculations in which K_{NOE} was varied slowly over each simulation, was employed,[61] improvements in the structural refinement were achieved. The final family of structures had RMSDs ranging from 1.4 to 1.7 Å (see Fig. 4.11).

4.3.2 Bovine pancreatic trypsin inhibitor (BPTI)

Bovine pancreatic trypsin inhibitor (BPTI) was the first protein for which the sequential assignments were carried out by Wüthrich and co-workers.[62] The protein has 58 residues ($M_r = 6500$), and the sequential assignments were made using combined COSY–NOESY experiments (see Fig. 4.12) which establish $d_{\alpha N}(i,i+1)$ connectivities as well as $^3J_{CH\alpha NH}$

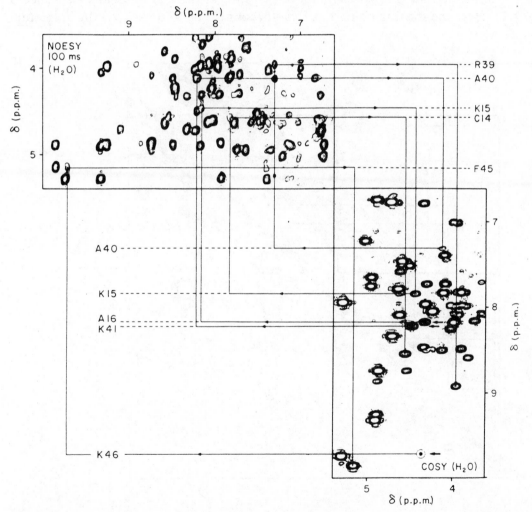

Fig. 4.12 Combined COSY–NOESY connectivity diagram for sequential resonance assignments between NH and the C$^\alpha$H of the preceding residue in BPTI. The straight lines and arrows indicate the sequential assignments obtained for segments 46–45, 41–39, and 16–14. Starting points are indicated by the arrows in the COSY spectrum. At this temperature of 68°C, the NH—C$^\alpha$H cross-peak for Lys-46 in the COSY spectrum was bleached out by the irradiation of the solvent, since the C$^\alpha$H resonance coincides with that of H$_2$O (although it is detectable at other temperatures). (Reprinted from ref. 62 with permission.)

couplings. The first step relies on the identification of a unique tripeptide segment at residues 43–45. Fig. 4.12 shows only those regions containing $C^\alpha H$—NH cross-peaks, and the horizontal line leading from Lys-46 $C^\alpha H$—NH COSY peak to the virtual diagonal position of the amide proton resonance. There is a NOESY cross-peak with the Lys-46 NH chemical shift, which is therefore assigned to $C^\alpha H$ of Phe-45, and a vertical line connects this peak with the diagonal position of Lys-46 NH. Continuing on from that peak, a horizontal line leads to the virtual diagonal position of $C^\alpha H$ of Phe-45, from where a vertical line connects with the already assigned $C^\alpha H$—NH COSY peak of Phe-45. Two other starting points, at the COSY peaks for Ala-16 and Lys-41, are shown in Fig. 4.12 and similar clockwise spirals define sequential assignments for those short segments. The $d_{NN}(i,i+1)$ connectivities are also useful for establishing sequential assignments, as shown in Fig. 4.13. Here the starting point is at Lys-46, connecting to Ser-47 via the horizontal

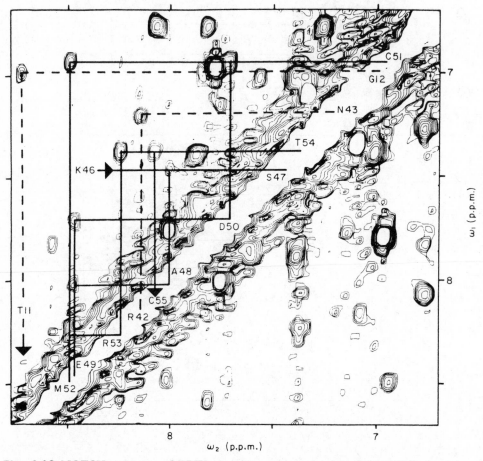

Fig. 4.13 NOESY spectrum of BPTI showing cross-peaks connecting NOEs between different amide protons. The solid lines indicate the sequential assignments for the segment 46–55. The arrows indicate the start and end of this sequence. The NOE cross-peak between the amide protons of Lys46 and Ser47 is not shown, since it is outside the sweep width of the spectrum. Connectivities between the amide protons of Thr11 and Gly12, and of Arg42 and Asn43 are indicated by broken lines. (Reprinted from ref. 62 with permission.)

line. Since Phe-45 was already established from Fig. 4.12, the only $d_{NN}(i,i+1)$ connectivity involving Lys-46 has to be with Ser-47. Because $d_{NN}(i,i+1)$ connectivities are symmetric with respect to the direction of the polypeptide chain, a second cross-peak linking Ser-47 with the NH of Ala-48 should be present. Since the Ala-48 NH might be either upfield or downfield, this second cross-peak could be either vertically above or horizontally to the left of the diagonal peak of Ser-47 NH. Fig. 4.13 shows that it is on the left, with a cross-peak with Ala-48 NH further downfield at 8.48 ppm, leading to Glu-49. The next two connectivities lead upfield, first to Asp-50 NH at 7.72 ppm, and then to Cys-51 NH at 6.95 ppm. This overall strategy was adopted for the whole polypeptide, checking against the known primary sequence. The NOE connectivities are summarized in Fig. 4.14.

The first NMR structure for BPTI appeared in 1987,[63] and several structure determinations have been carried out since then as tests for refinement algorithms and

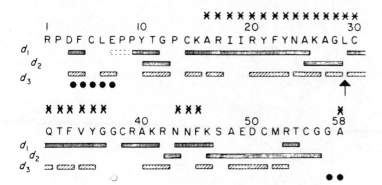

Fig. 4.14 NOE connectivities and amino acid sequence for BPTI. Sequential assignments were via d_1 ($d_{\alpha N}(i,i+1)$), d_2 ($d_{NN}(i,i+1)$), and d_3 ($d_{\beta N}(i,i+1)$) connectivities. The arrow indicates a location where all the resonances were assigned, but the connectivity between two neighbouring residues was not established. Stars (*) above the sequence identify the residues for which complete assignments had been established previously. (Reprinted from ref. 62 with permission.)

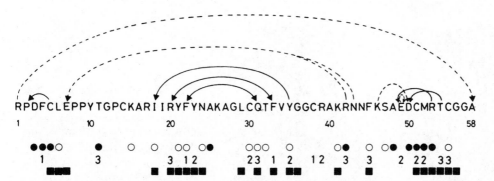

Fig. 4.15 Amino acid sequence of BPTI and supplementary conformational constraints. Continuous lines indicate H-bonds identified from NOEs, broken lines indicate H-bonds and salt-bridges identified from pH-titration studies. The arrows point from the H-bond donor to the acceptor. A double-headed arrow indicates a pair of H-bonds. The filled circles identify residues where $^3J_{HN\alpha}$ is <5.0 Hz and the torsion angle ϕ was constrained in the range [$-90°$, $-40°$]. Open circles identify residues where $^3J_{HN\alpha}$ is >8.0 Hz and ϕ was constrained in the range [$-160°$, $-80°$]. 1, 2, and 3 identify constraints on χ_1 of $60° \pm 60°$, and $-60° \pm 60°$ respectively. Filled squares indicate residues for which the peptide NH exchanges slowly, indicating involvement in an intramolecular H-bond. (Reprinted from ref. 63 with permission.)

assignment procedures. In the original structure, in addition to NOEs, $^3\mathcal{J}_{NH\alpha}$couplings, and disulphide bonds, hydrogen bonds and salt bridges established from pH titration and amide exchange data were also included, as shown in Fig. 4.15. The model was constructed using five random starting structures as input for each of the distance geometry algorithms DISGEO and DISMAN. Both sets of output were subjected to restrained energy minimization with up to 5000 steps, and the majority of the NMR constraints were fulfilled, with the average number of violations (N_{viol}) per structure around 30–40, and the residual distance constraint violations ($\sum \Delta r_{viol}$) always less than 1.0 Å, and in the best structures less than 0.5 Å. The RMSDs among the DISGEO

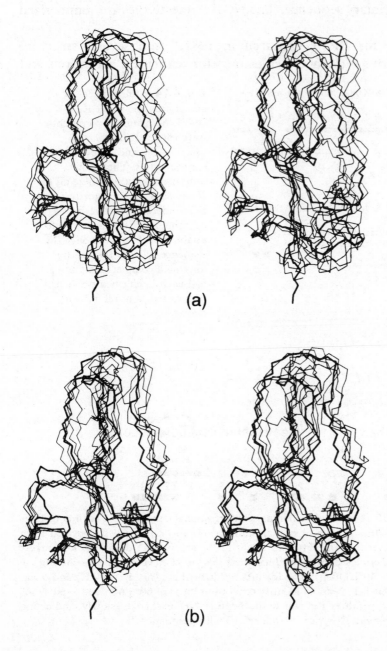

(a)

(b)

Fig. 4.16 Stereo view providing a comparison of the NMR structures calculated using (a) DISMAN or (b) DISGEO with the X-ray structure of BPTI. The heavy line represents the backbone conformation of the X-ray structure; the light lines correspond to the five conformers from the distance geometry. The NMR structures have been superimposed on the X-ray structure for minimal RMSD. (Reprinted from ref. 63 with permission.)

structures were 1.6 Å for the backbone residues, 2.3 Å for the backbone plus constrained side-chain residues, and 2.4 Å for the backbone plus non-constrained side-chain residues. The RMSDs among the DISMAN structures were 2.9 Å for the backbone residues, 3.2 Å for the backbone plus constrained side-chain residues, and 4.2 Å for the backbone plus non-constrained side-chain residues. The final structures from both sets of calculations are shown in Fig. 4.16.

4.4 EXAMPLES OF HIGH-RESOLUTION PROTEIN STRUCTURES

Until relatively recently, NMR protein structures were determined at a low level of detail, showing the global backbone fold typically with RMS deviations of at least 1.5–2 Å. The definition of side chains, if presented at all, was significantly poorer. With the use of stereospecific assignments and the availability of stronger magnetic fields, this situation has changed significantly. The introduction of heteronuclear three-dimensional and four-dimensional NMR techniques introduced in the last few years has also served to extend the molecular weight limit to much larger proteins. In this section, we will consider selected examples of these newer structures, starting with smaller proteins whose structures have been well-defined and proceeding to larger proteins for which heteronuclear multidimensional NMR methods were necessary. We will also concentrate on the refinement process rather than on the assignment procedure.

4.4.1 Tendamistat

The structure of the α-amylase inhibitor, Tendamistat (74 amino acids, $M_r = 8000$) was determined simultaneously and independently by the crystallographic group of Huber[64] and the NMR group of Wüthrich.[65,66] The backbone folds of the solution and crystal structures were virtually identical, although small differences in some of the side-chain conformations were observed between the two structures. The NMR structure was obtained using 842 NOEs and over 100 supplementary constraints from $\mathcal{J}$ couplings and identification of intramolecular hydrogen bonds. Stereospecific assignments were obtained for 41 of the 89 prochiral groups in the protein. The structure was calculated using the angle space distance geometry algorithm DISMAN, and nine separate structures were generated using all the constraints and random input (obtained by randomizing the dihedral angles within limited ranges, such that the NOE and ϕ angle constraints are satisfied). No further refinement by energy minimization was carried out on the output of the distance geometry calculations. RMS deviations between the NMR structures were 0.85 Å for the backbone, 1.04 Å for the backbone plus interior amino acid side chains, and 1.52 Å for all heavy atoms (i.e. C, O, N, S). The differences between solution and crystal structures were 1.0, 1.3, and 1.8 Å, respectively, and the structure is shown in Fig. 4.17. The quality of the definition of the side-chain conformations is shown in Fig. 4.18.

4.4.2 Antiviral protein BDS-I

The structure of the antihypertensive and antiviral protein BDS-I (43 residues, $M_r = 5000$) from the sea anemone was carried out in the laboratory of Clore and Gronenborn.[67] The structure was determined using 489 interproton and 24 hydrogen-

(a)

(b)

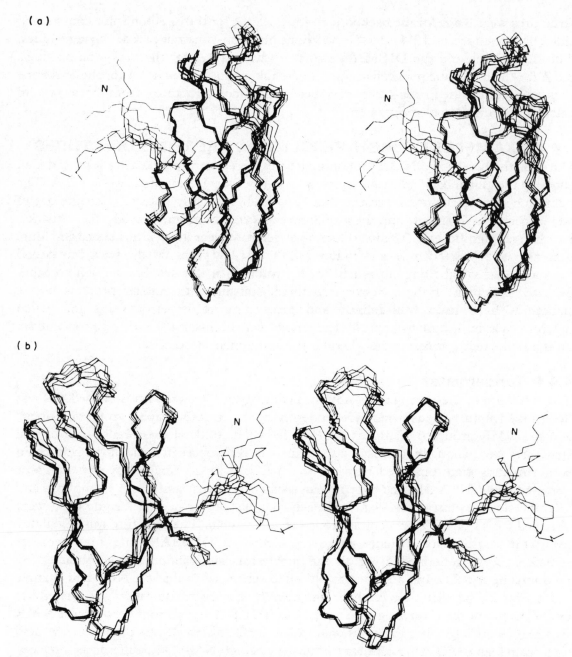

Fig. 4.17 Two stereoviews of Tendamistat showing a best-fit superposition of the backbone atoms of nine DISMAN structures in which (a) and (b) differ by 180°. N locates the N-terminus of the protein, which is virtually unconstrained in the NMR data. (Reprinted from ref. 66 with permission.)

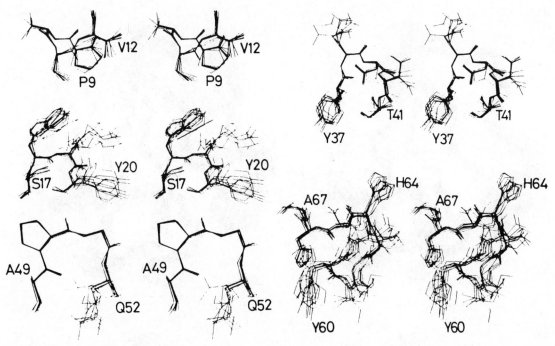

Fig. 4.18 Stereoviews of Tendamistat showing a best-fit superposition of selected side-chain atoms of nine DISMAN structures. (Adapted from ref. 66 with permission.)

Table 4.5 Structural statistics for BDS-1

	$\langle SA \rangle$	$\overline{SA}$	$(\overline{SA})_r$
rms deviations from experimental restraints (Å)			
all (513)	0.085 ± 0.002	0.059	0.079
interresidue short range ($|i-j| \leq 5$) (150)	0.083 ± 0.005	0.053	0.086
interresidue long range ($|i-j| > 5$) (105)	0.111 ± 0.007	0.082	0.100
intraresidue (234)	0.074 ± 0.006	0.053	0.065
hbond (24)	0.051 ± 0.010	0.038	0.043
F_{NOE} (kJ mol^{-1})	782 ± 46	385	669
F_{tor} (kJ mol^{-1})	109 ± 25	54	100
F_{repel} (kJ mol^{-1})	380 ± 29	10^4	301
F_{L-J} (kJ mol^{-1})	-431 ± 46	$>10^6$	-489
deviations from idealized geometry			
bonds (Å) (646)	0.014 ± 0.006	0.329	0.013
angles (deg) (1157)	2.910 ± 0.363	28.373	2.517
impropers (deg) (242)	0.830 ± 0.060	2.629	0.797

bonding distance restraints, supplemented with 44 torsion-angle restraints (ϕ, χ_1). This structure was refined using random input into the distance geometry program DISGEO, followed by simulated annealing, involving 200 steps of Powell minimization of the distance geometry substructures. The final 42 structures were subjected to 1000 steps of restrained energy minimization. The statistics for the structures are summarized in Table 4.5. The overall RMSDs were 0.67 Å for the backbone atoms and 0.90 Å for all the

(a)

(b)

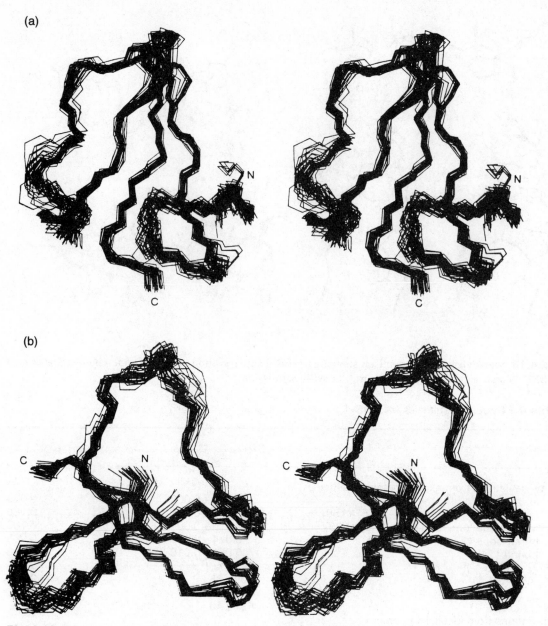

Fig. 4.19 Two stereoviews of BDS-I showing a best-fit superposition of the backbone (N, C^{α}, C′) atoms of 42 NMR structures after simulated annealing. (Reprinted from ref. 67 with permission.)

atoms. The structure is shown in Fig. 4.19. The excellent fit for selected side chains is illustrated in Fig. 4.20.

4.4.3 The inflammatory proteins C3a and C5a

The secondary structure of the human complement protein C3a (77 residues, $M_r = 8900$) was determined independently by the groups of Zuiderweg[68] and Wright.[69] This structure

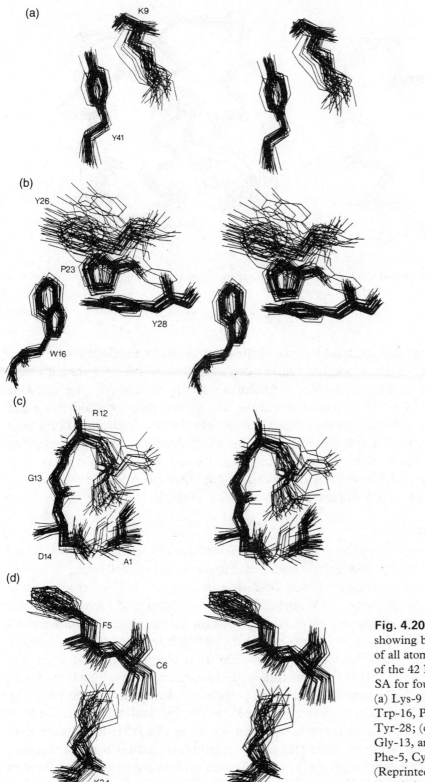

Fig. 4.20 Stereoviews of BDS-I showing best-fit superpositions of all atoms (excluding protons) of the 42 NMR structures after SA for four selected side chains. (a) Lys-9 and Tyr-41; (b) Trp-16, Pro-23, Tyr-26, and Tyr-28; (c) Ala-1, Arg-12, Gly-13, and Asp-14; and (d) Phe-5, Cys-6, and Lys-34. (Reprinted from ref. 67 with permission.)

Fig. 4.21 Stereoviews of human C5a showing best-fit superpositions of 11 equally converged NMR structures. (Reprinted from ref. 73 with permission.)

was not refined in any way, but substantial differences in the helical structure when compared to the X-ray crystal structure[70] have been attributed to crystal packing forces (see Section 4.5 for a detailed discussion). Analogous is the human complement C5a protein (74 residues, $M_r = 8500$), whose structure was determined by Zuiderweg and co-workers.[71-73] This structure was refined using 346 NOE, hydrogen-bond and disulphide constraints by the angle space distance geometry algorithm DISMAN starting with 14 random structures, from which 11 converged structures were obtained. The structure is shown in Fig. 4.21, which had an average RMSD of 1.6 Å for the α-carbons of the first 63 residues, and 0.8 Å for the α-carbons in the α-helices.

4.4.4 Plastocyanin

The structure of reduced (Cu^I) plastocyanin from French bean (99 residues, $M_r = 10$ kDa) was determined by Wright and co-workers[74] at high resolution, following a lower-resolution structure for the protein from *Scenedesmus obliquus*.[75] The structure was determined using 1120 distance, 103 dihedral angle (ϕ, χ_1), and 21 hydrogen-bond restraints. Stereospecific assignments were carried out for 26 methylene groups and 11 valine methyl groups. Additional restraints on the copper coordination were included in the restrained molecular dynamics calculations. The structure was refined using the distance geometry program DISGEO starting with 40 substructures, which led to 22 acceptable structures containing all the atoms. These were further refined using restrained energy minimization and then 'heated' to 1200 K and after equilibration allowed to cool and subjected to molecular dynamics using the AMBER[76] force field without solvent. A final cycle of 1000 steps of energy minimization was carried out. Sixteen MD-refined structures with the lowest restraint violation energies were used in the structure analysis. The RMSDs were 0.45 Å for all backbone heavy atoms, and 1.08 Å

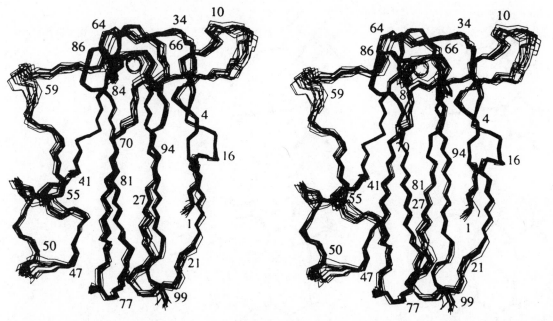

Fig. 4.22 Stereoview of plastocyamin showing best-fit superposition of the backbone atoms (N, C$^\alpha$, and C′) of 16 MD-refined structures. The position of the Cu atom is indicated by a circle (1.5 Å diameter). (Reprinted from ref. 74 with permission.)

for a side-chain heavy atoms. When compared with the X-ray structure of the reduced poplar plastocyanin, the RMSD between the 16 NMR structures and the X-ray structure was 0.76 Å for all backbone heavy atoms. Some of the side-chains are disordered in solution, and correlate with the side-chain temperature factors (*B*-factors) in the crystal structure, although no detailed backbone dynamics study using ^{15}N relaxation data has been carried out. The structure is shown in Fig. 4.22, with some examples of side-chain fits shown in Fig. 4.23.

4.4.5 Thioredoxin

The structure of reduced dithiol oxidoreductase thioredoxin from *E.coli* (108 residues, $M_r = 11.7$ kDa) was determined by Dyson, Wright, and co-workers[77] and independently the reduced human thioredoxin (105 residues, $M_r = 11.4$ kDa) structure was determined in the laboratory of Clore and Gronenborn.[78] The structure of the *E.coli* protein was determined using a total of 1244 interproton distance, 43 torsion-angle (ϕ, χ_1), and 17 hydrogen-bonding restraints. The structure was refined by starting with 28 structures calculated using the distance geometry program DISGEO, and further refined by restrained energy minimization and RMD with the AMBER force field, completing with 1000 steps of energy minimization. The best 12 structures had RMS deviations of 0.56 Å for the backbone residues 3–108 (0.77 Å for all the backbone residues) and 1.38 Å for side-chain heavy atoms. Confirmation of the disorder at the N- and C-termini has been

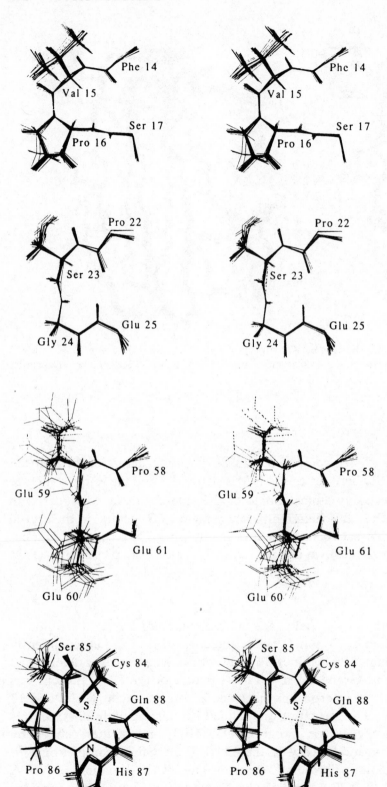

Fig. 4.23 Stereoviews of plastocyanin showing best-fit superposition of selected side-chain atoms of 16 MD-refined structures (Adapted from ref. 74 with permission.)

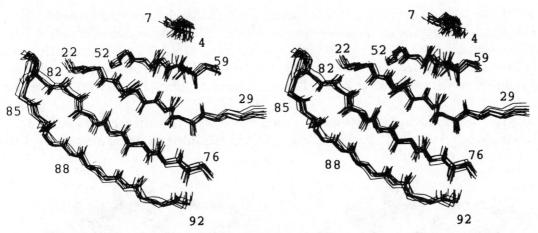

Fig. 4.24 Stereoviews of *E.coli* reduced thioredoxin showing the superposition of 12 refined NMR structures showing the central β-sheet. Carbonyl oxygens and amide protons involved in hydrogen bonds are shown as well as backbone C^{α}, C', and N atoms. (Reprinted from ref. 77 with permission.)

provided by work[79] showing that the S^2 order parameter for the whole protein is 0.86, whereas for the N- and C-termini it is 0.2–0.6. Part of the structure is shown in Fig. 4.24.

Human thioredoxin is only 25% homologous with the *E.coli* protein, and the structure was determined using 1983 interproton distance, 98 torsion-angle (ϕ, ψ, and χ_1), and 52 hydrogen-bonding restraints. Stereospecific assignments were made for 17 of the 53 non-degenerate β-methylene proton pairs, and eight of the 11 valine methyl groups. A total of 33 final simulated annealing structures was calculated. The procedure involved calculation of an initial set of 98 substructures using the real space distance geometry program DISGEO, followed by simulated annealing on 75 substructures with the program XPLOR, in an iterative manner. The structures had RMSDs of 0.40 Å for the backbone atoms and 0.78 Å for all the atoms. The structure is shown in Fig. 4.25.

4.4.6 Epidermal growth factor

For epidermal growth factor there was no crystal structure available when three groups independently determined very similar structures for the two domains of the protein by NMR. The groups of Scheraga and Wüthrich[80] and Inagaki[81] studied the murine (53 residues, $M_r \approx 5800$) protein, and Campbell[82] studied the human (53 residues although a 48-residue derivative was studied, $M_r \approx 5300$) protein. The structure of the murine protein determined by Scheraga, Wüthrich, and co-workers used 251 interproton distance, 36 hydrogen-bonding, 28 torsion-angle (ϕ), and 18 disulphide bond restraints. These were used in a distance geometry calculation with the angle space program DISMAN on 15 structures representing fragments 1–33 and ten structures for fragments 32–53. Five of these refined structural fragments were combined and subjected to further refinement with DISMAN and energy minimization. The RMSDs for the five final structures were 1.8 Å for the backbone residues. The study by Inagaki and co-workers used 186 tertiary interproton distances to assist in building a mechanical model! They also tested their model using 462 interproton distance, 40 hydrogen-bonding and 18 disulphide bond restraints as input for distance geometry calculations with DISMAN and found a similar structure to their mechanical model. Mechanical model-building was absolutely crucial in the determination of the structure of DNA by Watson and Crick in 1953, but it is extremely time-consuming for large molecules and its use is to be discouraged with the widespread availability of computers. Their final RMSDs were not reported.

The structure of the human protein was determined with 186 interproton distance, 23 torsion-angle (ϕ), nine hydrogen-bonding, and three disulphide bond restraints. The

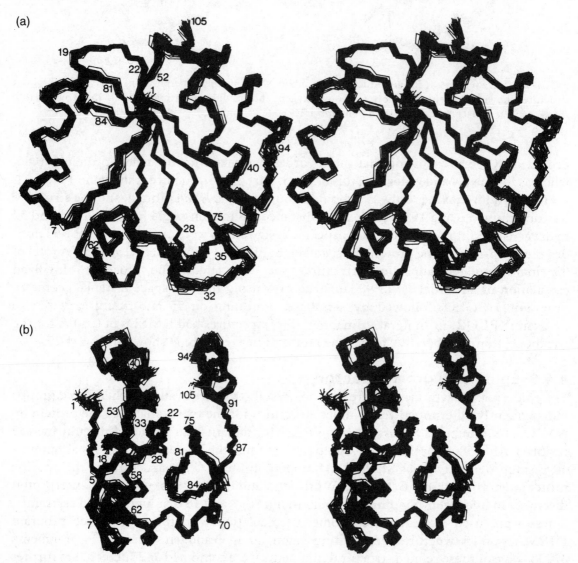

Fig. 4.25 Two stereoviews of human reduced thioredoxin showing best-fit superpositions of the backbone (N, C$^\alpha$, and C') atoms of the 33 refined NMR structures. (a) View towards the five strands of the twisted β-sheet with the surrounding helices. (b) View down the edge of the sheet, illustrating helices α_2 and α_4 interacting in a parallel orientation on one side and the other helices located in the opposite side. (Reprinted from ref. 78 with permission.)

Table 4.6 The energies of the hEGF structures[a]

| Structure | After distance geometry | | After energy minimization | | After molecular dynamics | |
	Potential energy	Restraints energy	Potential energy	Restraints energy	Potential energy	Restraints energy
I	$>10^5$	63	$-1,424$	211	$-2,222$	118
II	$>10^5$	33	$-1,097$	214	$-1,981$	140
III	$>10^5$	25	$-1,128$	187	$-2,262$	79
IV	$>10^5$	41	$-1,386$	154	$-2,142$	52
V	$>10^5$	36	$-1,494$	150	$-2,146$	125

The units are kJ mol^{-1}, but the zero is arbitrary. The potential energy includes contributions from geometric energy terms, van der Waals, and electrostatic interactions; the restraints energy arises from violations of the NMR-derived distance restraints. The force constant for the calculation of the restraints energy is 1000 kJ mol^{-1} nm^{-2}.
[a]Data from Ref. 82.

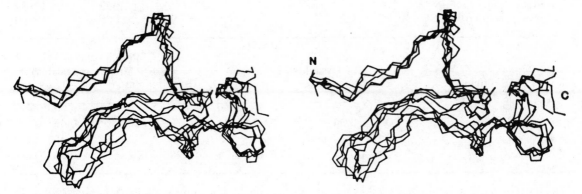

Fig. 4.26 Stereoviews of human epidermal growth factor showing the superposition of five distance geometry structures. Only bonds between main chain atoms (N, C^x, C$'$) are shown. (Reprinted from ref. 82 with permission.)

structure was calculated starting with five structures, which, after distance geometry using DISGEO, yielded RMSDs of 2.01 Å for the backbone atoms and 3.42 Å for all heavy atoms. The energies of the structures at this stage are given in Table 4.6. This was followed by restrained energy minimization and restrained molecular dynamics (see Table 4.6 for the energies of the structure after various stages of refinement) for 10–15 ps, although the RMSDs for the final structure were not reported. The final structure is shown in Fig. 4.26.

4.4.7 Interleukin 1β

The structure of interleukin 1β (IL-1β) has been solved in the laboratory of Clore and Gronenborn[83] by multidimensional heteronuclear NMR methods, and currently represents the largest protein whose structure has been determined (153 residues, $M_r = 17.4$ kDa), although there are a number of larger proteins whose structure

Table 4.7 Structural statistics and atomic RMS differences for Interleukin $1\beta^a$

Structural statistics	$\langle SA \rangle$	$(\overline{SA})_r$
RMS deviations from exptl distance restraints (Å)		
all (2780)	0.026 ± 0.0007	0.027
interproton distances		
interresidue sequential ($|i-j|=1$) (592)	0.027 ± 0.002	0.031
interresidue short range ($1 < |i-j| \leq 5$) (265)	0.024 ± 0.003	0.025
interresidue long range ($|i-j| > 5$) (848)	0.023 ± 0.002	0.024
intraresidue (923)	0.028 ± 0.002	0.027
H-bonds (114)	0.026 ± 0.003	0.000
RMS deviations from exptl dihedral restraints (deg) (366)	0.190 ± 0.024	0.199
F_{NOE} (kJ mol^{-1})	234 ± 13	247
F_{tor} (kJ mol^{-1})	3.72 ± 0.96	3.97
F_{repel} (kJ mol^{-1})	159 ± 8	238
E_{L-J} (kJ mol^{-1})	-2383 ± 46	-2345
deviations from idealized covalent geometry		
bonds (Å) (2474)	0.005 ± 0	0.005
angles (deg) (4469)	1.868 ± 0.002	2.109
impropers (deg) (929)	0.523 ± 0.006	0.556

Atomic RMS differences	Residues 2–151 Backbone atoms	All atoms	Internal residues ($\leq 40\%$ surface accessible) Backbone atoms	All atoms
$\langle SA \rangle$ vs $\overline{SA}$	0.41 ± 0.04	0.82 ± 0.04	0.28 ± 0.03	0.49 ± 0.03
$(\overline{SA})_r$ vs $\overline{SA}$	0.13	0.39	0.09	0.25
$\langle SA \rangle$ vs $(\overline{SA})_r$	0.43 ± 0.04	0.90 ± 0.04	0.29 ± 0.03	0.55 ± 0.04

^{a}The notation of the structures is as follows: $\langle SA \rangle$ are the final 32 simulated annealing structures; $\overline{SA}$ is the mean structure obtained by averaging the coordinates of the individual SA structures best fitted to each other (excluding residues 1, 52, and 153); $(\overline{SA})_r$ is the restrained minimized mean structure obtained from $\overline{SA}$. The number of terms for the various restraints is given in parantheses.

determination is in progress at the time of writing. We have already considered the heteronuclear three-dimensional and four-dimensional sequential assignment procedure for IL-1β in Chapter 2. The structure was determined using 3146 experimental restraints, including 2630 interproton distance restraints, 114 distance restraints for 57 hydrogen bonds, 36 distance restraints from hydrogen bonds involving seven bound water molecules, and 366 torsion-angle (ϕ, ψ, and χ_1) restraints. A total of 32 simulated annealing structures was calculated and the statistics are summarized in Table 4.7.

The procedure involved the calculation of an initial set of substructures using the real space distance geometry program DISGEO, followed by simulated annealing with the program XPLOR in an iterative manner. At the final stage of the refinement, the water molecules were added to the structure. The final structure is shown in Fig. 4.27. The atomic RMSDs were 0.41 Å for the backbone atoms and 0.82 Å for all the atoms excluding residue 1 at the N-terminus and residues 152 and 153 at the C-terminus, which

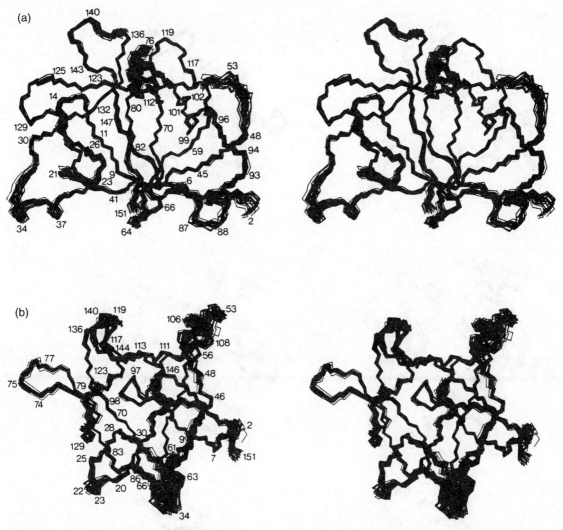

Fig. 4.27 Two stereoviews of interleukin-1β showing best-fit superpositions of the backbone (N, C$^{\alpha}$, C$'$) atoms of 32 SA structures. (Reprinted from ref. 83 with permission.)

are partially disordered. The fit for the side chains is illustrated in Fig. 4.28. We will consider the fit against the X-ray crystal structure in Section 4.5.

The backbone dynamics have been examined by ^{15}N relaxation studies[84] and all the residues exhibit very fast motions on a timescale of ≈ 20–50 ps, with a single-order parameter $S^2 = 0.82$. Thirty-two residues also display motions on a timescale of 0.5–4 ns, slightly less than the overall rotational correlation time of the protein (8.3 ns), requiring fitting using a two-parameter model-free approach.[85] Also, another 42 residues display some kind of motion on the 30 ns–10 ms timescale, as characterized by ^{15}N line broadening. There is also evidence for some slow conformational heterogeneity involving 19 residues. These dynamics data confirm that the C-terminus residues are conformationally disordered due to motion.

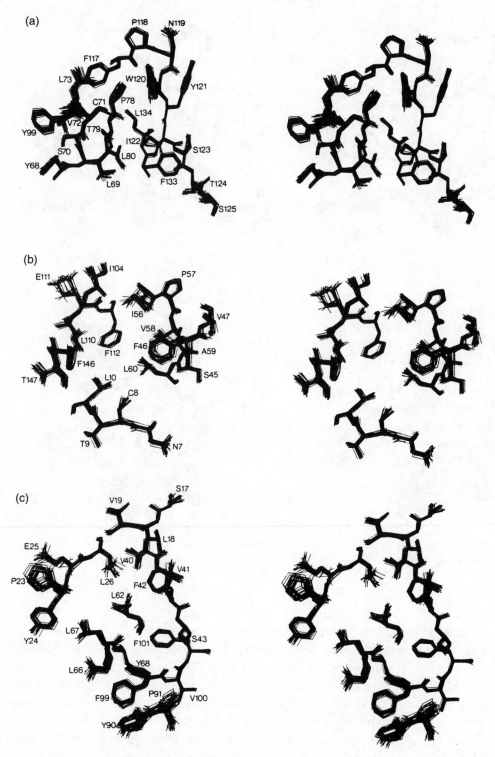

Fig. 4.28 Stereoviews of interleukin-1β showing best-fit superpositions of all atoms (excluding protons) of the 32 SA structures for three selected regions showing the side-chain fits. (Reprinted from ref. 83 with permission.)

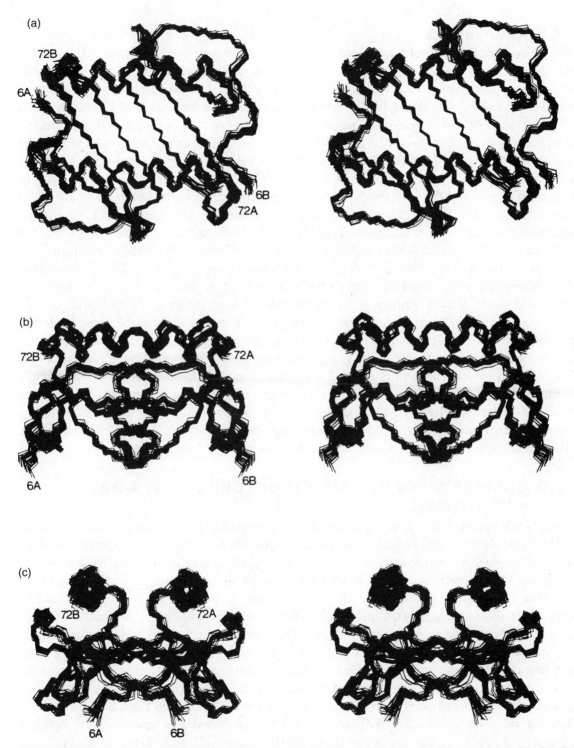

Fig. 4.29 Three stereoviews of interleukin-8 showing best-fit superpositions of the backbone (N, C^α, C') atoms of 30 SA structures. The molecule is viewed down the C_2 symmetry axis in (a), which is located in the centre of the β-sheet. The views in (b) and (c) are rotations about the C_2 axis which lies parallel to the plane of the paper. (Reprinted from ref. 86 with permission.)

4.4.8 Interleukin 8

The structure of the dimeric interleukin-8 (IL-8, 72 residues per monomer, $M_r = 16$ kDa per dimer) was solved in the laboratory of Clore and Gronenborn.[86] The structure was determined using 1694 interproton distance, 104 hydrogen-bonding and 362 torsion-angle (ϕ, ψ, and χ_1) restraints, for the two identical subunits combined and 70 interproton distance and 12 hydrogen-bonding restraints between the subunits. Stereospecific assignments were obtained for 38 of the 55 β-methylene protons per monomer and for the α-methylene protons of both glycine residues. A total of 30 final simulated annealing structures was calculated. The procedure involved calculation of an initial set of substructures using the real space distance geometry program DISGEO, followed by simulated annealing with the program XPLOR, in an iterative manner which involved incorporating an increasing number of restraints at each successive stage of the calculations. The final structure is shown in Fig. 4.29. The atomic RMSDs (excluding the disordered residues 1–5) are 0.41 Å for the backbone atoms and 0.9 Å for all the atoms. The presence of disordered residues 1–5 was confirmed by analysis of the ^{15}N backbone relaxation data,[87] which confirmed, using the 'model-free' analysis of Lipari and Szabo[88] (see Equation (3.50)), that S^2 for these residues range between 0.2 and 0.6 compared with the average of 0.88. The average angular RMSD for the ϕ and ψ angles is 8.7°, and the backbone torsion angles for all non-glycine residues lie in the allowed region of the Ramachandran ϕ,ψ plot. The internal side chains in the IL-8 structure are also well-defined and have atomic RMSDs of ≤ 0.5 Å. Eighteen side chains, however, have atomic RMSDs larger than 1 Å. All of these residues are highly surface accessible, and the $^3J_{\alpha\beta}$ coupling constants indicate multiple χ_1 side-chain conformations. A detailed study of the side-chain dynamics using ^{13}C relaxation data has yet to be carried out.

4.5 COMPARISONS OF NMR STRUCTURES WITH X-RAY STRUCTURES

Since NMR spectroscopy became established as a powerful method for the determination of the three-dimensional structures of protein (and DNA), there has been a revolution in the rather narrow-minded view that X-ray crystallography necessarily and solely generates the 'correct' structure. Such a view has its origins in the mistaken belief that proteins have rigid structures, which only methods such as X-ray crystallography can determine. This view is still prevalent among organic chemists, despite enormous physical organic evidence that small molecules are frequently dynamic and conformationally averaged. The situation was not helped by the fact that early NMR structures were not particularly precise and, occasionally, due to assignment errors, produced inaccurate structures such as the histidine-containing protein (HPr).[89] Having said this, of course the same is also true for the early X-ray structures, and the Brookhaven Data Bank contains a number of X-ray structures with errors and inaccuracies. For example, following a discrepancy between the low-resolution NMR structure[90] and X-ray structure[91] for metallothionein-2, the X-ray structure was reinvestigated and corrected.[92]

There have been a number of heated debates between NMR spectroscopists and X-ray crystallographers as to the physiological relevance of the structures that each have determined. Physiological relevance is a difficult argument, akin to those discussions among *aficionados* of health foods as to how 'natural' or even how 'organic' [*sic*] a particular food is. X-ray crystallographers will maintain that although crystals are in the solid state, they are 80–90% water and therefore approximate to the solution state, taking into account the effect of crystal packing forces. NMR spectroscopists, on the other hand, maintain that the whole molecule must be isotropically tumbling in order to approximate to the solution state, and that therefore their structures are more physiologically relevant. Of course, the truth is that while the cells do not generally contain crystalline proteins, only in certain circumstances are proteins at concentrations of 1–5 mM! Therefore in this context, physiological relevance is a specious argument. Each technique has its own distinctive advantages and disadvantages and clearly complements each other very well. X-ray crystallography does not suffer from the M_r limits that solution-state NMR does, but then it is sometimes impossible to obtain diffraction-quality crystals. NMR spectroscopy, on the other hand, can obtain quantitative estimates of protein dynamics, something which is not possible from the notoriously difficult to interpret crystallographic *B*-factors. Furthermore, as we shall consider in a moment, both methods give closely similar structures with a small number of exceptions. NMR spectroscopy has one great advantage, which is that it spans both the solution and solid states, and therefore can resolve discrepancies between solution-state and X-ray crystallographic data. An example of this is given for α-lytic protease in Section 7.1.1.

Two reviews have appeared comparing NMR and X-ray crystallographic structures.[93,94] Even the early NMR structures, which were relatively low precision, compared favourably with the corresponding X-ray structures in terms of their RMSDs. For example, bovine pancreatic trypsin inhibitor (BPTI), potato carboxypeptidase inhibitor (CPI), and barley serine proteinase inhibitor 2 (BSPI-2) were of the order of 2.0 Å. This is comparable to the differences between the NMR structures themselves. For BPTI, minor differences are found at the surface of the protein. For CPI, small deviations from the crystal structure were reported for two regions of the backbone. For BSPI-2, the small RMS difference (1.9 Å) for the backbone atom positions between crystal and solution structure appears not to be localized in particular regions of the protein.

The majority of the structures which we considered in Section 4.4 compare favourably with the X-ray structures where they are available. These are summarized in Table 4.8. Notable exceptions are the C3a and C5a inflammatory proteins, and interleukin-8. The C3a and C5a proteins are homologous (see Section 4.4.3) and exhibit a bundle of four helices containing residues 4–12, 18–26, 34–39, and 46–63 (in the C5a numbering). The crystal structure of C3a, however, lacks the N-terminal helix, and the crystal lacks the space to accommodate one. This helix is very well-defined in the NMR structure of C5a, which, on the basis of a large number of strong NOEs, folds back onto the core of the protein. In C3a, the helix appears to be much less stable, and the NOEs between the N-terminal helix and the rest of the molecule are much weaker. However, the NOEs are sufficient to determine that the position of the helix is the same as that of C5a. Interestingly, the mutation of residue 26 in C5a from an alanine to the methionine found in C3a destabilizes the N-terminal helix. These studies imply that crystallization destroys the labile N-terminal helix in C3a, whereas this helix is more stable in C5a, and this

Table 4.8 RMSD values from the mean NMR structure to the NMR conformers and to the X-ray structure

Protein[a]	Nr.	All residues[b] RMSD [Å][c] ⟨NMR⟩–NMR	⟨NMR⟩–X-ray	Nr.	Selected residues[b] RMSD [Å][c] ⟨NMR⟩–NMR	⟨NMR⟩–X-ray
α-amylase inhibitor	74	1.24; 1.69	2.17; 2.50	56; 43	0.40; 0.56	0.69; 1.04
Barnase	107	1.00; 1.44	1.76; 2.21	72; 48	0.62; 0.77	1.27; 1.39
Ca-binding protein	75	0.98; 1.44	1.31; 1.71	58; 44	0.49; 0.66	0.71; 1.10
Chymotrypsin inhib. 2	64	0.87; 1.43	1.19; 1.85	49; 29	0.47; 0.59	0.86; 1.09
ColE1 rop protein	56	0.90; 1.45	1.16; 1.95	42; 18	0.59; 0.65	1.01; 1.07
Interleukin 1β	153	0.48; 0.85	0.99; 1.52	118; 67	0.26; 0.29	0.64; 0.74
Interleukin 8	69	0.51; 1.04	2.00; 2.65	50; 27	0.24; 0.31	0.95; 1.26
Metallothionein: β	30	1.36; 1.96	2.05; 2.58	21; 15	0.86; 1.05	1.34; 1.73
α	31	1.16; 1.80	1.47; 2.36	23; 14	0.83; 0.87	1.28; 0.83
434 repressor (1–69)	63	0.61; 1.10	1.09; 1.62	49; 19	0.40; 0.45	0.66; 0.77
Ribonuclease A	124	0.77; 1.14	1.11; 1.58	94; 72	0.44; 0.52	0.76; 1.25
Thioredoxin	108	0.76; 1.13	1.01; 1.53	76; 39	0.37; 0.43	0.75; 1.05
Trypsin inhibitor	58	0.73; 1.07	1.16; 1.44	44; 24	0.32; 0.34	0.64; 0.81

[a]Data from Ref. 94.
[b]'All residues' are those residues which are found in both the NMR and the X-ray structure. 'Selected residues' are those that are less disordered.
[c]'⟨NMR⟩' denotes the mean NMR structure, 'NMR' the set of individual conformers describing the NMR structure, and 'X-ray' the X-ray structure. The first number is for the selected backbone fragments, the second for the selected heavy atoms.

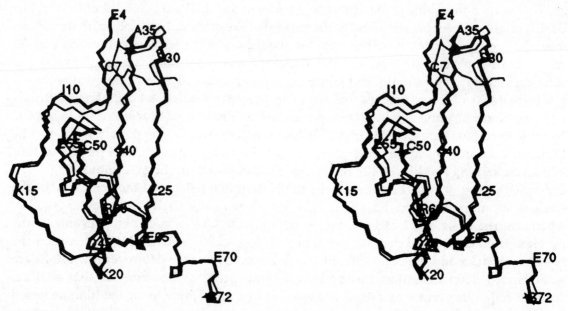

Fig. 4.30 Stereoview of interleukin-8 showing the backbone of the mean NMR structure (thick line) and of the X-ray structure (thin line). (Reprinted from ref. 94 with permission.)

stability may be one of the reasons why the protein will not crystallize in this space group. Another difference occurs at the C-terminal helix, which is the active site of the protein, and in solution is disordered, whereas in the crystal it extends farther out and ends in a β-turn. In the crystal, the C-terminal helix would appear to be stabilized by favourable contacts with neighbouring molecules in the unit cell. In NMR structures, active sites often appear disordered even though they are ordered in the X-ray structure. This is not so surprising, because mobility should facilitate recognition via induced fit. The ordered structure frozen out of the solution ensemble by crystallization may not correspond to the biologically active conformation. In such cases, the NMR ensemble should include not only the bound conformation, but also conformations involving recognition and binding.

The NMR structure of interleukin 8 has been extensively compared[95] to the X-ray crystal structure. The X-ray structure was solved in an unusual manner,[96] in which the NMR data was used to solve the phases for the X-ray structure determination.[97] The RMSD between the X-ray and NMR structures is shown in Table 4.8. Even in the well-defined region (residues 7–72), the RMSD between the two structures is sometimes as large as 3–4 Å. This is illustrated in Fig. 4.30, which shows the superposition of the backbones of the mean NMR structure and the X-ray structure. Residues 4–6 are disordered in the NMR structure but are well ordered in the X-ray structure, presumably because the salt bridge between Glu-4 of one subunit to Lys-23′ of the other subunit is replaced by one between Glu-29 and Lys-23′ in solution. Other differences include altered hydrogen bonding patterns, and a change in the overall quaternary structure such that the two helices sitting on top of the β-sheet structure are separated by 14.8 Å in solution, but only by 11.1 Å in the crystal. The crystal structure of the homologous platelet factor 4 (PF4) has the same interhelix distances as the NMR structure of IL-8, suggesting that these differences arise from crystal packing.

4.6 PEPTIDE AND PROTEIN STRUCTURES BY SOLID-STATE NMR

The introduction of the rotational resonance (R^2) and rotational echo double resonance (REDOR) discussed in Section 3.3.3 has been an important alternative to NOEs for the study of small peptides and potentially larger proteins that are insoluble or membrane-associated. These methods allow the determination of extremely accurate distance restraints which can be used to calculate structures. There are an increasing number of examples of the application of these methods to peptides, enzymes, and membrane proteins. We shall consider examples of the application of distance-based and torsion-angle-based methods to the latter two categories in Chapters 8 and 10. Currently there are three examples of peptide/protein structure determinations, two using REDOR and the other R^2 techniques, and we will consider them in some detail in this section.

4.6.1 Emerimicin

The REDOR technique was applied by Schaefer and co-workers[98] to a peptide fragment (1–9) of the antibiotic emerimicin, labelled with ^{13}C and ^{15}N. The sequence is shown in Fig. 4.33. The peptide was specifically labelled with ^{13}C at the backbone carbonyl of MeA-2 (where MeA = 2-methyl-alanine) and with ^{15}N at the amide nitrogen of Gly-6.

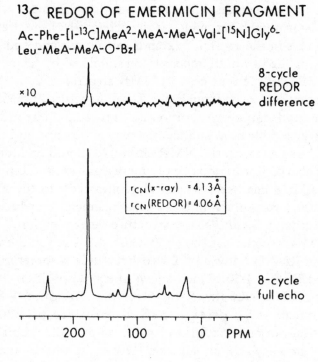

^{13}C REDOR OF EMERIMICIN FRAGMENT

Ac-Phe-[l-^{13}C]MeA^2-MeA-MeA-Val-[^{15}N]Gly^6-Leu-MeA-MeA-O-Bzl

×10

8-cycle REDOR difference

r_{CN}(x-ray) = 4.13 Å
r_{CN}(REDOR) = 4.06 Å

8-cycle full echo

200   100   0   PPM

Fig. 4.31 50.3 MHz REDOR ^{13}C NMR spectra obtained from 120 mg of [MeA^2]–^{13}C, Gly^6–^{15}N]emerimicin 1–9 benzyl ester. The REDOR difference spectrum (top, 10 ×) is the difference between rotational-echo ^{13}C NMR with and without 180° ^{15}N pulses. (Reprinted from ref. 98 with permission.)

These labels are across one turn of an α-helix, 4.13 Å apart in the X-ray structure of the peptide.[99] Fig. 4.31 shows the ^{13}C REDOR NMR spectrum (bottom) of microcrystalline peptide without ^{15}N dephasing pulses, and the difference spectrum (top) obtained by subtracting a spectrum collected with the dephasing pulses. The $\Delta S/S$ obtained was 0.0267.

In order to calculate the internuclear distance from $\Delta S/S$, the natural abundance backgrounds of ^{13}C (1.10%) and ^{15}N (0.37%) needed to be taken into account. The full S signal has a 7% contribution from the natural abundance ^{13}C of the peptide carbonyls that must be subtracted in order to obtain the intensity solely from the specific ^{13}C label. The ΔS signal has contributions both from the ^{13}C-labelled MeA-2 coupled to natural abundance ^{15}N and from ^{15}N-labelled Gly-6 coupled to natural abundance ^{13}C. The largest contributions to the ΔS signal arise from dipolar couplings to natural abundance carbons and nitrogens that are separated by one or two bonds from the labelled positions. The one- and two-bond distances are estimated to be 1.33 and 2.46 Å, which corresponds to couplings of 1.26 kHz and 200 Hz respectively. Despite the sizeable contribution of the natural abundance dipolar couplings to the ΔS signal when the coupling being measured is weak, the errors in the final distance determination are of the order of a few tenths of an Ångstrom. The natural abundance contribution to ΔS was estimated at more than 50% of the signal, leaving a residual $\Delta S/S$ of 0.0129 arising from the two labelled sites. Using Equation (3.7), with an eight-cycle evolution period and a spinning speed of 3205 Hz, the $\Delta S/S$ measurement leads to $D_{NC} = 44.1$ Hz, corresponding to an internuclear distance of 4.07 ± 0.1 Å. This is in good agreement with the 4.13 Å distance measured by X-ray crystallography (see Fig. 4.32). Since the 2–9 fragment is known to adopt a 3_{10}-helix in the

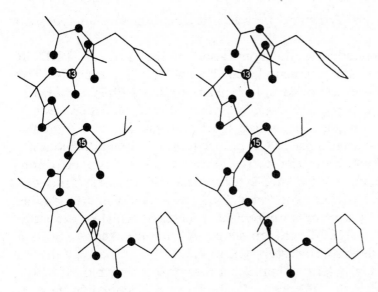

Fig. 4.32 Stereoview of the crystal structure of emerimicin 1–9 benzyl ester. Heteroatoms are indicated by filled circles; hydrogens have been omitted for clarity. Positions of ^{13}C and ^{15}N are indicated by numbers on filled circles. (Reprinted from ref. 98 with permission.)

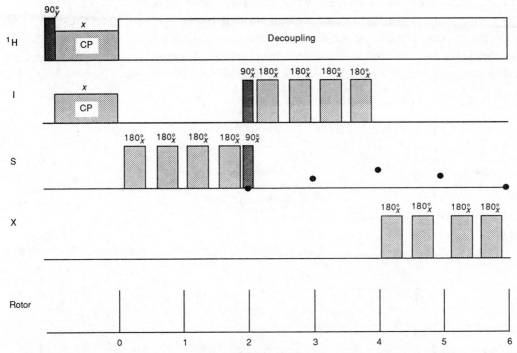

Fig. 4.33 Pulse sequence for the TEDOR–REDOR experiment, following initial cross-polarization transfer from abundant protons to generate initial magnetization of the I-spins. The TEDOR part of the experiment (which for this illustration extends over the first four rotor periods) selects S-spin magnetization by a coherence transfer from the I spins. The REDOR part of the experiment immediately follows the TEDOR preparation, and measures dipolar coupling of the selected S-spin to a third rare spin X. The solid circles represent observable magnetization.

crystal,[100] which would lead to a 5.87 Å distance between the labels, the formation of such a structure could be ruled out.

This method has been extended[101] to the application of the TEDOR–REDOR experiment to the (1–9) emerimicin fragment labelled with ^{19}F, ^{13}C, and ^{15}N. The TEDOR–REDOR experiment uses the quadruple resonance pulse sequence shown in Fig. 4.33. The pulse sequence involves establishing transverse ^{15}N magnetization through cross-polarization from the protons, and dephasing the ^{15}N magnetization by dipolar coupled ^{13}C spins through rotor-synchronized ^{13}C π pulses followed by simultaneous $\pi/2$ pulses to ^{13}C and ^{15}N. Next the coherence transferred to the ^{13}C spins is converted into observable ^{13}C transverse magnetization by rotor-synchronized ^{15}N π pulses. Up to this point, the pulse sequence is the TEDOR experiment, which is a rotational echo double resonance dephasing followed by a rotor-synchronized coherence transfer, the latter based on the same principles as an INEPT solution-state coherence transfer (see Section 1.1.7). Finally, the ^{13}C spins are dephased by rotor-synchronized ^{19}F π-pulses, and the ^{13}C signal can be observed synchronously (one sampling per rotor cycle) or as part of a two-dimensional experiment in which the full echo ^{13}C spectrum is obtained by normal acquisition starting at the beginning of a rotor cycle.

The result is that when used with high gyromagnetic ratio nuclei such as ^{19}F as a rare-spin label in background-free REDOR with ^{13}C observation, distances of the order of 10 Å can be measured with errors that are less than ± 0.5 Å. With [^{19}F, ^{13}C, ^{15}N]emerimicin (1–9), two rotor cycles (which is close to the optimum for the ^{15}N—^{13}C one-bond dipolar coupling) of ^{13}C dephasing pulses preceding the simultaneous $\pi/2$ ^{13}C, ^{15}N pulses, and two rotor cycles of ^{15}N refocusing pulses following the coherence transfer gave the spectrum shown in Fig. 4.34(a). The TEDOR spectrum gives one line from the

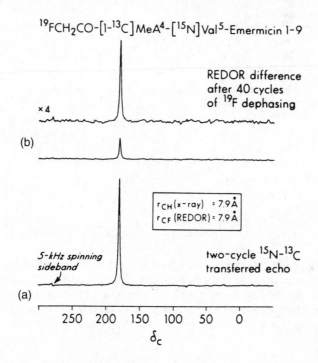

^{19}FCH$_2$CO-[1-^{13}C]MeA^4-[^{15}N]Val^5-Emermicin 1-9

REDOR difference after 40 cycles of ^{19}F dephasing

×4

(b)

r_{CH}(x-ray) = 7.9 Å
r_{CF} (REDOR) = 7.9 Å

5-kHz spinning sideband

two-cycle ^{15}N-^{13}C transferred echo

(a)

250 200 150 100 50 0

δ_C

Fig. **4.34** TEDOR–REDOR ^{13}C NMR spectra of a ^{19}F, ^{15}N, ^{13}C-triple-labelled emerimicin peptide: (a) the TEDOR spectrum; (b) the TEDOR–REDOR difference spectrum. The ratio of the difference signal to the TEDOR signal gives an r_{CF} of 7.9 $\pm$ 0.3 Å. (Reprinted from ref. 101 with permission.)

(a) $^{19}FCH_2CO-[1-^{13}C]MeA^4-[^{15}N]Val^5$-Emerimicin 1-9 *(diluted)*

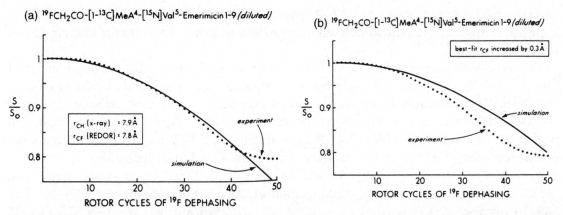

(b) $^{19}FCH_2CO-[1-^{13}C]MeA^4-[^{15}N]Val^5$-Emerimicin 1-9 *(diluted)*

Fig. 4.35 TEDOR–REDOR ^{13}C NMR data for a ^{19}F, ^{15}N, ^{13}C-triple-labelled emerimicin diluted ten-fold by natural-abundance peptide: (a) Best-fit data for r_{CF} of 7.9 Å; (b) best-fit data for r_{CF} of 8.1 Å. (Reprinted from ref. 101 with permission.)

^{13}C—^{15}N bond, without the natural abundance background. With ^{19}F dephasing, the difference spectrum shown in Fig. 4.34(b) corresponds to $\Delta S/S = 0.19$, which implies an internuclear distance of 7.9 Å, which is consistent with the X-ray determined value. The fit of these data to the simulated distance for 7.9 Å and 8.1 Å is shown in Fig. 4.35. This shows the sensitivity of the experimental data to the goodness of fit.

4.6.2 Alzheimer's β-amyloid peptide

The peptide fragment (34–42) of the 42-residue β-amyloid protein implicated in the formation of insoluble plaques in Alzheimer's disease and Down's syndrome has been studied using R^2 by Griffin, Lansbury Jr, and co-workers.[102] The synthetic peptide fragment (see Fig. 4.36) was ^{13}C-labelled at the α-carbon of Gly-37 and the carbonyl

H_2N-DAEFRHDSGYEVHHQKLVFFAEDVGSNKGAII**GLMVGGVVIA**-CO_2H

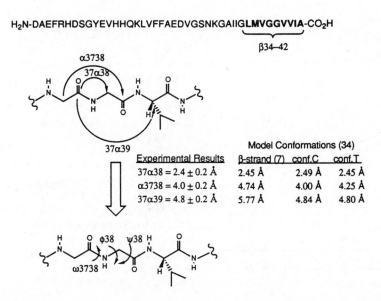

	Model Conformations (34)		
Experimental Results	β-strand (7)	conf.C	conf.T
37α38 = 2.4 ± 0.2 Å	2.45 Å	2.49 Å	2.45 Å
α3738 = 4.0 ± 0.2 Å	4.74 Å	4.00 Å	4.25 Å
37α39 = 4.8 ± 0.2 Å	5.77 Å	4.84 Å	4.80 Å

Fig. 4.36 Schematic representation of the Gly37—Gly38—Val39 region of β34–42 peptide, showing the distances measured by R^2 and those dihedral angles which have been determined. The experimentally determined distances are listed in tabular form along with the distances derived from models of three conformations: the idealized β-strand; and conformations C and T. The sequence of the whole protein is shown above, with the β34–42 sequence in boldface. (Reprinted from ref. 102 with permission.)

carbon of Gly-38 (abbreviated α37–38), and R^2 spectra are shown in Fig. 4.37. Also included in Fig. 4.37 are the calculated magnetization exchange curves for four simulated distances.

The relatively narrow linewidth of the resonance assigned to the amide carbonyl of Gly-38 (200 Hz) suggests that the peptide exists as a single conformer in the aggregate. The R^2 method was calibrated by measuring the distance between the carbonyl carbon of Gly-37 and the α-carbon of Gly-38, which was found to be 2.4 ± 0.2 Å, in agreement with typical peptide-bond geometry (2.45 Å for *trans* amide, 2.49 Å for *cis* amide). Rotational resonance was then used to measure two conformation-independent intercarbon distances in two different labelled peptides. The α37–38 distance was used to determine the dihedral angle ϕ38 and the configuration of the 37–38 amide bond (ω37–38). The α38–38 distance was measured to be 4.0 ± 0.2 Å, indicating that the Gly-37—Gly-38 amide bond has the unusual *cis* configuration (if ϕ38 $= \pm 121°$, then α37–38 $= 4.00$ Å) or that a strained conformation containing a *trans* amide is present (if ϕ38 $= 0°$, then α37 $- 38 = 4.25$ Å). The distance between the carbonyl of Gly-37 and the α-carbon of Val-39 (37–α39) depends on ϕ38, ψ38 and ω38–39. Given the two alternative

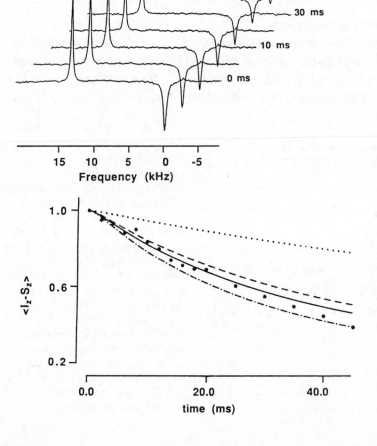

Fig. 4.37 (Top) Series of ^{13}C MAS spectra of β34–42 labelled at the α-carbon of Gly37 and the carbonyl of Gly38 (α37–38 distance). The times refer to the mixing time, τ_m. (Bottom) Comparison of magnetization exchange data (filled circles) for α37–38 along with the calculated curves for four distances: 4.74 Å (dotted line), 4.0 Å (dashed line), 3.9 Å (solid line), and 3.8 Å (dot-dashed line). The 4.74 Å distance would be expected for an idealized antiparallel β-strand. The best fit to the data is 3.9 Å. (Reprinted from ref. 102 with permission.)

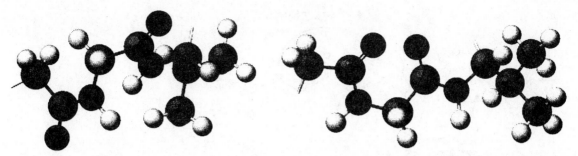

Fig. 4.38 Two possible conformations for the region of the β34–42 peptide backbone extending from the α carbon of Gly37 to the α carbon of Val39: conformation C (left) and conformation T (right). (Reprinted from ref. 102 with permission.)

conformations dictated by the α37–38 measurement, the experimental 37–α39 distance of 4.8 ± 0.2 Å constrains the dihedral angle ψ38 such that two representative conformations must be considered (see Fig. 4.38), designated C (ω37–38 $= 0°$, ϕ38 $= 121°$, ψ38 $= -25°$) and T (ω37–38 $= 180°$, ϕ38 $= 0°$, ψ38 $= \pm 134°$). Using semi-empirical MM2 molecular mechanics calculations, the relative energies of the two conformations in the relevant tripeptide segment were evaluated, and C was found to be 12 kJ mol^{-1} higher than an idealized antiparallel β-strand, whereas T was found to be 71 kJ mol^{-1} higher than a β-strand. It is therefore likely that the structure of the Gly-37—Gly-38 region of the peptide β34–42 in the amyloid probably resembles conformation C.

4.6.3 Melanostatin

The structure of the tripeptide melanostatin (Pro-Leu-Gly NH$_2$) has been determined using REDOR by Garbow and McWherter.[103] REDOR distances spanning the range 2.4–4.9 Å were measured for a series of selectively ^{13}C, ^{15}N-enriched melanostatins, and were found to be in good agreement with the X-ray structure. The REDOR distances were used to provide constraints on the values of the dihedral angles ϕ_{Leu}, ψ_{Leu}, and ϕ_{Gly} in the tripeptide. Together with conservative energetic considerations, the REDOR-measured distances determined a family of backbone structures for the tripeptide. ϕ_{Leu} was found to be limited to the range $-70°$ to $+20°$, ψ_{Leu} from 130° to 145°, and ϕ_{Gly} from $-5°$ to $+65°$. For a 14-atom backbone fragment, the pairwise RMSD between the family of REDOR structures and the X-ray conformation was found to be 0.437 ± 0.224 Å. Based on this work, Garbow and McWherter have proposed a general strategy for determining backbone conformations of selectively labelled sites in peptides and proteins, illustrated in Fig. 4.39.

The first step in the process is to choose the region of the peptide to be mapped, and place an [α-^{15}N]-labelled amino acid at residue i (see Fig. 4.39(a)). In the same molecule, ^{13}C labels are placed in the carbonyl position of residue $i-2$ and the α-carbon of residue $i+1$. Since these two carbon sites are well separated from one another and their NMR signals are resolved, two r_{CN} distances can be measured simultaneously in this sample in a single REDOR experiment. As shown in Fig. 4.39(a), the minimum and maximum distances between the labelled sites are within the range that can be measured by

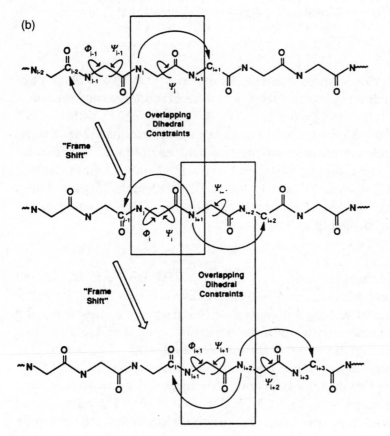

Fig. 4.39 (a) Isotopic labelling scheme for determining backbone dihedral angles using ^{13}C, ^{15}N REDOR NMR distance constraints; (b) General strategy for mapping the backbone conformation of peptides using ^{13}C, ^{15}N REDOR NMR. REDOR experiments provide constraints on the angles ϕ_{i-1} and ψ_{i-1}, and ψ_i. A frame-shift of the labelled triple along the backbone leads to overlapping dihedral angle constraints which can be used to map the backbone conformation. (Reprinted from ref. 103 with permission.)

REDOR. Each subsequent sample is then prepared by shifting the labelled ^{13}C, ^{15}N, ^{13}C triple, residue by residue, up or down the peptide chain, as illustrated in Fig. 4.39(b). The distances in these frame-shifted samples provide a series of overlapping dihedral constraints that allow the complete backbone conformation to be mapped.

4.7 PROTEIN STRUCTURES WITH $M_r > 30$ kDa—EXTENDING THE MOLECULAR WEIGHT LIMIT

It is tempting to think that simply increasing the dimensionality of multidimensional NMR experiments will continue to extend the molecular weight limit for the determination of NMR structures. This is a false hope, because of two crucial factors. First, the sensitivity of higher-dimensional experiments goes down as the number of dimensions increases, and the corresponding measuring time increases. Second, and more importantly, all the current multidimensional experiments rely on coherence or magnetization transfer, which can only occur when the linewidths do not exceed the J coupling between the atoms involved in the transfer. Another way of putting this is that if $T_2 < 1/(2J)$ then all these methods will fail. The source of the line broadening is of course significant, since for larger proteins the contributions to the shortening of T_2 come from both the increase in rotational correlation time τ_c and also other factors such as homonuclear dipolar relaxation. The latter problem has been addressed through employing fractional deuterium labelling, which reduces the ${}^1H$—${}^1H$ dipolar interactions (see Section 4.2.3). Although it has not yet been reported in the literature, fractional deuterium labelling in conjunction with ${}^{13}C/{}^{15}N$ uniform labelling and multidimensional NMR experiments may yet extend the limit somewhat beyond $M_r \approx 30$ kDa, since in addition to reducing the ${}^1H$—${}^1H$ dipolar interaction, the presence of ${}^2H$ reduces the heteronuclear linewidths by removing the ${}^1H$—${}^{15}N$ or ${}^1H$—${}^{13}C$ dipolar interactions. However, currently, the problem of avoiding conventional coherence transfer steps has been explored through alternative methods, for example involving cross-polarization in solution,[104] but none of these show much promise yet for breaking through the $M_r > 30$ kDa barrier. Instead, different strategies, aimed at obtaining some structural information without actually solving the three-dimensional structure have been adopted for a number of large molecules. Some of these strategies have been applied to the active sites of enzymes, and so are considered in Chapters 7 and 8. Here we consider studies on peptide antigen–antibody complexes and the use of Hahn spin echo methods for studying labile parts of large proteins.

4.7.1 Antibodies

Antibodies can have $M_r \approx 150$ kDa, which largely eliminates the kinds of NMR approaches outlined in Sections 4.1 and 4.2 (although, see Section 4.7.2 for counter-examples of this). Therefore probing the structure of the whole antibody is generally untenable. However, a number of approaches have been adopted to investigate the structure of the antigen when bound to the antibody, based on the binding affinity. These can be divided into three categories: (i) for low binding affinities (dissociation constants $K_d \geq 10^{-6}$M) and fast exchange between free and bound states, bound resonances can be studied by magnetization transfer and transferred NOE experiments; (ii) for tight binding ($K_d \leq 10^{-9}$M), bound resonances can be studied by ${}^1H$ COSY or isotope-edited ${}^1H\{{}^{15}N\}$ HMQC experiments; and (iii) for tight binding, the epitope can be mapped by hydrogen exchange experiments on antibody-bound antigen. We will consider examples of each of these approaches in turn.

An example of the application of the transferred NOE experiment to characterize antibody-bound peptides, is found in the work of Anglister and colleagues,[105] in which deuterated antibody fragments have been investigated interacting with short synthetic peptides. The choice of antibody is important. For example, the Fv antibody fragment (220 residues, $M_r \approx 25$ kDa) is the smallest fragment of antibody that retains its affinity for the antigen. In pioneering work, Dwek and co-workers investigated the interaction of the Fv fragment of the antibody MOPC315 complexed with dinitrophenyl hapten by NMR.[106,107] Unfortunately, the Fv fragment can be obtained by proteolytic cleavage only for a very limited number of antibodies. The Fab fragment (440 residues, $M_r \approx 50$ kDa) is more widely available and contains constant domains for each chain. Anglister and co-workers used the transferred NOE experiment, which will be considered in more detail in Section 6.5 but in brief involves measuring intramolecular NOEs for the antigen in the presence and absence of antibody, and, in some circumstances, intermolecular NOEs between antigen and antibody. Under conditions where the off-rate (k_{off}) for the ligand is fast relative to the T_1 relaxation time of both the Fab and the ligand protons, and to the mixing time of the NOESY experiment, the spectrum of the Fab in the presence of excess ligand contains TRNOE cross-peaks. The contribution of intramolecular NOEs between the Fab protons to the antigen–antibody complex spectrum can be removed by subtracting the NOESY spectrum of the Fab spectrum in the absence of ligand. They can be further reduced by growth of hybridoma cells producing the monoclonal antibody on a defined media containing specifically deuterated amino acids. The remaining TRNOEs in the difference spectrum arise from: (i) exchange between bound and free ligand; (ii) magnetization transfer between antibody protons and free protons via the bound state; and (iii) intramolecular magnetization transfer within the bound ligand via exchange with the free ligand. This means that cross-peaks from one resonance of the free ligand might

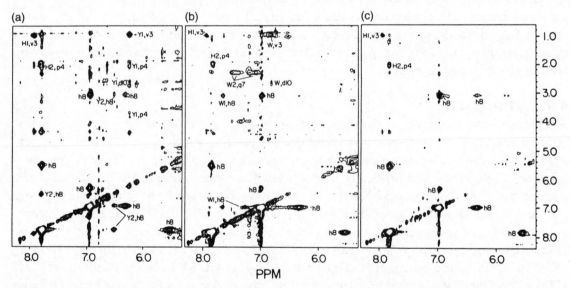

Fig. 4.40 Two-dimensional spectra between the NOESY spectrum of TE32 Fab with a four-fold excess of the peptide from the CTP3 cholera toxin and the NOESY spectrum of the peptide-saturated Fab, showing interactions of specific types of antibody aromatic residues with peptide residues. Assignment to antibody residues is marked by capital letters and arbitrary numbers, while assignment to peptide residues is marked by lower case letters and their location in the sequence. Interactions of antibody: (a) Tyr and His residues with peptide residues [antibody Phe and Trp residues are perdeuterated, and Tyr deuterated at 2,6-phenyl positions]; (b) Trp and His with peptide residues [antibody Phe and Tyr perdeuterated]; and (c) Phe and His residues with peptide residues [antibody Tyr and Trp residues perdeuterated]. (Reprinted from ref. 105 with permission.)

connect with a second resonance arising from the same proton of the bound ligand, an antibody proton or another proton of the free ligand. An example is shown in Fig. 4.40, which shows the interaction between the aromatic residues of three different deuterated TE32 Fab antibodies and the amino acid protons of the peptide from the CTP3 cholera toxin, which contains only one aromatic residue. The assignment of antibody resonances is based on specific deuteration, and the fact that cross-peaks between non-aromatic peptide resonances and aromatic resonances that do not have the same chemical shift as the peptide histidine must arise from the antibody.

The bound resonances of an antigen–antibody complex have been studied using COSY and DQF–COSY experiments by Cheetham and co-workers.[108] The method relies on the fact that the COSY cross-peaks for the peptide antigen in the presence of the antibody will broaden differentially such that unbound peptide residues will be unaffected, those residues bound at the periphery of the combining site will be weaker in intensity, and those bound at the centre of the combining site will disappear. In a sense this is the dynamic filtering analogue of isotope-edited filtering, which we will consider shortly. An example of the dynamic filtering approach is shown in Fig. 4.41, where the synthetic loop

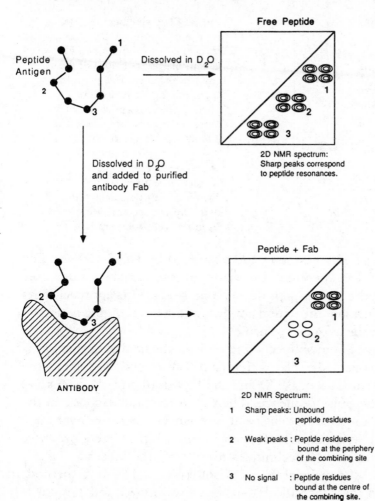

Fig. 4.41 Overall scheme for the study of high-affinity antibody–peptide complexes by two-dimensional ¹H COSY NMR. In the two-dimensional spectra, only cross-peaks on one side of the diagonal have been presented. (Reprinted from ref. 108 with permission.)

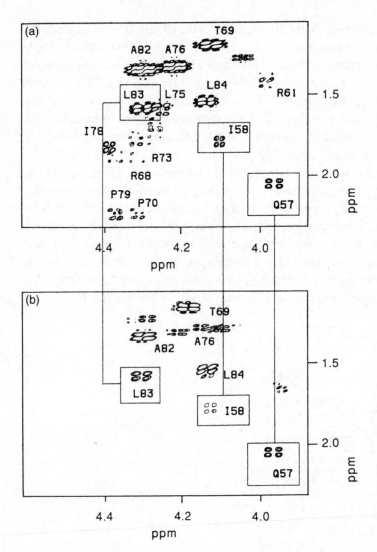

Fig. 4.42 Assignment of resonances in the COSY spectrum of the bound peptide. Chemical-shift positions of the cross-peaks in the free peptide spectrum (a) can be used to identify the cross-peaks of the corresponding resonances in the bound peptide spectrum (b). The boxed peaks highlight specific examples of the correlation between the chemical-shift positions in the two different spectra. (Reprinted from ref. 108 with permission.)

peptide is titrated with the Gloop1 Fab antibody and examined by DQF–COSY. The sequential assignments for the free peptide are made in the usual manner (see Section 4.1), and compared with the bound peptide (see Fig. 4.42). This approach does not provide structural information about the bound antigen, but does give information on the mobility of different parts of the antigen molecule.

An example of the application of Hahn spin echo methods in the field of antibodies is shown in Fig. 4.43. This work was carried out by Wright and co-workers[109] and shows the BPTI alone ($M_r = 6500$) and the complex of BPTI and the Fab' antibody ($M_r \approx 56$ kDa) obtained with the Hahn spin echo pulse sequence. There is a dramatic decrease in the spectral resolution as a result of the greater density of resonances (there are over 5000 protons) and a decrease in sensitivity as a result of increased linewidths. Therefore these workers have tried isotope-editing HMQC techniques to simplify the spectra.

An example[110] of the use of one-dimensional ^{1}H—^{15}N isotope-edited HMQC methods is illustrated in Fig. 4.44, in which a peptide ^{15}N-labelled at Gly-85 and Gly-86 is

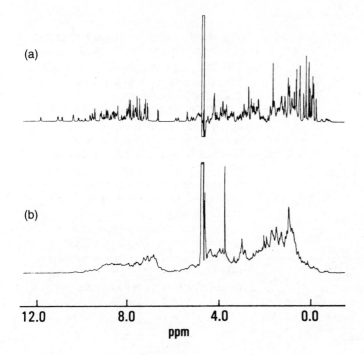

(a)

(b)

Fig. 4.43 500 MHz ¹H NMR spectra of 3mM BPTI (a) and 0.5 mM Fab′–peptide complex (b). Both spectra were acquired using the Hahn spin echo pulse sequence with solvent presaturation. (Reprinted from ref. 109 with permission.)

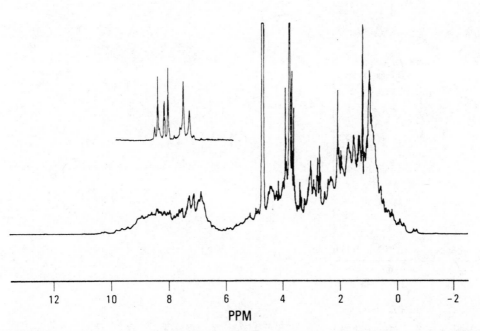

Fig. 4.44 ¹H spectra of a complex of 0.5 mM Fab′ and peptide (of sequence MHKDFLEKIGGL, corresponding to residues 76–87 of myohemerythrin, in a ratio 2:1 peptide:Fab) labelled with ¹⁵N at G85 and G86. The fill spectrum (lower) of the complex was acquired using a Hahn spin echo pulse sequence. The ¹⁵N-edited spectrum (upper), corresponding to the amide-aromatic region only, was acquired using an HMQC sequence with ¹⁵N-decoupling during acquisition. (Reprinted from ref. 109 with permission.)

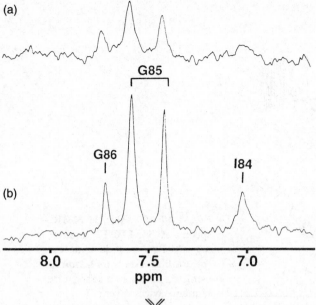

(a)

G85

G86   I84

(b)

8.0   7.5   7.0
ppm

Fig. 4.45 One-dimensional 600 MHz ^1H—^{15}N—HMQC–NOE spectra from two Fab′–peptide complexes recorded with a mixing time of 50 ms. (a) Complex with peptide MHKDFLEKIGGL labelled with ^{15}N at G85. (b) Complex of the same peptide ^{15}N-labelled at G85 and specifically deuterated at the adjacent residues, G86 and I84; the α protons of G85 were also replaced by deuterons. The resonances corresponding to the G85 amide doublet (no ^{15}N-decoupling was employed during acquisition) and NOEs assigned to the amide protons of G86 and I84 are indicated. (Reprinted from ref. 109 with permission.)

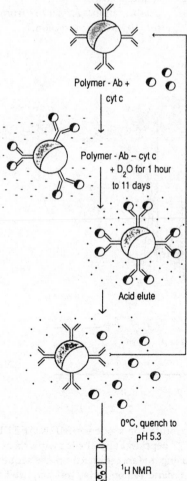

Polymer - Ab +
cyt c

Polymer - Ab -- cyt c
+ D$_2$O for 1 hour
to 11 days

Acid elute

0°C, quench to
pH 5.3

^1H NMR

Fig. 4.46 Procedure for hydrogen exchange on antibody-bound cytochrome *c*. (Reprinted from ref. 113 with permission.)

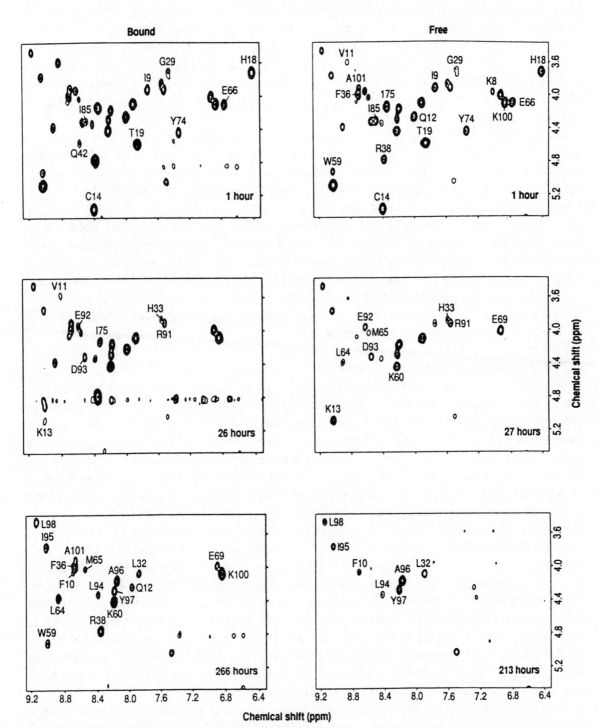

Fig. 4.47 Sections of two-dimensional COSY spectra of representative cytochrome *c* samples after partial hydrogen–deuterium exchange in the presence (left) and absence (right) of the E8 MAb antibody (Reprinted from ref. 113 with permission.)

complexed with the antibody Fab'. The full proton spectrum of the complex is shown, along with the amide proton–aromatic proton region of the ^{15}N isotope-edited spectrum containing only resonances from the free and bound forms of the peptide. These techniques can be used in conjunction with selective deuteration and NOESY experiments,[111] as shown in Fig. 4.45. These NOEs may be interpreted with reasonable accuracy for relatively short distances in the molecular weight and correlation time range of Fab–peptide complexes, and hence provide structural information on antigen–antibody complexes.[112]

An elegant use of hydrogen exchange data to probe antigen–antibody interactions was carried out by Paterson, Englander, and Roder.[113] The method involved immobilizing the monoclonal antibody (MAb) to horse cytochrome *c* on a solid support (see Fig. 4.46), complexing with its antigen, and then allowing the complex to exchange in D_2O for varying lengths of time before releasing the antigen and quenching exchange by adjusting the pH. Since the ^1H resonances for cytochrome *c* had already been assigned, the effect of the antibody on the exchange rates of the amide protons could be assessed by comparison with the exchange rates in cytochrome *c* alone (see Fig. 4.47), and these corresponded well with epitope mapping by chemical methods. This approach has been used with some success in studying protein folding (see Chapter 5). Further studies using this approach have uncovered the fact that antibody binding leads to a reduction of local antigen protein dynamics at sites remote from the epitope.[114,115]

4.7.2 Hahn spin echo methods

It is sometimes assumed by NMR spectroscopists that once $M_r > 30$ kDa, high-resolution ^1H NMR methods will not yield useful information about a protein. Whatever assumptions might be made as to the overall rotational correlation time calculated on the basis of the molecular weight, the degree of local librational freedom cannot be assumed *a priori*. For example, the *E.coli* pyruvate dehydrogenase multienzyme complex has $M_r \approx 4.5 \times 10^6$ and is one of the largest well-defined assemblies of proteins known, comprising multiple copies of three different enzymes. It has a correspondingly long rotational correlation time $\tau_c \approx 10^{-5}$ s, which suggests that the ^1H NMR spectrum of the complex should contain only very broad resonances with linewidths of the order of 10 kHz. Perham, Roberts, and co-workers[116] proved that this assumption is a hasty one, and Fig. 4.48(a) shows the remarkable ^1H NMR spectrum for pyruvate dehydrogenase complex. The enzyme complex is known to have swinging arms of lipoly–lysine residues, which carry substrate between the catalytic centres of the three enzymes and between lipoic acid residues attached to different subunits in the lipoate acetyltransferase core. It has been suggested that the lipoic acid-containing regions of the polypeptide chain might be flexible and this increased librational freedom gives rise to the sharper resonances in the spectrum. This has been confirmed by comparison of the ^1H NMR spectra of site-directed mutants[117] with mutations in the putative arms.

The resonances due to mobile domains of large proteins may be extracted from a broad envelope arising from an immobile core by the use of the Hahn spin echo experiment and related Carr–Purcell–Meiboom–Gill (CPMG) experiment. We saw an example of this in Section 4.7.1 for an antigen–antibody complex. A better example of this, in work by

(a)

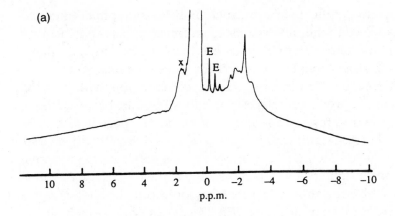

(b)

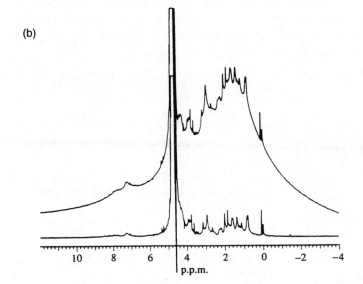

Fig. 4.48 ¹H NMR spectra of (a) *E.coli* pyruvate dehydrogenase multienzyme complex (70 mg mL⁻¹) *in toto* (the peaks marked E arise from EDTA); (b) gizzard myosin by conventional ¹H NMR spectroscopy (top), and by GPMG T_2 spin echo spectroscopy (bottom). ((a) Adapted from ref. 116 with permission; (b) Reprinted from ref. 118 with permission.)

Hartshorne, Sykes, and co-workers,[118] is shown in Fig. 4.48(b), in which the conventional ¹H NMR spectrum (Fig. 4.48(b) *upper*) of myosin (M_r = 470 kDa from turkey gizzard is compared with that obtained from CPMG ¹H NMR (Fig. 4.48(b) *lower*). The spin echo experiment takes advantage of the vastly different transverse relaxation rates ($1/T_2$) of the mobile and immobile components. For proteins with longer overall rotational correlation times, T_2 becomes very short and, from Equation (6.1), the linewidth becomes much greater. These resonances, not yet identified, allow mobile domains to be identified, and, although not really amenable to standard ¹H NMR, partial sequence-specific assignment strategies can be conducted when combined with site-directed mutagenesis studies.

4.8 OUTLOOK

NMR spectroscopy has clearly emerged as a structural technique rivalling X-ray crystallography for small- and medium-sized soluble proteins. Its greatest limitation is that it is unlikely that the $M_r \leq 30$ kDa barrier will be surpassable, since no matter how many dimensions are employed, no significant improvement in resolution can be obtained when the intrinsic linewidths exceeds the heteronuclear coupling constants. In other words, when there is no longer any coherence to be transferred, all the currently available pulse sequences, irrespective of dimensionality, will fail for large proteins. Alternative approaches, other than fractional deuterium labelling in combination with $^{15}N/^{13}C$ labelling, include the isolation of the protein of interest from thermophilic bacteria. Since these proteins are adapted for growth at high temperatures, their NMR spectra can also be recorded at high temperature, allowing significant increases in T_2 and therefore decreases in linewidths. Also, solid-state NMR methods are emerging as a particularly attractive alternative, although for a full three-dimensional structure determination, these methods are currently limited because of the large number of pairwise isotopically labelled samples required. The solid-state NMR equivalent of the NOESY experiment, in which all the distances can be obtained in one multidimensional experiment, is under development, and it is to be hoped that this will help solve the molecular weight barrier. Another alternative, in solution-state NMR, is the use of cross-polarization instead of coherence transfer for obtaining correlations between heteronuclei. Such approaches are also under development, and might help us reach the tantilizingly higher molecular weight structures.

REFERENCES

1. K. Wüthrich, *NMR of Proteins and Nucleic Acids*, John Wiley and Sons, New York (1986).

2. See the articles in *Computational Aspects of the Study of Biological Macromolecules by Nuclear Magnetic Resonance Spectroscopy*, eds. J. C. Hoch, F. M. Poulsen, and C. Redfield, NATO ASI Series Vol. 225, Plenum Press (1991) for an account of recent attempts at automated methods of sequential assignment.

3. S. W. Englander and A. J. Wand, *Biochemistry*, **26**, 5953 (1987).

4. D. Neuhaus and M. Williamson, *The Nuclear Overhauser Effect in Structural and Conformational Analysis*, VCH Publishers, New York (1989).

5. G. Wagner, *Quat. Rev. Biophys.*, **16**, 1–57 (1983).

6. K. Wüthrich, K. Strop, and M. P. Williamson, *Biochem. Biophys. Res. Commun.*, **122**, 1174–8 (1984).

7. C. E. Dempsey, *Biochemistry*, **25**, 3904–11 (1986).

8. D. A. Torchia, S. W. Sparks, and A. Bax, *Biochemistry*, **28**, 5509–24 (1989).

9. P. R. Gooley, M. S. Caffrey, M. A. Cusanovich, and N. E. MacKenzie, *Biochemistry*, **31**, 443–50 (1992).

10. R. S. Molday, S. W. Englander, and R. G. Kallen, *Biochemistry*, **11**, 150 (1972).

11. S. W. Englander and A. Poulsen, *Biopolymers*, **7**, 379 (1969).

12. G. Wagner, W. Braun, T. F. Havel, T. Schaumann, N. Gō, and K. Wüthrich, *J. Mol. Biol.*, **196**, 611–39 (1987).

13. N. J. Clayden and R. J. P. Williams, *J. Magn. Reson.*, **49**, 383 (1982).

14. L. Szilagyi and O. Jardetsky, *J. Magn. Reson.*, **83**, 441 (1989).

15. M. P. Williamson, *Biopolymers*, **29**, 1423–31 (1990).

16. A. Pastore and V. Saudek, *J. Magn. Reson.*, **90**, 165–76 (1990).

17. D. S. Wishart, B. D. Sykes, and F. M. Richards, *J. Mol. Biol.*, **222**, 311–33 (1991).

18. D. S. Wishart, B. D. Sykes, and F. M. Richards, *Biochemistry*, **31**, 1647–51 (1992).

19. R. Deslauriers and I. C. P. Smith, in *Biological Magnetic Resonance* (eds. L. J. Berliner and J. Reuben), vol. 2, pp. 243–344, Plenum Press, New York (1980).

20. F. Ni, D. R. Ripoll, and E. O. Purisima, *Biochemistry*, **31**, 2545–54 (1992).

21. H. Kessler, J. W. Bats, C. Griesinger, S. Koll, M. Will, and K. Wagner, *J. Am. Chem. Soc.*, **110**, 1033–49 (1988).

22. H. -R. Loosli, H. Kessler, H. Oschkinat, H. -P. Weber, T. J. Petcher, and A. Widmer, *Helv. Chim. Acta*, **68**, 682–703 (1985).

23. M. Jackson and H. M. Mantsch, *Biochim. Biophys. Acta*, **1078**, 231–5 (1991).

24. A. Bierzynski, P. S. Kim, and R. L. Baldwin, *Proc. Natl. Acad. Sci. USA*, **79**, 2470–4 (1982).

25. P. S. Kim and R. L. Baldwin, *Nature*, **307**, 329–34 (1984).

26. K. R. Shoemaker, P. S. Kim, D. N. Brems, S. Marqusee, E. J. York, I. M. Chaiken, *et al.*, *Proc. Natl. Acad. Sci.*, **82**, 2349–55 (1985).

27. K. R. Shoemaker, P. S. Kim, E. J. York, J. M. Stewart, and R. L. Baldwin, *Nature*, **326**, 563–7 (1987).

28. H. J. Dyson, M. Rance, R. A. Houghten, R. A. Lerner, and P. E. Wright, *J. Mol. Biol.*, **210**, 161–200 (1988).

29. H. J. Dyson, M. Rance, R. A. Houghten, P. E. Wright, and R. A. Lerner, *J. Mol. Biol.*, **210**, 201–17 (1988).

30. M. Mutter and K. -H. Altmann, *Int. J. Pept. Protein Res.*, **26**, 373–80 (1985).

31. F. D. Sonnichsen, J. E. Van Eyk, R. S. Hodges, and B. D. Sykes, *Biochemistry*, **31**, 8790–8 (1992).

32. S. W. Fesik and E. R. P. Zuiderweg, *J. Magn. Reson.*, **78**, 588–93 (1988).

33. W. M. Westler, B. J. Stockman, Y. Hosoya, Y. Miyake, M. Kainosho, and J. L. Markley, *J. Am. Chem. Soc.*, **110**, 6256 (1988).

34. B. H. Oh, W. M. Westler, P. Darba, and J. L. Markley, *Science*, **240**, 908–11 (1988).

35. A. Bax and S. Grzesiek, *Acc. Chem. Res.*, **26**, 131–8 (1993).

36. F. W. Studier, A. H. Rosenberg, J. J. Dunn, and J. W. Dubendorff, *Methods in Enzymol.*, **185**, 60–89 (1990).

37. M. Chamberlin, J. McGrath, and L. Waskell, *Nature (London)*, **228**, 227–31 (1970).

38. J. J. Dunn and F. W. Studier, *J. Mol. Biol.*, **166**, 477–535 (1983) and correction *Ibid.*, **175**, 111–12 (1984).

39. M. Golomb and M. Chamberlin, *J. Biol. Chem.*, **249**, 2858–63 (1974).

40. J. T. Moore, A. Uppal, F. Maley, and G. F. Maley, *Prot. Expr. Purif*, **4**, 160–3 (1993).

41. D. L. Wilkinson and R. G. Harrison, *Bio/technology* **9**, 443–8 (1991).

42. J. Sambrook, E. F. Fritsch, and T. Maniatis, *Molecular Cloning—A Laboratory Manual*, 2nd edn, Cold Spring Harbor Press (1989).

43. D. W. Hoffman and J. Spicer, *Techniques in Protein Chemistry II*, 409–16 (1991).

44. P. Zhang, G. F. Graminiski, and R. N. Armstrong, *J. Biol. Chem.*, **266**, 19475–9 (1991).

45. B. J. Bachmann, *Microbiol. Rev.*, **47**, 180–230 (1983).

46. L. McIntosh and F. W. Dahlquist, *Quat. Rev. Biophys.*, **23**, I–XX (1990).

47. P. Lodi and J. R. Knowles, *Biochemistry*, **30**, 6948–56 (1991).

48. W. A. Shuttleworth and J. N. S. Evans, unpublished results.

49. A. Bock, K. Forchhammer, J. Heider, W. Leinfelder, G. Sawers, B. Veprek *et al.*, *Mol. Microbiol.*, **5**, 515–20 (1991).

50. J. A. Ellman, B. F. Volkman, D. Mendel, P. G. Schultz, and D. E. Wemmer, *J. Am. Chem. Soc.*, **114**, 7959–61 (1992).

51. J. L. Markley, I. Putter , and O. Jardetsky, *Science*, **161**, 1249–51 (1968).

52. H. L. Crespi, R. M. Rosenberg, and J. J. Katz, *Science*, **161**, 795–6 (1968).

53. D. M. LeMaster and F. M. Richards, *Biochemistry*, **27**, 142–50 (1988).

54. D. M. LeMaster, *Quat. Rev. Biophys.*, **23**, 133–74 (1990).

55. C. H. Arrowsmith, R. Pachter, R. B. Altman, S. B. Iyer, and O. Jardetsky, *Biochemistry*, **29**, 6332–41 (1990).

56. E. R. P. Zuiderweg, R. Kaptein, and K. Wüthrich, *Eur. J. Biochem.*, **137**, 279–92 (1983).

57. E. R. P. Zuiderweg, R. Kaptein, and K. Wüthrich, *Proc. Natl. Acad. Sci. USA*, **80**, 5837–41 (1983).

58. R. Kaptein, E. R. P. Zuiderweg, R. M. Scheek, R. Boelens, and W. F. van Gunsteren, *J. Mol. Biol.*, **182**, 179–82 (1985).

59. E. R. P. Zuiderweg, R. M. Scheek, R. Boelens, W. F. van Gunsteren, and R. Kaptein, *Biochimie*, **67**, 707–15 (1985).

60. R. Kaptein, R. Boelens, R. M. Scheek, and W. F. van Gunsteren, *Biochemistry*, **27**, 5389–95 (1988).

61. J. de Vlieg, R. M. Scheek, W. F. van Gunsteren, H. J. C. Berendsen, R. Kaptein, and J. Thomason, *Proteins: Struct., Funct. & Genet.*, **3**, 209–18 (1988).

62. G. Wagner and K. Wüthrich, *J. Mol. Biol.*, **155**, 347–66 (1982).

63. G. Wagner, W. Braun, T. F. Havel, T. Schaumann, N. Gō, and K. Wüthrich, *J. Mol. Biol.*, **196**, 611–39 (1987).

64. J. W. Pflugrath, G. Wiegand, and R. Huber, *J. Mol. Biol.*, **189**, 383–6 (1986).

65. A. D. Kline, W. Braun, and K. Wüthrich, *J. Mol. Biol.*, **189**, 377–82 (1986).

66. A. D. Kline, W. Braun, and K. Wüthrich, *J. Mol. Biol.*, **204**, 675–724 (1988).

67. P. C. Driscoll, A. M. Gronenborn, L. Beress, and G. M. Clore, *Biochemistry*, **28**, 2188–98 (1989).

68. D. G. Nettesheim, R. P. Edalji, K. W. Mollison, J. Greer, and E. R. P. Zuiderweg, *Proc. Natl. Acad. Sci. USA*, **85**, 5036–40 (1988).

69. W. J. Chazin, T. E. Hugli, and P. E. Wright, *Biochemistry*, **27**, 9139–48 (1988).

70. R. Huber, H. Scholze, E. P. Paques, and J. Deisenhofer, *Hoppe-Seyler's Z. Physiol. Chem.*, **361**, 1389 (1980).

71. E. R. P. Zuiderweg, K. W. Mollison, J. Henkin, and G. W. Carter, *Biochemistry*, **27**, 3568–80 (1988).

72. E. R. P. Zuiderweg, J. Henkin, K. W. Mollison, G. W. Carter, and J. Greer, *Proteins: Struct. Funct. & Genet.*, **38**, 139–45 (1988).

73. E. R. P. Zuiderweg, D. G. Nettesheim, K. W. Mollison, and G. W. Carter, *Biochemistry*, **28**, 172–85 (1989).

74. J. M. Moore, C. A. Lepre, G. P. Gippert, W. J. Chazin, D. A. Case, and P. E. Wright, *J. Mol. Biol.*, **221**, 533–55 (1991).

75. J. M. Moore, W. J. Chazin, R. Powls, and P. E. Wright, *Biochemistry*, **27**, 7806–16 (1988).

76. S. J. Weiner, P. A. Kollman, D. T. Nguyen, and D. A. Case, *J. Computat. Chem.*, **7**, 230–52 (1986).

77. H. J. Dyson, G. P. Gippert, D. A. Case, A. Holmgren, and P. E. Wright, *Biochemistry*, **29**, 4129–36 (1990).

78. J. D. Forman-Kay, G. M. Clore, P. T. Wingfield, and A. M. Gronenborn, *Biochemistry*, **30**, 2685–98 (1991).

79. M. J. Stone, K. Chandrekhar, A. Holmgren, P. E. Wright, and H. J. Dywon, *Biochemistry*, **32**, 426–35 (1993).

80. G. T. Montelione, K. Wüthrich, E. C. Nice, A. W. Burgess, and H. A. Scheraga, *Proc. Natl. Acad. Sci. USA*, **84**, 5226–30 (1987).

81. D. Kohda, N. Go, K. Hayashi, and F. Inagaki, *J. Biochem.*, **103**, 741–3 (1988).

82. R. M. Cooke, A. J. Wilkinson, M. Baron, A. Pastore, M. J. Tappin, I. D. Campbell, H. Gregory, and B. Sheard, *Nature*, **327**, 339–41 (1987).

83. G. M. Clore, P. T. Wingfield, and A. M. Gronenborn, *Biochemistry*, **30**, 2315–23 (1991).

84. G. M. Clore, P. C. Driscoll, P. T. Wingfield, and A. M. Gronenborn, *Biochemistry*, **29**, 7387–401 (1990).

85. G. M. Clore, A. Szabo, A. Bax, L. E. Kay, P. C. Driscoll, and A. M. Gronenborn, *J. Am. Chem. Soc.*, **112**, 4989–91 (1990).

86. G. M. Clore, E. Appella, M. Yamada, K. Matsushima, and A. M. Gronenborn, *Biochemistry*, **29**, 1689–96 (1990).

87. B. L. Grasberger, A. M. Gronenborn, and G. M. Clore, *J. Mol. Biol.*, **230**, 364–72 (1993).

88. G. Lipari and A. Szabo, *J. Am. Chem. Soc.*, **104**, 4546–59 (1982).

89. R. E. Klevit and B. E. Waygood, *Biochemistry*, **25**, 7774–81 (1986).

90. P. Schultze, E. Wörgötter, W. Braun, G. Wagner, M. Básak, J. H. R. Kägi, *et al.*, *J. Mol. Biol.*, **203**, 251–68 (1988).

91. W. F. Furey, A. H. Robbins, L. L. Clancy, D. R. Winge, B. C. Wang, and C. D. Stout, *Science*, **231**, 704–10 (1986).

92. A. H. Robbins, D. E. McRee, M. Williamson, S. A. Collett, N. H. Xuong, W. F. Furey, *et al.*, *J. Mol. Biol.*, **221**, 1269–93 (1991).

93. G. Wagner, S. G. Hyberts, and T. F. Havel, *Annu. Rev. Biophys. Biomol. Struct.*, **21**, 167–98 (1992).

94. M. Billeter, *Quart. Rev. Biophys.*, **25**, 325–77 (1992).

95. G. M. Clore and A. M. Gronenborn, *J. Mol. Biol.*, **217**, 611–20 (1991).

96. A. T. Brünger, R. L. Campbell, G. M. Clore, A. M. Gronenborn, M. Karplus, G. A. Petsko, *et al.*, *Science*, **235**, 1049–53 (1987).

97. E. T. Baldwin, T. T. Weber, R. St Charles, J. -C. Xuian, E. Appella, M. Yamada, *et al.*, *Proc. Natl. Acad. Sci. USA*, **88**, 502–6 (1991).

98. G. R. Marshall, D. D. Beusen, K. Kociolek, A. S. Redlinski, M. T. Leplawy, Y. Pan, *et al.*, *J. Am. Chem. Soc.*, **112**, 963–6 (1990).

99. G. R. Marshall, E. E. Hodgkin, D. A. Langs, G.

D. Smith, J. Zabrocki, and M. T. Leplawy, *Proc. Natl. Acad. Sci. USA*, **87**, 487–91 (1990).

100. C. Toniolo, G. M. Bonora, A. Bavoso, E. Benedetti, B. DiBlasio, V. Pavone, *et al.*, *J. Biomol. Struct. Dyn.*, **3**, 585 (1985).

101. S. M. Holl, G. R. Marshall, D. D. Beusen, K. Kociolek, A. S. Redlinski, M. T. Leplawy, *et al.*, *J. Am. Chem. Soc.*, **114**, 4830–3 (1992).

102. R. G. S. Spencer, K. J. Halverson, M. Auger, A. E. McDermott, R. G. Griffin, and P. T. Lansbury Jr., *Biochemistry*, **30**, 10382–7 (1991).

103. J. R. Garbow and C. A. McWherter, *J. Am. Chem. Soc.*, **115**, 238–44 (1993).

104. M. H. Levitt, *J. Chem. Phys.*, **94**, 30–8 (1991).

105. J. Anglister and F. Naider, *Methods in Enzymol.*, **203**, 228–41 (1991).

106. S. K. Dower and R. A. Dwek, in *Biological Applications of Magnetic Resonance* (ed. R. G. Schulman), p. 271, Academic Press, New York (1979).

107. R. A. Dwek, J. C. A. Knott, D. Marsh, A. C. McLaughlin, E. M. Press, N. C. Price, *et al.*, *Eur. J. Biochem.*, **53**, 25 (1975).

108. J. C. Cheetham, C. Redfield, R. E. Griest, D. P. Raleigh, C. M. Dobson, and A. R. Rees, *Methods in Enzymol.*, **203**, 202–28 (1991).

109. P. Tsang, M. Rance, and P. E. Wright, *Methods in Enzymol.*, **203**, 241–61 (1991).

110. P. Tsang, T. M. Fieser, J. M. Ostresh, R. A. Lerner, and P. E. Wright, *Peptide Res.*, **1**, 87 (1988).

111. P. S. Tsang, P. E. Wright, and M. Rance, *J. Am. Chem. Soc.*, **112**, 8183 (1990).

112. A. Odaka, J. I. Kim, H. Takahashi, I. Shimada, and Y. Arata, *Biochemistry*, **31**, 10686–91 (1992).

113. Y. Paterson, S. W. Englander, and H. Roder, *Science*, **249**, 755–9 (1990).

114. L. Mayne, Y. Paterson, D. Cerasoli, and S. W. Englander, *Biochemistry*, **31**, 10678–85 (1992).

115. D. C. Benjamin, D. C. Williams Jr, S. J. Smith-Gill, and G. S. Rule, *Biochemistry*, **31**, 9539–45 (1992).

116. R. N. Perham, H. W. Duckwork, and G. C. K. Roberts, *Nature (London)*, **292**, 474–7 (1981).

117. F. L. Texter, S. E. Radford, E. D. Laue, R. N. Perham, J. S. Miles, and J. R. Guest, *Biochemistry*, **27**, 289–96 (1988).

118. L. E. Sommerville, G. D. Henry, B. D. Sykes, and D. J. Hartshorne, *Biochemistry*, **29**, 10855–64 (1990).

5 · Protein folding

The way in which a given primary sequence of amino acids in a polypeptide chain determines how that chain folds up to give a fully functional protein is one of the enigmas of late twentieth-century biomolecular science. Clearly, as the protein folds up, there are a number of conformational states that form *en route* to the final native structure. Early models for protein folding suggested that although transient secondary structural elements may form, there are essentially only two states involved: the folded and the unfolded, and folding is heterogeneous, following several alternative pathways. A more recent model is that native-like structural domains form in discrete, highly transient steps, and this has been called[1,2] the *molten globule*, which is thought to have a high content of secondary structure, but little or no tertiary structure. In the early studies in this field, NMR was used to follow the amide exchange rates in folding intermediates, and some of the evidence pointed to the existence of discrete native-like intermediates (or, in some examples, non-native intermediates, such as was proposed for BPTI). Now, more recent evidence for the molten globule intermediate is emerging, which displays a marginally retarded rate of amide proton exchange and a compactness which is somewhat less than that of the native but considerably greater than that of the fully unfolded form. Although if detectable, partially folded forms would be a powerful source of mechanistic information, such states are not significantly populated at equilibrium for most small proteins. Furthermore, those which are detectable have extremely short lifetimes, making the acquisition of detailed structural information difficult on transient intermediates. NMR has been particularly useful in unravelling the structural details of these folding-unfolding transitions.

5.1 THERMAL UNFOLDING

The interconversion between folded and unfolded forms is usually slow on the NMR timescale, giving rise to separate lines for the native and denatured states at equilibrium. Fig. 5.1 illustrates the thermal unfolding of a trypsin inhibitor (HPI) from the snail, *Helix pomatia*. The ^{1}H NMR spectrum of the native protein at 43°C displays a number of resolved and assigned resonances characteristic of the folded conformation. In particular, some of the groups in the hydrophobic interior (Ala-9, Tyr-23, and Phe-45) exhibit significant ring-current shifts and three C$^\alpha$H located in the central β sheet are shifted downfield. The spectrum recorded at 65°C, where the inhibitor is about 90% unfolded, resembles the random coil spectrum expected on the basis of the chemical shifts presented in Table 4.2. Since a spectrum recorded after lowering the temperature back to 43°C is indistinguishable from the original one, the folding–unfolding interconversion is clearly reversible. The spectrum shown at the intermediate temperature, is near the midpoint of the thermal unfolding transition, and is essentially the superposition of the spectra at 43°C and 65°C. This means that there are no discrete partially folded states with sufficient lifetime and population to give rise to separate resonances, which might be seen as evidence for the two-state model. Any existing intermediates would have to be in rapid

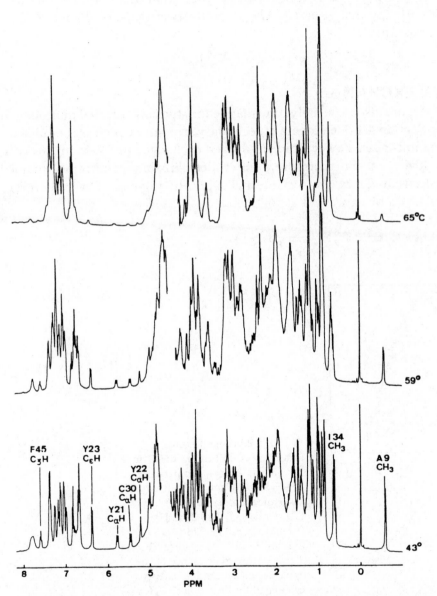

Fig. 5.1 Thermal denaturation of HPI, a 58-amino acid trypsin inhibitor related to BPTI, monitored by ^{1}H NMR. The spectra were recorded in D_2O (pD 2.5) after complete exchange of all labile protons. At 50°C the protein is 50% unfolded, and at 65°C it is about 90% unfolded. (Reprinted from H. Roder, *Methods in Enzymol.*, **176**, 446–73 (1989) with permission.)

exchange with either the folded or unfolded conformations. This can be tested by examining the transition curves for three resolved resonances, which, as shown in Fig. 5.2, are identical within experimental error.

The normalized intensity of a resolved native protein resonance yields the fraction of molecules in the native form, f_n, and the equilibrium constant for unfolding is given by

$$K_u = (1 - f_n)/f_n \tag{5.1}$$

The unfolding free energy, $\Delta G_u = -RT \ln K_u$, is best expressed in terms of the midpoint temperature (T_m), the enthalpy change at T_m (ΔH_m), and the heat capacity change (ΔC_p), by the following relationship:

$$\Delta G_u = \Delta H_m (1 - T/T_m) - \Delta C_p [T_m - T + T \ln(T/T_m)] \tag{5.2}$$

5.2 HYDROGEN EXCHANGE

Hydrogen exchange rates are also extremely informative for studying internal motions of proteins. As we saw in Section 4.1.3, elements of secondary structure which are stabilized by hydrogen bonds exhibit much slower amide hydrogen exchange rates than in parts of the protein lacking structure. Furthermore, under destabilizing conditions, major structural unfolding leads to exchange for internal labile hydrogens. The structural unfolding model, which can be expressed as follows:

$$F(H) \underset{k_f}{\overset{k_u}{\rightleftharpoons}} U(H) \underset{D_2O}{\overset{k_c}{\rightleftharpoons}} U(D) \underset{k_u}{\overset{k_f}{\rightleftharpoons}} F(D) \tag{5.3}$$

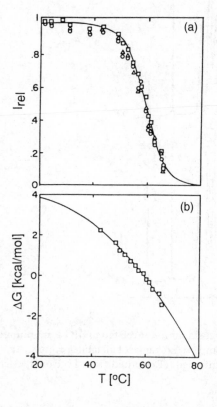

Fig. 5.2 Thermal folding-unfolding equilibrium of HPI in D_2O at pD 2.5 (a) Temperature dependence of the normalized ^1H resonance intensities for Ala-9 CH_3 (squares), Try-21 $C^\delta H$ (triangles) and Try-23 $C^\delta H$ (circles) in the folded form. (b) Temperature dependence of the apparent unfolding free energy, calculated by Equation (5.1) for the Ala-9 data. (Reprinted from H. Roder, *Methods in Enzymol.*, **176**, 446–73 (1989) with permission.)

is particularly useful when structural unfolding occurs via a cooperative transition between an ensemble of folded states, F, and unfolded conformations, U. The opening transition can then be characterized by an unfolding rate, k_u, and a refolding rate k_f, or an opening equilibirium, $K_{op} = k_u/k_f$. It is generally thought that a hydrogen atom initially buried within the native protein structure can exchange only after it is brought into contact with water by a structural unfolding transition, at an intrinsic rate k_c, which is given by

$$k_c = k_H[H^+] + k_{OH}[OH^-] \tag{5.4}$$

Exchange of amide NH hydrogens does not occur spontaneously—the reactions are catalysed by solvent acids or bases. Having said this, there has been a certain amount of debate on the solvent penetration model, in which it is hypothesized that solvent can penetrate protein secondary structure and catalyse exchange. Different amino acid side chains and neighbouring amino acids affect the magnitude of k_c. This is summarized in Fig. 5.3, which shows the exchange behaviour of the NH group in a random chain of polyalanine.

Under conditions in which the folded state is energetically favoured ($k_f \geq k_u$), Equation (5.3) leads to the following expression for the observed exchange rate, k_{ex}:

$$k_{ex} = k_u k_c/(k_f + k_c) \tag{5.5}$$

In the limit of structural fluctuations which are rapid compared to the intrinsic exchange ($k_f \geq k_c$), then Equation (5.5) becomes:

$$k_{ex} = K_{op} k_c \tag{5.6}$$

The exhange rate is then determined by the relative concentration of unfolded protein, which is in turn determined by the unfolding equilibrium constant, K_{op}, and the rate-limiting intrinsic exchange rate. This exchange mechanism, known as the EX2 mechanism, is second order with the rate being dependent on the catalyst concentration, and is typically encountered in proteins under conditions where the folded state is stable and intrinsic exchange is relatively slow. The quantity K_{op} contains information on the

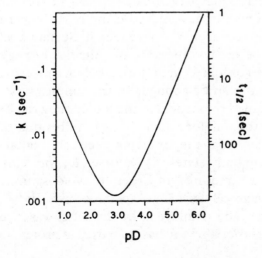

Fig. 5.3 Exchange behaviour for random poly-D, L-alanine (in D_2O at 20°C). (Reprinted from S. W. Englander and L. Mayne, *Ann Rev. Biophys. Biomol. Struct.*, **21**, 243–65 (1992) with permission.)

structural contribution to hydrogen exchange. With $\Delta G_{op} = -RT\ln K_{op}$, Equation (5.6) can be rewritten as

$$\Delta G_{op} = -RT\ln(k_{ex}/k_c) \tag{5.7}$$

In the limit of slow structural fluctuations relative to the intrinsic exchange rate ($k_f \leq k_c$), the measured exchange rate is determined by the rate of unfolding, and Equation (5.5) becomes:

$$k_{ex} = k_u \tag{5.8}$$

In this case, known as the EX1 mechanism, exchange occurs with each structural opening and the measured rate gives the rate of unfolding directly. Conditions favouring exchange due to limited opening are basic pH, where k_c is fast, or destabilizing conditions, where folding rates tend to be slow.

The distinction between these two kinetic exchange mechanisms is crucial for the interpretation of hydrogen exchange data. Indications for exchange due to limited opening can be obtained from the pH dependence of exchange. In the EX2 limit the pH dependence is determined by the strong variation of the intrinsic exchange rate (see Equation (5.4)), modulated by any pH dependence of the structural equilibrium. In the EX1 limit, only the unfolding rate contributes to the pH dependence. Only in the EX1 limit can exchange of different protons occur in a structurally correlated way, whereas in the EX2 limit the protein has to undergo many unfolding transitions before a particular proton is exchanged, giving rise to a random distribution of exchanged sites. Under most circumstances the EX2 mechanism can be expected to prevail, but the EX1 mechanism may occur under the conditions of some protein refolding experiments. According to Equation (5.6) exchange rates measured in the EX2 regime, together with the known values for k_c, provide equilibrium information on structural folding transitions.

5.3 STRUCTURAL AND KINETIC CHARACTERIZATION OF THE UNFOLDED STATE

NMR spectroscopy is one of the few techniques which can be used to correlate the unfolded and folded states structurally. This can be achieved using NMR experiments which involve magnetization (or saturation) transfer. We have seen in Section 1.3.5 how magnetization transfer experiments can be used to measure the kinetics of enzymatic reactions. The kinetics of interconversion of unfolded and folded states of a protein are no different, and the same type of experiments can be applied. In the simplest case, one-dimensional saturation transfer of the assigned resonances in the folded state can be used to assign those unassigned resonances in the unfolded state. This requires that there is slow exchange on the NMR timescale, and that the spectrum therefore contains the superposition of resonances arising from both states. Of course, for proteins with $M_r \geq 15$ kDa, two-dimensional methods are required in order to avoid saturation of overlapping resonances. Here, as in chemical exchange spectroscopy, the EXSY, ROESY, or TOCSY pulse sequences may be employed. In the latter case, relayed magnetization transfer by isotropic mixing is possible and has been used in protein folding problems.[3] Fig. 5.4 illustrates the use of the two-dimensional EXSY experiment, showing the spectrum of bovine pancreatic trypsin inhibitor (BPTI) modified by selective reduction and carboxyaminomethylation (RCAM–BPTI) of the 14—38 disulphide bond (because of its lower thermal stability), at the midpoint of the thermal unfolding transition.

The kinetics of folding can be determined under equilibrium conditions from such two-dimensional spectra, or by one-dimensional saturation transfer, varying the delay τ between the selective excitation pulse and the non-selective $90°$ read pulse. An example of this is shown in Fig. 5.5, again with RCAM–BPTI. The $C^{\varepsilon}H$ ring proton of Tyr-23 in the unfolded state is irradiated and the normalized intensity of the corresponding resonance in the folded state is plotted as a function of saturation time τ. At each temperature the intensity decays exponentially from the equilibrium value, M_0, to a limiting value, M_∞, indicating that a two-state model is adequate.

The decay rate is then given by

$$M_\infty/M_0 = R_f/(R_f + k_u) \tag{5.9}$$

where k_u is the unfolding rate, R_f is the relaxation rate of the irradiated resonance in the folded state. From an exponential fit to the data, $(R_f + k_u)$ and M_∞ may be determined. Furthermore, k_f and R_f may be determined independently, for example, k_f may be measured by irradiating a resonance in the folded state and observing the unfolded state.

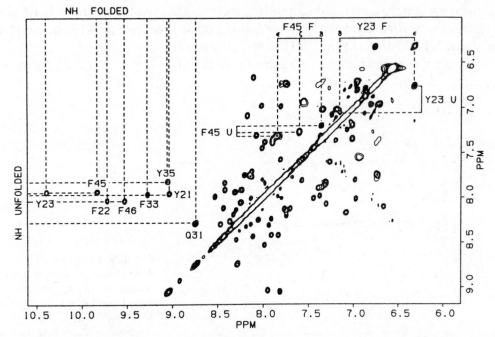

Fig. 5.4 Two-dimensional EXSY spectrum (mixing time 100 ms) of 15 mM RCAM–BPTI correlating the folded and unfolded forms of the enzyme. The majority of the cross-peaks arise from magnetization transfer between the two states. Selected examples of NH and aromatic resonance assignments are shown. Dashed lines indicate the resonance positions in the folded form (vertical) and the corresponding shifts in the unfolded form (horizontal). (Reprinted from H. Roder, *Methods in Enzymol.*, **176**, 446–73 (1989) with permission.)

Or, of course, both k_u and k_f may be obtained simultaneously from an analysis of the cross-peak intensities in an EXSY experiment.

5.4 HYDROGEN EXCHANGE LABELLING OF FOLDING INTERMEDIATES—pH COMPETITION

As we have seen in Section 5.2, competition between amide hydrogen exchange and folding can be used to probe the existence of folding intermediates during the formation of the native state. In this type of experiment, protein refolding and amide hydrogen exchange are initiated simultaneously. When refolding is performed at low pH where exchange is slow, the refolded native structure traps the initial isotope. At higher pH, hydrogen exchange proceeds more rapidly, and some exchange occurs before hydrogen bonds can be reformed. Half of a given site will be labelled when its hydrogen exchange rate and hydrogen-bond formation rate are equal. If the intrinsic amide hydrogen exchange rate (k_c) is known (from Equation (5.4)), then the folding rate can be determined. Experimentally, a rapid mixing apparatus is used, as outlined in Fig. 5.6. The protein is initially unfolded in H_2O in the presence of a suitable denaturant or at extreme pH and placed in syringe S1. Refolding and H-D exchange are induced simultaneously in a first mixer (M1) by rapid dilution of the denaturant (or adjustment of the pH) with a D_2O buffer solution (in syringe S2), causing a jump to native conditions. After a reaction time τ, which is adjustable, the exchange process is quenched by rapid lowering of the pH in the second mixer (M2). This stops further H-D exchange but allows refolding to proceed to completion, thereby protecting the remaining unexchanged amide hydrogens against further exchange. The protein is then recovered and examined by two-dimensional NMR in which the NH cross-peak volumes are analysed.

Relative proton occupancies, P, at each site are calculated by normalizing the measured signal intensities, I_m, as follows:

$$P = [(I_m/I_0) - f_h]/(1 - f_h) \tag{5.10}$$

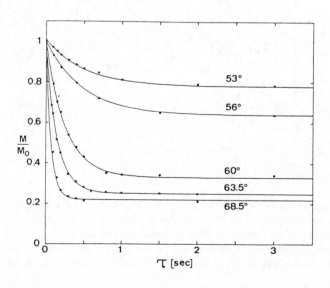

Fig. 5.5 Temperature dependence of the time-resolved saturation transfer between folded and unfolded forms of RCAM–BPTI induced by selective irradiation at the resonance position of Tyr-23 $C^\varepsilon H$ in the unfolded form. (Reprinted from H. Roder, *Methods in Enzymol.*, **176**, 446–73 (1989) with permission.)

where I_0 is the signal intensity of the fully protonated group, and f_h is the residual fraction of H_2O present in the reaction mixture.

As an illustration of the method, consider a hypothetical folding pathway in which an exchangeable proton (1) is protected at a rate k_i by rapid formation of an intermediate I containing a helical segment (see Fig. 5.7). Another proton (2) is protected at a slower rate k_f as it forms a hydrogen bond in a later stage of refolding. In this example, pH is used to vary the intrinsic amide exchange rate (k_c), and the limiting degree of protonation, P_∞, is measured at long reaction times ($\tau \gg 1/k_c$). Assuming that refolding is essentially irreversible, then P_∞ measured at site (1) is given by:

$$P_\infty = k_i/(k_i + k_c) \tag{5.11}$$

and for site (2):

$$P_\infty = k_f/(k_f + k_c) \tag{5.12}$$

Assuming that the pH dependence of P_∞ is dominated by k_c according to Equation (5.4), both protons are trapped at low pH where $k_c < k_f$. As k_c is increased, the proton occupancy of site (2) drops since the more rapidly formed intermediate structure protects site (2) and $k_c < k_i$. Finally, at more basic pH where exchange is fast ($k_c \approx k_i$) site (2) competes equally with site (1). The effective rate of protection for each site may be estimated from the value of k_c at $P_\infty = 0.5$. The values for k_c need to be determined in an independent experiment, for example using exchange measurements in the fully unfolded protein.

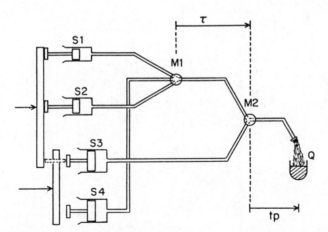

Fig. 5.6 Schematic diagram of a commercial perparative rapid mixing apparatus for quenched-flow, adapted for competition and pulse labelling experiments. The solutions from syringes S1 and S2 are mixed in the first mixing chamber, M1. After a variable time, τ, the mixture is combined in M2 with a third reagent from syringe S3. Aging times from about 5 to 200 ms are achieved by continuous flow, and longer times by interrupted flow, in which the sample is allowed to rest between M1 and M2 before being driven out with S4 and mixed with reagent from S3. For competition experiments, H-exchange is quenched in M2 by mixing with an acidic buffer from S3. In pulse-labelling experiments, H or D labelling is induced in M2 and after delay, t_p, quenching is achieved by injecting the mixture through a nozzle into excess buffer in an external container. (Reprinted from H. Roder, *Methods in Enzymol.*, **176**, 446–73 (1989) with permission.)

A variation on the competition method is shown in Fig. 5.7(c), in which the reaction time (τ) is varied at a fixed pH value, e.g. at pH 8.8, where $P_\infty \approx 0.5$ for site (1). The relative protonation decays exponentially at a rate ($k_c + k_i$) from $P = 1$ at short τ, to the value for P_∞ given by Equation (5.11). Exchange and folding rates can thus be measured independently from the decay rate and P_∞ can be determined from an exponential fit.

A serious problem with the pH competition approach is that the change in pH may also perturb the folding pathway and the stability of the intermediates. An experimental advantage is that only one mixing step is necessary, and that the early stages of folding may be investigated.

5.5 HYDROGEN EXCHANGE LABELLING OF FOLDING INTERMEDIATES—pH PULSE

The pH-pulse labelling method effectively uses many hydrogen exchange probes in a protein, in order to unravel the details of protein folding. Unlike the pH competition experiment, the entire time-dependent folding process is monitored under constant solution conditions. Such conditions may be chosen to help stabilize intermediates, for example by using low temperatures or neutral pH, or even including chaperone proteins that assist in protein folding. It is particularly useful for studying the later stages of folding. In this approach the protein is initially unfolded in denaturant in D_2O, so that all amide hydrogens are already exchanged with deuterium. In a first mixing step, the solution is diluted to initiate folding. The first dilution can be into the H_2O buffer at a pH and temperature that are sufficiently low that folding is initiated but exchange labelling

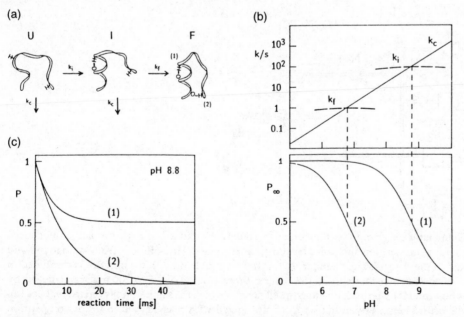

Fig. 5.7 Schematic illustration of the competition method for a hypothetical folding pathway (a). The principle of competition experiments at long reaction times and variable pH is shown in (b). The expected results of competition experiments at constant pH and variable reaction times are illustrated in (c). (Reprinted from H. Roder, *Methods in Enzymol.*, **176**, 446–73 (1989) with permission.)

does not yet occur. After some folding time, τ, a labelling pulse by jumping to higher pH is applied by the addition of a small volume of high pH buffer. Now the exchange of deuterium with hydrogen occurs rapidly, and those sites still available for exchange become labelled with hydrogen. Sites already hydrogen bonded retain their deuterium atoms. The labelling pulse is terminated by quenching at low pH (through another small addition of low pH buffer). The protein folds to completion, and at each site trapped by hydrogen bonding, the H/D ratio can be determined by two-dimensional NMR analysis of the NH cross-peak intensities as outlined in the previous section. Results at a series of τ values can then trace out the history of hydrogen-bond formation for each amide NH trapped in the native state.

Three-stage rapid mixing can be used, and has the advantage that the length of the labelling pulse can be set such that the intensity of the pulse can be controlled. Pulse intensity (I_p) is given by

$$I_p = \Delta t_p / \tau_{ex,p} = k_{ex,p} \times \Delta t_p \tag{5.13}$$

where Δt_p is the pulse time and $\tau_{ex,p}$ is the hydrogen exchange lifetime at the time of the pulse. Note that, here,

$$k_{ex,p} = 1/\tau_{ex,p} = k_c \tag{5.14}$$

At pH 9 and 10°C, the intrinsic amide hydrogen lifetime is ≈ 3 ms, so that a 30-ms pulse corresponds to an intensity $I_p = 10$, which is in the range that is large enough to ensure labelling of exposed NHs but small enough to minimize labelling of hydrogen-bonded NHs. Such a short pulse is unlikely to perturb the structure labelled in its duration. Varying I_p can be used to test the stability of various intermediates and their change in stability monitored as folding progresses. When the pulse intensity is reduced, say $I_p \approx 2$, at a constant $\tau = 30$ ms, at pH 8.3 and 10°C, then labelling falls off and the fraction not labelled is $\exp(-I_p)$. If the pulse intensity is increased, sites already hydrogen bonded will begin to be labelled when $k_{op} I_p \approx 0.15$, and from Equation (5.6) K_{op} can be estimated and the stability (ΔG_{op}) of local folded structure in a folding intermediate can be estimated from Equation (5.7). This energy analysis necessarily assumes that the hydrogen exchange in the intermediate is determined in an EX2 mechanism, an assumption which may not always be justified.

To illustrate the results of a pulse labelling experiment, consider Fig. 5.8, which plots the course of proton labelling as a function of folding time τ. As τ increases, protection against labelling increases, and the proton occupancies measured by two-dimensional NMR decrease. The example shows possible results for a protein with four hydrogen bonds, with time plotted logarithmically to cover a realistically broad time span. Folding may proceed from the unfolded state (U) through a simple linear sequence of intermediates (I_i) to the native state (N), as follows: $U \rightarrow I_1 \rightarrow I_2 \rightarrow N$. Folding may be heterogeneous, with different fractions of U following alternate, parallel paths: $U_1 \rightarrow N$; $U_2 \rightarrow I \rightarrow N$. Although such steps are reversible, this can be neglected given the short duration of the labelling pulse.

Folding may appear to be concerted as illustrated in Fig. 5.8(a), in which all the probe sites achieve 100% protection at the same rate, in a single first-order kinetic step. The protection rate observed represents the rate at which molecules cross the rate-limiting barrier preceding N. Although other barriers and intermediates will occur, they are not detectable under these conditions. In Fig. 5.8(b), a simple sequential pathway (U→I→N) is represented, in which an intermediate that has formed hydrogen bonds A and B is populated before hydrogen bonds C and D are formed. A and B attain 100% protection with the same time constant, and the presence of a partially folded intermediate is revealed when different probes show different degrees of protection at any given folding time. Such labelling results also reveal the identity of the residues involved in hydrogen bonding in the populated intermediate.

For intermediates to be well-populated, they must occupy an energy well in the free-energy profile for the reaction which is lower than the preceding wells, and the barrier to the next well must be larger than all preceding barriers. Since this is relatively rare, most folding intermediates are not detected. This means that intermediates can appear to form simultaneously, and formation of hydrogen bonds A and B, for example, may reflect independent folding events which are kinetically indistinguishable.

In searching for intermediates, therefore, the presence of population heterogeneity with independent parallel pathways needs to be taken into account. Heterogeneity is

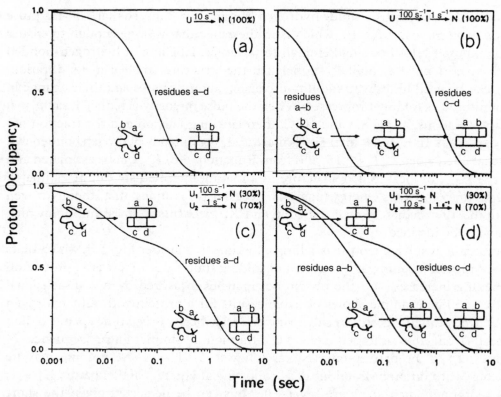

Fig. 5.8 Idealized HX pulse-labelling results for a four-hydrogen bond protein that folds as in the pathways indicated: (a) concerted; (b) sequential with one intermediate state; (c) two heterogeneous, parallel paths with concerted folding; (d) two parallel paths with one intermediate in the slower path. (Reprinted from S. W. Englander and L. Mayne, *Ann. Rev. Biophys. Biomol. Struct.*, **21**, 243–65 (1992) with permission.)

indicated when less than 100% of a given probe is protected in a given first-order step ($U_1 \rightarrow I_1$ at k_1; $U_2 \rightarrow I_1$ or I_x at k_2; where $k_1 \neq k_2$). In different fractions of the protein population, a given amino acid residue forms hydrogen bonds at different rates. In the example shown in Fig. 5.8(c), the system is heterogeneous, but each folding step involves all the NHs and intermediates are not detectable. In Fig. 5.8(d), the earliest step involves all of the NHs in 30% of the molecules, and appears concerted. In 70% of the molecules, the presence of a partially folded intermediate is revealed by the fact that different probes follow a different time course. It is also possible to interpret folding heterogeneity in terms of the molten globule, which would represent the superposition of kinetically indistinguishable secondary structural elements.

5.6 EXAMPLES OF PROTEIN FOLDING STUDIES BY NMR
5.6.1 Ribonuclease A

The first NMR study on ribonuclease A folding was by Adler and Scheraga,[4] which employed a novel approach to the detection of transient folding intermediates known as the continuous recycled flow method. In this experiment, the protein solution is continously pumped through a closed circuit containing first a high-temperature bath, a low-temperature bath, and a sample cell located in the NMR magnet. The RNase A is unfolded in the high-temperature ($60°C$) bath, rapidly cooled in the low-temperature bath ($< 42°C$) to initiate folding, and then pumped into the NMR probe for observation. The sample is recycled back to the high-temperature bath in order to allow signal averaging. Since at pH 2.4, RNase A unfolding is reversible, the same protein solution may be recycled repeatedly. The dead-time, that is, the time elapsed between temperature jump and NMR spectroscopy, of this approach is around 15 s, and therefore can only be used to study slow-folding intermediates. The result from the continuous recycled flow experiments, together with far UV CD data, suggest that the intermediate has a rapidly fluctuating structure with an overall conformation close to that of the thermally denatured protein. The N-terminal helix is present and constitutes the only obvious element of native secondary structure. The apparent molten globule formed by the bulk of the protein may provide a hydrophobic surface on which the amphipathic helical peptide can dock.

Baldwin and co-workers[5,6] used pH pulse labelling to study the refolding of RNase A. This protein has a large β-sheet structure and three short helices (see Fig. 5.9) and is stabilized by four disulphide bonds. In their initial experiments,[5] these investigators initiated refolding of the denatured, disulphide-intact protein (dissolved in 2.6M guanidine D_2O buffer) by using a ten-fold dilution in D_2O (pD ≈ 9, $5°C$) and a long (10 s) labelling pulse (six-fold dilution into H_2O buffer at pH 9). In their subsequent experiments,[6] the initial dilution was into H_2O (pH 4, $10°C$) and a short labelling pulse (37 ms) was used (1.5-fold dilution into pH 9 or 10 H_2O buffer). The final two-dimensional COSY experiments followed 27 probe NHs, based on previous assignments for RNase.[7,8] It was already known that there was three-part population heterogeneity, with a fast component ($U_F \rightarrow N$; $\tau = 40$ ms; 20% of population) and a slow component ($U_S I \rightarrow N$; $\tau = 50$ ms; 15% of the population), and that both these components appeared to

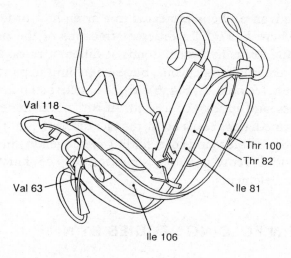

Fig. 5.9 Tertiary structure of RNase A. The three α-helices and the β-sheet are clearly depicted in this ribbon diagram (courtesy of Jane Richardson). The locations of selected amide protons are indicated. (Reprinted from ref. 5 with permission.)

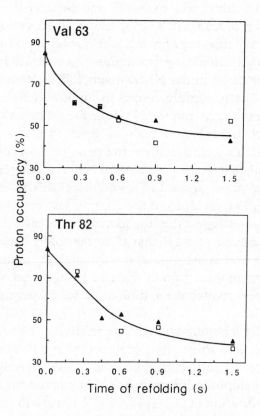

Fig. 5.10 Kinetics of protection of backbone amide deuterons from exchange with solvent protons for RNase A. The accessibility to exchange (proton occupancy) of a selection of two amide proton sites is plotted versus time after initiation of refolding. The intensities of the $C^\alpha H$—NH cross-peaks are followed in the COSY spectrum, at pH 9 (closed symbols) and pH 10 (open symbols) after dilution into buffer. Evidence for an early intermediate is clear. (Reprinted from ref. 5 with permission.)

fold into the native state in kinetically two-state reactions. A major component (65%) displayed more kinetic structure, with an early hydrogen-bonded intermediate and a late native-like intermediate ($U_SII \rightarrow I_1 \rightarrow I_N \rightarrow N$).

The pH pulse-labelling studies revealed the existence of population heterogeneity very clearly. The very fast component (or early intermediate) was easily detectable by NMR, as shown in Fig. 5.10. The slow component was detected by the change in UV absorbance for a tyrosine residue. The major (65%), multistep pathway was elucidated by using two probe NHs in the N-terminal helix of native RNase (helix 1), which were found to be protected against exchange labelling in one kinetic step ($\tau = 1$ s). On the same timescale, changes in the UV-absorption of a tyrosine residue indicate the formation of the intermediate I_N. The presence of a partially folded intermediate is established by the fact that other residues, in the extended β-sheet, are protected in a different manner. These latter NHs fold in two heterogeneous components: 40% of the population in a faster group ($\tau \approx 20$ ms) and 25% in a slower group ($\tau \approx 1$ s). Therefore the behaviour of the major 65% phase is similar to that shown in Fig. 5.8(d), except that in RNase A the phase with the common rate is the slower one. The implication of such results is that a partially folded intermediate (I_1) forms with the extensive β-sheet structure, and perhaps also with helices 2 and 3 largely intact, but with helix 1 missing at this stage.

The stability of the I_1 structure changes during the 0–400 ms time interval. In experiments that varied the intensity of the labelling pulse, a 3 ms pulse was used at pH 9 and pH 10. The β-sheet structure in I_1 is resistant to labelling at pH 9, whereas at pH 10, incomplete protection is seen initially ($p \approx 250$), but stability increases to full protection ($P > 1000$) over the 400 ms period. The investigators tested the stability of the structure present at $\tau = 400$ ms by varying the pulse pH from 7.6 to 11 systematically. The labelling achieved falls off for all amide NHs. Above pH 10, some already unstructured sites become labelled presumably because of local unfolding, whereas others are not labelled, suggesting that the intermediate structure present ($I_1 + I_N$) has great overall stability ($P > 10^4$).

5.6.2 Cytochrome *c*

Roder, Englander, and co-workers[9] used pH pulse labelling to study the refolding of cytochrome *c*. Folding was inititated by dilution from a guanidine buffer (in D_2O) into H_2O at pH 6, 10°C. A 50 ms labelling pulse was used at pH 9 (with a small dilution). The two-dimensional NMR COSY analysis followed 35 assigned[10–12] amide NHs (see Fig. 5.11). An early phase of the refolding takes place within the ≈ 3 ms dead-time of the mixing apparatus, and involves $\approx 20\%$ of every NH, suggesting that there is rapid folding of this component to a near-native state. At longer times, several heterogeneous phases were detected. Sites in the N- and C-terminal helices gain protection simultaneously, in three parallel steps with timescales of about 20 ms (40%), 400 ms (20%), and 4 s (20%). Similar time constants were detected by fluorescence spectroscopy of Trp-59 which is quenched by proximity to the haem group. Population heterogeneity is also observed in the 60S helix, the short 70S helix, and the centrally buried indole NH (see Fig. 5.12). These sites become protected in at least two different phases (≈ 80 ms and 4 s). This behaviour is analogous to that in Fig. 5.8(d), with more steps.

These results show that an early intermediate occurs in which the N- and C-terminal helices are largely formed before the rest of the structure (≈ 20 ms; 40% of the population). In the native protein these helices interact, and therefore the early intermediate may be native-like and rate-limited by the productive docking of the two helices. Concerted formation of the terminal helices is followed by the 60S and the 70S helices together with the tertiary changes at Trp-59 at longer times.

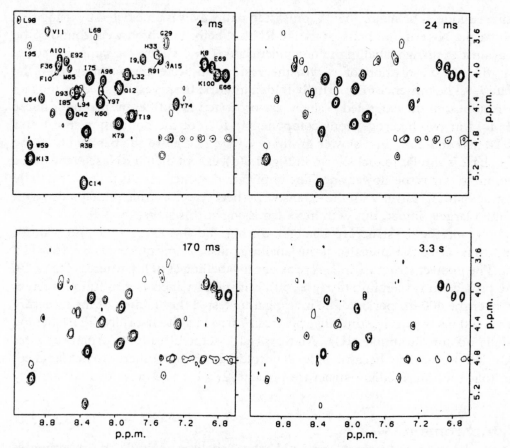

Fig. 5.11 Two-dimensional COSY spectra of representative cytochrome *c* samples prepared by the pulse labelling method. The sections of the COSY spectra contain the majority of the NH—C$^{\alpha}$H cross-peaks, shown for samples labelled at four different refolding times. (Reprinted from ref. 9 with permission.)

5.6.3 Barnase

In a *tour de force*, Fersht and co-workers have investigated[13] *Bacillus amyloliquefaciens* ribonuclease (barnase) by a plethora of methods, including pH pulse labelling.[14,15] These workers used two-stage mixing with a long (≈ 10 s) labelling pulse (four-fold dilution into H$_2$O buffer, pH 8.5) to study barnase (in urea D$_2$O buffer at pD 6.3, 25°C) and quenching at pD 3.5 in D$_2$O buffer. Of the 106 assigned[16,17] amide hydrogens, 29 NHs exchanged slowly enough to be measured under the trapping and NMR conditions used, with 25 being involved in secondary structural elements (such as α-helix, β-sheet, and β-turn) and four were involved in tertiary interactions.

A very slow phase ($t_{1/2} = 8.5$ s) that involves 20% of every NH was not followed in this work. All the secondary structure probes are protected in two faster phases (as in Fig. 5.8(c)), with the first phase (≈ 20–40% of each secondary structural NH) proceeding within the mixing dead-time. The second phase (≈ 30–60% of each secondary structural NH) proceeds more slowly ($k \approx 10$–30 s^{-1}). Several tertiary NH probes and one NH in a reverse turn achieve protection at a common, slower rate ($k \approx 5$ s^{-1}).

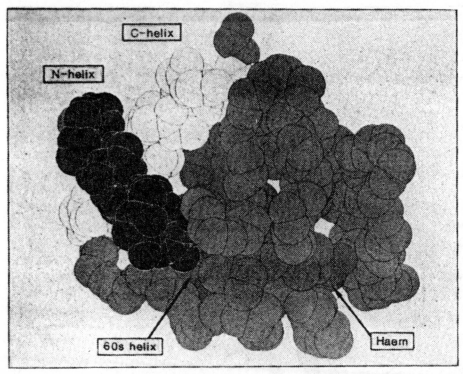

Fig. 5.12 Space-filling model of cytochrome c showing the location of various structural elements followed in the refolding experiments. Side chains are omitted. The view was chosen to emphasize the contact between the N-terminal (dark) and C-terminal (light) α-helices, which are formed early in the folding. Also identified are the 60 s helix, an example of a structural feature formed late in folding, and the haem. Trp 59 is behind the lower right edge of the haem. (Reprinted from ref. 9 with permission.)

The interpretation of these results is that extensive secondary structure formation is an early folding event. All parts of the three α-helices (see Fig. 5.13) that can be monitored, as well as all antiparallel β-strands and several β-turns, are formed rapidly. This species represents a transient intermediate, possibly analogous to the molten globule. Evidence of an interaction close to the 3_{10}-helix suggests that the helix is formed only late in the folding pathway. Furthermore, with one exception, many of the tertiary interactions form slowly.

5.6.4 Apomyoglobin

Apomyoglobin (apoMb) has been studied by pH pulse labelling by Wright, Baldwin, and co-workers.[18] Removal of the haem group renders apoMb less stable than Mb, and reduces its helical content. Earlier studies on the acid-induced unfolding of apoMb indicated the presence of an intermediate, perhaps a molten globule. Pulse labelling experiments were applied to probe the structure of both native apoMb and the acid intermediate. In an ingenious modification of the procedure outlined in Section 5.5, haem is added to the apoprotein solution in order to quench exchange. This creates the holomyoglobin species which is readily analysed by NMR spectroscopy, based on previous ^{1}H NMR assignments.[19] The amide protons in the A-, G-, and H-helices (see Fig. 5.14) are significantly protected from exchange with solvent in the intermediate state.

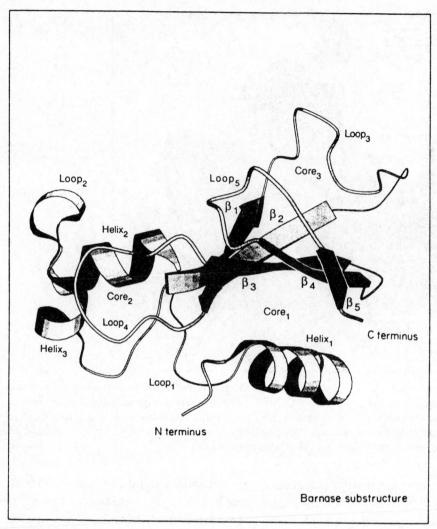

Fig. 5.13 General overview of barnase structure. (Reprinted from ref. 13 with permission.)

Fig. 5.14 Ribbon drawing model of myoglobin, showing the proposed region of the structured subdomain in the apomyoglobin intermediate. Helices B–E are unshaded and are proposed to be predominantly unfolded in the I state. The F helix is lightly shaded to denote the lack of amide proton probes—and, consequently, structural information—for this region. The A, G, and H helices are darkly shaded and are proposed to be folded in a native-like conformation in the I form. (Reprinted from ref. 18 with permission.)

This leads to the hypothesis that apoMb may fold through a species with a hydrophobic core on which stable amphipathic helices are docked. More evidence[20] for this hydrophobic core has been presented by Baldwin and co-workers, who find strong support for the notion that a molten globule forms as the major kinetic intermediate near the start of the folding pathway.

5.6.5 α-Lactalbumin

Baum and co-workers[21] investigated α-lactalbumin by magnetization transfer NMR techniques under conditions of high temperature or low pH. This protein appears to form the so-called molten globule at pH 7.4 and at the thermal transition (68°C) exchange between the native conformation and the molten globule could be monitored by EXSY, thus correlating assignments in the two states. Alternatively the molten globule state also forms at pH 2 or with intermediate amounts of denaturants such as urea. Additional experiments, used hydrogen exchange and pH pulse-labelling, albeit with an extremely long pulse, since the protein was labelled by dissolving the protein in D_2O buffer at pH 2 for 10 h before adjustment to pH 5.4. A number of amide protons were found not to have exchanged even over the course of 10 h, as illustrated in Fig. 5.15. The results show that a

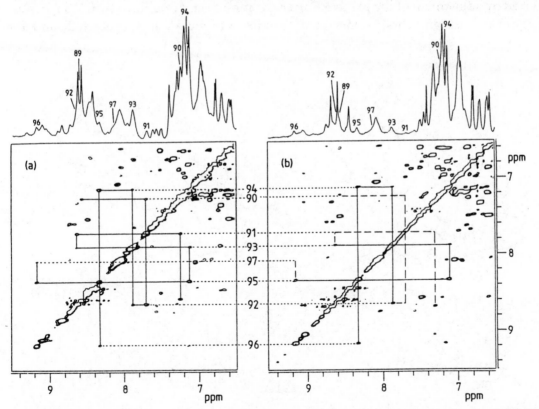

Fig. 5.15 Low-field regions of the NOESY spectra of α-lactalbumin at 35°C freshly dissolved (a) in ^2H_2O at pH 5.4 following the pH-jump experiment. The sequential connectivities between residues 90 and 97 are shown in both cases. The broken lines in (b) indicate connectivities from (a) that are not observed in the spectra following the pH jump; these are between residues 90–91, 91–92, and 96–97. (Reprinted from ref. 25 with permission.)

tertiary structure in the native protein is conserved in the molten globule. This consists of an amphipathic helical region including residues 89–96, bordering on a hydrophobic box region.

More recent work by Baum and co-workers[22] has established from pH-jump measurements that fewer than 20 of the amides are involved in highly persistent structure in the molten globule, despite CD evidence for significant amounts of unstable structural elements. Also, this structure was found to be native-like, with a hydrophobic core persisting in the helical domain of the molten globule.

5.6.6 Lysozyme

The folding of hen lysozyme has been investigated[23] by Miranker, Karplus, and co-workers using pH competition methods and by Radford and co-workers[24] using pH pulse-labelling methods. Also, T4 lysozyme has been studied by pH pulse-labelling by Lu and Dahlquist.[25] In the first study, Miranker and co-workers performed competition experiments at pH 7.5 in which the 65 slowly exchanging amide hydrogens in the native protein were used to probe the structure. The protein was denatured in guanidinium hydrochloride in H_2O at pH 7.5, and diluted (25-fold) in D_2O buffer at pH 7.5, exchange quenched by adjustment of the pH to 3.8 after one minute and examined by COSY NMR, using previous assignments.[26] One set of amides was measurably protected, and for

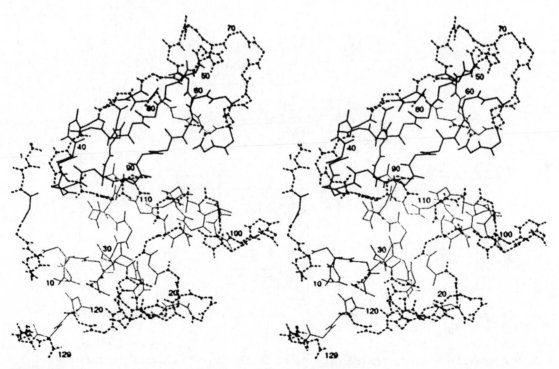

Fig. 5.16 Stereoviews of the structure of hen lysozyme. Dark shaded residues indicate amides having exchange $>30\%$, and light shaded residues indicate $<30\%$. Dashed residues are those for which there is no information. These represent mainly surface amides and proline residues. (Reprinted from ref. 23 with permission.)

another set, exchange competed with protection. The first set involved residues at the N- and C-termini, corresponding to the recognizable structural α-domain (see Fig. 5.16). The second set involved the central part of the sequence and formed the β-domain. The termini of the protein, which fored the major α-helix and interact in the native structure, was also protected rapidly ($t_{1/2} \approx 10$ ms). These results support the notion that, like α-lactalbumin, hen lysozyme forms a transient intermediate in folding, but that its formation occurs much more rapidly with $k_f \approx 1$ s^{-1}, and $P \approx 100$, rather like cytochrome c.

In the pH pulse-labelling studies, about half of the total 126 amide sites were used to monitor the refolding reaction. The protein in guanidine—D_2O buffer, pH 6.0, was diluted (ten-fold in H_2O buffer, pH 5.5) to initiate folding for a variable time (3.5–2000 ms) and then the pH labelling pulse (by adjustment of the pH to 9.5 for 8.4 ms) and termination by five-fold dilution into acidic H_2O buffer (final pH ≈ 4). Distinct biphasic loss of proton occupancy with refolding time was observed for all the probes (see Fig. 5.17). The faster phase occurs within a few milliseconds and results in the protection

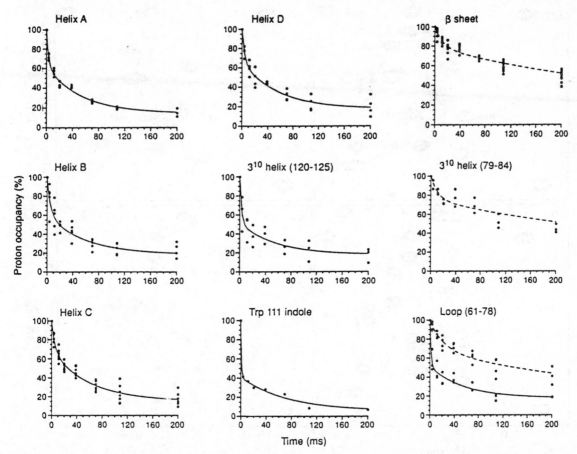

Fig. 5.17 Time courses for the protection of amides from exchange during the refolding of hen lysozyme. The curve drawn represents the average of a two exponential fit to the time course for individual amides in each native secondary structural element. In the case of the loop region, two averages are drawn, one for Trp 63, Cys 64, and Ile 78, and one for the remainder.

of the α-domain amides in half of the population, and the β-domain amides in one-third of the population. The slower phase, which is due to the folding of the remaining population, distinguishes the α-domain (<100 ms) and the β-domain (>100 ms). These results are consistent with parallel pathways (as in Fig. 5.8(d)) of domain segregation with heterogeneity in the unfolded state or in the intermediate state, formed within the first 2 ms of refolding. However, they are also consistent with the formation of a single molten globule.

In contrast to hen lysozyme, the structurally homologous T4 lysozyme appears to refold via a single channel. Unfolded [^{15}N]protein in which the amide protons had been exchanged for deuterium was allowed to refold in 50% D_2O pH 5.1 for a variable length of time (8–400 ms), exposed to a 41-ms proton-labelling pulse at pH 9.0 and quenched by adjustment to pH 3.2. The samples were then examined by ^{1}H—^{15}N HMQC NMR (see Fig. 5.18), using previous assignments and NMR structure[27] for interpretation. A rapid

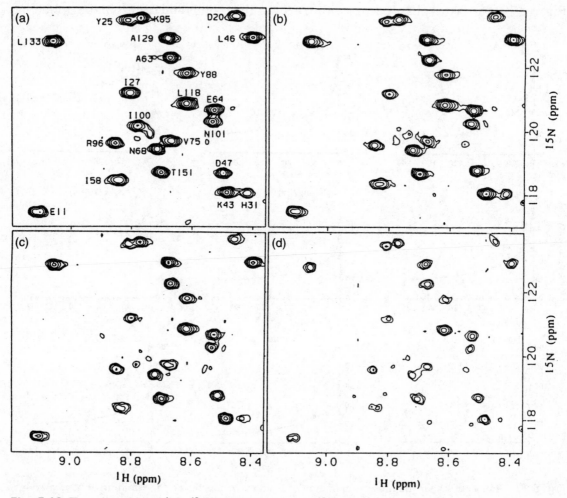

Fig. 5.18 Two-dimensional ^{1}H—^{15}N HMQC spectra of ^{15}N-labelled T4 lysozyme prepared by pulsed hydrogen-exchange methods: (a) reference sample; (b) 8 ms sample; (c) 100 ms sample; (d) 250 ms sample. (Reprinted from ref. 25 with permission.)

phase (occurring within the 8 ms of the instrument deadtime) first produces intermediates with several amides that are poorly but measurably protected from exchange. A slower phase (after 170 ms) leads from the intermediates to a native-like structure with well-protected amides. In spite of the structural homology between these two lysozymes, domain segregation is not observed with the T4 phage enzyme. The presence of four disulphide bonds in hen lysozyme may enhance the structure and stability of the early intermediates which play key roles in establishing parallel folding routes.

5.6.7 Staphylococcal nuclease A

Wild-type nuclease from *Staphylococcus aureus* exists in two slowly interconverting forms, N and N'. Magnetization transfer experiments[28,29] detected exchange between N and N', and also demonstrated the existence of corresponding unfolded forms, U and U', at the midpoint of thermal unfolding. The reason for this heterogeneity has been attributed to *cis-trans* isomerism at the amide bond preceding Pro-117. Site-directed mutagenesis was used to replace Pro-117 with Gly, and the alternative forms of the protein were suppressed in the NMR spectra,[30] thereby implicating P117 as the source of the heterogeneity. In some elegant experiments using isotopically labelled staphylococcal nuclease, Torchia and co-workers[31] (protein liganded with Ca^{2+} and thymidine-3', 5'-bisphosphate) and Bolton, Gerlt, and co-workers[32] (with unliganded protein), provided conclusive proof by showing that P117 is exclusively *cis* in the liganded protein (in keeping with the X-ray structure) and largely *cis* in the unliganded protein.

Analysis of the one- and two-dimensional NMR data obtained at the midpoint of the thermal denaturation (40°C, pH 5.3) provides a quantitative description of the equilibria among two native and two unfolded species in the wild-type protein. Four rate constants for unfolding and refolding, corresponding to N⇌U and N'⇌U', and two equilibrium constants for proline isomerization, corresponding to N⇌N' and U⇌U', were obtained. The interconversion rates in this case ranged between 0.1 and 4 s^{-1}. On the basis of the X-ray structure and the relative rates of interconversion, it was concluded that the predominant form N contains a *cis* 116—117 bond, whereas the predominant form U' contains a *trans* 116—117 bond. It was also possible to discriminate among several limiting energy profiles for the folding/unfolding reaction. Stabilization of the *cis* bond in the N-form seems to occur at a late stage in folding.

In another study by Markley and co-workers,[33] magnetization transfer experiments were applied to a number of mutant staphylococcal nucleases, demonstrating that mutations at sites remote from Pro-117 perturb the N⇌N' equilibrium. At high (>3 mM) concentrations of protein, a third native form, N'', is detectable, and is presumably a dimeric species.

5.6.8 Ubiquitin

Ubiquitin is a simple, small (76-residue) protein lacking prosthetic groups or disulphide bonds, which was studied through pH pulse-labelling methods by Briggs and Roder.[34] This protein has an extended β-sheet and an α-helix surrounding a pronounced hydrophobic core (see Fig. 5.19). The protein was dissolved in guanidine D_2O buffer at pD 3.5, allowed to exchange for 30 minutes, refolding was initiated by rapid six-fold

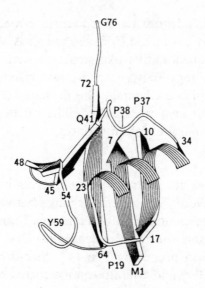

Fig. 5.19 Ribbon diagram of the ubiquitin backbone, based on the X-ray coordinates. (Reprinted from ref. 34 with permission.)

dilution into H_2O buffer at pH 5.0, and a labelling pulse was initiated by two-fold dilution into H_2O buffer at pH 9.4 and terminated after 31 ms by two-fold dilution into H_2O buffer at pH 3.4. Protection from exchange was measured for two backbone amide protons that form stable hydrogen bonds upon refolding and exchange slowly under native conditions. Amide protons in the β-sheet and the α-helix, as well as protons involved in hydrogen bonds at the helix–sheet interface, become 80% protected in an initial 8 ms folding phase, indicating that the two elements of secondary structure form and associate in a common cooperative folding event. Somewhat slower protection rates for selected residues provide evidence for the subsequent stabilization of a surface loop. Most of the amide probes also exhibit two minor phases with time constants of about 100 ms and 10 s. Only two of these residues, Gln-41 and Arg-42, display significant slow folding phases, with relative amplitudes of 37% and 22% respectively, and this can be attributed to native-like folding intermediates containing *cis* Pro-37 and/or Pro-38. Compared with cytochrome *c*, ribonuclease, and barnase, the initial formation of hydrogen-bonded structure in ubiquitin occurs at a more rapid rate and slow-folding species are less prominent.

5.6.9 Bovine pancreatic trypsin inhibitor (BPTI)

BPTI was one of the first proteins whose folding pathway was studied. This is a small single-domain protein stabilized by three disulphide bonds (see Fig. 5.20), designated [30–51; 5–55; 14–38] to indicate the cysteine residues involved. Native BPTI can be unfolded by simply reducing its disulphides and efficiently refolds under conditions where disulphide formation is favoured. During the 1970s, Creighton elucidated the pathway of reductive and oxidative refolding by isolating and characterizing partially disulphide-bonded intermediates.[35] The folding pathway is outlined in Fig. 5.20. One of the striking results of this study was the discovery that the kinetic mechanism involved intramolecular rearrangements involving intermediates with non-native disulphide bonds, and that these were well populated. In particular, two species containing non-native disulphides ([30–51; 5–14] and [30–51; 5–38]) were detected. This finding raised

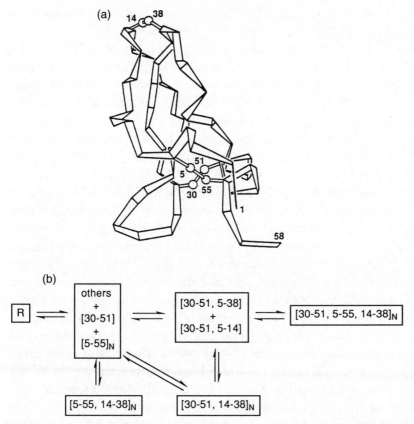

Fig. 5.20 (a) Schematic representation of the crystal structure for BPTI. The first and last residues and the cysteine residues involved in disulfide bonds are labelled. The disulfide bond 14–38 is accessible to solvent, exposing 48% of its total surface area, whereas the disulfide bonds 30–51 and 5–55 are inaccessible, exposing 0% of their total surface area. (b) Schematic diagram of the folding pathway of BPTI as proposed by Creighton.[42] ((a) Reprinted from ref. 37 with permission.)

the possibility that the amino acid sequence might have to specify the structures of folding intermediates which have different conformations from that of the final folded protein. Such a scenario would add new complications to the already formidable problem of predicting the three-dimensional structure of a protein from the primary sequence alone. Recent studies[36] from the laboratory of Peter Kim have suggested, through a combination of HPLC analysis and NMR spectroscopy, that all the well-populated folding intermediates contain only native disulphide bonds, and that in Creighton's original study, the concentrations of the species containing non-native disulphides were overestimated. NMR evidence from Kim's laboratory suggests[37] that even early intermediates such as the [5–55] disulphide corresponds to an essentially completely folded protein. Based on NMR studies[38] of a synthetic peptide model for the intermediate [30–51], it appears that this too has native-like secondary and tertiary structure. This is also true for later intermediates, such as the [30–51; 14–38] intermediate. The implication from this work was that the species containing non-native disulphides are kinetically irrelevant and probably have to undergo significant conformational rearrangement to reach the final native structure.

The relative importance of non-native intermediates in the folding of BPTI is still a matter of considerable debate.[39–41] However, it is clear that NMR spectroscopy is playing an important role in resolving it. For example, the partially folded conformation of the [30–51] intermediate has been determined in the laboratories of Creighton and Neuhaus[42] by ^{1}H and ^{15}N NMR methods, including a detailed analysis of ^{15}N T_1 and T_2 relaxation data to assess conformational flexibility in regions lacking NOEs (see Section 3.10). According to Creighton and co-workers, this intermediate does not readily form the [5–55] disulphide directly, but rather the [5–14], [5–38] or [14–38] disulphides prior to rearrangement to the native disulphides. The NMR structure of the [30–51] intermediate provides insights into this point, because it is precisely the regions that are flexible that also contain Cys5, Cys14 and Cys38, explaining how non-native disulphides could readily form in a folding intermediate whose conformationally ordered regions are native-like. In a less detailed study, Creighton, Neuhaus, and co-workers[43] used NMR (among other methods) to show that the two predominant intermediates containing non-native disulphides, [30–51, 5–14] and [30–51, 5–38], also contain regions with stable structure comparable to the [30–51] intermediate, also with similar regions of conformational flexibility. The same groups[44,45] have also used two-dimensional NMR methods to determine the structures of the [14–38, 30–51] and [5–55] intermediates, and again these have essentially fully folded, native-like conformations. Thus the implication from this work suggests that the presence of non-native disulphide bonds does not mean that the protein needs to adopt a non-native conformation. In fact the non-native disulphide bond formation occurs in flexible regions of the folding intermediates. Indeed recent work from Kim's laboratory[46] also show that intermediates with non-native disulphide bonds are involved in the pathway leading to native BPTI, although Kim believes that the formation of such species are not obligatory for productive folding to the final native structure.

The earlier stages of folding were studied[47] by Roder and Wüthrich in the first example of pH competition by NMR. Eight of the amide hydrogens were followed, some of which are located in the β-sheet and one in the C-terminal α-helix. For all eight, the pH at which $k_c \approx k_f$ lies around 6.0. All appear to be protected at about the same rate, with k_fs of the order of 15–60 s^{-1} at 70°C. This corresponds to the fast folding rates detected by other techniques, and implies that there is a high degree of cooperativity and the formation of an early intermediate with native-like secondary structure.

5.7 OUTLOOK

One of the difficulties in the field of protein folding has been the structural characterization of transient folding intermediates. Since these hold the key to unlocking the secrets of how the primary structure determines the final three-dimensional structure of a protein, work in the future will need to focus on methods of trapping and characterizing these intermediates. Although to date, the amide exchange method has been used with great success to characterize a number of intermediates, it depends on the existence of relatively stable intermediates accumulating to a reasonable extent. In this respect, the study of protein folding is reminiscent of the study of enzyme intermediates (see Chapter 6), in that transient intermediates need to be trapped for sufficient periods of time, at sufficient populations, to be detectable by NMR spectroscopy. Here it seems likely that methods such as time-resolved solid-state NMR (see Section 6.4.2), coupled with solid-state distance measurement techniques (see Section 3.3), could be gainfully employed in probing specific structural aspects of the structure of thermally or chemically trapped folding intermediates.

REFERENCES

1. O. B. Ptitsyn, *J. Protein Chem.*, **6**, 272–93 (1987).

2. K. Kuwajima, *Proteins*, **6**, 87–103 (1989).

3. Y. Feng and H. Roder, *J. Magn. Reson.*, **78**, 597–602 (1988).

4. M. Adler and H. A. Scheraga, *Biochemistry*, **27**, 2471–80 (1988).

5. J. B. Udgaonkar and R. L. Baldwin, *Nature*, **335**, 694–9 (1988).

6. J. B. Udgaonkar and R. L. Baldwin, *Proc. Natl. Aca. Sci. USA*, **87**, 8197–201 (1990).

7. A. D. Robertson, E. O. Purisima, M. A. Eastman, and H. A. Scheraga, *Biochemistry*, **28**, 5930–8 (1989).

8. H. Rico, M. Ruix, J. Santoro, C. Gonzalez, J. L. Neira, and J. Herranz, *Eur. J. Biochem.*, **183**, 623–38 (1989).

9. H. Roder, G. A. Elöve, and S. W. Englander, *Nature*, **335**, 700–4 (1988).

10. A. J. Wand and S. W. Englander, *Biochemistry*, **25**, 1100–6 (1986).

11. A. J. Wand, D. L. DiStefano, Y. Feng, H. Roder, and S. W. Englander, *Biochemistry*, **28**, 186–94 (1989).

12. Y. Feng, H. Roder, S. W. Englander, A. J. Wand, and D. L. DiStefano, *Biochemistry*, **28**, 195–203 (1989).

13. A. R. Fersht and co-workers, *J. Mol. Biol.*, **224**, 771–859 (1992).

14. M. Bycroft, M. Matouschek, J. T. Kellis, L. Serrano, and A. R. Fersht, *Nature*, **346**, 488–90 (1990).

15. M. Matouschek, L. Serrano, E. M. Meirin, M. Bycroft, and A. R. Fersht, *J. Mol. Biol.*, **224**, 837–45.

16. M. Bycroft, R. N. Sheppard, F. T.-K. Lau, and A. R. Fersht, *Biochemistry*, **29**, 7425–32 (1990).

17. M. Bycroft, S. Ludvigsen, A. R. Fersht, and F. M. Poulsen, *Biochemistry*, **30**, 7425–32 (1991).

18. F. M. Hughson, P. E. Wright, and R. L. Baldwin, *Science*, **249**, 1544–8 (1990).

19. T. C. Pochapsky, C. Dalvit, Y. Thériault, and P. E. Wright, unpublished results.

20. D. Barrick and R. L. Baldwin, *Protein Science*, **2**, 869–76 (1993).

21. J. Baum, C. Dobson, P. A. Evans, and C. Hanley, *Biochemistry*, **28**, 7–13 (1989).

22. C.-L. Chyan, C. Wormald, C. M. Dobson, P. A. Evans, and J. Baum, *Biochemistry*, **32**, 5681–91 (1993).

23. A. Miranker, S. E. Radford, M. Karplus, and C. M. Dobson, *Nature*, **349**, 633–6 (1991).

24. S. E. Radford, C. M. Dobson, and P. A. Evans, *Nature*, **358**, 302–7 (1992).

25. J. Lu and F. W. Dahlquist, *Biochemistry*, **31**, 4749–56 (1992).

26. C. Redfield and C. M. Dobson, *Biochemistry*, **27**, 122–36 (1988).

27. L. P. McIntosh, A. J. Wand, D. F. Lowry, A. G. Redfield, and F. W. Dahlquist, *Biochemistry*, **29**, 6341–62 (1990).

28. R. O. Fox, P. A. Evans, and C. M. Dobson, *Nature*, **320**, 192–4 (1986).

29. P. A. Evans, R. A. Kautz, R. O. Fox, and C. M. Dobson, *Biochemistry*, **28**, 362–70 (1989).

30. P. A. Evans, C. M. Dobson, G. Hatfull, R. A. Kautz, and R. O. Fox, *Nature*, **329**, 266–8 (1987).

31. D. A. Torchia, S. W. Sparks, P. E. Young, and A. Bax, *J. Am. Chem. Soc.*, **111**, 8315–17 (1989).

32. S. M. Stanczyk, P. H. Bolton, M. Dell'Acqua, and J. A. Gerlt, *J. Am. Chem. Soc.*, **111**, 8317–18 (1989).

33. A. T. Alexandrescu, E. L. Ulrich, and J. L. Markley, *Biochemistry*, **28**, 204–11 (1989).

34. M. S. Briggs, and H. Roder, *Proc. Natl. Acad. Sci USA*, **89**, 2017–21 (1992).

35. T. E. Creighton, *Prog. Biophys. Mol. Biol.*, **22**, 221–98 (1978).

36. J. S. Weissman and P. S. Kim, *Science*, **253**, 1386–93 (1991).

37. J. P. Staley and P. S. Kim, *Proc. Natl. Acad. Sci USA*, **89**, 1519–23 (1992).

38. T. G. Oas and P. S. Kim, *Nature*, **336**, 42–8 (1988).

39. D. P. Goldenberg, *Trends in Biochem. Sci.*, **17**, 257–61 (1992).

40. T. E. Crieghton, *Science*, **256**, 111–12 (1992).

41. J. S. Weissman and P. S. Kim, *Science*, **256**, 112–14 (1992).

42. C. P. M. van Mierlo, N. J. Darby, J. Keeler, D. Neuhaus, and T. E. Creighton, *J. Mol. Biol.*, **229**, 1125–46 (1993).

43. N. J. Darby, C. P. M. van Mierlo, G. H. E. Scott, D. Neuhaus, and T. E. Creighton, *J. Mol. Biol.*, **224**, 905–11 (1992).

44. C. P. M. van Mierlo, N. J. Darby, D. Neuhaus, and T. E. Creighton, *J. Mol. Biol.*, **222**, 353–71 (1991).

45. C. P. M. van Mierlo, N. J. Darby, D. Neuhaus, and T. E. Creighton, *J. Mol. Biol.*, **222**, 373–90 (1991).

46. J. S. Weissman and P. S. Kim, *Proc. Natl. Acad. Sci. USA*, **89**, 9900–4 (1992).

47. H. Roder and K. Wüthrich, *Proteins*, **1**, 34–42 (1986).

Part III · *Enzymes*

6 · Enzyme function

Since the first reported NMR study[1] of an enzyme, ribonuclease, in 1957, NMR spectroscopy has been applied to an increasingly wide range and number of enzymes. Although the introduction and development of multidimensional ^{1}H NMR techniques over the last ten years has revolutionized the structural elucidation of proteins with $M_r \leq 35$ kDa, the majority of enzymes fall outside this molecular weight range. Indeed, in our discussion of protein structures in Part II, only three enzyme structures were covered: lysozyme, staphylococcal nuclease, and thioredoxin. For larger enzymes, the complete structure cannot currently be determined by NMR. But one of the distinguishing features of NMR spectroscopy as a technique for structural biology is its ability to provide crucial structural information about a specific part or region of a molecule. This sets NMR apart from other structural techniques such as X-ray crystallography, in which generally the whole structure must be completely solved before a selected region can be investigated. This ability to focus on structural regions of interest is no more important than in the field of protein function, where the relationship between structure and function is intimately connected at the enzyme active site. The use of heteronuclear NMR, in particular nuclei such as ^{13}C, ^{15}N, ^{19}F, and ^{31}P, has been important in providing the means to focus on enzyme active sites for proteins whose $M_r = 25-100$ kDa.

6.1 WHICH TYPES OF ENZYMES CAN BE STUDIED?

Not all enzymes are amenable to study by NMR spectroscopy. The limitations arise because of a number of factors, the two most important being: molecular weight, M_r, and also the k_{cat} of the enzyme. We will consider the effects of these limitations in sections 6.2 and 6.4. The types of functional information that can be obtained on enzymes may be summarized as follows:

1. The flux of isotopically labelled substrates to products through an enzyme-catalysed pathway may be obtained by solution-state NMR—this can lead to the identification of enzyme-free intermediates. An example of this is shown in Fig. 6.1, which shows the turnover of porphobilinogen (PBG, structure **6.1** in Fig. 6.2) to the product uroporphyringen I (uro'gen I, structure **6.3**) catalysed by hydroxymethylbilane synthase (EC 4.3.1.8, also known as PBG deaminase—see Section 8.4.2) by ^{1}H NMR spectroscopy.

The transient appearance of two lower intensity resonances at 4.4 and 6.4 ppm, corresponding to the HO-CH_2-pyrrole and pyrrole-H of hydroxymethylbilane (HMB, structure **6.2**). It had already been established that HMB is the product of the enzymatic reaction, and chemically cyclizes to uro'gen I.[2]

2. The structure of an isotopically labelled inhibitor bound to an enzyme may be determined by solution-state ($M_r \leq 50$ kDa) or solid-state ($M_r \leq 150$ kDa) NMR. This can be used in conjunction with protein structural methods by NMR (see Part II) or X-ray crystallography, for the identification of functional residues in the enzyme active site and a possible rôle in catalysis. We shall see examples of this in Sections 7.1.1 and 8.4.3.

3. The structure of an enzyme-substrate or enzyme-intermediate complex of a multi-substrate enzyme with an insufficient number of substrates for the reaction to proceed can be carried out by solution-state or solid-state NMR. An example of this is given in Section 8.4.1.

Fig. 6.1 An example of a transient reaction intermediate which is released into solution. The 300 MHz ^{1}H NMR spectra of PBG (**6.1**), in the presence of PBG deaminase are shown as a function of time. The low-intensity resonances that appear fleetingly at 4.2 and 6.3 ppm arise from the product of the enzyme, hydroxymethylbilane (**6.2**), which is converted chemically to uro'gen I (**6.1**). (Reprinted from ref. 2 with permission.)

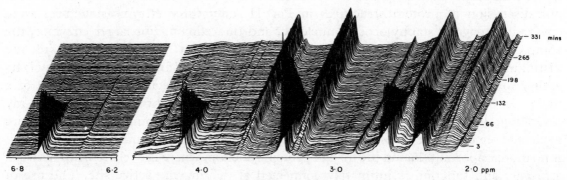

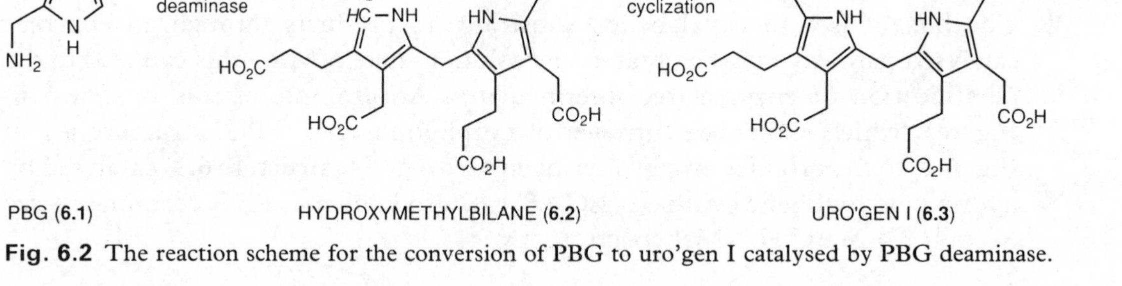

Fig. 6.2 The reaction scheme for the conversion of PBG to uro'gen I catalysed by PBG deaminase.

4. The structure of an enzyme-substrate, or enzyme-intermediate complex of a freely reversible enzyme, can be determined by solution-state or solid-state NMR by balancing the concentrations of substrate and product, and allowing maximum accumulation of intermediate (depending on k_{cat}, K_M and exchange rates). An example of this is given in Section 8.4.3.

5. The structure of an enzyme-substrate or enzyme-intermediate complex can be determined at low temperature by cryoenzymological solution-state NMR and solid-state NMR. We will discuss these methods in more detail in Section 6.4, and examples are given in Sections 7.1.3 and 8.4.3.

In the foregoing list, the word 'structure' is used frequently, and it is important to realize the limits of the structural information which can be obtained. Clearly, the molecular weight will affect the detail of the information obtainable, or, put more formally, above $M_r = 35$ kDa there is an inverse relationship between the number of distance restraints and molecular weight. The same sorts of considerations as were outlined in the section for proteins in general are, of course, applicable to enzymes. Therefore alternative strategies have to be adopted to obtain structural information. The use of isotopic labels introduced at specific locations in the enzyme substrate or inhibitor molecule can be used for heteronuclear NMR studies. But the detection of the resonance from such an isotope provides, in the first instance, only a chemical shift. This can be highly diagnostic—for example if the enzyme catalyses the cleavage of an amide bond via a tetrahedral sp^3 carbon, then the ^{13}C resonance would shift from around 200 ppm to around 100 ppm. But chemical shifts a structure do not make! By analogy with the chemical shifts of model compounds, together with pH titration behaviour, relaxation data, and J coupling data (e.g. for heteronuclei the spectrum can be obtained with and without ^1H decoupling), a structure can be deduced. However, the information that is most useful concerns distance or torsion-angle restraints of some kind, either from NOEs (see Section 3.1), transferred NOEs (see Section 6.5), paramagnetic relaxation enhancements (see Section 3.4), or dipolar couplings (see Section 3.3). Only with this information coupled with the other data can the structure of a substrate, inhibitor, or intermediate be determined with any certainty. The relationship of that ligand to the rest of the biomolecule poses a different type of problem, and yet again distance or torsion-angle restraints are the data required to resolve it. Sometimes distance or torsion-angle restraints can be used in conjunction with a low-resolution X-ray crystal structure, or an X-ray structure of the protein in the absence of a ligand, in docking experiments. Above all it is the quality rather than the quantity of restraints that helps to define good structures for larger enzyme-substrate or enzyme-intermediate complexes. This is in contrast to smaller protein structures, where we saw (Section 4.1) that the quantity of relatively poor quality distance restraints can generate remarkably accurate, and in some cases precise, structures.

6.2 LINEWIDTHS IN ENZYME COMPLEXES

In any NMR investigation of an enzyme complex, either heteronuclear studies of isotopically enriched substrates or intermediates, or homonuclear studies of the protein itself, the single most important factor crucial to successful detection of the resonance of interest is its linewidth. The linewidth of an NMR resonance is given by:

$$v_{1/2} = \frac{1}{\pi T_2^*} \tag{6.1}$$

where $v_{1/2}$ is the linewidth at half-height, and T_2^* is the apparent transverse relaxation time. Linewidths in enzyme complexes are determined essentially by two factors: (i) the overall rotational correlation time, τ_c, of the macromolecule or (ii) chemical exchange. The rotational correlation time for globular proteins may be calculated using the Stokes–Einstein equation:

$$\tau_c = \tfrac{4}{3}\pi r^3 \frac{\eta}{k_B T} \tag{6.2}$$

where r is the molecular radius, η is the viscosity of the medium, k_B is Boltzmann's constant, and T is the temperature in Kelvin.

Assuming $M_r \propto r$,

$$\tau_c \approx M_r \times C \tag{6.3}$$

where C is typically 1.3×10^{-12} (calculated from T_1 values for backbone α carbons at field strengths ≤ 2.4 T).[4] Note that this latter relation is *very* approximate.

Such an estimate of rotational correlation time is based on the assumption that the overall tumbling rate dominates the relaxation for nuclei in the molecule, even if it is a small molecule complexed with a large one such as an enzyme-substrate or enzyme-intermediate complex. There are exceptions to this, in which the local librational motion is greater than the tumbling. In these cases, as was considered in Section 4.7.2, mobile domains are detectable even in very large proteins.

6.3 ENZYME-INHIBITOR COMPLEXES

The study of enzyme-inhibitor complexes by NMR relies on certain basic features of the inhibition. When the inhibitor structure resembles postulated intermediate structures, the inhibitor is sometimes called a 'transition state analogue'. The use of this name, originally introduced by Jencks[3] and promoted by Wolfenden,[4,5] derives from the premise suggested by Pauling[6] that enzymes achieve their remarkable catalytic efficiency by energetic stabilization of transition states along the reaction pathway. While that is undoubtedly one of the most important mechanisms available to enzymes as a motive for catalysis, it should be noted that not all enzymes do this: destabilization of enzyme-substrate and enzyme-product complexes also produce the same net effect. Furthermore, structures of transition states can only be inferred, and even then only with a considerable body of data (e.g. from kinetic isotope studies) on the internal microscopic rate constants, and hence thermodynamics, for the reaction. These data are required in order to predict whether the transition state resembles the substrate or intermediate, according to the Hammond Postulate. Thus to the structural biophysicist, the use of the term 'transition state analogue' is something of a misnomer, since in general almost all the rationally designed enzyme inhibitors are analogues of the substrate, product, or postulated intermediate, all of which are ground-state structures that, at least in principle, can be determined with certainty.

The types of inhibitors amenable to study by NMR (using ^1H, ^{19}F, ^{31}P, ^{13}C, or ^{15}N NMR of appropriately labelled inhibitor, or in some cases, enzyme) can be separated into the following categories:

1. Covalent or 'suicide' inhibitors. An example of this will be given in Section 7.1.1.

2. Electrostatically bound inhibitors whose $K_i \leq \mu M$, and whose exchange is slow on the NMR timescale. An example of this is given in Section 8.4.3.

The structure of an isotopically labelled inhibitor bound to an enzyme may be determined by solution-state (for enzymes with $M_r \leq 50$ kDa) or solid-state (for enzymes with $M_r \leq 150$ kDa) NMR, which, when used in conjunction with high-resolution ^1H NMR described in Chapters 2 and 4, or with paramagnetic ions, can lead to identification of functional residues in enzyme-active sites. Studies of the pH behaviour of functional groups on the enzyme-bound inhibitor molecule can also be very informative as to the identity of enzyme-active site functional groups in the vicinity, which can have perturbed pK_a values. Saturation transfer experiments, inversion transfer, or two-dimensional EXSY experiments can be used to determine k_{on} and k_{off} and therefore K_d (or K_i) values for inhibitors.

6.4 DIRECT OBSERVATION OF ENZYME–SUBSTRATE/INTERMEDIATE COMPLEXES

The quest for an understanding of the molecular details of enzymatic catalysis has been a major motivation for much biochemical and biophysical research during the last decade. A crucial part of this endeavour has been the structural investigation of intermediates, which has played an important rôle in the elucidation of enzymatic reaction mechanisms. An enzymatic reaction mechanism can be said to be understood when (i) the enzyme-bound intermediate(s) can be detected and its structure determined; (ii) the kinetic competence of the enzyme has been demonstrated (i.e. the rate of formation and degradation are consistent with the forward and reverse reaction rates);[7] and (iii) the internal microscopic rate data are consistent with a reaction pathway in which all thermodynamic and kinetic states have been described. The use of ^{31}P, ^{13}C, and ^{15}N NMR in solution is a well-established technique for the study of enzymatic reaction mechanisms.[8-11]

The elucidation of the structure of an enzyme-substrate complex at various points along the reaction coordinate is one of the most sought after goals of enzyme chemistry. Currently there are two approaches to stabilizing enzyme-intermediate complexes so that they can be examined by NMR, by low-temperature solution-state cryoenzymological NMR, and by the recently introduced method of time-resolved solid-state NMR spectroscopy.

6.4.1 Cryoenzymology

There are a considerable number of enzymatic reactions for which the catalytic rate is too fast for significant accumulation of intermediates during the reaction. In this case, the idea

Table 6.1 Effect of temperature on relative reaction rates[a]

| °C | Activation energy (kJ mol^{-1}) | | | |
	40	60	80	100
+20	1	1	1	1
0	3.3	6.1	11	20
−20	13	49	180	660
−30	29	160	860	4600
−40	69	570	4700	39 000
−50	173	2280	30 000	3 900 000
−80	4900	350 000	2.4×10^7	1.7×10^9

[a]From Ref. 9.

of lowering the temperature to slow the rate is straightforward. The kinds of reduction in reaction rates are shown in Table 6.1.

To prevent freezing of an aqueous solution of an enzyme and substrate below 0°C, cryosolvents are employed. In general the cryosolvent needs to be inert, and commonly 40–50% DMSO-water is used. Unfortunately, the viscosity of this cryosolvent also increases at lower temperatures, resulting in increases in linewidth. For example there is a four-fold increase in linewidth in going from 0°C to −20°C in 40% DMSO-water. The use of a cryosolvent and low temperatures requires a careful kinetic characterization of the enzymatic reaction under the new conditions, to ensure that the reaction pathway has not been altered. Furthermore, the current molecular weight limit is around 50 kDa.[9] This method has been applied to a number of enzymes (see Sections 7.1.2 and 7.1.3 for examples), and application of solution-state NMR spectroscopy to the study of enzyme-intermediate complexes at subzero temperatures can provide valuable data (chemical shift, relaxation times, J coupling, and NOEs).[12,13]

6.4.2 Time-resolved solid-state NMR spectroscopy

One way to circumvent some of the disadvantages of solution-state cryoenzymology is to use the recently developed method of time-resolved solid-state NMR spectroscopy.[14–16] In the past, study on intermediate species has depended upon their isolation from the reaction mixture and is therefore limited to relatively stable, long-lived species. With the advent of spectroscopic techniques, it has become possible to investigate these intermediate species *in situ* during the reaction. However, the techniques that have gained the widest application have been those which furnish the least structural information, namely UV-visible spectroscopy, IR spectroscopy, Raman spectroscopy, and ESR spectroscopy. The two techniques that generate the most structural information, X-ray crystallography and NMR spectroscopy, are also among the slowest and most insensitive methods for the data collection. In the case of X-ray crystallography, the technique is wholly dependent upon obtaining suitable crystals, which is not a trivial undertaking. To date the best time resolution achieved with cryoenzymological methods[17] has been of the order of 0.5 h, although with the recent developments in Laue diffraction methods,[18–21] time-resolved kinetic studies of enzymatic reactions have been reported[22] with time resolutions of 1–3 s, and in theory it should be possible to carry out kinetics at time resolutions in the millisecond regime.[23] However, there are intrinsic limitations to the time-resolved Laue diffraction methods, since diffusion of the substrate results in a kinetically inhomogeneous population of enzyme-bound intermediate states and transient lattice disorder during substrate entry results in diffuse electron densities just at the time when the reaction takes place. These latter problems have been mitigated to some extent by employing caged substrate molecules, where a photo-releasable version of the substrate is co-crystallized with the enzyme, and photo-release of the substrate is synchronized with X-ray irradiation.[24,25] Even with this approach, the Laue method only provides somewhat diffuse electron density maps for the substrate or intermediate, to which it is difficult to fit a precise structure. This is not too surprising, since during catalysis it is to be expected that the enzyme-bound substrate or intermediate is dynamic and undergoing significant molecular motion.

Although NMR spectroscopy has been used to study rapid reactions in solution, in particular by continuous-flow and stopped-flow methods,[26,27] the time resolution is at very best ≈ 20 ms and in practice of the order of 200 ms to > 10 s. Laboratory introduced time-resolved solid-state NMR spectroscopy, in which the time resolution achievable is of the order of ≈ 2 ms.

An example of the application of time-resolved solid-state NMR spectroscopy is to a well-characterized enzyme, 5-enolpyruvyl-shikimate-3-phosphate (EPSP) synthase. We will consider a variety of applications of NMR experiments to this enzyme in greater detail in Section 8.4.3. In contrast to cryoenzymological solution-state NMR, with solid-state NMR spectroscopy,[28] as we saw in Section 1.2, line-broadening due to chemical-shift anisotropy and dipolar coupling can be reduced by CP–MAS, so that there is theoretically no known molecular weight limit, since the rotational correlation time governs only the nuclear relaxation properties under magic-angle spinning. The key, therefore, is the use of solid-state NMR methods in conjunction with rapid freeze-quench methods, which were developed[29] for ESR studies in the 1960s, so that short-lived enzyme intermediates can be trapped. For this method, while knowledge of the pre-steady-state kinetics for the enzyme can be helpful, it is not a requirement, since the method can be used without prior knowledge in much the same way that other physical methods used to measure pre-steady-state kinetics start with initial guesses at the reaction time. Indeed, time-resolved solid-state NMR seems to be emerging as an alternative strategy for carrying out pre-steady-state kinetic analyses which are superior to monitoring substrate consumption or product appearance by chromatographic techniques, since with these approaches there is always uncertainty as to the precise origin of the species being analysed. Time-resolved solid-state NMR spectroscopy essentially provides 'snapshots' of all the species, free and enzyme-bound, present at a particular time in the reaction. Furthermore, without the need for lengthy kinetic studies in cryosolvents, the method may be simpler and easier to use than cryoenzymological solution-state NMR. The disadvantages are that considerable quantities of protein are required—of the order of 100 mg per time point, although with modern overexpression systems (see Section 4.2.2), this does not need to be a limiting factor, and also a reaction with a k_{cat} of 10^{-5} s and a large activation enthalpy (e.g. -25 kJ mol^{-1}) will not be quenched even at $-75°$C. The latter limitation is not too severe, given the number of enzymatic reactions that do not fall into this category.

The rapid freeze-quench method involves rapidly mixing enzyme and substrate together and freezing by spraying the mixture directly into a secondary cryogen such as liquid propane cooled to ~ 85 K (see Fig. 6.3). Concerns about the fate of the protein under these conditions of freezing have been addressed,[30] and at the protein concentrations and freezing rates (10^5 K s^{-1}, or in the millisecond time regime) usually employed, there is significant dispersal of the solute in the frozen water.[31] Furthermore the frozen water is probably largely amorphous,[31] with the protein itself acting in a similar manner to a cryoprotectant,[32] thereby reducing the formation of hexagonal ice that is detrimental to the protein. For the example given in Fig. 6.4 the specific activity of EPSP synthase was assayed before and after rapid freezing and was found to be unchanged at the high protein concentrations that were used.

Fig. 6.4 shows time-resolved solid-state ^{13}C CP–MAS NMR spectra of EPSP synthase mixed with [^{13}C]-substrate under pre-steady-state conditions where t indicates the time elapsed from the start of the reaction. The intermediate (E·I) is clearly visible at 104 ppm and its build-up is demonstrated as the reaction proceeds. It is worth noting that the intensities of the E·I resonance correlate well with the concentrations observed by chemical quench methods.[33] On allowing the pre-steady-state reaction to proceed for a few minutes, the turnover of intermediate (E·I) to product (E·EPSP) is evident. In

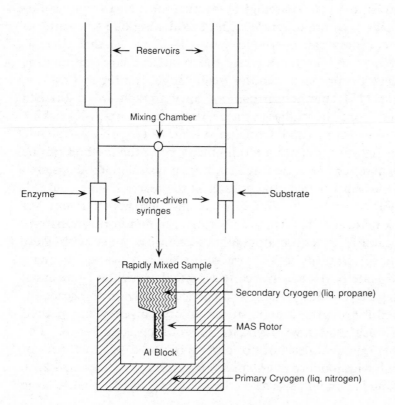

Fig. 6.3 The apparatus used for rapid freeze-quench of intermediates in time-resolved solid-state NMR.

addition to the resonance due to the E · I complex, the resonance due to the free product (EPSP) builds up at 155 ppm, and the free substrate (PEP) decays at 151 ppm.

Time-resolved solid-state NMR spectroscopy provides a major technological advance in the study of enzymatic reaction mechanisms. One important consideration in any attempt to detect transient intermediates in addition to their lifetimes is their pre-steady-state concentrations. This is dependent upon the kinetics and thermodynamic stability of intermediates of each individual enzyme. Some enzymes, such as EPSP synthase, have unusually stable intermediates (see Section 8.4.3). However, following the notions of Albery and Knowles,[34] we would expect that the majority of enzymes are evolving towards 'perfection', and stabilize highly unstable intermediates. For example, with triosephosphate isomerase (TIM) at equilibrium, the concentration of the postulated intermediate bound to the enzyme is approximately 5% of the total substrate concentration (with enzyme in excess). While the intermediate of TIM is probably beyond the limits of detection by time-resolved solid-state NMR spectroscopy, a labelled intermediate which is enzyme-bound (at concentrations of 1–4 mM) at greater than or equal to 10% of the protein concentration (with substrate in excess) can be detected using careful difference spectroscopy. While that excludes the 'perfect' enzymes, there remain a large number of enzymes for which this approach will still be very powerful. Furthermore, when coupled with the elegant solid-state NMR distance measurement methods that have been introduced recently (see Section 3.3), time-resolved solid-state NMR could be used to 'map out' molecular conformations of intermediates and enzyme-active site-intermediate

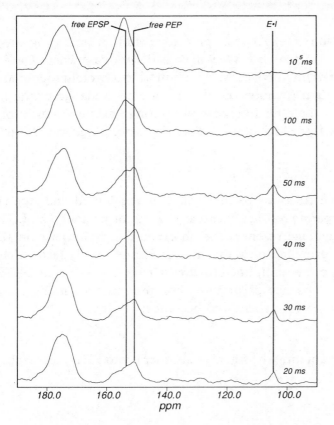

Fig. 6.4 Time-resolved solid-state NMR spectra of an enzyme (EPSP synthase) plus [^{13}C]-labelled substrate (PEP), showing the transient formation enzyme-intermediate (E · I) and the formation of product (EPSP) as a function of reaction time.

distances as an enzymatic reaction proceeds. This would provide the crucial missing structural details which Laue X-ray diffraction and allied techniques cannot provide, and enable the complete definition of the molecular events of enzyme catalysis.

6.5 THE TRANSFERRED NUCLEAR OVERHAUSER EFFECT

The transferred NOE or TRNOE[35,36] (which was considered briefly in Section 4.7.1) is a particular example of the NOE in the presence of exchange, for example between an enzyme-bound ligand and a ligand free in solution.[37-39] The aim of the TRNOE is to obtain conformational information on the bound ligands. The experiment relies on weakly binding ligands that remain on the enzyme for a relatively short time, so that NOEs between nuclei in the bound ligand are transferred to the more easily detected nuclei in the ligand free in solution (which should be present in excess) by virtue of chemical exchange between the free and bound states. In other words intramolecular cross-relaxation for the bound state will be retained in the free state provided the ligand dissociates fast enough. The negative NOEs arising from proximal nuclei in the protein-ligand complex with a correlation time in the spin diffusion limit (i.e. it works best for large proteins) are readily distinguished from the positive NOEs encountered in the ligand, which will possess a correlation time in the extreme narrowing limit.

In order to obtain quantitative distance information from TRNOEs, we need to express the NOE as $f_I(S)$, the fractional change in intensity of the resonance of proton I when proton S is saturated:

$$f_I(S) = \sigma_{IS}/\rho_I \tag{6.4}$$

where σ_{IS} is the cross-relaxation rate between protons I and S, and ρ_I is the total spin-lattice relaxation rate $(1/T_1)$ of proton I. We will consider the case where I and S are protons on a ligand that is exchanging between a state bound to a protein (denoted as B) and a second state free in solution (denoted F). If the rate of exchange of the ligand between the two states is greater than the difference in proton relaxation rates, protons I and S will each be characterized by relaxation rates that are the weighted mean of the two states. Therefore the NOE is:

$$f_I(S) = (X_F\sigma_{IS,F} + X_B\sigma_{IS,B})/(X_F\rho_{I,F} + X_B\rho_{I,B}) \tag{6.5}$$

where X_B and X_F are the mole fractions of the ligand in the bound and free states. Equation (6.5) is valid when separate resonances are observed for protons I and S in the two states, or when a single, average resonance is observed (i.e. in either slow or fast exchange—see Section 1.3.2), provided only that the exchange rate is faster than the relaxation rates. Equation (6.5) can be simplified further when $\sigma_{IS,F} = 0$, i.e. no NOEs are observed for the ligand in the absence of protein. For proteins, assuming isotropic tumbling so that $\sigma_{IS,B} \neq 0$ and $\rho_{I,B} \gg \rho_{I,F}$, then (6.5) becomes:

$$f_I(S) = \sigma_{IS,B}/\rho_{I,B} \tag{6.6}$$

If the effect on resonance I of irradiating the signal of another proton T is measured, $f_I(T)$ will be given by:

$$f_I(T) = \sigma_{IT,B}/\rho_{I,B} \tag{6.7}$$

Thus,

$$f_I(S)/f_I(T) = \sigma_{IS,B}/\sigma_{IT,B} = r_{IT}^6/r_{IS}^6 \tag{6.8}$$

where r_{ij} are internuclear distances. Equation (6.8) holds provided that the correlation times of the I-S and I-T vectors are the same, and that indirect cross-relaxation effects are negligible. Under these circumstances the TRNOE can provide quantitative information about the conformation of the ligand bound to a protein without direct observation of the signals of the bound species.

The principal of the simple one-dimensional TRNOE experiment is illustrated schematically in Fig. 6.5. An example of the use of the TRNOE is shown for NAD^+ building to yeast alcohol dehydrogenase ($M_r = 150$ kDa) in Fig. 6.6. On the basis of these measurements, the model for the conformation of NAD^+ when bound to the enzyme was proposed as shown in Fig. 6.7.

Experimentally TRNOEs can be obtained in either of the two ways that NOEs are measured. Nowadays the two-dimensional TR–NOESY, which is of course the same pulse sequence as the simple NOESY experiment (see Section 2.1.7), is generally employed.[40,41] There are a number of examples of the application of one-dimensional and two-dimensional TRNOE methods to quite complex enzyme-bound ligands such as oligopeptides.[42,43] Recently a complete relaxation matrix analysis (see Section 3.8) of transferred NOEs has been presented,[44] which will be extremely useful for the refinement of the structures of bound ligands.

6.6 STEREOCHEMISTRY OF ENZYMATIC REACTIONS

The stereochemistry of an enzymatic reaction provides crucial information as to its mechanism. Methods have been devised which employ labelling of substrates with stable

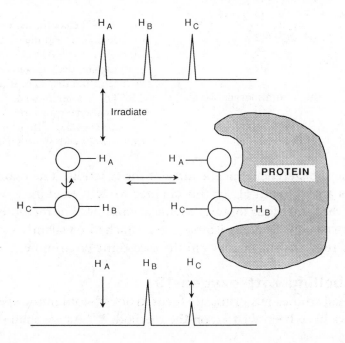

Fig. 6.5 Illustration of the TRNOE experiment, depicting a slowly exchanging ligand with three resolved resonances (H_A, H_B, and H_C). If H_A is saturated in the free state, and then the ligand binds and undergoes a conformation change such that H_A and H_C are now close to one another, an NOE will be observed (loss of intensity in the resonance due to H_C) which would be absent in the spectrum of the ligand on its own. (Redrawn from D. J. Craik, K. A. Higgins, M. M. Kneen, S. L. A. Munro, and K. J. Waterman, *J. Chem. Ed.*, **68**, 259–61 (1991) with permission.)

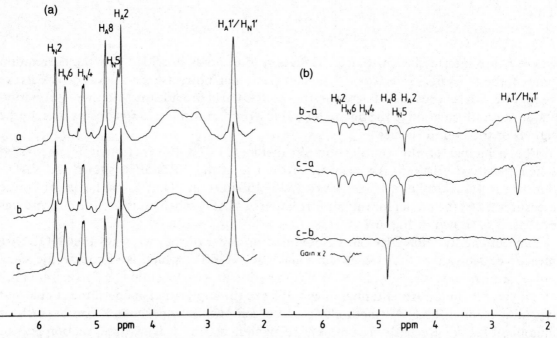

Fig. 6.6 The 270 MHz ^{1}H NMR spectrum of NAD$^+$ in the presence of yeast alcohol dehydrogenase (A) spectrum with (a) no irradiation; (b) irradiation for 0.4 s at a control frequency (in the protein envelope); and (c) irradiation of the H_A1'/H_N1' resonance at 1.00 ppm for 0.4 s. (B) Difference spectra of (b)–(a); (c)–(a); and (c)–(b). (Reprinted from ref. 39 with permission.)

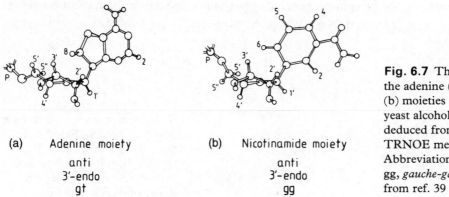

(a)   Adenine moiety

anti

3'-endo

gt

(b)   Nicotinamide moiety

anti

3'-endo

gg

Fig. 6.7 The conformation of the adenine (a) and nicotinamide (b) moieties of NAD$^+$ bound to yeast alcohol dehydrogenase deduced from distances from TRNOE measurements. Abbreviations: gt, *gauche-trans*; gg, *gauche-gauche*. (Reprinted from ref. 39 with permission.)

isotopes in order to determine the stereochemical course of the reaction. The two most frequently employed isotopes are ^{18}O and ^{2}H. In this chapter we will examine selected examples of the application of these methods to stereochemical analysis. The methods rely on the property that nuclei to which different isotopes are attached covalently exhibit different chemical shifts, as a result of the changes in the electronic environment.

6.6.1 Using isotopic labelling with oxygen-18

The use of ^{18}O in stereochemical studies has primarily focused on ^{31}P-containing species. A number of excellent reviews have been written on the method.[45–49] As we shall see in

Chapter 7, the study of phosphoryl transfer enzymes has relied to a large extent on the examination of ^{18}O (and ^{17}O) substitutions during the enzymatic reaction. Fig. 6.8 shows the ^{31}P NMR spectrum of inorganic phosphate in which ^{18}O has been substituted for ^{16}O. Each resonance represents the ^{18}O and ^{16}O composition of the species present in solution at an enrichment of 78% ^{18}O. Clearly, the ^{31}P chemical shifts are sensitive to the attached ^{18}O, and this can be used to probe the transfer of oxygen atoms or labelled phosphate groups between substrates and products in an enzyme-catalysed reaction.

The magnitude of the ^{18}O-induced isotope shift in a phosphoryl group is approximately 0.02 ppm downfield per ^{18}O, and a number of phosphate-containing species have been studied as listed in Table 6.2. It is clear that the magnitude of the shift correlates with the bond order of the phosphorus–oxygen bond, with a larger shift for a P=O double bond than for the P—O single bond. The effect arises because the presence of a heavier isotope such as ^{18}O results in a perturbation of the zero-point vibrational levels, which causes an electrostatic change in shielding at phosphorus. Essentially the absolute shielding at the observed nucleus is diminished by the electrical fields from the bonded atom. These fields arise from oscillations of the bonded atoms in the applied magnetic field. When a heavier isotope is substituted for a lighter one, its average distance to the observed nucleus is slightly shorter, and its vibratory amplitude is smaller. This gives rise to smaller electric fields, which in turn do not deshield the observed nucleus as strongly as the lighter isotope. This explanation is also consistent with the observation of the dependence of isotope shift on bond order.

The chemical-shift differences for ^{16}O- and ^{18}O-containing phosphate species have been used in a number of biological systems to determine the stereochemical course of an enzymatic reaction. This has been achieved using chiral phosphate-containing compounds and subsequent analysis by ^{31}P NMR. The chirality has been introduced using thiophosphates[50] in which one ^{16}O has been replaced with ^{18}O, or, alternatively, with phosphates containing[51,51] ^{16}O, ^{17}O, and ^{18}O. These have been used to evaluate the stereochemical course of a variety of reactions involving phosphate group transfers, for example by replacement of the β-phosphate of nucleotide triphosphates with a chiral phosphate moiety. Since in nucleotide triphosphates the ribose ring is also chiral, the resulting diastereomers can often be resolved by ^{31}P NMR. Fig. 6.9 shows a comparison

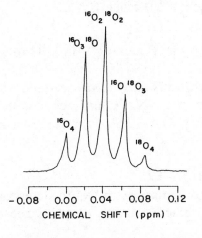

Fig. 6.8 145.7 MHz ^{31}P NMR spectrum of randomly ^{18}O-labelled inorganic phosphate. (Reprinted from M. Cohn and A. Hu, *Proc. Natl. Acad. Sci. USA*, 75, 200–3 (1978) with permission.)

Table 6.2 Chemical shifts of ^{31}P NMR resonances due to ^{18}O in various phosphates[a]

Compound	$\Delta\delta/^{18}$O (ppm)
$HP^{18}O_4^{2-}$	0.021
$(CH_3O)_3P{=}^{18}O$	0.036
$(CH_3O)_2P^{18}O_2{-}$	0.029
$(CH_3O)P^{18}O_3^{2-}$	0.023

$$AMP{-}^{18}O{-}\overset{\overset{\displaystyle O}{|}}{\underset{\underset{\displaystyle {}^{18}O}{|}}{P}}{-}O \qquad 0.0205$$

$$AMP{-}O{-}\overset{\overset{\displaystyle {}^{18}O}{|}}{\underset{\underset{\displaystyle {}^{18}O}{|}}{P}}{-}O \qquad 0.0227$$

$$ADP{-}^{18}O{-}\overset{\overset{\displaystyle O}{|}}{\underset{\underset{\displaystyle {}^{18}O}{|}}{P}}{-}O \qquad 0.0206$$

$$ADP{-}O{-}\overset{\overset{\displaystyle {}^{18}O}{|}}{\underset{\underset{\displaystyle {}^{18}O}{|}}{P}}{-}O \qquad 0.0225$$

$$AMP{-}^{18}O{-}\overset{\overset{\displaystyle {}^{18}O}{|}}{\underset{\underset{\displaystyle {}^{18}O}{|}}{P}}{-}^{18}O$$

0.0166 (a,β bridge);
0.0215 (non-bridge)

$$AMP{-}^{18}O{-}\overset{\overset{\displaystyle {}^{18}O}{|}}{\underset{\underset{\displaystyle {}^{18}O}{|}}{P}}{-}O{-}\overset{\overset{\displaystyle O}{|}}{\underset{\underset{\displaystyle O}{|}}{P}}{-}O$$

0.0172 (a,β bridge, a-P)
0.0281 (non-bridge)

[a]Data from J. Villafranca [*Methods in Enzymology*, **177**, 392 (1989)].

of the two diastereomeric forms of chiral labelled ATP containing S and ^{18}O to varying extent.

Oxygen-18 has not just been used with ^{31}P. Use with ^{15}N and ^{13}C has been reported and reviewed.[53] For example, the ^{15}N NMR spectrum of the [^{15}N, ^{18}O]nitrate ion is shown in Fig. 6.10. Similarly the time-dependent exchange of β-D-[1-^{13}C]erythrofuranose in 60% $H_2^{18}O$ is shown in Fig. 6.11.

6.6.2 Using isotopic labelling with oxygen-17

This nucleus has been used in an exactly analogous manner to ^{18}O to determine the stereochemical course of a variety of biochemical reactions. The disadvantage of this

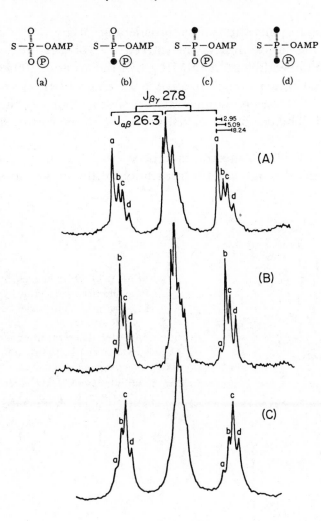

Fig. 6.9 P^β signals (29.8 ppm downfield from H_3PO_4) of the 145.7 MHz ^{31}P NMR spectra of the (R_p)-ATPβS obtained from $[^{18}O_3]P_{si}$ (a) and from two chiral P_{si} (b, c). The coupling constants and isotope shifts are expressed in Hertz. (Reprinted from M. -D. Tsai, *Biochemistry*, **19**, 5310–16 (1980) with permission.)

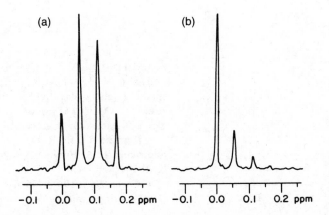

Fig. 6.10 ^{15}N NMR spectra of the $[^{15}N,^{18}O]$nitrate ion: (a) 50% enrichment with ^{18}O; and (b) 15% enrichment with ^{18}O. (Reprinted from ref. 53 with permission.)

nucleus is that, since it is quadrupolar, it causes broadening of the ^{31}P through quadrupolar relaxation. This has also limited the direct use of the nucleus in stereochemical studies, since the linewidths in ^{17}O NMR are of the order of 100 Hz and the sensitivity is 2.91×10^{-2} (at 100% enrichment) relative to ^{1}H. The sensitivity problem is in part mitigated by the fact that ^{17}O relaxation times are very short (<5 ms) so that very short recycle times may be used. The use of ^{17}O NMR gives rise to the sort of spectra shown in Fig. 6.12.

^{17}O isotope-induced shifts in ^{31}P NMR spectra have been used[54] to investigate the stereochemical course of reactions, through careful integration of the intensities of the

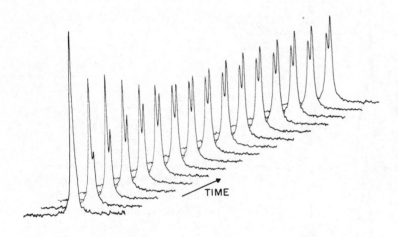

Fig. 6.11 Oxygen exchange at the anomeric carbon in D-erythrose. The series of ^{13}C NMR spectra of β-D-[1-^{13}C]erythrofuranose anomer of D-erythrose during the oxygen exchange in unbuffered 60% $H_2{}^{18}$O. (Reprinted from ref. 53 with permission.)

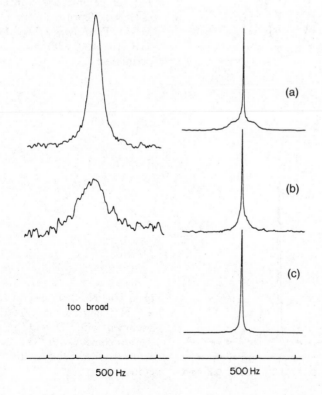

Fig. 6.12 ^{17}O NMR spectra (left) and ^{31}P (^{17}O) NM,R spectra (right) of $H_3P^{17}O^{16}O_3$ (50 atom % ^{17}O) in D_2O (a), or 1:1 glycerol/water (b), or glycerol (c). (Reprinted from R. D. Sammons, P. A. Frey, K. Bruzik, and M. -D. Tsai, *J. Am. Chem. Soc.*, **105**, 5455–61 (1983) with permission.)

various lines that arise in the spectrum. Fig. 6.13 shows data for ATPαS derived from the reaction of acetyl-CoA synthetase which catalyses the reaction shown in Fig. 6.14. The ^{31}P resonances broaden in the presence of ^{17}O, and the consequential loss of intensity is monitored through analysis of the integrated intensities as opposed to measuring changes in peak heights, a parameter far more subject to error. The enrichment of the ^{17}O-acetate was determined to be 19%, and in the synthetic material (Fig. 6.13(b)) the P_α signal decreased to $67 \pm 1\%$ and the P_β signal to $83 \pm 4\%$. The ATPαS from the enzymatic reaction (Fig. 6.13(c)) showed a P_α signal decrease to $80 \pm 4\%$ and a P_β signal decrease to $82 \pm 5\%$, and since both P_α and P_β decreased, this implies that ^{17}O must be located at the bridging position and therefore the reaction proceeds with the inversion of configuration.

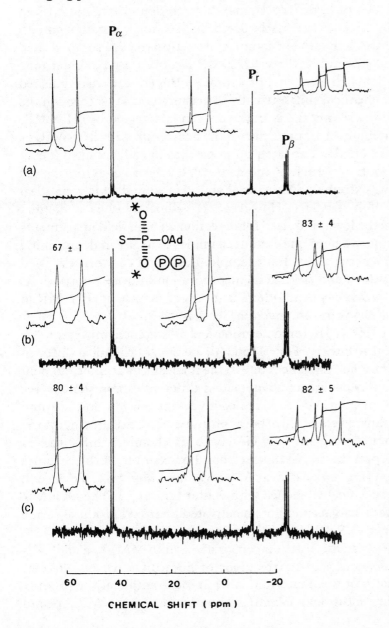

Fig. 6.13 ^{31}P NMR spectra (at 32.2 MHz) showing the results of acetyl-CoA synthetase. (a) Unlabelled (S_p)-ATPαS. (b) Synthetic (S_p)-[α-^{17}O, αβ-^{17}O]ATPαS. (c) (S_p)-ATPαS from [^{17}O$_2$]acetate. The insets represent the integrations of the corresponding signals. (Reprinted from ref. 54 with permission.)

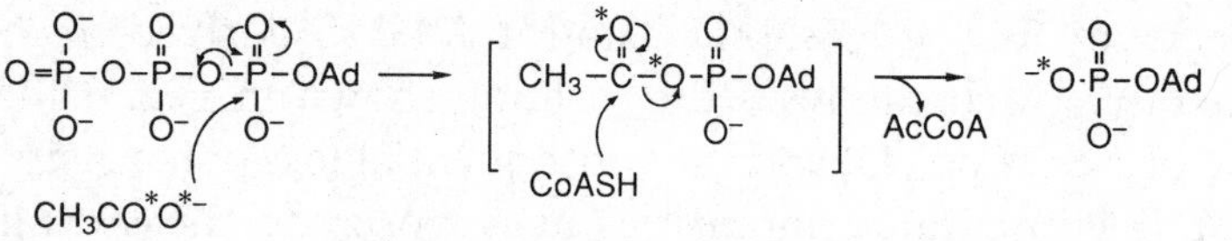

$$O=\overset{O^-}{\underset{O^-}{P}}-O-\overset{O}{\underset{O^-}{P}}-O-\overset{O}{\underset{O^-}{P}}-OAd \longrightarrow \left[CH_3-\overset{*O}{C}-*O-\overset{O}{\underset{O^-}{P}}-OAd \right] \xrightarrow[AcCoA]{} \quad *O-\overset{O}{\underset{O^-}{P}}-OAd$$

$$CH_3CO^*O^{*-} \qquad\qquad CoASH$$

Fig. 6.14 The mechanism of acetyl-CoA synthase, showing the fate of the oxygens (marked with '*').

6.6.3 Using isotopic labelling with deuterium

Of course the change in stereochemistry accompanying a biological transformation at phosphorus represents an important, and yet relatively small, proportion of all the biochemical reactions in living cells. Changes in stereochemistry at carbon have been evaluated through tracing the fate of hydrogen atoms, through deuterium and tritium labelling. Such approaches have been particularly useful in secondary metabolism,[55,56] where the number of chiral centres in relatively complex molecules require sophisticated methods for stereochemical analysis. While direct ^{1}H NMR has been used to determine the stereochemical fate of particular hydrogens, prochiral hydrogens are distinguished only by virtue of their different chemical shifts (usually in diastereomers), or through the use of deuterium labelling, in which only the chiral hydrogen is detected by ^{1}H NMR. Alternatives to this approach include direct detection of the deuterium with ^{2}H NMR, or direct detection of tritium by ^{3}H NMR. The latter approach has the advantage of being the most sensitive nucleus (spin$=\frac{1}{2}$, nuclear moment$=2.96$, relative gyromagnetic ratio$= 1.067$), and in view of its extremely low natural abundance might make an ideal nucleus for tracing the fate of hydrogens. However, its radioactivity is a distinct disadvantage, particularly since the levels required for detection on high field instruments can be in the mCi range. There have been some elegant examples of its use in determining the stereochemistry of biochemical reactions. For example, Crout and co-workers[57] used chirally labelled $[4\text{-}^2H_1, 4\text{-}^3H_1]$valine and studied its incorporation into cephalosporin C in whole cells of *Cephalosporium acremonium*. Other examples of the use of ^{3}H NMR in studying enzymatic reaction mechanisms are given in Section 8.4.2.

The disadvantages of direct ^{2}H NMR for determination of stereochemistry are its inherently low sensitivity relative to either ^{1}H or ^{3}H, poor dispersion of chemical shifts (in Hz, about 15% that of ^{1}H), and broad resonances due to the nuclear quadrupole moment of ^{2}H. Of course some of these aspects are an advantage in different circumstances (see Chapter 10). It is relatively inexpensive, non-radioactive, and has a low natural abundance (0.016%) allowing low levels of incorporation to be detected. Furthermore, since ^{2}H chemical shifts in ppm correspond very closely to ^{1}H chemical shifts, the ^{1}H assignments can be used to interpret the ^{2}H NMR spectrum. For example, in their studies on the formation of the β-lactam ring in nocardicin A, by *Nocardia uniformis*, Townsend and Brown[58] used serine stereospecifically labelled with deuterium at C-3. The ^{2}H NMR results suggest that there is direct displacement of an activated hydroxyl group of serine by the nitrogen atom. If either $(2S,3R)$-$[2,3\text{-}^2H_2]$serine (see Fig. 6.15) or $(2S,3R)$-$[2,3\text{-}^2H_2]$-2,3-diaminopropanoic acid was incorporated into N^2-L-alanyl-N^3-fumaramoyl-L-2,3-diaminopropanoic acid (a precursor of nocardicin A), by *Streptomyces collinus*, the stereochemistry of the deuterium label in each sample of the product was identical. With help from additional labelling experiments, ^{2}H NMR spectra demonstrated that the conversion of serine into 2,3-diaminopropanoic acid proceeds through retention of configuration at the β-carbon. Such an enzymatic reaction is likely to be catalysed by a pyridoxal-phosphate-dependent enzyme.

An alternative to the direct detection of deuterium is indirect detection by ^{1}H NMR, in which the absence of a proton is followed. For example, in work on the mechanism of isopenicillin N synthase, Baldwin and co-workers[59] examined the product from incubation of two diastereomeric peptides derived from (2R,3R)- and (2R,3S)-2-amino[3-^{2}H$_1$]butyrate when incubated separately with the enzyme (see Fig. 6.16). The ^{1}H NMR spectra of the product and its derivatives (derived from chemical degradation) showed that the identical pencillin analogue had been formed from each of the precursors. This was interpreted in terms of a radical mechanism, implying that there is rotation about the C^α—C^β bond of the valine moiety, for which there is a substantial isotope effect.

By far the most popular method for tracing the fate of hydrogens has been the indirect detection of ^{2}H by ^{13}C NMR. The advantage of ^{13}C NMR is the large dispersion of chemical shifts that help in situations where ^{1}H or ^{2}H spectra are very crowded, and, furthermore, double labelling of ^{13}C—^{2}H bonds can be used to establish unambiguously the stereochemistry of an enzymatic reaction. For directly attached deuterium atoms, the so-called α-isotope shift (upfield by 0.3–0.6 ppm) and quadrupolar splitting of the ^{13}C resonance is highly diagnostic for the integrity of the ^{13}C—^{2}H bond. The disadvantage of examining α-isotope shifted ^{13}C resonances is that even when broadband ^{1}H-decoupled, the ^{13}C NMR spectrum of a carbon to which one deuterium is attached appears as a triplet of lines of equal intensity, a CD$_2$ group as a quintet (1:2:3:2:1), and a CD$_3$ group as a septet (1:3:6:7:6:3:1). This multiplicity is compounded further by quadrupolar broadening as well as a loss of the {^{1}H}^{13}C NOE. As a result, ^{2}H-coupled, ^{1}H-decoupled ^{13}C NMR spectra can be very complex and difficult to interpret when the sample is derived biosynthetically and consists of a mixture of different labelled species. Note again, however, that while these are disadvantages for stereochemical analyses, they can be advantages in studying enzyme-substrate complexes (see Chapter 8). The use of a triple-resonance probe for simultaneous decoupling of ^{1}H and ^{2}H (with an ^{19}F lock) greatly facilitates this type of experiment.

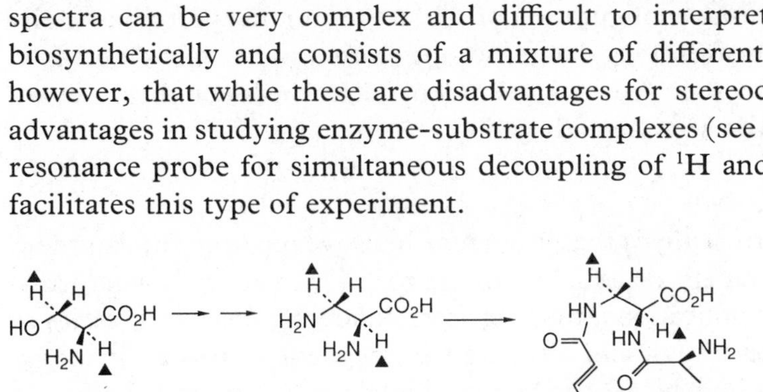

Fig. 6.15 The biosynthesis of a nocardicin A precursor, N^2-L-alanyl-N^3-fumaramoyl-L-2,3-diaminopropanoic acid, from serine with ▲ = ^{2}H.

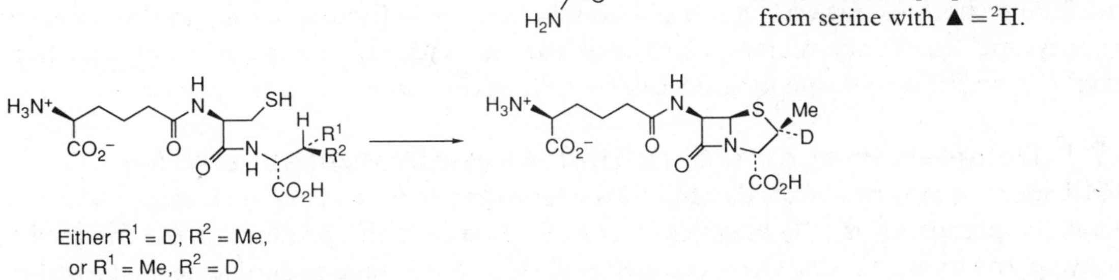

Either R^1 = D, R^2 = Me,
or R^1 = Me, R^2 = D

Fig. 6.16 The reaction catalysed by isopenicillin N synthase from two diastereomeric peptide precursors.

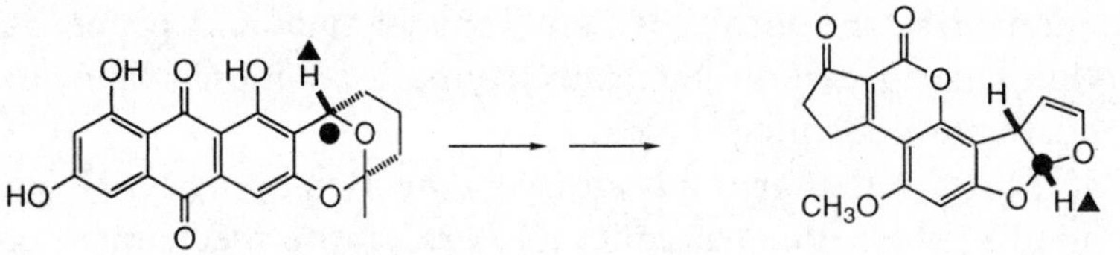

Fig. 6.17 The biosynthesis of aflatoxin, B$_1$ from averufin ($\blacktriangle = {}^2H$, $\bullet = {}^{13}C$).

An example of the use of ^2H α-isotope shifted ^{13}C NMR spectroscopy comes from the work of Townsend and co-workers,[60,61] who demonstrated that doubly labelled [1′–^{13}C, 1′–^2H]averufin ($\blacktriangle = {}^2H$, $\bullet = {}^{13}C$) was converted into [13–^{13}C, 13–^2H]aflatoxin B$_1$ by *Aspergillus parasiticus* without any deuterium being lost (see Fig. 6.17) by observing the α-shifted species in the ^{13}C NMR spectrum.

A simple alternative to the use of ^1H, ^2H-decoupled ^{13}C triple resonance NMR spectroscopy was reported by Abell and Staunton,[62] who demonstrated that the retention of deuterium in 6-methylsalicylic acid that had been derived from sodium [1–^{13}C,$^2H_3$]acetate (see Scheme 4.4) could be detected by the changes that were induced in the chemical shifts of ^{13}C atoms that were two bonds away from deuterium (β-isotope shift). Problems of line broadening and reduced NOE are largely avoided, and the ^2H—^{13}C coupling over two bonds is negligible (<1 Hz). Furthermore this method permits the detection of two adjacent bonds which have remained intact during biosynthesis, analogous to the use of ^{13}C—^{13}C double labelling in biosynthetic studies.[63,64] Since isotope shifts are additive, the ^{13}C resonances due to the various isotopic species can be resolved. The problem, however, is that β-isotope shifts are most frequently upfield (≈ 0.01–0.1 ppm per deuterium), but they can also be downfield to zero. This is especially likely for carbonyl groups. Also, the magnitude of both β-isotope shifts and γ- (three bond) isotope shifts depends on the configuration of the deuterium in conformationally constrained systems. In special cases, long-range isotope shifts can become detectable, especially if the ^{13}C and ^2H are close in space. Thus independent checks on the observed β-isotope shifts by, say, ^2H NMR are often advisable.

6.7 ENZYMES *IN VIVO*

A unique feature of NMR spectroscopy is that it is a non-invasive method: the sample is returned at the end of the experiment essentially unchanged. This feature is what lends NMR spectroscopy to *in vivo* applications, ranging from following enzyme kinetics *in vivo*, to delineating metabolic pathways, and ultimately to magnetic resonance imaging. In this section we shall consider how NMR can be used to delineate metabolic pathways and follow the biosynthesis of simple metabolites (for discussions on more complex primary and secondary metabolism the reader is referred to a number of excellent texts[65–67]), and how enzyme kinetics can be followed *in vivo*.

6.7.1 Delineation of metabolic pathways—biosynthetic NMR

NMR spectroscopy can be used to follow the fate of a particular isotopic label present in a metabolic precursor as it is chemically transformed into the product of the metabolic pathway. In circumstances where the metabolic intermediates accumulate to a reasonable degree (mM concentrations), then novel metabolites can be identified and structurally characterized. Also the flux of metabolites in the pathway can be determined. The isotopic labels commonly employed in biosynthetic studies include ^1H, ^3H, ^{13}C, ^{15}N, ^{19}F, and ^{31}P. As we considered in Section 6.6, when coupled with ^2H, ^{17}O, or ^{18}O, the stereochemistry of a particular biosynthetic pathway can also be investigated.

Let us consider a simple example of a novel metabolic pathway which was delineated in the archaebacterium *Methanobacterium thermoautotrophicum*. These organisms produce methane by reduction of carbon dioxide with hydrogen, and grow at elevated temperatures. It was found by ^{31}P NMR that these bacteria accumulate a novel metabolite from *in vivo* ^{31}P NMR studies of the whole cells[68] (see Fig. 6.18). The resonances for this novel metabolite, which accumulate in the cell at concentrations up to 10 mM, were consistent with 2,3-cyclopyrophosphoglycerate (CPP), whose structure is also shown in Fig. 6.18. The biosynthesis of CPP was shown to have a role in carbohydrate metabolism[69] from labelling studies with $^{13}CO_2$, [^{13}C]acetate, [^{13}C]pyruvate, and [^{13}C]glucose, all of which were incorporated into CPP specifically. The catabolism of CPP was also shown to involve incorporation into cell wall carbohydrate.[70] However the integrity of labelling of a C_2-unit can be followed using $^{13}C_2$ precursors.[71] Fig. 6.19 shows the result when CPP is labelled with [$^{13}C_2$]-acetate and pyruvate, and in this case is examined by ^{13}C NMR after ethanol extraction from whole cells. Interestingly, although the resonances display $^1J_{CC}$ coupling (and a smaller, $^2J_{CP}$ coupling), the resonance assigned to C3 shows additional intensity in the centre of the doublet. This arises from the presence of ^{12}C at C2, which implies that the carbon–carbon bond of acetate or pyruvate is cleaved to some extent prior to incorporation into CPP. This is because the organism possesses an asymmetric CO_2 fixation pathway, as outlined in Fig. 6.20.

When the organism is fed unlabelled CO_2 and the double labelled precursor, the [$^{13}C_2$]acetate, for instance, can be cleaved to a methyl group and an enzyme-bound carbon monoxide, the latter being in oxidative exchange with unlabelled CO_2. In this way, the ^{13}C label at the carboxyl of acetate (or carbonyl of pyruvate) can be partially lost. Since C2 of CPP arises from the carboxyl of acetate, then ^{13}C labelling of C2 of CPP is

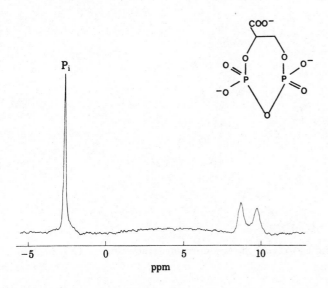

Fig. 6.18 109.3 MHz ^{31}P NMR spectrum of *Mb.thermo-autotrophicum* cells in a growth media in the NMR tube. The two resonances centred at 9.3 ppm arise from a novel cyclic pyrophosphate metabolite, 2,3-cyclopyrophosphoglycerate (CPP, whose structure is shown). (Reprinted from ref. 68 with permission.)

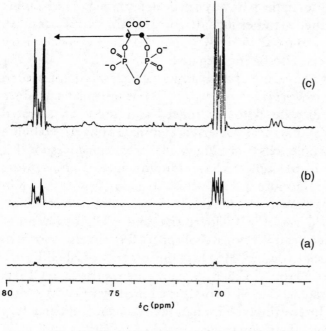

Fig. 6.19 ^{1}H-decoupled 100.6 MHz ^{13}C NMR spectra of ethanol extracts of *Mb. thermoautotrophicum* fed for six hours with (a) $^{12}CO_2/H_2$; (b) $^{12}CO_2/H_2$ plus [1,2-$^{13}C_2$]acetate; (c) $^{12}CO_2/H_2$ plus [2,3-$^{13}C_2$]-pyruvate. Spectral intensities in (a) and (b) are normalized to those in (C).

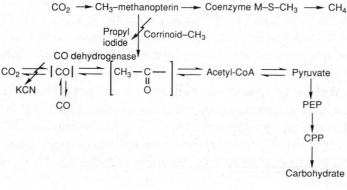

Fig. 6.20 The pathway by which CPP is asymmetrically labelled from CO_2. Note that KCN appears to inhibit the conversion of CO_2 to carbon monoxide dehydrogenase-bound CO, and propyl iodide inhibits the corrinoid methyl group transfer.

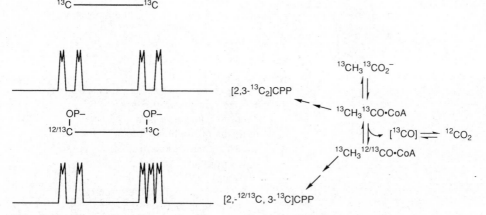

Fig. 6.21 Illustration of how the $^1J_{CC}$ coupling can reveal the presence of an asymmetric CO_2 fixation pathway in methanogens.

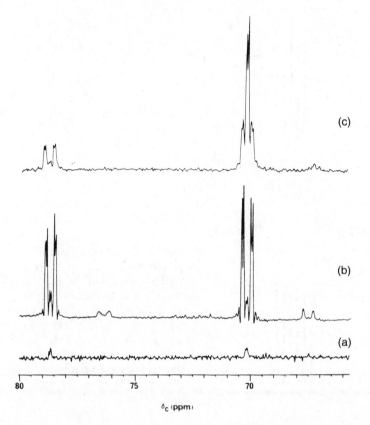

Fig. 6.22 [1]H-decoupled 100.6 MHz ^{13}C NMR spectra of ethanol extracts of *Mb. thermoautotrophicum* fed for six hours with [12]CO$_2$/H$_2$ plus [1,2-[13]C$_2$]-acetate, and (a) no additions (control); (b) cyanide; and (c) propyl iodide.

correspondingly reduced, giving rise to a lack of a spin–spin coupling partner for C3, and resulting in the presence of a ^{13}C uncoupled resonance for C3. This is shown schematically in Fig. 6.21. This situation can be altered, since cyanide is known to inhibit carbon monoxide dehydrogenase, the enzyme responsible for the formation of carboxyl of acetyl CoA, and the assembly of the molecule. Also, propyl iodide is known to inhibit methanogenesis, through alkylation of the methyl corrinoid transferase that transfers the methyl group formed from reduction of CO$_2$ to the active site of CO dehydrogenase for incorporation into the methyl of acetyl CoA. The effect of these inhibitors of this part of the pathway on the incorporation of [[13]C$_2$] acetate into CPP is illustrated in Fig. 6.22. Clearly, cyanide abolishes the exchange of the [13]C=O of acetate with unlabelled [12]C=O. Furthermore, propyl iodide actually promotes the exchange, presumably because of kinetic sequestering of the methyl group and carboxyl of acetyl CoA at the active site of CO dehydrogenase, prior to carbon–carbon bond formation.

6.7.2 Measuring metabolic fluxes and enzyme kinetics *in vivo*

The flux of metabolites can be probed by NMR. This can be achieved, for example, by following the fate of an isotopic label in a metabolite as a function of the number of turns of a particular cycle. An example[72] is shown in Fig. 6.23, which shows the ^{13}C resonance for succinate present in a cell-free extract of *E.coli* fed with [[13]C$_6$]glucose under anaerobic conditions. The complex pattern of resonances, simplified using one-dimensional double-quantum coherence spectroscopy, can be analysed by simulating the observed

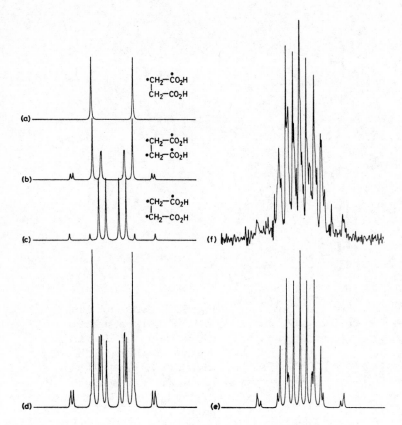

Fig. 6.23 Simulated coupling pattern for the C2 resonance of enriched succinate. (a) [1,2-^{13}C]; (b) [U-^{13}C]; (c) [1,2,2'-^{13}C]; (d) 1:1:1 addition of (a), (b) and (c); (e) 1:5:25 addition of (a), (b), and (c) respectively, with addition of an uncoupled centre line; (f) observed multiplet. (Reprinted from ref. 72 with permission.)

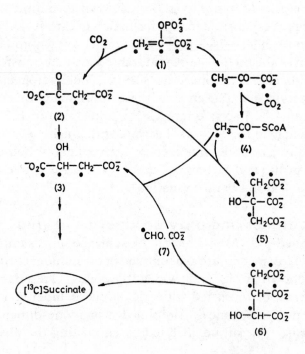

Fig. 6.24 The formation of ^{13}C-enriched succinate from [U-^{13}C]phosphoenolpyruvate (derived from D-[U-^{13}C]-glucose). (Reprinted from ref. 72 with permission.)

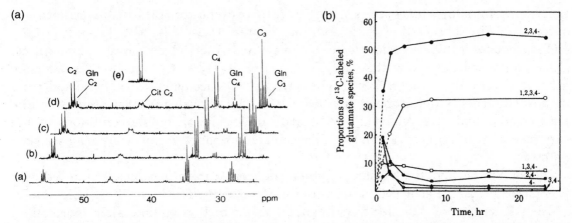

Fig. 6.25 (a) ^{13}C NMR spectra of cell-free perchloric acid lysates of *S. cerevisiae* after incubation with [2-^{13}C]acetate. Spectra (a)–(d) refer to lysates made after 1, 2, 4, and 8 hours of sporulation. Spectrum (e) (*inset*) shows the multiplet corresponding to citrate C2 obtained for the 8 hr sample after addition of EDTA. C_2-C_4 refer to glutamate C2, C3, and C4; Gln C_2-C_4 refer to glutamine C2, C3 and C4; Cit C_2 refers to the C2 of citrate. (b) Proportions of ^{13}C-labelled glutamate species in cell lysates as a function of time. (Reprinted from ref. 73 with permission.)

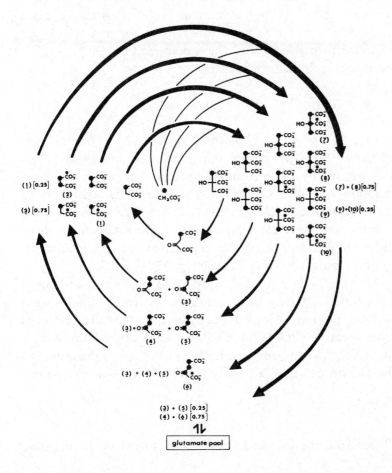

Fig. 6.26 ^{13}C isotopomer distribution for incorporation of [2-^{13}C]acetate on sequential turns of the tricarboxylic acid cycle. Independent operation of the glyoxalate cycle sequentially gives rise to [2-^{13}C]citrate, [2,4-^{13}C$_2$]- and [2,3-^{13}C$_2$]citrate, and [2,3,4-^{13}C$_3$]citrate exclusively. (Reprinted from ref. 73 with permission.)

spectrum using addition of spectra arising from three of the possible labelling patterns for succinate. The observed 1:5:25 addition of the [1,2–^{13}C]-, [^{13}C$_4$]- and [1,2,2'–^{13}C]-succinates provides clues as to the ways in which glycolysis is taking place. Quantitative analysis shows that [1,2,2'–^{13}C]-, [^{13}C$_4$]-, [1,2–^{13}C)- and [2–^{13}C)-succinate are present in the ratio 8:3:2:1. The [1,2,2'–^{13}C]-succinate probably arises from carboxylation of [^{13}C$_3$]phosphoenolpyruvate to [1,2,3–^{13}C]oxaloacetate and reversal of the tricarboxylic acid cycle via [1,2,3–^{13}C]malate under the anaerobic conditions of the experiment (see Fig. 6.24). The [^{13}C$_4$]succinate probably forms via the glyoxalate pathway.

In a similar fashion, the labelling of glutamate from *Saccharomyces cerevisiae* fed with [2–^{13}C]acetate can be analysed in terms of isotopomer distributions.[73] Fig. 6.25A shows ^{13}C NMR spectra of perchloric acid extracts as a function of time after sporulation. Fig. 6.25B shows the kinetics of isotopomer formation, which arise from successive turns of the tricarboxylic acid cycle as shown in Fig. 6.26, resulting in [^{13}C]oxaloacetate labelled to different degrees, which is transaminated to give [^{13}C]glutamate. First-order analytical expressions[74] describing the distribution of ^{13}C intensity within the multiplets can be solved to give a quantitative estimate of the fluxes. Fluxes are defined unambiguously by the analytical expression, according to the accuracy of the model used. Because of line broadening *in vivo*, such analyses are best performed on cell-free extracts. For example experiments on perfused liver[75] fed with [2–^{13}C]pyruvate and [1,2–^{13}C]ethanol lead to labelling of C4 of glucose, which occurs as a doublet and a singlet superimposed (see Fig. 6.27), and can be used to infer information about the metabolic flux.

The intensity ratio for C4 doublet:singlet can be written as:

$$\frac{I_{\text{doublet}}}{I_{\text{singlet}}} = y/[y(1-z)+(1-y)+(1-y)(1-z)] \tag{6.8}$$

where $y = f([1,2–^{13}\text{C}]\text{acetyl-CoA})$, $(1-y) = f([1–^{13}\text{C}]\text{acetyl-CoA})$, $(1-z) = f(\text{methylene}\,^{13}\text{C}$ in oxaloacetate derived from [2–^{13}C]pyruvate), and f denotes probability. The formation of [1–^{13}C]acetyl-CoA occurs by the action of pyruvate dehydrogenase on [2–^{13}C]pyruvate, and the formation of [1,2–^{13}C]acetyl-CoA occurs via oxidation of [1,2–^{13}C$_2$]ethanol. In this way a set of simultaneous equations can be generated and solved in general form. When the measured intensity ratios are substituted into these expressions, relative fluxes under the conditions of the experiment can be estimated. Similar types of experiment can be employed using ^{15}N/^{14}N or ^{1}H/^{2}H labelling.

Alternatively, saturation transfer methods can be used to measure fluxes, and more generally, kinetics through a particular enzyme in a metabolic pathway. There are a number of excellent reviews of this approach.[76] Consider the example shown in Fig. 6.28 of the interconversion of phosphocreatine (PCr) and ATP in the reaction catalysed by creatine kinase (also see Section 7.2.3), with subsequent hydrolysis of the γ-phosphate of ATP to yield inorganic phosphate. This can be analysed in terms of three-site exchange:

$$\text{PCr} \underset{k_{\text{rev}}}{\overset{k_{\text{for}}}{\rightleftharpoons}} \gamma\text{-ATP} \underset{k_{\text{syn}}}{\overset{k_{\text{hyd}}}{\rightleftharpoons}} \text{P}_{\text{i}}$$

where $k_{\text{for}}[\text{PCr}]$ and $k_{\text{rev}}[\text{ATP}]$ describe the forward and reverse fluxes in the creatine

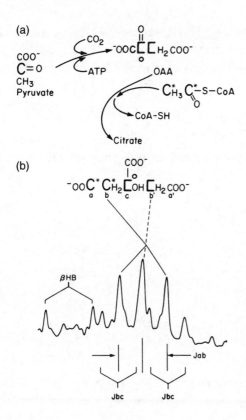

Fig. 6.27 (a) Simplified pathway of incorporation of [2-¹³C]pyruvate and [1,2-¹³C]-ethanol into citrate. The main flux of pyruvate into the tricarboxylic acid cycle is via pyruvate carboxylase. The ¹³C label is randomized in mitochondrial oxaloacetate. The small open circle follows the original label at pyruvate C2. The main flux into the mitochondrial acetyl-CoA pool is via the oxidation of [1,2-¹³C]ethanol (boldface Cs with '*'). (b) The citrate methylene region of the ¹³C spectrum of the perchloric acid extract of freeze-clamped rat liver fed with [2-¹³C]pyruvate and [1,2-¹³C]ethanol under steady-state conditions. (Reprinted from ref. 75 with permission.)

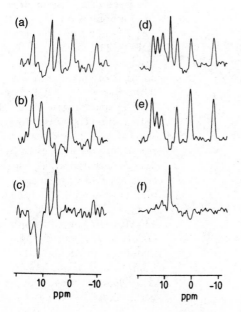

Fig. 6.28 Saturation transfer measurements of creatine kinase fluxes in the human brain. ³¹P spectra were acquired at 32.7 MHz using a coil which surrounded the entire head. In (b) the ATP-γ-phosphate resonance was selectively saturated, while in (a) a control irradiation was performed elsewhere in the spectrum away from the resonances of interest. The spectrum (c) is the difference (a–b). The spectra show the transfer of saturation between the ATP and phosphocreatine resonances. In (e) the phosphocreatine resonance was selectively saturated, with (d) as the control irradiation spectrum. Spectrum (f) is the difference (d–e), in which no transfer of saturation between ATP and phosphocreatine can be observed. (Reprinted from ref. 76 with permission.)

kinase reaction, respectively, and $k_{hyd}[ATP]$ and $k_{syn}[P_i]$ describe the sum of the fluxes for all those reactions catalysing the interconversion of ATP and P_i. The modified Bloch equations describing the transfer of z-magnetization of the ATP-γ-phosphate can be written as:

$$\frac{d\mathbf{M}_\gamma}{dt} = \frac{(\mathbf{M}_\gamma^0 - \mathbf{M}_\gamma)}{T_{1\gamma} - k_{rev}\mathbf{M}_\gamma - k_{hyd}\mathbf{M}_\gamma + k_{syn}\mathbf{M}_i + k_{for}\mathbf{M}_{PCr}} \tag{6.9}$$

If at $t=0$ the PCr z-magnetization is inverted, the initial rate of change of the ATP-γ-phosphate magnetization is given by:

$$\frac{d\mathbf{M}_\gamma}{dt} = -(k_{rev}\mathbf{M}_\gamma^0 - k_{hyd}\mathbf{M}_\gamma^0 + k_{for}\mathbf{M}_{PCr}^0) + k_{syn}\mathbf{M}_i^0 \tag{6.10}$$

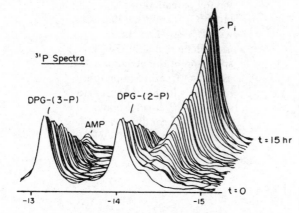

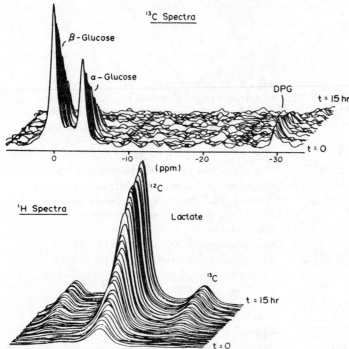

Fig. 6.29 Alternate ^{31}P, ^{13}C, and ^{1}H NMR spectra of erythrocytes metabolizing [1-^{13}C]glucose. The ^{31}P spectra show resonances from inorganic phosphate, AMP and 2,3-diphosphoglycerate (DPG). The ^{13}C spectra show resonances from the α and β anomers of [1-^{13}C]glucose and 2,3-diphospho[3-^{13}C]glycerate. The ^{1}H spectra were obtained with a spin echo pulse sequence ($\tau = 136$ ms). The water resonance was suppressed by selective saturation. The spectra show resonances from the methyl group of [3-^{12}C]lactate and [3-^{13}C)lactate ($^1J_{CH} = 130$ Hz). (Reprinted from S. T. Oxley, R. Porteous, K. M. Brindle, J. Boyd, and I. D. Campbell, *Biochim. Biophsy, Acta*, **805**, 19 (1984) with permission.)

Since $k_{hyd}\mathbf{M}_\gamma^0 = k_{syn}\mathbf{M}_i^0$ and $k_{for}\mathbf{M}_{PCr}^0 = k_{rev}\mathbf{M}_i^0$, this simplifies to:

$$\frac{d\mathbf{M}_\gamma}{dt_{t=0}} = -2k_{rev}\mathbf{M}_\gamma^0 = -2k_{for}\mathbf{M}_{PCr}^0 \tag{6.11}$$

The same initial rate is obtained for changes in the PCr z-magnetization following inversion of the ATP γ-phosphate resonance. This represents one way in which fluxes may be measured. Rather than using saturation transfer or inversion transfer, the two-dimensional EXSY experiment (see Section 1.3.6) can be used, which has a number of advantages, as has already been discussed.

Of course, a simple metabolic pathway such as glycolysis can be monitored using single isotopic labels or multiple labels of differing nuclei, as illustrated in Fig. 6.29. Here the metabolism of [1–^{13}C]glucose by erythrocytes is illustrated in a special experiment in a triple resonance probe, in which the spectrometer frequency alternates between ^{1}H, ^{13}C, and ^{31}P, collecting spectra for each of the three nuclei. The decrease of the ^{13}C resonances for the anomeric glucose carbons correlates with the transient labelling of 2,3-diphosphoglycerate (DPG). The fact that it is the flux of ^{13}C through the DPG pool, rather than a change in the DPG pool concentration, is illustrated in the ^{31}P spectra, which show that the DPG resonances remain at roughly the same intensity during the period in which the [^{13}C] appears. The ^{1}H spectra show the production of both ^{12}C- and ^{13}C-labelled lactate, the ratio of the centre resonance to the satellites giving the fractional labelling.

REFERENCES

1. M. Saunders, A. Wishnia, and J. G. Kirkwood, *J. Am. Chem. Soc.*, **79**, 3289 (1957).

2. J. N. S. Evans, R. C. Davies, A. S. F. Boyd, I. Ichinose, N. E. Mackenzie, A. I. Scott *et al.*, *Biochemistry*, **25**, 896–904 (1986), and references cited therein.

3. W. P. Jencks in *Current Aspects of Biochemical Energetics* (eds N. O. Kaplan and E. P. Kennedy), Academic Press, New York, pp. 273–98 (1966).

4. R. Wolfenden, *Annu. Rev. Biophys. Bioeng.*, **5**, 271–306 (1976).

5. R. Wolfenden and L. Frick in *Enzyme Mechanisms* (eds M. I. Page and A. Williams), Royal Society of Chemistry, London, pp. 97–122 (1987).

6. L. Pauling, *Chem. Eng. News*, **24**, 143 (1946).

7. W. W. Cleland, *Biochemistry*, **29**, 3194–7 (1990).

8. J. N. S. Evans, G. Burton, P. E. Fagerness, N. E. Mackenzie, and A. I. Scott, *Biochemistry*, **25**, 905–12 (1986).

9. J. P. G. Malthouse, *Prog. in NMR Spectroscopy*, **18**, 1–59 (1985).

10. P. Rösch, *Ibid.*, **18**, 123–69 (1986).

11. K. Kanamori and J. D. Roberts, *Acc. Chem. Res.*, **152**, 35 (1983).

12. J. P. G. Malthouse, M. P. Gamcsik, A. S. F. Boyd, N. E. Mackenzie, and A. I. Scott, *J. Am. Chem. Soc.*, **104**, 6811 (1982).

13. N. E. Mackenzie, J. P. G. Malthouse, and A. I. Scott, *Biochem. J.*, **219**, 437 (1984).

14. R. J. Appleyard and J. N. S. Evans, *Bull. Magn. Reson.*, **14**, 81–5 (1992).

15. J. N. S. Evans, R. J. Appleyard, and W. A. Shuttleworth, *J. Am. Chem. Soc.*, **115**, 1588–90 (1993).

16. J. N. S. Evans, Time-resolved solid-state NMR of enzyme-substrate interactions, in *Encyclopaedia of NMR*, John Wiley & Sons (1995).

17. P. Douzou and G. Petsko, *Adv. in Protein Research*, **36**, 245 (1984).

18. J. Hajdu, P. A. Machin, J. W. Campbell, T. J. Greenhough, I. J. Clifton, S. Zurek, *et al.*, *Nature*, **329**, 178–81 (1987).

19. G. K. Farber, P. Machin, S. C. Almo, G. A. Petsko, and J. Hajdu, *Proc. Natl. Acad. Sci. USA*, **85**, 112–15 (1988).

20. J. Hajdu and L. N. Johnson, *Biochemistry*, **29**, 1669–78 (1990).

21. L. N. Johnson, *Protein Sci.*, **1**, 1237–4 (1992).

22. J. Hajdu, K. R. Acharya, D. I. Stuart, P. J. McLaughlin, D. Barford, N. G. Oikonomakos, *et al.*, *EMBO J.*, **6**, 539–46 (1987).

23. K. Moffat, *Annu. Rev. Biophys. Chem.*, **18**, 309–32 (1989).

24. E. M. H. Duke, A. Hadfield, J. L. Martin, I. J. Clifton, J. Hajdu, L. N. Johnson, *et al.*, in *Protein Conformation*, Ciba Foundation Symposium, John Wiley and Sons, Chichester, pp. 75–86 (1991).

25. B. L. Stoddard, P. Koenigs, N. Porter, K. Petratos, G. A. Petsko, and D. Ringe, *Proc. Natl. Acad. Sci. USA*, **88**, 5503–7 (1991).

26. C. A. Fyfe, M. Cocivera, and S. W. H. Damji, *Acc. Chem. Res.*, **11**, 277 (1978).

27. J. J. Grimaldi and B. D. Sykes, *J. Amer. Chem. Soc.*, **97**, 273 (1975).

28. S. O. Smith, I. Palings, V. Copie, D. P. Raleigh, J. Courtin, J. A. Pardoen, *et al.*, *Biochemistry*, **26**, 1606 (1987).

29. R. C. Bray, *Biochem. J.*, **81**, 180–95 (1961).

30. See, for example, *Proteins at low temperatures* (ed. O. Fennema), *Am. Chem. Soc. Adv. in Chem. Series* No. 180 (1979); F. Franks, *Biophysics and Biochemistry at Low Temperatures*, Cambridge University Press (1985).

31. R. J. Appleyard and J. N. S. Evans, *J. Magn. Reson., Series B*, **102**, 245–52 (1993).

32. L. Bachmann and E. Mayer, in *Cryotechniques in Biological Electron Microscopy* (eds R. A. Steinbrecht and K. Zierold), Springer-Verlag, Berlin p. 3. (1987); P. Douzou, *Cryobiochemistry: An Introduction*, Academic Press (1977).

33. R. J. Appleyard, W. A. Shuttleworth, and J. N. S. Evans, *Biochemistry*, **33**, 6812–21 (1994).

34. W. J. Albery and J. R. Knowles, *Biochemistry*, **15**, 5631 (1976).

35. A. A. Bothner-By and R. Gassend, *Ann. N. Y. Acad. Sci.*, **222**, 668 (1972).

36. T. L. James and M. Cohn, *J. Biol. Chem.*, **249**, 2599 (1972).

37. J. P. Albrand, B. Birdsall, J. Feeney, G. C. K. Roberts, and A. S. V. Burgen, *Int. J. Biol. Macromol.*, **1**, 37 (1979).

38. G. M. Clore and A. M. Gronenborn, *J. Magn. Reson.*, **48**, 402–17 (1982).

39. G. M. Clore and A. M. Gronenborn, *J. Magn. Reson.*, **48**, 423–42 (1983).

40. F. Ni, Y. Konishi, and H. A. Scheraga, *Biochemistry*, **29**, 4479–89 (1990).

41. C. P. J. Glaudemans, L. Lerner, G. D. Daves Jr, P. Kováč, R. Venable, and A. Bax, *Biochemistry*, **29**, 10906–11 (1990).

42. G. M. Clore, A. M. Gronenborn, G. Carlson, and E. F. Meyer, *J. Mol. Biol.*, **190**, 259–67 (1986).

43. K. Wakamatsu, A. Okada, M. Suzuki, T. Higashijima, Y. Masui, A. Sakakibara *et al.*, *Eur. J. Biochem.*, **154**, 607–15 (1986).

44. F. Ni, *J. Magn. Reson.*, **96**, 651–6 (1992).

45. J. J. Villafranca, *Methods in Enzymology*, **177**, 390–403 (1989).

46. F. M. Raushel and J. J. Villafranca, *CRC Crit. Rev. Biochem.*, **23**, 1–26 (1988).

47. M. -D. Tsai, *Methods in Enzymology*, **87**, 235–79 (1982).

48. S. L. Buchwald, D. E. Hansen, A. Hassett, and J. R. Knowles, *Methods in Enzymology*, **87**, 279–301 (1982).

49. C. W. DeBrosse and J. J. Villafranca, in *Magnetic Resonance in Biology* (J. S. Cohen, ed.), pp. 1–52, Wiley, New York (1983).

50. D. A. Usher, D. I. Richardson, and F. Eckstein, *Nature (London)*, **228**, 663 (1970); D. A. Usher, E. S. Erenirch, and F. Eckstein, *Proc. Natl. Acad. Sci USA*, **69**, 115 (1972).

51. S. J. Abbott, S. R. Jones, S. A. Weinmann, and J. R. Knowles, *J. Am. Chem. Soc.*, **100**, 2558 (1978); S. J. Abbott, S. R. Jones, S. A. Weinman, F. M. Bockhoff, F. W. McLafferty, and J. R. Knowles, *J. Am. Chem. Soc.*, **101**, 4323 (1979).

52. P. M. Cullis and G. Lowe, *J. Chem. Soc., Chem. Commun.* p. 512 (1978).

53. J. M. Risley and R. L. Van Etten, *Methods in Enzymology*, **177**, 376–89 (1989).

54. M. -D. Tsai, *Biochemistry*, **18**, 1468 (1979).

55. M. J. Garson and J. Staunton, *Chem. Soc. Rev.*, **8**, 539–61 (1979).

56. J. C. Vederas, *Nat. Prod. Rep.*, 277–337 (1987).

57. E. P. Abraham, C. -P. Pang, R. L. White, D. H. G. Grout, M. Lutstorf, P. J. Morgan, and A. E. Derome, *J. Chem. Soc. Chem. Commun.*, 723–4 (1993).

58. C. A. Townsend and A. M. Brown, *J. Am. Chem. Soc.*, **104**, 1748 (1982).

59. J. E. Baldwin, E. P. Abraham, R. M. Adlington, J. A. Murphy, N. B. Green, H. H. Ting, *et al.*, *J. Chem. Soc., Chem. Commun.*, 1319 (1983).

60. C. A. Townsend and S. B. Christensen, *Tetrahedron*, **39**, 3575 (1983).

61. C. A. Townsend, S. B. Christensen, and S. G. Davis, *J. Am. Chem. Soc.*, **104**, 6152 (1982).

62. C. Abell and J. Staunton, *J. Chem. Soc., Chem. Comm.* 856 (1981).

63. J. C. Vederas, *Nat. Prod. Rep.*, 277–337 (1987).

64. T. J. Simpson, *Chem. Soc. Rev.*, **16**, 123–60 (1987).

65. J. Staunton, *Primary Metabolism*, Oxford University Press (1978).

66. J. Mann, *Secondary Metabolism*, 2nd edn, Oxford University Press (1987).

67. E. Haslam, *Metabolites and Metabolism*, Oxford University Press (1985).

68. S. Kanodia and M. F. Roberts, *Proc. Natl. Acad. Sci. USA*, **80**, 5217–21 (1983).

69. J. N. S. Evans, C. J. Tolman, S. Kanodia, and M. F. Roberts, *Biochemistry*, **24**, 5693–8 (1985).

70. J. N. S. Evans, D. P. Raleigh, C. J. Tolman, and M. F. Roberts, *J. Biol. Chem.*, **261**, 16323–31 (1986).

71. J. N. S. Evans, C. J. Tolman, and M. F. Roberts, *Science*, **231**, 488–91 (1986).

72. N. E. Mackenzie, R. L. Baxter, A. I. Scott, and P. E. Fagerness, *J. C. S. Chem. Comm.*, 145–7 (1982).

73. J. R. Dickinson, I. W. Dawes, A. S. F. Boyd, and R. L. Baxter, *Proc. Natl. Acad. Sci. USA*, **80**, 5847–51 (1983).

74. S. M. Cohen, *J. Biol. Chem.*, **258**, 14294 (1983).

75. S. M. Cohen, *Methods in Enzymol*, **177**, 417–35 (1989).

76. K. M. Brindle, *Prog. NMR Spectros.*, **20**, 257–93 (1988).

7 · Acyl and phosphoryl transfer enzymes

In this chapter we will consider selected examples from two extremely important classes of enzymes: the acyl transfer and the phosphoryl transfer enzymes. The selection was based on the breadth of chemistry which the enzymes catalyse and the, often ingenious, ways in which NMR spectroscopy has been used to address specific mechanistic and structural problems.

7.1 ACYL TRANSFER ENZYMES

The proteases are an important class of digestive enzymes, and are among the best characterized acyl transfer enzymes. Every major textbook of biological chemistry contains a chapter on the structure and mechanism of these enzymes. However, this chapter is still being written, and NMR is playing a key rôle in determining the outcome. The serine/cysteine proteases are the best understood (e.g. α-lytic protease, chymotrypsin, trypsin, papain) of the proteases, although the aspartic proteases (e.g. pepsin) and the zinc proteases (e.g. thermolysin) are under active investigation.

7.1.1 α-Lytic protease

An example of the use of NMR to study a labelled enzyme interacting with an inhibitor is the work of Bachovchin and co-workers[1,2] with [^{15}N-His]-labelled α-lytic protease. α-Lytic protease ($M_r = 19.8$ kDa) is a serine protease containing only one histidine (His-57, to use the chymotrypsin numbering), which forms part of the characteristic catalytic triad (see structure **7.1**).

7.1   **7.2**

^{15}N chemical shifts are exquisitely sensitive to the type of nitrogen involved in the structure and whether it participates in a hydrogen-bonding network, as shown in Table 7.1. Hydrogen bonding typically induces an 8–10 ppm chemical-shift change for these nitrogens, moving the hydrogen-bond donor NH and +NH type nitrogens *downfield* while moving the hydrogen-bond acceptor =N-type nitrogen *upfield*.

For α-lytic protease in solution at pH 4.0, where His-57 is essentially fully protonated ($pK_a = 7.0$), $N^{\delta 1}$ and $N^{\epsilon 2}$ resonate at 191.6 ppm and 204.2 ppm, respectively, as in

Table 7.1 ^{15}N chemical shifts of [^{15}N]histidine by nitrogen type[a]

Nitrogen type[b]	δ^{15}N (ppm)	δ^{15}N (ppm) in H-bond
NH	210	200
=N−	128	138
+NH	201	191

[a]From Ref. 2.
[b]NH and =N− refer to the protonated and non-protonated nitrogens, respectively, within a neutral imidazole ring, while +NH refers to both nitrogens within a protonated imidazole ring.

structure **7.3**. Here both ring nitrogens must be +NH type nitrogens. The low-field position of $N^{\delta 1} \approx 10$ ppm downfield from that typical of +NH nitrogens, reveals the presence of the hydrogen bond to Asp-102. At pH 9.5, where His-57 is essentially fully neutral, $N^{\delta 1}$ and $N^{\varepsilon 2}$ resonate at 199.4 ppm and 138 ppm respectively (structure **7.4**). There are two possible explanations for these ^{15}N chemical shifts, one involving tautomeric exchange and consequent ^{15}N chemical-shift averaging, the second involving a pure $N^{\delta 1}$—H tautomer strongly hydrogen bonded at both ring nitrogens, $N^{\delta 1}$ as a donor to Asp-102, $N^{\varepsilon 2}$ as an acceptor from Ser-195 (e.g. structure **7.4** with ?=H-bond).

This latter interpretation has been shown to be correct in experiments in which the possibility of His–Ser hydrogen-bonding was eliminated by specific sulphonylation with the covalent inhibitor phenylmethane-sulphonyl fluoride (PMSF) or phosphorylation with diisopropylfluorophosphate (DIFP) at Ser-195. In the derivatized enzymes $N^{\varepsilon 2}$ moves 10 ppm downfield to 128 ppm (structure **7.6**, cf. structure **7.4**) while $N^{\delta 1}$ remains at 200 ppm (structure **7.6** cf. structure **7.4**). This finding provides compelling evidence for a strong His–Ser hydrogen bond in solution, a conclusion which remained in conflict with the X-ray diffraction data on α-lytic protease.[3] Solid-state NMR work on the lyophilized ^{15}N-his-labelled enzyme by Bachovchin, Griffin, and co-workers,[4] established that the hydrogen-bonding network is maintained in the resting enzyme without any significant change in the pK_a. In an elegant experiment with crystalline ^{15}N-his-labelled α-lytic protease, Bachovchin, Griffin, and co-workers[5] showed that in the crystal the pK_a of His-57 increases to 7.9, nearly one pH unit higher than in solution (see Fig. 7.1). This result explains the discrepancy in the X-ray data, since the latter was performed at pH 7.2, where His-57 would be mostly protonated and unable to act as a hydrogen acceptor for the serine hydroxyl proton.

One of the interesting additional results of this work is that the Asp-102—His-57 hydrogen bond is disrupted in the sulphonylated enzyme at low pH, i.e. when the imidazolium anion is formed (structure 7.5). Since the latter is strongly implicated in the catalytic mechanism of the enzyme, this suggests that it is the combination of tetrahedral

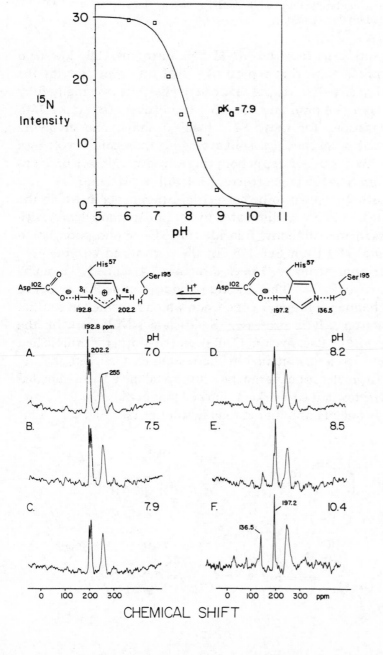

Fig. 7.1 ^{1}H-decoupled ^{15}N CP–MAS NMR spectra of [^{15}N-His]-labelled α-lytic protease crystals as a function of pH (lower). The 192.8 ppm resonance moves to 197.2 ppm, while the 202.2 ppm resonance moves to 136.5 ppm as the pH is increased. The differences in intensity arise from differences in efficiency of cross-polarization as a function of the ^{1}H exchange rate (which is affected by pH). The plot of $N^{\delta 1}$ signal intensity as function of pH (upper). The solid line is a calculated titration curve for a pK_a of 8.0. (Reprinted from ref. 5 with permission.)

adduct to Ser-195 and a protonated imidazolium ion which signals the rupturing of the hydrogen bond to move the imidazolium into position to donate its proton to the nitrogen atom in the leaving group of the peptide substrate. This moving histidine mechanism is yet another modification of the classic catalytic mechanism for serine proteases, so often reported incorrectly in major biochemistry textbooks.[6]

7.1.2 Chymotrypsin and trypsin

Chymotrypsin ($M_r = 25$ kDa) and trypsin ($M_r = 24$ kDa) are very similar to α-lytic protease, although they have been studied by NMR by employing isotopically labelled substrates or inhibitors. The earliest study was carried out by Richards and co-workers,[7] who recorded one of the first enzyme–substrate complex spectra, of chymotrypsin complexed with N-acetyl-DL-*p*-fluorophenylalanaine. Another study was conducted by Sykes, who measured the forward and reverse rates of binding for the inhibitor trifluoroacetyl-D-phenylalanine.[8] The suicide inhibitor[9] 3-benzyl-6-chloro-2-pyrone (structure **7.7**) covalently modifies chymotrypsin. Using ^{13}C NMR of [2,6-[13]C$_2$]-3-benzyl-6-chloro-2-pyrone with the enzyme, Gelb and Abeles[10] established the mechanism for inhibition (see structures **7.7–7.9**).

Four resonances are detectable (see Fig. 7.2), two sharp lines at 181.7 and 176 ppm due to C-2 and C-6 resonances of the product (structure **7.9**), respectively, and two broad lines at 177.5 and 173.1 ppm due to a carboxylate and ester group respectively. The latter was established by [18]O exchange studies. With mono-labelled [6-^{13}C]-inhibitor, of the two broad resonances only the 177.5 ppm line was observed, establishing that the assignments are enzyme-bound C-6 carboxylate at 177.5 ppm and enzyme-bound C-2 ester at C-2. The migration of the double bond to the ester (structure **7.8**) was established by similar studies with [3,5-[13]C$_2$]-inhibitor. This inhibitor bound to the enzyme was also characterized by X-ray crystallography[11] at 1.9 Å resolution, and the formation of a covalent adduct to Ser-195 was confirmed.

The direct observation of the acyl enzyme intermediate was first reported by Niu *et al.*,[12] who incubated the non-specific substrate [1-^{13}C]-*p*-nitrophenyl acetate with chymotrypsin at pH 5.1 and then slowed down the general base-catalysed deacylation step by lowering the pH to 3.0 (see structures **7.10–7.12**). In addition to the sharp resonances due to substrate (structure **7.10**, $\delta = 170.4$ ppm) and product (structure **7.12**, $\delta = 175.1$ ppm), an additional broad resonance at 174 ppm was observed (see Fig. 7.3). This broad resonance was assigned to the acyl enzyme intermediate (structure **7.11**). After prolonged incubation at pH 3.0 (for three days), the resonances due to substrate and product were present but the broad resonance due to acyl enzyme was not detected, implying that it had hydrolysed to products.

A number of workers have examined the interaction of α-chymotrypsin with mechanism-based inhibitors. Abeles and co-workers have reported[13] ¹H, ¹³C, and ¹⁹F NMR studies of the complex of chymotrypsin with N-acetyl-L-leucyl-1-[1-¹³C] phenylalanyl trifluoromethyl ketone, for which evidence was presented that the ketone group forms an ionized hemiketal. Groutas and co-workers[14] studied the complex of the enzyme with 3-benzyl-N-((methylsulphonyl)oxy)-[5-¹³C]succinimide and proved that

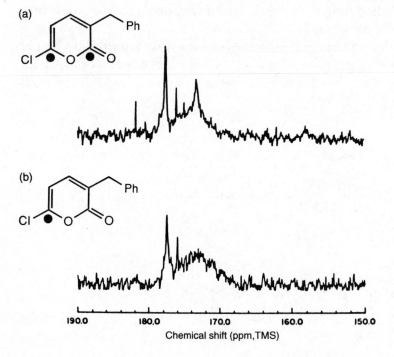

Fig. 7.2 ¹H-decoupled ¹³C NMR spectra of chymotrypsin inactivated with (a) [2,6-¹³C₂]-3-benzyl-6-chloro-2-pyrone and (b) [6-¹³C]-3-benzyl-6-chloro-2-pyrone. (Reprinted from ref. 10 with permission.)

this is mechanism-based, proceeding via a Lossen rearrangement. Jordan and co-workers[15,16] have studied boronic acid inhibitors by ^{11}B NMR, and found evidence for various modes of binding. Malthouse and co-workers[17] used ^{13}C NMR to investigate the structure of the enzyme-inhibitor complex formed from α-chymotrypsin and tosyl-[2-^{13}C]phenylalanine chloromethyl ketone (Tos-Phe-CH$_2$Cl), and detected two resonances. One was at 204.8 ppm, did not titrate, and was assigned to alkylated Met-192. The other tirated from 99.1 to 103.4 ppm, with a $pK_a = 8.67$, and was assigned to the tetrahedral adduct with Ser-195. It was argued that the resultant oxyanion could be stabilized through interaction with the imidazolium of His-57. There was evidence also for the formation of the inhibitor adduct with His-57 upon protein denaturation. In an additional study, Malthouse and co-workers[18] measured the relaxation properties of the enzyme alkylated with the same inhibitor, and found that there is no significant librational freedom for the enzyme-bound species. Using [^{2}H,[13]C]-labelled inhibitor in combination with a spin echo pulse sequence, the broad envelope from the natural abundance protein background could be attenuated. This result contrasts with a result from Williams, Gerig,

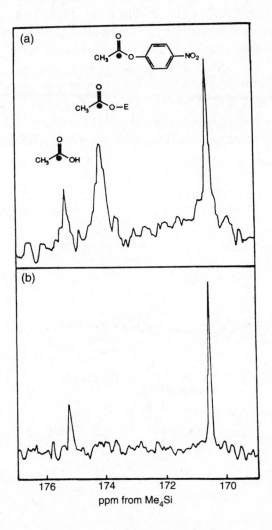

Fig. 7.3 ^{1}H-decoupled ^{13}C NMR spectra of (a) α-chymotrypsin (2 mM) and p-nitrophenyl-[1-^{13}C]acetate (3 mM) at room temperature, pH 5.1; (b) p-nitrophenyl-[1-^{13}C]acetate (3.7 mM) under the same conditions after three days. (Reprinted from ref. 12 with permission.)

and co-workers,[19] who studied the dynamics of the tosyl group of chymotrypsin tosylated at Ser-195 by ^{3}H NMR and found that it is highly mobile.

An example of a covalent inhibitor for the serine protease, trypsin, is the chloromethylketone inhibitor derived from Z-lys, known as Z-lys-CMK (structure **7.13**). Z-lys-CMK ^{13}C enriched in the ketone ([2-^{13}C]-Z-lys-CMK) has been used by Malthouse, MacKenzie, Scott, and co-workers[20] to determine whether a tetrahedral adduct is formed and to attempt to assess its ionization state. The ^{13}C NMR spectra obtained are shown in Fig. 7.4. Using [2-^{13}C]-Z-lys-CMK it was shown that in aqueous solution, both the ketone (structure **7.14**, $\delta_C = 204.7$ ppm) and its hydrate (structure **7.15**, $\delta_C = 95.4$ ppm) are present (Fig. 7.4(a)). At pH 3.2 no inhibition or alkylation of trypsin was detected and the spectrum of [2-^{13}C]-Z-lys-CMK was unperturbed (Fig. 7.4(c)), demonstrating that at low pH there is no detectable tetrahedral adduct formation prior to alkylation. At pH 6.9, there was rapid irreversible inhibition with the resonances due to the free inhibitor carbonyl and hydrate being replaced by a single new resonance at 98.0 ppm (structure **7.16**, Fig. 7.4(d)). Addition of excess [2-^{13}C]-inhibitor led to the reappearance of the free ketone and hydrate resonances (Fig. 7.4(e)) which could then be removed either by adding more trypsin (Fig. 7.4(f)) or by gel filtration (Fig. 7.4(g)). This clearly demonstrates that the carbonyl carbon is sp^3 hybridized and that a covalent tetrahedral adduct is formed. X-ray crystallographic studies[21,22] and solid-state NMR studies[22] have confirmed that the tetrahedral adduct is a hemiketal and not a hydrate.

On denaturation, the resonance at 98 ppm decreased in intensity and new resonances appeared at 205.5 and 95.1 ppm (Fig. 7.4(h)), which were assigned to species (structure **7.17**) and (structure **7.20**). The other two species shown (structures **7.18** and **7.19**) were detected by altering the pH of the sample, and assigned as shown. The titration

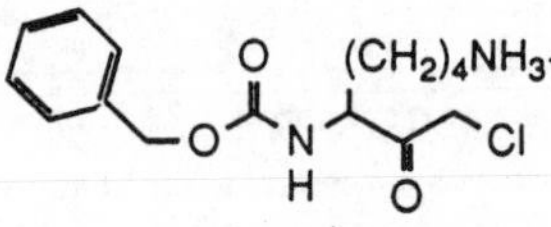

7.13

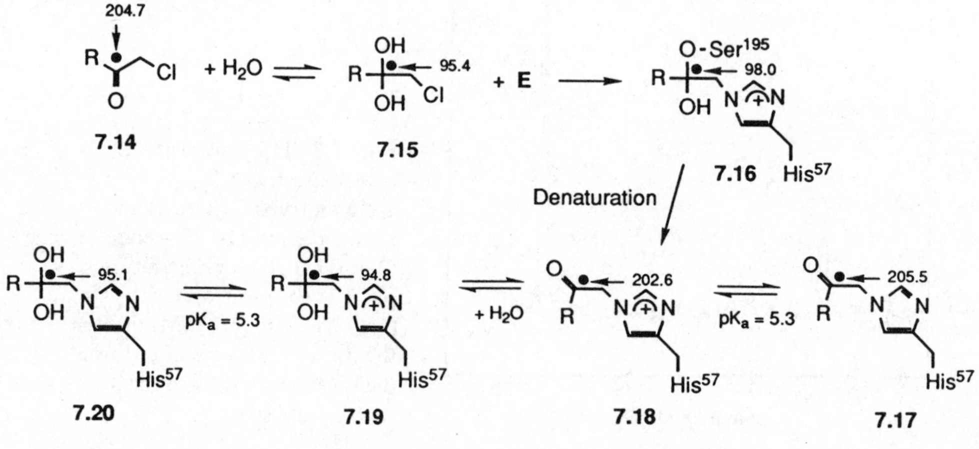

7.14   7.15   7.16   His^{57}

Denaturation

7.20   7.19   7.18   7.17

shifts of the resonances provided evidence for the ionization state of the histidine imidazole in the complex.

Scott and co-workers[23] have examined the interaction of the inhibitor leupeptin (N-acetyl-Leu-Leu-Arginal) ^{13}C-labelled at the arginal carbonyl with trypsin, using solution-state ^{13}C and ^{1}H 1D HMQC–NOESY and solid-state ^{13}C NMR experiments. Two pH-dependent covalent complexes were detected, and these were interpreted as arising from two interconverting diastereomers at the new asymmetric tetrahedral centre created by covalent addition of Ser-195 to either side of the ^{13}C-enriched aldehyde of the inhibitor. At pH 7, two ^{13}C resonances are present, at 98.8 and 97.2 ppm in a 84:16 ratio, while at pH 3, the latter signal predominates. In the selective proton ^{13}C-edited NOE spectrum of the major diastereomer at pH 7, a strong NOE is observed between the hemiacetal proton of the inhibitor and the C2 proton of His-57 of the enzyme, thereby

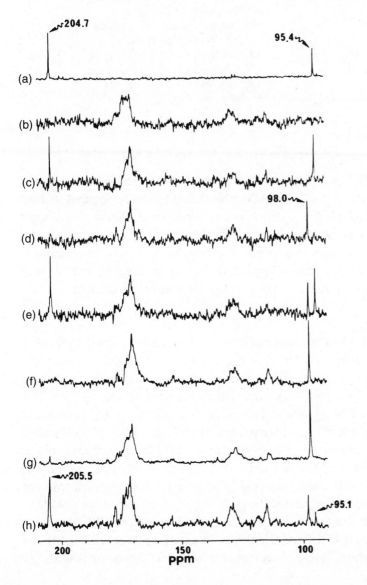

Fig. 7.4 ^{1}H-decoupled ^{13}C NMR spectra of (a) Z-lys-CMK (47.6 mM), pH 3.1; (b) trypsin (0.3 mM), pH 3.2; (c) trypsin (0.3 mM) plus Z-lys-CMK (0.38 mM), pH 3.2; (d) trypsin (0.3 mM) plus Z-lys-CMK (0.38 mM), pH 6.9; (e) trypsin (0.3 mM) plus Z-lys-CMK (0.74 mM), pH 6.9; (f) denatured trypsin (0.4 mM) plus Z-lys-CMK (0.6 mM), pH 6.9; (g) sample from (f) after gel filtration and concentration, pH 6.1; (h) sample from (g) after incubating for 30 mins at pH 11, and adjustment back to pH 6.1. (Reprinted from ref. 20 with permission.)

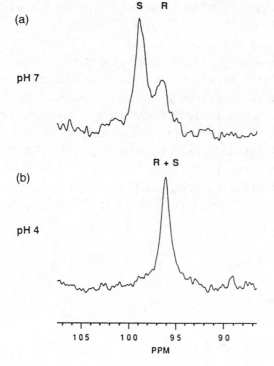

Fig. 7.5 75.4 MHz CPMAS ^{13}C NMR spectra of the trypsin–leupeptin complex after soaking commercial trypsin crystals for 15 days with 5 mM [^{13}C]leupeptin in 3M ammonium sulphate and 0.1 M phosphate buffer at pH 7 (a) and then at pH 4 (b). (Reprinted from ref. 23 with permission.)

defining the stereochemistry to the *S* configuration, in which the hemiacetal oxygen resides in the oxyanion hole. pH titration studies established that the ^{13}C chemical shift of the *S* diastereomer titrates with a p$K_a = 4.69$, which is consistent with direct titration of the hemiacetal oxygen. Using ^{13}C CP–MAS solid-state NMR on crystalline enzyme, the same dependence on chemical shift was observed (see Fig. 7.5). These experiments establish the existence of the same diastereomeric equilibrium in the solid state, and permit a model of the inhibitor–active site complex to be derived (see Fig. 7.6).

7.1.3 Papain

Papain is a thiol protease ($M_r = 23.35$ kDa), in which there is a catalytic triad including cysteine-25, histidine-159 and asparagine-175. The mechanism is thought to be closely similar to the serine proteases, except that the thiolate ion of cysteine is acylated during catalysis, possibly via a tetrahedral species (although this has been questioned[24,25]). The first evidence for the formation of a tetrahedral species was obtained by Lowe and co-workers,[26,27] using saturation transfer experiments by ^{1}H NMR. In a sample containing papain and excess inhibitor, *N*-benzoylaminoacetaldehyde, irradiation at 6.19 ppm led to a significant decrease in the intensity of the free aldehyde resonance at 9.61 ppm. This effect was highly specific and most likely to arise from saturation transfer from the corresponding proton in the enzyme-inhibitor complex. Lowe *et al.* were able to estimate the chemical shift of the thiohemiacetal proton (6.51 ppm) and J_{CH} coupling constant, by observing the residual splitting of the thiohemiacetal carbon with inhibitor enriched in ^{13}C at the aldehyde carbon. Saturation transfer of the ^{1}H resonances with the

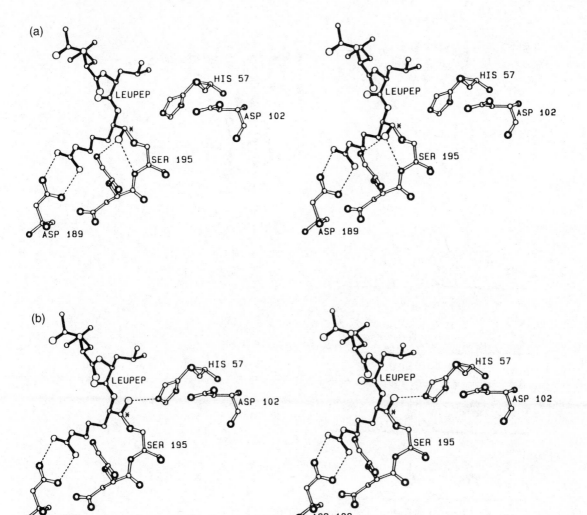

Fig. 7.6 Stereo structures of the two diastereomers of the leupeptin–trypsin complex showing the hemiacetal in the (a) *S* configuration and (b) the *R* configuration. Bonds of the inhibitor are filled. Hydrogen bonds are shown as dashed lines. (Reprinted from ref. 23 with permission.)

[13]C enriched inhibitor confirmed the assignment of the resonance due to the enzyme-bound species. The formation of such a thiohemiacetal has been confirmed by ^{13}C NMR studies of the inhibitor *N*-acetyl-L (and D)-phenylalanyl-glycinal with papain.[28] This work demonstrated that both stereoisomers form the same diastereometric thiohemiacetal, and that the thiohemiacetal hydroxyl oxygen is not within hydrogen-bonding distance of the so-called 'oxyanion hole'. Since then, this work has been corrected,[29] and it has been found that each diastereomer forms a *different* diastereomeric thiohemiacetal. However, it is still evident that there is no conformational preference in the oxyanion hole. One interpretation of this result[29,30] suggests that papain does not stabilize a tetrahedral intermediate during catalysis. This is an example of the danger of

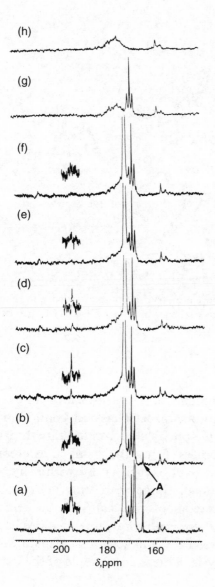

Fig. 7.7 ^{1}H-decoupled ^{13}C NMR spectra of papain (1.7 mM) plus [[13]C= O]benzoylimidazole (23.6 mM) in 25% DMSO-water, pH 4.1 $T = 267$ K at times (a) 6; (b) 51; (c) 96; (d) 141; (e) 186; and (f) 231 mins after addition of benzoylimidazole; the insert at 200–192 ppm is a three-fold vertical expansion; (g) papain (2 mM) containing formate buffer (0.1 mM); (h) papain (2 mM) without formate. (Reprinted from ref. 35 with permission.)

interpretation of limited structural data, since it is equally possible that the tetrahedral species forms but that the inhibitor is not the 'transition-state analogue' that it is claimed[31] to be. Other inhibitors of papain have been investigated by ^{13}C NMR, such as nitriles, which form thioimidate esters,[32] epoxysuccinyl peptides,[33] and peptidyl acyloxy-, aryloxy- and chloromethyl ketones.[34]

The direct observation of an acyl intermediate of papain by ^{13}C NMR using the non-specific substrate, N-benzoylimadazole (structure **7.21**), was carried out by Malthouse, Scott, and co-workers[35] under cryoenzymological conditions. Papain was incubated with [^{13}C=O]-N-benzoylimidazole at 0°C to achieve maximum acylation, and then cooled to -6°C to reduce the rate of deacylation (see Fig. 7.7). A resonance at 196 ppm characteristic of a thioacyl species (structure **7.22**) was observed. All other resonances in the spectrum could be assigned to natural abundance background from papain (Fig. 7.7(h)), formate buffer (Fig. 7.7(g)), substrate, product (structure **7.23**), and benzoic anhydride (formed by reaction of product with the substrate). Since formate occurs as a partially decoupled triplet centred at 169.3 ppm (pH 3.8 Fig. 7.7(a–f)) or 170.1 ppm (pH 4.1, Fig. 7.7(g)), the substrate ([^{13}C=0]-N-benzoylimidazole) resonance ($\delta = 168.7$ ppm) is superimposed on the formate resonance.

7.1.4 Pepsin

Pepsin is an aspartic protease ($M_r = 35$ kDa) which is 39% homologous with human renin, the enzyme that plays an important rôle in the regulation of blood pressure. A tetrahedral adduct of the enzyme with a ^{13}C-labelled pepstatin analogue was reported by Rich and co-workers.[36] They also found evidence from ^{1}H NMR for an enzyme conformational change accompanying pepstatin binding.[37] Instead of using heteronuclear NMR to study the interaction of an enzyme with substrate or inhibitor, Fesik and co-workers[38] were the first to apply two-dimensional HMQC spectroscopy, originally introduced by Müller,[39] to ^{1}H NMR studies of large complexes. They used the [^{1}H, ^{15}N] and [^{1}H, ^{13}C]–HMQC–NOESY experiment to detect intermolecular or intramolecular NOEs for a ^{15}N (or ^{13}C)-labelled tripeptide inhibitor (see Fig. 7.8) bound to pepsin. This greatly simplifies the spectrum compared with the conventional NOESY, which would contain over 1000 cross-peaks for the same spectral region. The ^{15}N-labelled tripeptide exhibited intramolecular NOEs (Fig. 7.8(a)) whose assignment was simplified by employing different partially deuterated peptides (Figs. 7.8(b and c)). Intermolecular NOEs between the inhibitor and the enzyme-active site were observed using ^{13}C-labelled tripeptide and analysed by comparison with X-ray crystallographic data. The intense NOEs observed between the amide and α protons of adjacent amino acid residues of the enzyme-bound inhibitor in conjunction with the absence of NOEs between the amide protons are characteristic of an extended chain conformation of the main chain of the inhibitor. Using the partially refined, unpublished X-ray crystal strucure of the pepsin/inhibitor complex, Fesik and co-workers were able to identify NOEs from the inhibitor to Phe-111, Glu-13, Thr-77, Try-114, Met-290, and Thr-218, and generate a computer model for the structure (Fig. 7.8(d)). For a complete identification of the active site contacts, additional experiments will be necessary.

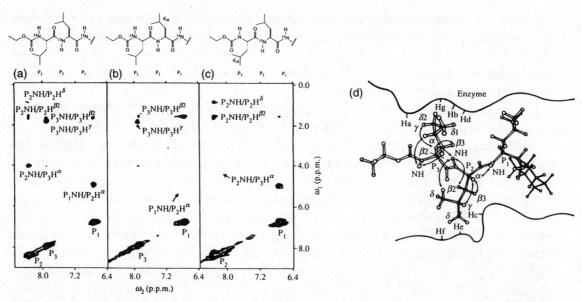

Fig. 7.8 ^{1}H NMR spectra (a–c) obtained with HMQC ^{15}N-isotope-editing of two-dimensional NOE experiments on pepsin-inhibitor complex (1:1) using the labelled inhibitors shown at the top of the spectra, and acquired with a mixing time of 50 ms. In (a) diagonal peaks corresponding to the P_1–P_3 amide protons are labelled, and in (b) and (c) arrows indicate the location in the spectra where cross-peaks are absent due to deuterium labelling. (d) Computer-generated model of the bound pepsin inhibitor that is consistent with the NMR data. The arrows indicate NOEs between proton pairs of the ligand. Inhibitor/enzyme NOEs are designated by letters corresponding to the chemical shifts of the enzyme protons and have been interpreted on the basis of a model built from the partially refined X-ray crystal structure of a similar inhibitor complexed with pepsin (Ha Phe-111; Hb Tyr-114; Hc Thr-218; Hd Glu-13; He,f Met-290; Hg Thr-77). (Reprinted from ref. 38 with permission.)

7.1.5 Thermolysin

Thermolysin is a thermostable zinc-containing endopeptidase ($M_r = 34.6$ kDa) related to peptidases such as angiotensin converting enzyme and enkephalinase that play important rôles in cellular and hormonal metabolism. The mechanism is thought to involve general base catalysed attack of water on the scissile carbonyl, and subsequent protonation of the peptide nitrogen leading to a tetrahedral intermediate,[40] which collapses with cleavage of the C—N bond. Potent inhibitors, said to be 'transition-state analogues',[41] have been synthesized in which the scissile peptide linkage has been replaced by a tetrahedral phosphorus ester (phosphonate) or amide (phosphonamidate) moiety.[42] The tightest binding inhibitor, carbobenzoxy-Phe^p-L-leucyl-L-alanine (structure **7.24**, abbreviated ZF^{p}LA, where Phe^p or F^p represents replacement of the trigonal carbonyl by a

7.24

phosphonamidate group), has a $K_i = 0.068$ nM. From X-ray crystallographic studies,[43] it has been postulated that the tightest binding inhibitors are those which (i) have optimal interactions with the enzyme amino acid residues in their hydrophobic clefts, (ii) can form a tight bidentate complex with thermolysin's zinc ion; and (iii) possess a long P—N bond and a reduced P—N—C bond angle. These features are consistent with the transition state suggested by the proposed mechanism, in which cleavage of the C—N bond and protonation of the amide nitrogen occur concomitantly. For the ZF^{P}LA-thermolysin complex, the X-ray data suggests[32] that the P—N bond is lengthened to a distance of ≈ 0.19 nm.

A variation on one of the homonuclear distance measurement methods (see Section 3.3.2), called an extended dipolar modulation experiment, and introduced by Schaefer and co-workers, employs a two-dimensional pulse sequence graphically described[44] as XDMR4. This was used by Griffin and co-workers[45] to determine the P—N bond length in crystalline ZF^{P}LA-thermolysin complex. In contrast to the X-ray crystallographic data already described, a P—N distance of $r_{\text{P-N}} = 0.168$ nm was obtained for both free and enzyme-bound inhibitors (see Fig. 7.9). The accuracy of these measurements was found to be within 0.04 nm by comparison of X-ray crystallographic data ($r_{\text{P-N}} = 0.1605$ nm) on a model compound, diphenyl-N-methylphosphoramidate, with that obtained by the solid-state NMR method ($r_{\text{P-N}} = 0.165$ nm). Furthermore, for the enzyme-inhibitor complex the isotropic chemical shift $\sigma_{\text{iso}} = 36.7$ ppm (similar to that obtained in solution, where $\delta = 37.8$ ppm), some 20 ppm downfield from the free inhibitor, $\sigma_{\text{iso}} = 17.9$ ppm. This change originates from a 30 ppm downfield shift in σ_{11} and σ_{22} elements of the chemical-shift tensor, whereas σ_{33} remains almost unchanged. Recall from Equations (1.61) and (1.62) that the principal elements of the shift tensor, the shielding anisotropy $\Delta\sigma$, and the asymmetry parameter η may be calculated from the shape of the spectrum. The principal elements of the shift tensor can be obtained by rotational side-band intensity analysis.[46] The parameter $\Delta\sigma$ provides an estimate of the departure from spherical symmetry of the electrons around the ^{31}P nucleus. In the case of the enzyme-inhibitor complex, $\Delta\sigma$ changes from 76.5 ppm to 54.0 ppm, which suggests that

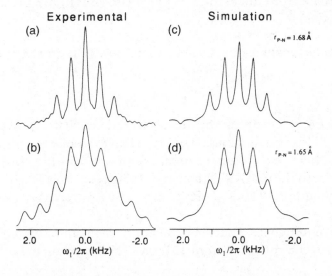

Fig. 7.9 XDMR4 ω_1 projection of free ZF^{P}LA inhibitor and thermolysin–ZF^{P}LA complex with spinning rate $= 2$ kHz. (a) Experimental XDMR4 ω_1 projection of free ZF^{P}LA inhibitor. (b) Experimental XDMR4 ω_1 projection of thermolysin–ZF^{P}LA complex at $T = 182$ K. (c) Simulated XDMR4 ω_1 projection of free ZF^{P}LA inhibitor with $r_{\text{P-N}} = 0.168$ nm. (d) Simulated XDMR4 ω_1 projection of thermolysin–ZF^{P}LA complex with $r_{\text{P-N}} = 0.165$ nm. (Reprinted from ref. 45 with permission.)

the electronic distribution around the phosphorus atom becomes more symmetric on binding to the enzyme. This reduction in chemical-shift anisotropy could originate from charge neutralization of the phosphonamidate by the protein zinc ion, and/or by changes in bond angles or distances. The asymmetry parameter η describes the cylindrical symmetry of the electronic distribution around the ^{31}P nucleus, with $\eta = 0$ representing axial symmetry about σ_{33}. For the enzyme-inhibitor complex, η changes from 0.69 to 1.00, which suggests that the phosphorus tensor becomes fully axially asymmetric. The interpretation of $\Delta\sigma$ and η must be qualitative, since the theoretical foundations as to the structural origins of chemical-shift changes, particularly for phosphorus (see Section 7.2), are poorly understood.

The ^{15}N isotropic chemical shift of the inhibitor was determined at natural abundance, and found to shift 4 ppm in the presence of the enzyme. This suggests that the nitrogen is not protonated in the complex. As with papain (Section 7.1.3), the NMR data suggests that either the mechanism of the enzyme is not properly understood, and a tetrahedral species is not an intermediate, or that the synthetic ZF^{P}LA inhibitor is not really a 'transition-state analogue'. Yet again this underscores the importance of detailed structural information about the enzyme intermediates in an enzyme-catalysed reaction.

7.2 PHOSPHORYL TRANSFER ENZYMES

Since almost all of the chemical energy derived from metabolic processes is stored in the form of adenosine triphosphate (ATP) or other nucleoside triphosphates, the enzymes which transfer phosphoryl groups, often to and from ATP, are clearly extremely important. The phosphoryl transfer enzymes include the kinases, which are probably the single most important group of this class of enzymes, and involve the phosphorylation of a substrate.

The other important class of phosphoryl transfer enzymes are the phosphatases. These are enzymes which hydrolyse phosphoester bonds and are very common in Nature, since nucleoside triphosphates such as ATP, the storage form of energy for the cell, are hydrolysed by cleavage of the terminal phosphate–phosphate ester bond. Formally this is a phosphoryl transfer of phosphate from nucleoside triphosphate (NPT) to water:

$$NTP + H_2O \rightleftharpoons NDP + P_i$$

As with most phosphoryl transfer enzymes, a divalent metal ion is required for this process.

7.2.1 Adenylate kinase

Adenylate kinase (AK, EC 2.7.4.3) catalyses the interconversion between MgATP + AMP and MgADP + ADP. This small kinase (AK1, from cytosolic mammalian muscle, 194 residues, $M_r = 21.7$ kDa) has become something of a model for kinases in general, despite the fact that between the mid-1970s and mid-1980s, the substrate binding sites deduced from X-ray crystallography[47] and NMR[48,49] spectroscopy were very much at odds (see Fig. 7.10).

The picture of the substrate binding site, and the detailed mechanism of the enzyme that is now emerging turns out to be different from either of these two models.[50] The original X-ray crystal structure turned out to be incorrect, because of three factors: (i) the substrate sites were determined by soaking the crystals with a variety of substrates and substrate analogues, from which it was assumed that the site where salicylate was found to bind corresponds to the ATP binding site, despite their lack of structural similarity; (ii) the site occupied by MnATP was assigned to the AMP site (which was probably an artifact[51]); (iii) the measurements of the complexes were made at 6 Å resolution. The NMR structure was determined by Mildvan and co-workers,[52] and turned out to be

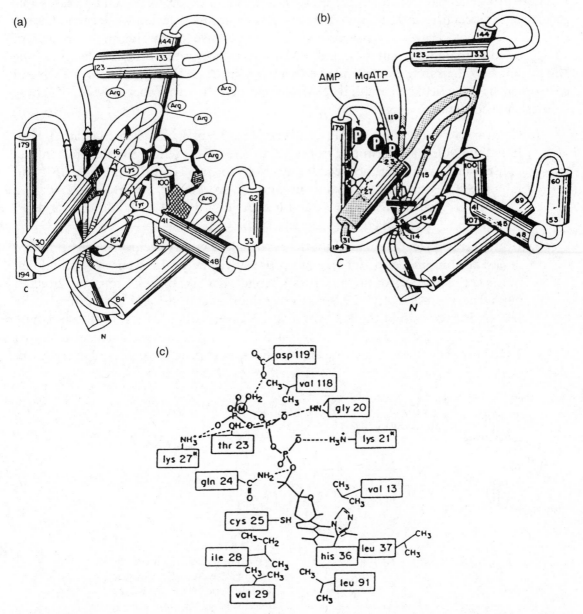

Fig. 7.10 Sketches of adenylate kinase (AK1p) and its substrate binding sites deduced from X-ray crystallography (a) and from NMR (b). The detailed interactions in the NMR model are shown in (c). (Reprinted from ref. 50 with permission.)

incorrect for three reasons: (i) the X-ray structure of AK was only at 3 Å resolution that was used to dock the substrates using NMR data derived from NOEs and paramagnetic measurements (which also contained large errors), making the structure in Fig. 7.10(c) highly questionable; (ii) the distance measurements required assignments for the resonances AK, which had not been carried out completely (or correctly); (iii) the MgATP binding fragment later used by Mildvan and co-workers[53] for two-dimensional NMR and other physical studies turns out to bind ATP or MgATP very weakly.[50]

These problems have been addressed by a number of laboratories, and a reasonable picture of the structure and function of AK is now emerging. Elegant work carried out in the laboratory of Tsai and co-workers[50] has made use of an iterative process of site-directed mutagenesis, kinetics, and NMR spectroscopy to probe the details of the AK1 from chicken muscle cloned and overexpressed in *E.coli*. Independently, Rösch and co-workers have carried out careful ^{1}H NMR assignments for a number of AKs.[54,55] These studies showed the following:

1. Contrary to the earlier NMR evidence, His-36 plays a structural rather than functional role, as determined from site-directed mutagenesis studies. In contrast to the transferred NOE between bound MgATP and His-36 of porcine AK1 and rabbit AK1 detected by Mildvan and co-workers,[56,57] work by Tsai and co-workers[58] with chicken AK1, and independently by Rösch and co-workers[59] with porcine AK1, failed to detect any such NOEs.

2. Of those lysines postulated to be essential, Lys-21 occurs in the putative nucleotide binding loop, but plays an important structural and possibly functional rôle (see Fig. 7.10(c) for the structural changes that take place). From site-directed mutagenesis studies, Lys-27 is non-essential for catalysis, and, interestingly, the ^{1}H NMR spectrum of the K27M mutant is essentially identical to the wild-type (see Fig. 7.11(b,c)). The laboratory of Rösch and co-workers[60] has also shown NOEs between the adenine ring and the side-chain protons of Phe-19 in the *E.coli*

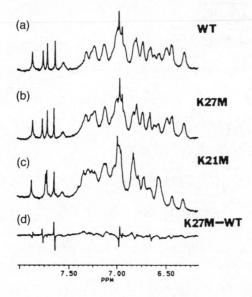

Fig. 7.11 500 MHz ^{1}H NMR spectra of adenylate kinase showing the aromatic regions of (a) wild-type (WT); (b) K27M mutant; (c) K21M mutant; and (d) K27M mutant—wild-type differences. (Reprinted from ref. 50 with permission.)

AK (which corresponds to Lys-27 in AK1). Similar cross-peaks were observed[61] for Trp-210 and His-143 in the yeast AK (corresponding to Phe-183 in AK1 and His-134 in *E.coli* AK, respectively).

3. Consistent with the X-ray data, Arg-132, -138, and -149 are important in transition-state stabilization, as illustrated in Fig. 7.12, which shows the effect of various ligands on the ^{1}H NMR spectrum of both wild-type and the R138K mutant. Similar studies were performed with R132M and R149M mutants.

4. Arg-44 and Arg-97 interact with AMP specifically, suggesting that the X-ray structure, which had assigned these residues to the MgATP site, was in error. Fig. 7.13 shows the result of incubating wild-type and R44M mutant enzymes with the phosphorothiate substrate analogue of AMP, AMPS, and analysis via ^{31}P NMR. The wild-type enzyme converts AMPS to $[S_p\text{-}\alpha S]$ADP at the AMP site and subsequently converts it to $[S_p\text{-}\alpha S]$ATP at the MgATP site. With the R44M mutant, not only is $[\alpha S]$ATP not produced, but the $[\alpha S]$ADP that is generated has the opposite stereochemistry (R_p) from the wild-type enzyme product. This latter effect presumably arises because the preferred orientation of the P—S bond changes with the changes in the AMP binding site when R44 is replaced with methionine. Arg-97 has been shown to interact with the phosphoryl group of AMP in the binary ground-state complex and also in the transition-state.[62]

5. Asp-93 is critical for Mg^{2+} binding and does not bind to the enzyme in the absence of substrates, confirming the earlier NMR evidence. Figure 7.14 shows the effects of added Mg^{2+} ions on the ^{1}H NMR spectra of various complexes of wild-type and D93A mutant enzymes and the potent inhibitor AP_5A (P^1,P^5-bis(5′-adenosyl)-pentaphosphate), which is considered to be a static mimic of the enzyme intermediates, occupying both AMP and MgATP sites. Clearly, the

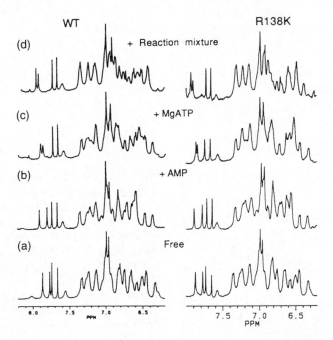

Fig. 7.12 500 MHz ^{1}H NMR spectra of adenylate kinase showing the aromatic regions of wild-type (left) and R138K mutant (right). (Reprinted from ref. 50 with permission.)

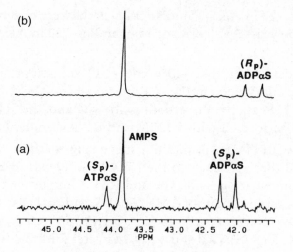

Fig. 7.13 121.5 MHz ³¹P NMR analysis of stereospecificity of adenylate kinase at neutral pH. Only the regions of P$_\alpha$ are shown. AMPS plus (a) wild-type (WT) enzyme; and (b) R44M mutant. The right half of the doublet of (S_p)-ATPαS overlaps with the singlet of AMPS. (Reprinted from ref. 62 with permission.)

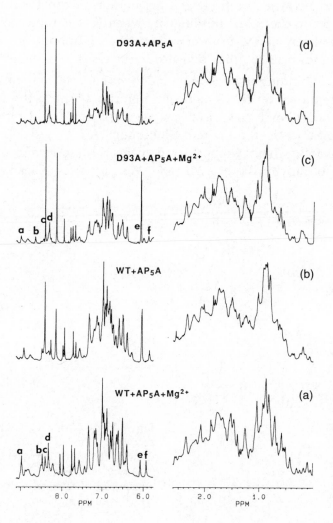

Fig. 7.14 500 MHz ¹H NMR spectra of adenylate kinase showing the effects of Mg²⁺ ions on the AP₅A complexes of wild-type (WT) and D93A mutant enzymes. (a) WT + AP₅A + Mg²⁺; (b) WT + AP₅A; (c) D93A mutant + AP₅A + Mg²⁺; and (d) D93A mutant + AP₅A. (Reprinted from ref. 50 with permission.)

ability of AP_5A to bind to the enzyme depends on Mg^{2+}, which in turn depends on D93. This fact is corroborated by the ^{31}P NMR studies in Fig. 7.15, which show the effects of added Mg^{2+} on the ^{31}P resonances of AP_5A in the presence of wild-type and D93A mutant enzymes.

More recent X-ray studies on the complex of $MgAP_5A$ at 2.6 Å resolution[63] confirm many of these findings, as shown in Fig. 7.16, in particular the arrangement of the AMP and MgATP binding sites.[64]

There is a hard lesson to be learnt from the somewhat tortuous history of adenylate kinase, which is that of consistent overinterpretation of data by many of the investigators involved over the years. The paramagnetic metal ion binding studies of Mildvan and co-workers, while undoubtedly representing the best available technology of the time, made crucial and evidently flawed interpretations as to the location of the bound metal ions and, by inference, the nature of the residues in close proximity. This highlights one of the weaknesses of the paramagnetic approach, since without detailed knowledge of location of the binding site, and rigorous elimination of adventitious or irrelevant binding patterns, the data can be very difficult to interpret. Mildvan and co-workers made an enormous contribution to the field, and, as we shall see in other examples of phosphoryl transfer enzymes, applied it to a number of difficult systems. However, a better approach nowadays is the use of the dipolar coupling between two spins that we discussed in Section 3.3. This has the great advantage that the probe used to measure distances can be an isotopic substitution made on the substrate molecule itself, so that in general, there can

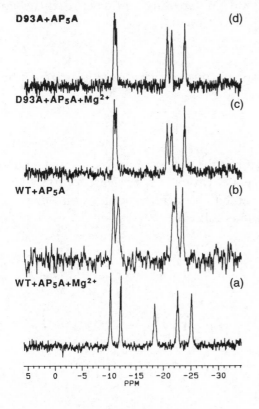

Fig. 7.15 121.5 MHz ^{31}P NMR spectra of adenylate kinase showing the effects of Mg^{2+} ions on AP_5A bound to wild-type (WT) and D93A mutant enzymes. (a) WT + AP_5A + Mg^{2+}; (b) WT + AP_5A; (c) D93A mutant + AP_5A + Mg^{2+}; and (d) D93A mutant + AP_5A. (Reprinted from ref. 50 with permission.)

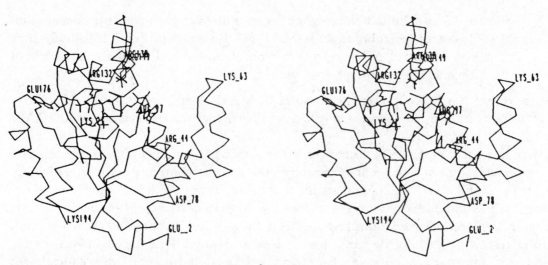

Fig. 7.16 Stereoview of the backbone of the 2.1 Å structure of free AK1p with side chains of Lys-21, Arg-44, Arg-97, Arg-132, Arg-138, and Arg-149. Using the chain fold overlay of AKy M·gAP$_5$A and free AK1p as a guide, AP$_5$A has been placed into the structure to demonstrate the relationship of the functionally important side chains in the free enzyme and the possible position of the bisubstrate analogue inhibitor deduced from the structure of AKy·MgAP$_5$A. (Reprinted from ref. 50 with permission.)

be no doubt as to where it binds. Alternatively, when the complete resonance assignments are available, the solution-state NMR studies will have a much firmer footing. Rösch and co-workers[65] have employed fractional deuterium labelling to assist in this endeavour, and Tsai and co-workers[66] have employed ^{15}N and ^{13}C labelling and have recently reported virtually complete three-dimensional NMR assignments.

7.2.2 Pyruvate kinase

Pyruvate kinase (PK, EC 2.7.1.40) catalyses the formation of pyruvate from phosphoenol-pyruvate (PEP):

$$PEP + ADP \rightleftharpoons Pyr + ATP$$

The mechanism is outlined in Fig. 7.17.

The mechanism involves nucleophilic attack of the β-phosphoryl oxygen of ADP on the PEP phosphorus atom, displacing enolpyruvate, which is not released into solution but rapidly rearranges to give pyruvate. At catalytic concentrations of the enzyme, the pyruvate kinase has an equilibrium constant as follows:

$$K_{eq} = \frac{[ADP][PEP]}{[ATP][Pyr]} \cong 3 \times 10^{-4}$$

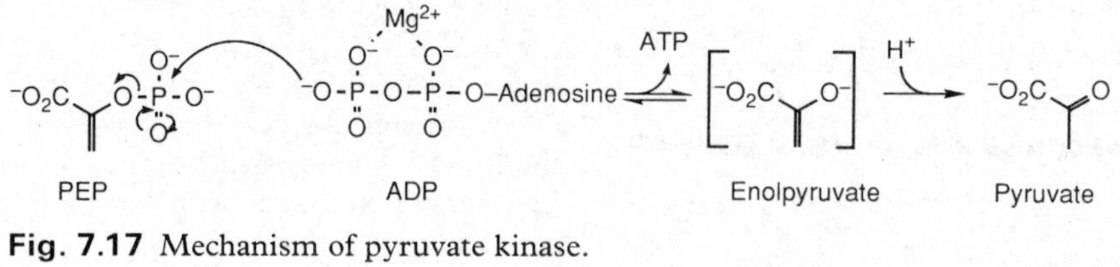

Fig. 7.17 Mechanism of pyruvate kinase.

This is clearly in favour of ATP production. One of the interesting results of the study of the bound substrates in this reaction (despite that PK is tetrameric and large, with $M_r = 59$ kDa per subunit) by ^{31}P NMR was the discovery[67] that in an equilibrium mixture with enzyme at concentrations in excess over substrate, the equilibrium constant $K'_{eq} \cong 1$. Another reaction catalysed by pyruvate kinase is between glycolate, ATP, on the one hand, and phosphoglycolate, ADP, on the other, and was known to have $K_{eq} \geq 50$ in favour of ADP, and yet again the equilibrium constant on the enzyme $K'_{eq} \cong 1$. Thus for the two reactions catalysed by pyruvate kinase for which the values of K_{eq} differ by around 10^5, both values of K'_{eq} are 1 regardless of whether K_{eq} is strongly in favour of one side of the reaction or the other. We now know of course that this is a general phenomenon for many enzymes, and simply reflects the fact that the free-energy change in the interconversion steps is very small, and the main contribution to free-energy changes comes from association and dissociation steps.

PK requires two metal ions, one of which is probably complexed directly with the enzyme and the other bound to the enzyme-bound nucleotide. The interaction of these two metals was studied in the complex $PK \cdot Mn^{2+} \cdot ATP \cdot Cr^{3+}$. The influence of the paramagnetic ions on the water relaxation rate was determined[68] in one of the first studies on the conformation of bound substrates using distances calculated from paramagnetic relaxation enhancements. The distance between the two paramagnetic metals was estimated to be around 5 Å. In further studies by ^{31}P, ^{13}C, and ^{1}H NMR, 15 separate distances were measured, allowing construction of a composite picture[69] from all the intersubstrate distances, as shown in Fig. 7.18. This model was put to the test with a 2.6 Å resolution X-ray crystal structure of the enzyme from cat muscle,[70] in which the Mg^{2+} site was identified as interacting with Glu-271 and the carboxylate oxygen of Ala-292 and Arg-293, together with a putative monovalent cation binding site at Glu-328 and Glu-363. PEP appears to interact with the phospho group close to Ser-242, and Lys-269 is the probable acid/base catalyst. The distances involved in the proposed substrate and metal binding to the uncomplexed crystal structure are consistent with Mildvan's NMR data.

The resonances of the nucleotide substrates were found to be unchanged when

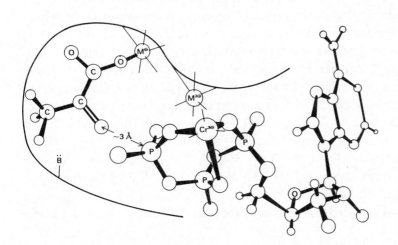

Fig. 7.18 Active-site geometry for pyruvate kinase, with chromium-ATP and pyruvate from proton relaxation enhancement measurements. (Reprinted from ref. 69 with permission.)

complexed to the enzyme, unlike the PEP resonances which shifted ≈ 1 ppm downfield from free PEP upon binding to the enzyme. Addition of Mg^{2+} resulted in a further ≈ 2 ppm shift downfield, and considerable linebroadening. When ADP is added, the PEP ^{31}P resonance is shifted a further 1.3 ppm downfield and narrowed somewhat. Addition of K^+ has no discernible effect on the signal. Changes in chemical shift are not easy to interpret, since they can represent changes in the local structure of the binding site as each substrate binds, or changes in ionization state at the phosphate groups. The contributions to ^{31}P chemical-shift changes in enzyme-substrate complexes will be discussed in some more detail in Section 8.4.3, and are nicely reviewed by Gorenstein.[71]

The binding of Mg^{2+} ions to the $PK \cdot ATP$ complex was monitored through changes in the chemical shift of the resonance due to the β-phosphate of bound ATP, which is always the resonance affected by metal complexation. Qualitatively, the apparent dissociation constant for the $E \cdot ATP \cdot Mg^{2+}$ complex was found to be larger than for the enzyme free $ATP \cdot Mg^{2+}$ complex, implying that Mg^{2+} may also bind to a secondary site in fast exchange with the primary one.

The C2H proton resonances of six out of the total of 14 histidine resonances per subunit could be detected and their pK_a values determined. The pK_a of one of the histidines was found to decrease upon formation of the $PK \cdot PEP$ complex.[72]

The stereochemical course of the PK-catalysed reaction was shown to proceed with inversion of configuration of the phospho group transferred from $[\gamma\text{-}^{16}O, {}^{17}O, {}^{18}O]ATP$, which is consistent with an in-line transfer mechanism without the formation of a phosphorylated enzyme intermediate.[73] Pyruvate kinase was shown to catalyse the reaction via an associative transfer mechanism (i.e. S_N2), by reaction with the substrates except pyruvate, and using $[\alpha,\beta\text{-}^{18}O, \beta\text{-}^{18}O_2]ATP$ with ^{16}O in the $\beta-\gamma$ bridging position, and also $[\gamma\text{-}^{16}O, {}^{17}O, {}^{18}O]ATP$. No scrambling was observed.[74]

Conformational changes in the active site of PK were probed using ^7Li relaxation of a LiCl solution in the presence of the Li^+-activated enzyme-Mn^{2+} complex.[75] It was concluded that the monovalent and divalent cation sites were separated by about 6 Å. Having said this, it was also deduced[76] that there are two structures of the $PK \cdot Mn^{2+} \cdot Li^+ \cdot PEP$ complex, with different Li^+-Mn^{2+} distances.

More recently ^{19}F NMR was used[77] to determine the stereochemistry of the PK-catalysed protonation of enolpyruvate using (Z)-fluorophosphoenolpyruvate, which is a known substrate for the enzyme.[78,79] The reaction was performed in D_2O, thereby yielding chiral $[3,3\text{-}^1H,^2H]$fluoropyruvate, which was coupled to pyruvate transcarboxylase and malate dehydrogenase, both enzymes whose stereochemical course was already known. If the deuteron transfer takes place with some degree of specificity, then an excess of C-3 protonated over C-3 deuterated $(2R,3R)$-fluoromalate would be expected to form in the coupled reaction. On the basis of the $^{19}F—^1H$ J couplings, for which it was known that trans couplings are larger than cis couplings, it was inferred that $^3J_{HF} \approx 25$ Hz corresponds to $(2R,3R)$-fluoromalate and $^3J_{HF} \approx 35$ Hz corresponds to $(2R,3S)$-fluoromalate.[80] The product from this reaction was indeed the $(2R,3R)$-fluoromalate, confirming earlier work with alternative substrates[81,82] or with PEP itself.[83] This established that proton addition to PEP takes place from the 2-*si* face, like many other PEP-utilizing enzymes.

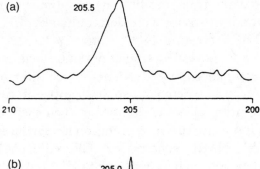

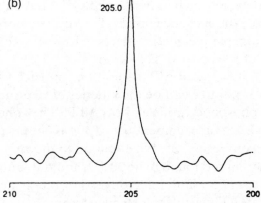

Fig. 7.19 Expansion in the region of C2 of pyruvate of the ^{13}C NMR spectra of pyruvate kinase plus (a) [2-^{13}C]pyruvate; and (b) sample (a) plus the inhibitor, oxalate. (Reprinted from ref. 84 with permission.)

Burbaum and Knowles have attempted to address the question of the nature of pyruvate when bound to PK by ^{13}C NMR spectroscopy.[84] They incubated PK (4 mM) with [2-^{13}C]- or [3-^{13}C]-pyruvate (2 mM) in the presence of ATP and ADP (see Fig. 7.19). Addition of the competitive inhibitor oxalate resulted in the displacement of bound [2-^{13}C]pyruvate and the sharpening of the resonance at $\delta_C = 205$ ppm. The bound resonance of C-2 of pyruvate had a linewidth of 100 Hz as compared with the free pyruvate resonance of 5 Hz, and was shifted downfield by 0.5 ppm. This was interpreted in terms of polarization of the keto carbonyl group when bound to the enzyme, presumably through proximity to the cations. Such polarization would make mechanistic sense, since electrophilic catalysis is expected during the formation of the enol or enolate intermediate or during the phospho group transfer to the C-2 oxygen.

7.2.3 Creatine kinase

Creatine kinase (CK, EC 2.7 3.2) makes use of the reservoir of phosphocreatine (PCr) in vertebrates via transfer of the phosphate group of phosphocreatine to ADP (see Fig. 7.20).

As with all kinase reactions, a divalent metal such as Mg^{2+} is required for activity. All

Fig. 7.20 Transfer of the phosphate group of phosphocreatine to ADP.

NMR studies have focused on the enzyme from rabbit muscle (dimer with total $M_r = 83$ kDa). The chemical shifts of bound substrates were determined by ^{31}P NMR.[85] Quaternary complexes $CK \cdot ADP \cdot Mg^{2+} \cdot creatine \cdot NO_3^-$ and $CK \cdot ADP \cdot Mg^{2+} \cdot creatine \cdot HCO_2^-$, which were thought to be so-called transition-state analogues (see Section 6.3), showed upfield shifted β-resonances by ^{31}P NMR, which split into two resonances at low temperature. This was interpreted as an indication of two non-equivalent active sites (with substrate in intermediate exchange between them at room temperature).[86] However, the evidence supporting this interpretation is very flimsy. The effects of linebroadening on the ^{31}P NMR spectra of $CK \cdot ADP \cdot M^{2+}$ and $CK \cdot ATP \cdot M^{2+}$, where M is a paramagnetic ion such as Mn^{2+} and Co^{2+}, have also been examined. The linebroadening was found to be exchange limited for Mn^{2+}, and therefore unsuitable for estimating distances using paramagnetic effects. However, for Co^{2+}, ^{31}P-metal ion distances were estimated, of the order of 2.4–4.6 Å.[87]

Campbell, Radda, and co-workers[88] have used ^{31}P saturation transfer and the ACCORDION experiment (see Section 1.3.6) to examine the kinetics of creatine kinase catalysed phosphoryl exchange between phosphocreatine and MgATP. They obtained a pseudo first-order rate constant of about 0.3 s^{-1} at a temperature of 304 K for the transfer in the PCr direction. The substrate interconversion rate when bound to the enzyme was estimated from a complete lineshape analysis, and found to be 90 s^{-1}, and the equilibrium constant was determined to be $K_{eq} = 1$ at 277 K. The interconversion rate is nearly 20 times faster than the overall reaction rate. Thus the transfer step cannot be the rate limiting factor.

Creatine kinase was shown to catalyse an associative reaction mechanism (S_N2), in which bond formation to the second substrate occurs prior to bond cleavage. ATP with ^{18}O in three β-positions (one $\alpha\beta$-bridging and two non-bridging) was incubated with CK and the competitive inhibitors L-arginine and tauromycin. A dissociative reaction scheme (S_N1) would be expected to lead to bond cleavage of the enzyme-bound substrate and thus scrambling of the ^{18}O in the β-position observed in the free [^{18}O$_3$]ATP. No such scrambling was observed by ^{31}P NMR.[89,90]

Creatine kinase is rather large for ^{1}H NMR studies, although evidence was obtained on the CK-formate complex that a lysyl residue is involved in the enzyme-substrate complex.[91] One-dimensional NOE experiments[92] also implicated an arginine residue when the protein resonances at 0.9 and 1.7 ppm were irradiated and NOEs were measured to C2H of the adenine ring in the $CK \cdot ADP$ complex. Attempts to observe this NOE by irradiation of the C2H (and also C8H) of the adenine ring led to negative NOEs for several protein resonances in the aliphatic region including those at 0.9 and 1.7 ppm, and several aromatic resonances.[93] However, these results need to be regarded with caution, since spin diffusion for a protein of this size will be extensive.

There are 16 histidine residues per subunit in CK, and one C4H and six imidazole C2H resonances were detected by ^{1}H NMR. The pK_a of one of these was found to be 7.6 in the presence of phosphorylated substrates and 7.0 in the presence of creatine. This was interpreted as implying that the observed histidine may be responsible for acid/base catalysis of the phosphotransfer. Distances of three histidine residues from the paramagnetic Cr^{3+} of bound $ATP \cdot Cr^{3+}$ could be estimated to be from 12 to 14 Å by analysis of linebroadening effects. Furthermore the off-rate was determined to be 350 s^{-1} directly from ADP · Mg^{2+} exchange linebroadening.[94]

7.2.4 3-Phosphoglycerate kinase

The enzyme 3-phosphoglycerate kinase (PGK, EC 2.7.2.3) catalyses the interconversion of 3-phosphoglyceric acid (3-PG) and 1,3-diphosphoglycerate (1,3-DPG):

$$3\text{-PG} + \text{ATP} \rightleftharpoons 1,3\text{-DPG} + \text{ADP}$$

The mechanism is very similar to pyruvate kinase. The enzyme which has been studied by NMR is the yeast PGK ($M_r = 47$ kDa), although there has been a study on the enzyme from halibut.[95] ^{31}P NMR has been used to measure the equilibrium constant for the substrates bound to the enzyme ($K'_{eq} \approx 1$), which compares with the equilibrium constant of the overall reaction, $K_{eq} = 3 \times 10^{-4}$. Forward and reverse reaction rates on the enzyme were also measured to be 100 s^{-1}. Given that the ratio of the concentrations of the bound substrates [E · Mg^{2+} · ADP · 1,3-DPG]/[E · Mg^{2+} · ATP · 3-PG] = 0.8, and the corresponding ratio of free substrate concentrations was around 3000, this implies that the phosphotransfer is not rate-limiting. The ^{31}P chemical shifts for ATP, MgATP, and ADP on binding to the enzyme did not change significantly, unlike MgADP, whose α- and β-phosphate shifted 1.1 and 1.6 ppm upfield respectively.[96] Changes in the chemical shifts of approximately 2 ppm were induced for 3-PG and 1,3-DPG upon binding. Two resonances for the β-phosphate of enzyme-bound ATP in the presence of Mg^{2+} were observed, one at the position expected for the metal complex, and one at the position of the metal-free compound, suggesting the presence of a second ATP binding site with a larger dissociation constant for the metal complex. This has been confirmed by subsequent ^{31}P NMR studies[97] on yeast PGK, in which it was found that for Mg · ATP binding K_2(1st site) = 65 μM and K'_2(2nd site) = 5 mM; and for ATP binding K_S(1st site) = 70 μM and K'_S(2nd site) = 0.5 mM. There is also evidence that ADP binds to two sites, presumably the same as the ATP sites.[98] Also it was found that none of the ^{31}P resonances of bound species titrated in the pH range 6.4–9.0. As part of the same study, Nageswara Rao and co-workers[99] used paramagnetic relaxation enhancements of ^{31}P to measure distances in the E · CoADP and E · CoATP complexes, which for Co(II)–^{31}P were all in the range of 2.7–4.1 Å, and in the quaternary complex E · MnADP · 3-PG complex, which for Mn(II)–^{31}P was 11.1 ± 0.3 Å. The latter distance suggests that in E · MnADP · 3-PG, the enzyme is in an open conformation.

From the X-ray structure, PGK has a distinctly bilobal appearance (see Fig. 7.21), with the MgADP binding site located in one domain and the 1,3-DPG binding site in the other domain separated, as we have seen, by about 11 Å. Upon binding of the substrates, it was hypothesized[100,101] that the two domains swing together about a hinge region, closing down on the substrates and excluding water. Despite the high molecular weight of PGK, Williams and co-workers have used ^{1}H NMR in conjunction with site-directed mutagenesis, in an attempt to investigate the large changes in protein conformation postulated to occur during substrate binding. As an initial step they attempted[102] to assign the ^{1}H NMR resonances of some of the residues and probe the triose and anion binding sites. Highly conserved residues in a 'basic-patch' region located at the hinge between the two domains were mutated, and the ^{1}H NMR one-dimensional and two-dimensional NOESY spectra were compared. The conserved Arg-168, was replaced with lysine (R168K) and methionine (R168M), and the non-conserved His-170 was replaced with aspartate (H170D). The spectra revealed that only minor structural changes accompanied these mutations. These studies also led to robust assignments for three histidine residues. NMR was used to measure the K_d values for 3-PG binding to the mutants, which was found to decrease 15-fold and four-fold for mutants R168M and H170D respectively. This reduction was rationalized in terms of the compromised ability of the mutant enzymes to undergo the necessary conformational changes proposed to accompany substrate binding. It was further concluded that His-170 is not of major importance in substrate binding, despite the implication from the X-ray structure that this residue is equally important as His-62 in substrate binding. Further studies by Sherman and co-workers[103] on the R38A and H62A mutants suggest that R38 is catalytically essential, whereas His-62 is important for anion binding, but not for catalytic activity. NMR studies also suggest that conformational changes which accompany substrate binding are required for anion- and substrate-dependent activation.

The relationship between the two domains was investigated[104] by genetically engineering the expression of the N- and C-domains separately. ^{1}H NMR of the two

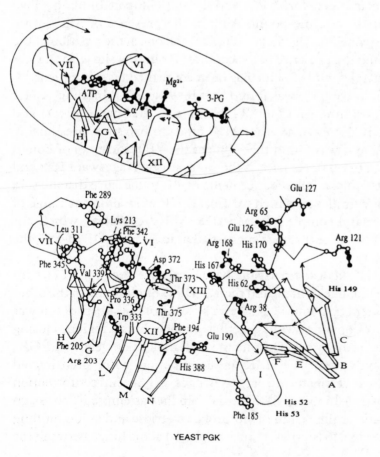

Fig. 7.21 Schematic drawing of the active-site cleft of yeast phosphoglycerate kinase showing the binding site for $Mg^{2+} \cdot ATP$ and 3-phosphoglycerate as determined from X-ray crystallography. The helical segments are denoted by cylinders and the β-sheet strands by arrows. Residues Phe 194, Phe 205, and His 388 in the interdomain region of the protein are also shown. (Reprinted from ref. 106 with permission.)

separate domains suggested that both folded with regions of native-like structure, with the N-domain showing greater conformational flexibility than the C-domain (for example see Fig. 7.22). The 'basic patch' region of the N-domain is perturbed by the removal of the C-domain, manifested in changes in pK_a values for all the observable histidine residues. This change is presumably because of the absence of stabilizing interactions between the two domains. The C-domain was also found to bind nucleotides only three times less than that of the native protein.

When the interdomain residue His-388 was replaced with glutamine (H388Q) and the ^{1}H NMR spectra was compared,[105] it was found that the nature of the conformational

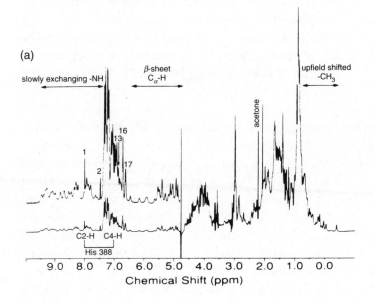

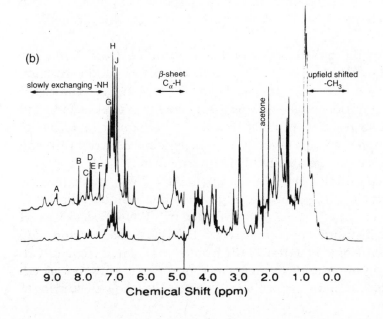

Fig. 7.22 500 MHz ^{1}H NMR spectra of (a) the C-terminal domain and (b) the N-terminal domain of yeast phosphoglycerate kinase in D$_2$O. (Reprinted from ref. 104 with permission.)

changes induced by 3-PG binding was different for the wild-type versus the mutant proteins. These differences were not only in the hinge region but also in the long-range regions. Since this observation was accompanied by a 30-fold lowering of the K_d for 3-PG with the mutant, these results suggest that an aromatic residue at position 388 plays an important role in the hinge-bending process. Investigation of site mutants in the interdomain region were carried out[106] for F194L and F194W, and from the NMR data, Phe-194 is clearly important for the interdomain mobility, and F194W reduces the mobility, whereas F194L increases it.

The anion binding site was investigated[107] using paramagnetic probes ($[Cr(CN)_6]^{3-}$ and $[Fe(CN)_6]^{3-}$) and it was found that Arg-65 and Arg-168, located in the 'basic patch' region, interact with the probes in the absence of substrate, and less strongly with His-62 (or His-167, His-170). The anions were found to be directly competitive with 3-PG, confirming the earlier observation of the involvement of His-62 and Arg-168 in triose binding. The conclusion from these data suggest that the X-ray structure represents an 'open' conformation, containing a general anion binding site comprising Arg-65, Arg-168, and His-62.

7.2.5 Phosphoenolpyruvate-dependent phosphotransferase system

The phosphoenolpyruvate-dependent phosphotransferase system (PTS) found in anaerobic and facultatively anaerobic bacteria is a multienzyme system transferring a phosphate group from PEP to a carbohydrate. This is performed during transport through the cell membrane, and in the case of *Staphylococcus aureus* the enzyme system consists of four separate proteins: Enzyme I (E I), Heat-stable Protein (HPr), Factor 3 (F III) and Enzyme 2 (E II). E II is membrane-bound, whereas the other three are located in the cytosol. The reaction scheme is outlined in Fig. 7.23. E I and HPr are non-specific, whereas F III and E II are sugar-specific proteins. F III is substituted by a multifunctional protein called enzyme 3 (E III) in *E.coli* and *S.typhimurium*. E III is believed to have additional regulatory properties, in that the unphosphorylated E III can stimulate adenylate cyclase activity, whereas when phosphorylated this property is switched off.

The stereochemical course of the phosphoryl transfer of the PTS from *E.coli* was investigated starting with $[^{16}O, ^{17}O, ^{18}O]PEP$. The stereochemistry of the reaction product, methyl-α-D-glucopyranoside, was analysed by ^{31}P NMR. As Knowles[108] suggested, an inversion of configuration at phosphorus is expected to take place for every in-line phosphoryl transfer. The reaction product was found to exhibit inversion of configuration. This, in turn, implies an odd number of transfer steps between PEP and product, suggesting that there is a phosphorylated E II intermediate.[109]

NMR studies initially focused on HPr, since this is small ($M_r = 7650$) and extremely stable. ^1H NMR studies suggested that N^δ of the single histidine residue (His-15) is phosphorylated in the *S.aureus* enzyme.[110] This was confirmed by ^{31}P NMR as well as ^1H NMR data for model phosphohistidine compounds.[111,112] Robillard, Bachovchin, and co-workers[113] have used ^{15}N NMR of $[^{15}N$-His-15$]HPr$ from *E.coli*, which contains two histidines (His-15 and His-76) in order to determine which nitrogen is phosphorylated. HPr was labelled with either $[N^{\delta 1}, N^{\varepsilon 2}]His$ or $[N^{\delta 1}]His$. Their results show that, prior to phosphorylation, His-15 has an abnormally low $pK_a(5.8)$, and $N^{\varepsilon 2}$ is strongly hydrogen bonded, probably to a carboxylate. It was postulated that this hydrogen bond would strengthen the nucleophilic character of the deprotonated $N^{\delta 1}$, resulting in a good acceptor for the phosphoryl group. In the phosphorylated HPr, the pK_a of His-15 shifts to 8.0, and the hydrogen bond to the $N^{\varepsilon 1}$ is disrupted (see Fig. 7.24). The chemical shifts

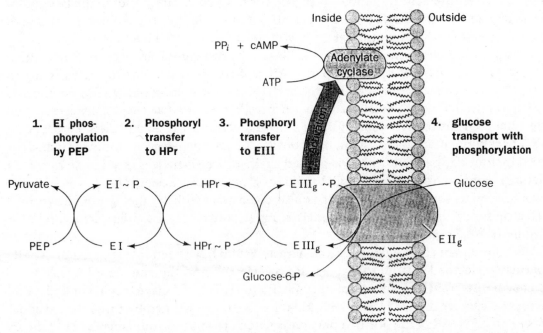

Fig. 7.23 Transport of glucose by the PEP-dependent phosphotransferase system (PTS). HPr and E I are cytoplasmic proteins common to all sugars transported. E II$_g$ and E III$_g$ refer to E II^{Glc} and E III^{Glc}, respectively, and are proteins specific for glucose. Adenylate cyclase is activated in the presence of E III$_g$–P. (Adapted from D. Voet and J. G. Voet, *Biochemistry*, John Wiley and Sons, New York, p. 449 (1990) with permission.)

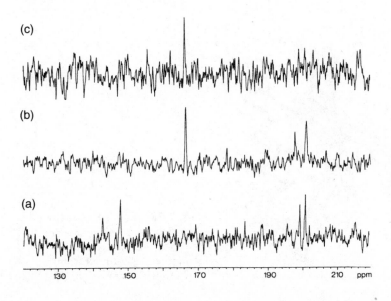

Fig. 7.24 ^{15}N NMR spectra of HPr and PHPr at pH 6.7: (a) HPr containing doubly $^{15}N^{\delta 1}$, $^{15}N^{\varepsilon 2}$-labelled histidine in the presence of all the components required for phosphorylation except PEP; (b) PHPr containing doubly labelled histidine; and (c) PHPr containing $^{15}N^{\delta 1}$-labelled histidine. (Reprinted from ref. 113 with permission.)

were interpreted according to Table 7.1 (p. 269). Changes in the chemical shifts for His-76 upon phosphorylation suggest that there is a conformational change extending beyond the active site when phosphorylation occurs. Wittekind and co-workers[114] found that the same conformational change that accompanies phosphorylation can be achieved through site-directed mutagenesis of Ser-46 to Asp.

The first structure determinations of *E.coli* HPr have been carried out in the laboratory of Delbaere and co-workers by X-ray crystallography[115] and in the laboratory of Klevit and co-workers by two-dimensional ^{1}H NMR spectroscopy.[116] These two structures are considerably different, and it appears that both contain significant errors. Subsequent work by Klevit's laboratory showed that their first structure contained assignment errors,[117] and since then complete ^{1}H, ^{13}C, and ^{15}N three-dimensional NMR assignments have been carried out by Robillard and co-workers,[118] and the structure firmly established using 1200 NOE restraints, which for the nine final structures had an RMSD = 1.2 Å for the C^α backbone atoms. The structure consists of an open-face β sandwich which is formed by four antiparallel β strands packed against three α-helices (see Fig. 7.25). This structure is consistent with the overall topology of the structure determined by Klevit and co-workers, with some differences. It is also consistent with (i) the X-ray structure[119] NMR structure[120] of the *Bacillus subtilis* HPr, and (ii) the NMR secondary structure[121] for the *Staphylococcus aureus* HPr.

Other components of the phosphotransferase system have been investigated by NMR. In particular, factor III from the lactose PTS (III^{Lac})[122] and the glucose PTS (III^{Glc}),[123] and enzyme II from the mannitol-specific PTS (E II^{Mtl})[124] have been studied. The secondary structure of III^{Glc} (M_r = 18.1 kDa) has been determined by three-dimensional NMR. The ^{31}P NMR of the purified phosphorylated cytoplasmic domain of E II^{Mtl} shows two resonances consistent with phosphohistidine and phosphocysteine. Addition of mannitol plus membranes containing the N-terminal mannitol-binding domain of E II^{Mtl} resulted in the formation of mannitol-1-phosphate and the disappearance of the two

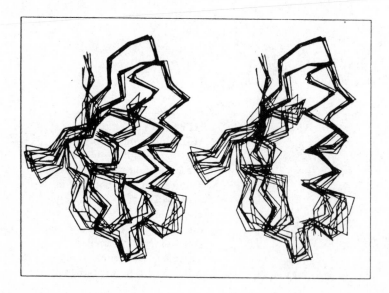

Fig. 7.25 Fold of *E.coli* HPr after final refinement, showing the superposition of nine structures. (Reprinted from ref. 118 with permission.)

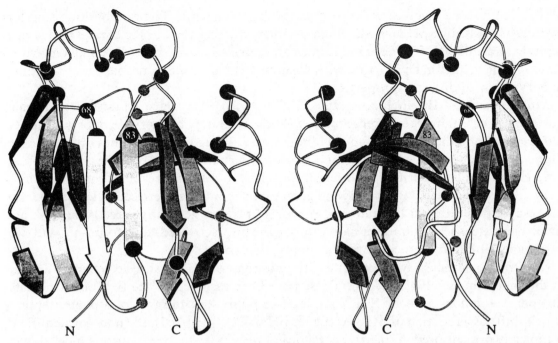

Fig. 7.26 Schematic representation of the solution structure of the IIA^{Glc} domain. The two views differ by a 180° rotation. The positions of amino acids conserved in six sequences are indicated by spheres at the C$^{\alpha}$ positions. Conserved glycines and prolines, which may be expected to fulfil a structural role, are represented by small grey spheres. All other totally conserved residues are represented by larger black spheres. The site of phosphorylation, His83, is labelled, as is the site of His68, which has been implicated in the phosphoryl transfer mechanism by site-directed mutagenesis experiments. (Reprinted from ref. 127 with permission.)

resonances. Homology comparisons suggest that cysteine may be the important phosphorylated residue. These results also confirm Knowles' prediction as to the involvement of a phosphorylated E II. The domain that HPr binds to on II^{Mtl} has been mapped using [^{15}N]HPr, in which it was found that 14 out of 85 amino acid residues are involved.[125] The ^{1}H, ^{13}C, and ^{15}N NMR spectra of glucose permease IIA^{Glc} have been completely assigned by three-dimensional NMR methods by Wright and co-workers,[126] a low-resolution structure was reported (see Fig. 7.26),[127] and the binding interface between HPr and IIA^{glc} was mapped onto both proteins,[128] in which it was found to involve predominantly hydrophobic surfaces near the active site His-15 of HPr and the phosphoryl acceptor His-83 of IIA^{Glc}.

7.2.6 Alkaline phosphatase

Alkaline phosphatase (AP, EC 3.13.1) is a metalloenzyme with a maximal activity above pH 7.5 AP (dimer, $M_r = 48$ kDa per subunit) catalyses the non-specific hydrolysis of phosphomonoesters, and it is widely accepted that a covalent phosphoserine intermediate (E—P) and a non-covalent E·P complex play an important role in catalysis:

$$\text{ROP} + \text{E} \rightleftharpoons \text{E·ROP} \rightleftharpoons \text{E–P} + \text{ROH} \rightleftharpoons \text{E–P} + \text{ROH} \rightleftharpoons \text{E} + \text{P} + \text{ROH}$$

The serine residue involved is Ser-102 in the *E.coli* enzyme. The rate limiting step for hydrolysis at alkaline pH is the $E \cdot P$ dissociation, whereas the E—P cleavage step is rate limiting at acidic pH. The enzyme is normally purified as the zinc (II) metalloenzyme, although the metal can be replaced with cadmium, cobalt, copper, or manganese, leading to enzymes with different properties.

Multinuclear NMR, including ^{13}C, ^{19}F, ^{31}P, and ^{113}Cd, have addressed two main questions concerning the number of covalently bound phosphoryl groups per active dimeric unit, and the number of metal binding sites. There has been some confusion as to the correct number of metal ions in this protein, principally arising because enzymes from different sources and purified in different ways lead to different results. This situation is quite common for metalloenzymes, and great care needs to be exercised when comparing the results from different laboratories. However, it is now generally agreed that there are six divalent metal ions bound per native unit in the fully activated state, four Zn^{2+} and two Mg^{2+} ions, and that there are two phosphorylated sites.[129]

On titration of the Zn_4Mg_2–enzyme with phosphate at pH 5.5 three resonances appear in the spectrum (see Fig. 7.27(a)). These have been assigned to the E—P intermediate (8.4 ppm), the non-covalent $E \cdot P$ complex (4.2 ppm), and the free phosphate (0 ppm, which shifts as a function of pH). At higher pH values (pH > 6.5), the non-covalent $E \cdot P$ complex is present predominantly; at values of pH < 5.0, the E—P form is most stable, and no $E \cdot P$ can be observed in the spectra.[130-133] These measurements correlate well with earlier studies on the pH stability of the intermediates. The chemical shifts for the bound species are anomalously downfield, and this has been attributed to bond strain introduced by the enzyme. In studies of the Cd_4Mg_2–enzyme, the covalent E—P appeared at 8.4 ppm, identical to the Zn enzyme, but was stable over a wider range of alkaline pH values. Enzyme-bound phosphate cannot be observed for the Co- and Mn-substituted enzymes at an enzyme:phosphate ratio of 1:1. When more phosphate is added, a resonance attributable to $E \cdot P$ appears, suggesting that the first phosphate is bound very closely to the rigidly held metal ion. In additional studies with the ^{113}Cd-substituted enzyme, the $E \cdot P$ but not the E—P, appears as a doublet due to ^{113}Cd—^{31}P $\mathcal{J}$ coupling (see Fig. 7.27(c)).[131]

Analysis of the linewidth of the $E \cdot P$ complex indicated that dissociation of the non-covalent phosphate is the rate-limiting step in the turnover of the enzyme.[132] A study of the phosphorylated apoenzyme revealed that the linewidth observed was much narrower than for the metal-containing enzyme. This implies considerable flexibility for the residue, in contrast to the holoenzyme, where the observed linewidth has been interpreted as being held rigidly. pH titrations showed that the residue in the apoenzyme cannot be protonated until the enzyme dissociates and unfolds at low pH. The phosphorus nucleus in the apoenzyme is coupled to the methylene protons of the serine with a coupling constant of 13 Hz, indicating that the phosphate group is *syn* to the C^{α} of serine, rather than *anti*.

The kinetics of the reaction were investigated using saturation transfer measurements.[131] The results of these experiments are shown in Fig. 7.28, in which saturation of the E—P resonance reduced the intensities of $E \cdot P$ and P_i by 24 and 12% respectively. From such effects, given the values for T_1 relaxation times, the kinetic information can be extracted, assuming the following kinetic scheme:

$$E + P_i \underset{k_{-1}}{\overset{k_1}{\rightleftharpoons}} E \cdot P \underset{k_{-2}}{\overset{k_2}{\rightleftharpoons}} E - P + H_2O$$

It was found that $k_{-1} > 10 \text{ s}^{-1}$, whereas k_2 and k_{-2} are 0.19 and 0.23 s^{-1} respectively.[131] These values are in excellent agreement with those obtained by other methods.

The phosphoryl group transfer from [R-^{16}O, ^{17}O, ^{18}O]phenylphosphate to the product 1-[^{16}O, ^{17}O, ^{18}O]phosphopropane-1,2-diol was found to proceed with retention of configuration,[134] consistent with a pair of S_N2 reactions, or, less likely, an S_N1 (A-1) mechanism. This aspect was resolved by examination of the ^{18}O-induced isotope shift of the ^{31}P resonance of the [^{16}O$_3$, ^{18}O]phosphate liberated by alkaline phosphatase hydrolysis of benzyl phosphate in [^{18}O]H$_2$O. This indicated that fission of the P—O bond had taken

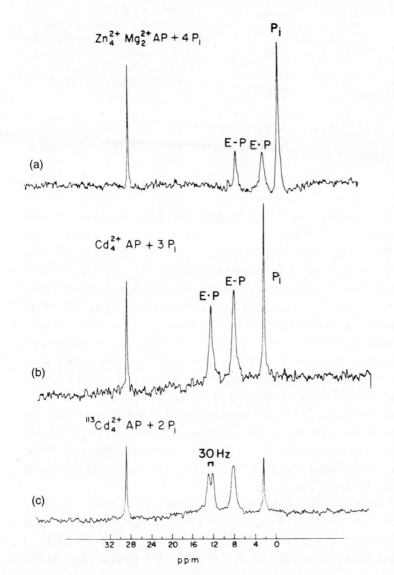

(a)

(b)

(c)

Fig. 7.27 ^{31}P NMR spectra of alkaline phosphatases (a) $Zn_4^{2+}Mg_2^{2+}$ enzyme plus 9.6 mM phosphate, pH 5.5; (b) Cd_4^{2+} enzyme (natural abundance ^{113}Cd) plus 5.25 mM phosphate, pH 8.9; (c) $^{113}Cd_4^{2+}$ enzyme (enriched in ^{113}Cd) plus 5.8 mM phosphate, pH 9.1. (Reprinted from ref. 131 with permission.)

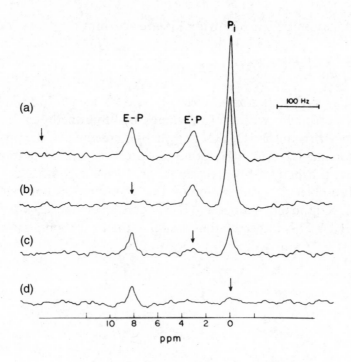

Fig. 7.28 ^{31}P NMR saturation transfer spectra of $Zn_4^{2+}Mg_2^{2+}$ alkaline phosphatase plus 9.6 mM phosphate at pH 5.5. The arrows indicate the frequencies at which the saturating pulse was applied, with (a) off resonance; (b) at the E–P resonance; (c) at the E·P resonance; and (d) at the P_i resonance. (Reprinted from ref. 131 with permission.)

place, implicating a pair of S_N2 reactions, as might be expected from the kinetic scheme outlined above.

7.2.7 ATPase

The ATPases are rather large proteins (EC 3.6.1.3, $M_r = 250$ kDa) and membrane-embedded. In general they catalyse the active transport of ions across a membrane with concomitant hydrolysis of ATP. In spite of its size, ^{31}P NMR has been used to observe the catalytic phosphoaspartate intermediate in detergent-solubilized Na$^+$/K$^+$ ATPase.[135] The resonance can be observed for low protein concentrations, in an unusually short time, probably owing to a remarkably short $T_1 = 0.2$ s. The resonance appears at -17 ppm, which is ≈ 15 ppm upfield from the region where acyl phosphate resonances are usually observed. The support for the assignment came from the following: (i) the resonance appeared both after addition of the substrate ATP or the product P_i; (ii) the resonance disappeared on addition of K$^+$ (which had been reported to destabilize the intermediate); (iii) a similar chemical shift was observed for a model compound, although the latter was not purified and was poorly characterized. Curiously, no resonances were observed for the phospholipids known to co-purify with these enzymes. Grisham and co-workers[136] have used paramagnetic relaxation enhancements to infer the geometry of the ATP triphosphate side chain, and, in conjunction with TRNOEs, to infer the conformation of the bound ATP; their model is shown in Fig. 7.29.

The Ca^{2+}-ATPase of rabbit skeletal muscle also exhibited a similar resonance arising from phosphorylated aspartate. The Ca^{2+}-ATPase from rabbit sarcoplasmic reticulum (SR) has been studied extensively by a variety of NMR experiments. This enzyme is responsible for the active transport of calcium across the SR membrane, and exists in the

SR as a 115 kDa protein with associated, essential phospholipids. Clore and co-workers[137] determined the conformations of bound nucleotide (MgATP, MgADP, and MgAMPPNP) with TRNOE experiments. It was found that intramolecular NOEs between H8 and H1′ occurred in the bound state. Therefore the conformation of the bound nucleotide could be inferred, which was *anti* about the glycosidic bond (see Section 9.1.1). Grisham and co-workers[138] have used Gd^{3+} ions to examine the proton relaxation times for two high-affinity Ca^{2+} transport sites to which the Gd^{3+} binds. These sites are occluded with a very low accessibility to solvent water. In the presence of MgATP or $Co(NH_4)ATP$, a new state for bound Gd^{3+} with one less water molecule of hydration is observed, together with two additional states, one of which corresponds to the E—P intermediate.

The ^{18}O exchange reaction catalysed by gastric H^+/K^+ ATPase has been examined by ^{31}P NMR spectroscopy. It was found that ^{18}O exchange occurs when the enzyme is incubated with [^{18}O]phosphate, and the resultant E—P intermediate is detected by ^{31}P NMR. From the time dependence of the relative concentration of [^{18}O]phosphate, the pseudo first-order rate constant for the cleavage step could be estimated. It was also found that ^{18}O exchange was strictly dependent upon the presence of Ca^{2+} ions. This allowed the intermediacy of a phosphoryl enzyme to be firmly established.[139]

The F_1F_0–ATPase differs from the previous ATPases, which hydrolyse ATP in order to pump ions across membranes, in that it is the enzyme responsible for ATP synthesis, and as such is not strictly a phosphatase. However, given its close relationship to the aforementioned ATPases, we will consider it briefly. NMR studies have been mainly concerned with the F_1–ATP synthase ($M_r = 371$ kDa) which is the peripheral membrane protein where ATP synthesis occurs. Clore, Gronenborn, and co-workers[140] studied the mitochondrial F_1–ATPase, which contains three exchangeable nucleotide binding sites. Using TRNOE measurements they determined the conformation of bound nucleotides ATP, ADP, and nucleotide analogues, 8-azido-ATP, 8-azido-ADP, and AMPPNP. These analogues are predominantly in the *syn* conformation in solution, and were found to

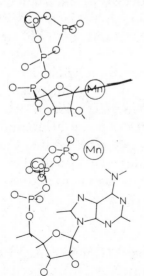

Fig. 7.29 Conformation of CoATP bound at the active site of Na^+,K^+-ATPase, based on paramagnetic relaxation studies and two-dimensional TRNOE measurements. (Reprinted from ref. 136 with permission.)

be in the *anti* conformation when bound to the F_1 catalytic sites. This was found from TRNOEs observed between the H8 ring proton and the H2′ and H3′ sugar protons, irrespective of nucleotide and whether Mg^{2+} was present. The *anti* conformation is indicative of the 2′-*endo* sugar pucker (see Section 9.1.1).

The F_1–ATPase from chloroplasts (CF_1, $M_r = 400$ kDa, five subunits with stoichiometry $\alpha_3\beta_3\gamma\delta\epsilon$) has been examined by ^1H and ^{31}P NMR by Grisham and co-workers,[141] and, independently, by Haddy and co-workers,[142,143] who examined the interaction of nucleotides and nucleotide analogues and their metal complexes with Mn^{2+} bound to the enzyme. They used kinetics, water proton relaxation enhancements (PRE), and ^1H and ^{31}P relaxation measurements on the substrates. CF_1 has three nucleotide binding sites, one of which is thought to be a regulatory ADP binding site, the second binds MgATP with a dissociation half-life of several days, and the third is believed to be the catalytic site for ADP/ATP turnover. There are 1–3 metal binding sites in the absence of added ATP and an additional class of weaker non-interacting metal sites in the presence of ATP. Titration of activated Mn^{2+}-CF_1 complexes with ATP, ADP, P_i, $Co(NH_3)_4ATP$, $Co(NH_3)_4ADP$, and β,γ-bidentate $Co(NH_3)_4AMPPCP$ (adenosine 5′-[β,γ-methylene-triphosphate]) leads to increases in the water relaxation enhancement, consistent with enhanced metal binding. The ^1H and ^{31}P relaxation measurements were used to determine the conformation of the bound nucleotide analogue $Co(NH_3)_4AMPPCP$, and its arrangement with respect to the high- and low-affinity sites of Mn^{2+}. The average Mn^{2+}–P distance was found to be 5.29 Å, in moderate agreement with EXAFS data. Distances were also derived from the Mn^{2+} to H2 and H8 of the adenine ring, and H1′, H3′, H4′, and H5′ of the sugar. On the basis of these distances, molecular modelling was used to determine the nucleotide conformation, which was found to be *syn*, in contrast to the other ATPases. In more recent work, Campbell and co-workers[144] examined nucleotide binding to active and 4-chloro-7-nitrobenzofurazan (Nbf-Cl) inhibited form of CF_1. The latter inhibitor modifies a single tyrosine and at least two cysteine residues, although pre-treatment of CF_1 with N-ethylmaleimide (NEM) reduced the modification to Tyr-328 on the β-subunit. ^{31}P and ^1H NMR were used to identify a single tight-binding site for the Ca^{2+} complex of the ATP analogue 5′-adenylylimidodiphosphate (AMPPNP). The Nbf-Cl modified enzyme was found to bind only one ADP and one ATP, although resonances from the bound ATP were not detectable.

The F_0 part of F_1F_0—ATP synthase has three kinds of subunits, designated a, b, and c. The c subunit is a transmembrane protein, which is thought to be important in the formation of the proton channel. In the *E.coli* form of the enzyme, it consists of 79 residues ($M_r = 8290$), and, using ^1H and ^{15}N heteronuclear three-dimensional NMR methods, Campbell and co-workers[145] have assigned the spectrum and determined the secondary structure. The protein contains two extended regions of predominant α-helical character separated by an eight-residue segment which displays little evidence of ordered secondary structure.

7.2.8 Ribonuclease

As we mentioned in the introduction to Part III, ribonuclease A (RNase A, EC 3.1.27.5, 124 residues, $M_r = 14$ kDa) was the first enzyme to be examined by NMR spectroscopy, and that original spectrum was reproduced in Fig. 1.1. It was also one of the first enzymes whose protein folding was examined by NMR (see Section 5.6.1). Despite this, and the considerable amount of NMR work that has been carried out on the enzyme, the three-dimensional structure of ribonuclease A by NMR has only just been published,[146,147] presumably because the X-ray structure has been available[148] since 1971. Other ribonucleases whose three-dimensional structures by NMR have been reported include ribonuclease F1[149], H[150], and T1.[151] In this section we will consider mainly RNase A. The earliest studies, by Jardetsky and co-workers,[152] proposed a structure for the binding of the inhibitor 3'-CMP to RNase on the basis of changes in the titration behaviour of the histidine ^{1}H resonances in the protein. For example, the pK_a for His-12 and His-119 changed from 5.8 and 6.2, respectively, in the enzyme alone, to 7.4 and 8.0, respectively, in the enzyme-inhibitor complex. On the basis of proposed charge states of those resonances analysed, the inhibitor was manually 'docked' into the enzyme 2 Å X-ray crystal structure. Remarkably, the agreement was quite good with subsequent X-ray structures of the 3'-CMP-RNase complex, even though it turned out that the original assignment of the two histidine resonances was in fact reversed. With 2'-CMP it was established that the inhibitor phosphate group is dianionic when bound to RNase in solution.[153,154] The difference between the binding modes for 2'-CMP and 3'-CMP was probed[155] by ^{1}H and ^{15}N NMR, and the formation of a hydrogen bond for 3'-CMP is the primary reason for their relative specificity.

The so-called transition-state analogue, uridine vanadate, when bound to RNaseA, has been examined[156] by ^{1}H and ^{51}V solution-state NMR and neutron diffraction in the solid state (see Fig. 7.30). This work established that in the complex, the resonance of His-12 and His-119 do not titrate. The diffraction data support the notion that His-12 lies in an equatorial position while Lys-41 lies in an axial position with respect to the pentacoordinate phosphorus analogue, in contrast to the implications from X-ray diffraction studies which place both histidines in the axial position. The NMR data implied 58% protonation of His-12 and 26% protonation of His-119, whereas the neutron diffraction data suggested that both histidines are doubly protonated. This discrepancy might arise because of differences between the crystal and solution.

In a more recent study,[157] this issue was resolved by solution-state ^{31}P NMR of the 2'-CMP-RNaseA complex (see Fig. 7.31), compared with the solid-state NMR of RNase

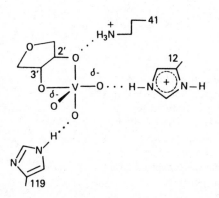

Fig. 7.30 Schematic representation of the RNase–U–V complex on the basis of neutron diffraction and NMR studies. The protonation state of the two active-site histidine residues was determined from the NMR studies. The neutron diffraction results indicated a greater degree of protonation of His119 and showed that Lys41 was apical and His12 equatorial. (Reprinted from ref. 156 with permission.)

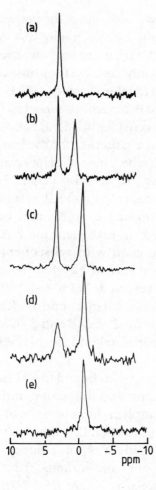

Fig. 7.31 202.4 MHz ^{31}P NMR spectra of 30% ethanol/70% water solutions containing 2'-CMP and RNase at pH (a) 8.13; (b) 6.40; (c) 5.41; (d) 4.18; and (e) 3.15. (Reprinted from ref. 157 with permission.)

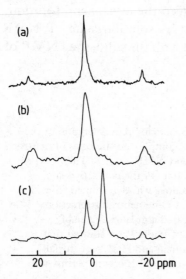

Fig. 7.32 80.96 MHz CPMAS ^{31}P NMR spectra of 2'-CMP in RNase crystals obtained by soaking at pH (a) 7.1; (b) 5.5; and (c) 5.5 using crystals prepared in the presence of pyrophosphate. (Reprinted from ref. 157 with permission.)

crystals soaked in 2'-CMP (see Fig. 7.32). This established that 2'-CMP bound to RNase is dianionic both in the solution and in the solid, which in turn implies that the hydrogens in the hydrogen bonds to the phosphate oxygens must lie closer to the histidine residues. This therefore implies that the histidine residues should be largely protonated. ^{15}N NMR studies[158] on the RNase S' fragment (21–124) implied that His-12 is involved in a hydrogen bond upon binding of 2'-CMP, with $N^{\delta 1}$ and $N^{\epsilon 2}$ both protonated (due to the observation of $^1J_{NH}$ coupling at low temperature). These studies did not address the ^{15}N shifts of His-119.

7.2.9 Staphylococcal nuclease

Staphylococcal nuclease (S.Nase, EC 3.1.4.7, 136 residues, $M_r = 18$ kDa) is another enzyme, like ribonuclease, that has been extensively studied since the early studies by Jardetsky and co-workers,[159] particularly with respect to its folding pathway (see Section 5.6.7). The full sequential assignments and secondary structure by heteronuclear two-dimensional methods have been reported by Torchia, Bax, and co-workers,[160] although the enzyme has been studied by Markley and co-workers by heteronuclear and isotope-edited two-dimensional NMR methods.[161,162] The backbone dynamics have also been studied[163] in the presence of deoxythymidine 3',5'-diphosphate (pdTP) and Ca^{2+}, by investigation of the T_1, T_2, and NOE values for over 100 assigned backbone ^{15}N amides detected indirectly by a modified CPMG-reverse INEPT pulse sequence. The order parameter S^2 (see Section 3.8) for the majority of the nitrogens was 0.86, implying that there is no correlation between rapid small-amplitude motions and the secondary structure of S.Nase. However, an analysis of the ^{15}N linewidths suggests a possible correlation between secondary structure and motions on the millisecond timescale, with a loop between residues 42 and 56 exhibiting considerably more flexibility than the rest of the protein.

The conformation of the enzyme-bound substrate deoxythymidine-deoxyadenine (dTdA) was investigated by Mildvan and co-workers.[164,165] Using paramagnetic relaxation enhancements of Co^{2+} in the ternary complex on T_1 and T_2 of the phosphorus, and on the T_1 of six proton resonances of dTdA, seven metal–nucleus distances and nine lower-limit metal–nucleus distances were calculated. The long Co^{2+} to ^{31}P distance of 4.1 ± 0.9 Å measured is intermediate between that expected for direct coordination (3.3 Å) and a second sphere complex with an intervening water ligand (4.8 Å), implying either a distorted inner sphere complex or rapid averaging between an inner sphere and second sphere complex. In addition, 17 interproton distances and 108 lower-limit interproton distances were measured by NOESY experiments in the $E \cdot La^{3+} \cdot dTdA$ complex. Using all these restraints, the conformation of dTdA was found to be highly extended, with no base stacking and high *anti* glycosidic torsion angles for dT ($64° \leq \chi \leq 73°$) and dA ($66° \leq \chi \leq 68°$) and predominantly C-2'-endo sugar puckers for both nucleosides. There is also evidence that in the ternary complex the presence of the metal induces a conformational change in relation to the binary $E \cdot dTdA$ complex. Using this substrate conformation, the recent X-ray structure[166] for the enzyme $\cdot Ca^{2+} \cdot 3',5'$-pdTp complex, and further three-dimensional NOESY–HMQC data,[164] Mildvan and co-workers 'docked' the substrate into the X-ray structure using the NMR restraints

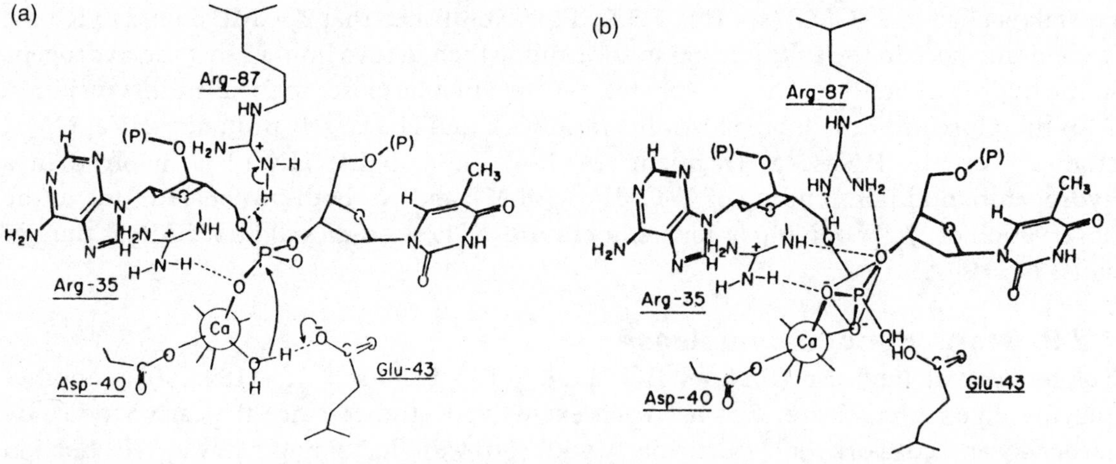

Fig. 7.33 Active site and mechanism of staphylococcal nuclease (a), and proposed transition state (b). (Reprinted from ref. 165 with permission.)

followed by restrained energy minimization. The final active site structure is shown in Fig. 7.33.

The implication for the enzymatic mechanism from this structural model is that one of eight water molecules coordinated to Ca^{2+} (in the inner and second coordination spheres) is adjacent to the phosphodiester bond of the substrate, and positioned for nucleophilic attack on phosphorus, based on three criteria. First, the oxygen of this water is 3 Å from the carboxylate of Glu-43, permitting Glu-43 to act as a general base. Second, this water oxygen is within 4.9 Å (in fact 4.31 Å) from the reactive phosphorus atom, consistent with an associative (S_N2) nucleophilic displacement at phosphorus. Third, the angle of attack defined by the water oxygen, the phosphorus, and the leaving 5′-deoxyadenosyl oxygen approximates (131°) the expected value of 180° based on the observed stereochemical inversion at phosphorus.[167] The less than optimal angle of attack of this water probably explains the low catalytic activity generally observed for the hydrolysis of dinucleotides.

REFERENCES

1. W. W. Bachovchin and J. D. Roberts, *J. Am. Chem. Soc.*, **100**, 8041–7 (1978).

2. W. W. Bachovchin, *Biochemistry*, **25**, 7751–9 (1986) and references therein.

3. For example: M. Fujinaga, L. T. J. Delbaere, G. D. Brayer, and M. N. G. James, *J. Mol. Biol.*, **183**, 479–502 (1985).

4. T. -H. Huang, W. W. Bachovchin, R. G. Griffin, and C. M. Dobson, *Biochemistry*, **23**, 5933–7 (1984).

5. S. O. Smith, S. Farr-Jones, R. G. Griffin, and W. W. Bachovchin, *Science (Washington DC)*, **244**, 961–4 (1989).

6. For example: L. Stryer, *Biochemistry*, 3rd edn, W. H. Freeman and Co., New York, p. 224 (1988).

7. T. McL. Spotswood, J. M. Evans, and J. H. Richards, *J. Am. Chem. Soc.*, **89**, 5052–4 (1967).

8. B. D. Sykes, *J. Am. Chem. Soc.*, **91**, 949–55 (1969).

9. R. B. Westkaemper and R. H. Abeles, *Biochemistry*, **22**, 3256 (1983).

10. M. H. Gelb and R. H. Abeles, *Biochemistry*, **23**, 6596 (1984).

11. D. Ringe, J. M. Mottonen, M. H. Gelb, and R. H. Abeles, *Biochemistry*, **25**, 5633–8 (1986).

12. C. -H. Niu, H. Shindo, and J. S. Cohen, *J. Am. Chem. Soc.*, **99**, 3161 (1977).

13. T. C. Liang and R. H. Abeles, *Biochemistry*, **26**, 7603–8 (1987).

14. W. C. Groutas, M. A. Stanga, and M. J. Brubaker, *J. Am. Chem. Soc.*, **111**, 1931–2 (1989).

15. F. Adebodun and F. Jordan, *J. Am. Chem. Soc.*, **110**, 309–10 (1988).

16. S. Zhong, F. Jordan, C. Kettner, and L. Polgar, *J. Am. Chem. Soc.*, **113**, 9429–35 (1991).

17. M. D. Finucane, E. A. Hudson, and J. P. G. Malthouse, *Biochem. J.*, **258**, 853–9 (1989).

18. J. P. G. Malthouse and M. D. Finucane, *Biochem. J.*, **280**, 649–57 (1991).

19. T. M. O'Connell, J. T. Gerig, and P. G. Williams, *J. Am. Chem. Soc.*, **115**, 3048–55 (1993).

20. J. P. G. Malthouse, N. E. MacKenzie, A. S. F. Boyd, and A. I. Scott, *J. Am. Chem. Soc.*, **105**, 1685–6 (1983).

21. J. Walter and W. Bode, *Hoppe Seyler's Z. Physiol. Chem.*, **364**, 949 (1983).

22. A. I. Scott, N. E. MacKenzie, J. P. G. Malthouse, W. U. Primrose, P. E. Fagerness, A. Brisson, L. Z. Qi, W. Bode, C. M. Carter, and Y. J. Yang, *Tetrahedron*, **42**, 3269–76 (1986).

23. C. Ortiz, C. Tellier, H. Williams, N. J. Stolowich, and A. I. Scott, *Biochemistry*, **30**, 10026–34 (1991).

24. B. Asboth, E. Stokum, I. U. Khan, and L. Polgar, *Biochemistry*, **24**, 606 (1985).

25. N. E. MacKenzie, S. Grant, A. I. Scott, and J. P. G. Malthouse, *Biochemistry*, **25**, 2293–8 (1986).

26. P. I. Clark, G. Lowe, and D. Nurse. *J. Chem. Soc. Chem. Comm.*, 451 (1977).

27. M. R. Bendall, I. L. Cartwright, P. I. Clark, G. Lowe, and D. Nurse, *Eur. J. Biochem.*, **79**, 201, (1977).

28. M. P. Gamcsik, J. P. G. Malthouse, W. U. Primrose, N. E. MacKenzie, A. S. F. Boyd, R. A. Russell, and A. I. Scott, *J. Am. Chem. Soc.*, **105**, 6324 (1983).

29. N. E. MacKenzie, S. K. Grant, A. I. Scott, and J. P. G. Malthouse, *Biochemistry*, **25**, 2293–8 (1986).

30. J. P. G. Malthouse, *Prog. NMR Spectros.*, **18**, 1–59 (1986).

31. J. O'C. Westerik and R. Wolfenden, *J. Biol. Chem.*, **247**, 8195 (1972).

32. J. B. Moon, R. S. Coleman, and R. P. Hanzlik, *J. Am. Chem. Soc.*, **108**, 1350–1 (1986).

33. Y. Yabe, D. Guillaume, and D. H. Rich, *J. Am. Chem. Soc.*, **110**, 4043–4 (1988).

34. V. J. Robinson, H. W. Pauls, P. J. Coles, R. A. Smith, and A. Krantz, *Bioorg. Chem.*, **20**, 42–54 (1992).

35. J. P. G. Malthouse, M. P. Gamcsik, A. F. S. Boyd, N. E. MacKenzie, and A. I. Scott, *J. Am. Chem. Soc.*, **104**, 6811–13 (1982).

36. D. H. Rich, M. S. Bernatowicz, and P. G. Schmidt, *J. Am. Chem. Soc.*, **104**, 3535–6 (1982).

37. P. G. Schmidt, M. S. Bernatowicz, and D. H. Rich, *Biochemistry*, **21**, 6710–16 (1982).

38. S. W. Fesik, J. R. Luly, J. W. Erickson, and C. Abad-Zapatero, *Biochemistry*, **27**, 8297–301 (1988).

39. L. Müller, *J. Am. Chem. Soc.*, **101**, 4481 (1979).

40. D. G. Hangauer, A. F. Monzingo, and B. W. Matthews, *Biochemistry*, **23**, 5730–41 (1984).

41. P. A. Bartlett and C. K. Marlowe, *Biochemistry*, **22**, 4618–24 (1983).

42. P. A. Bartlett and C. K. Marlowe, *Science (Washington DC)*, **235**, 569–71 (1987).

43. H. M. Holden, D. E. Tonrud, A. F. Monzingo, L. H. Weaver, and B. W. Matthews, *Biochemistry*, **26**, 8642–553 (1987).

44. T. Gullion, M. D. Poliks, and J. Schaefer, *J. Mgn. Reson.*, **80**, 553 (1988).

45. V. Copié, A. C. Kolbert, D. H. Drewry, P. A. Bartlett, T. G. Oas, and R. G. Griffin, *Biochemistry*, **29**, 9176–84 (1990).

46. J. Herzfeld and A. E. Berger, *J. Chem. Phys.*, **73**, 6021–30 (1980).

47. E. F. Pai, W. Sachsenheimer, R. H. Schirmer, and G. E. Schulz, *J. Mol. Biol.*, **114**, 37–45 (1977).

48. G. G. McDonald, M. Cohn, and L. Noda, *J. Biol. Chem.*, **250**, 6947–54 (1975).

49. D. C. Fry, S. A. Kuby, and A. S. Mildvan, *Biochemistry*, **24**, 4680–94 (1985).

50. M.-D. Tsai and H. Yan, *Biochemistry*, **30**, 6806–18 (1991).

51. K. Diederichs and G. E. Schulz, *Biochemistry*, **29**, 8138–44 (1990).

52. A. S. Mildvan and D. C. Fry, *Adv. Enzymol. Relat. Areas Mol. Biol.*, **58**, 241–313 (1987).

53. D. C. Fry, M. Byler, H. Susi, E. M. Brown, S. A. Kuby, and A. S. Mildvan, *Biochemistry*, **26**, 1645–55 (1987).

54. W. Klaus, M. Scharf, S. Zimmerman, and P. Rösch, *Biochemistry*, **27**, 5407–11 (1988).

55. I. Bock, J. Reinstein, A. Wittinghofer, and P. Rösch, *J. Mol. Biol.*, **200**, 745–8 (1988).

56. G. M. Smith and A. S. Midvan, *Biochemistry*, **21**, 6119–23 (1982).

57. D. C. Fry, S. A. Kuby, and A. S. Mildvan, *Biochemistry*, **24**, 4680–94 (1985).

58. H. Yan, Z. Shi, and M.-D. Tsai, *Biochemistry*, **29**, 6385–92 (1990).

59. P. Rösch, W. Klaus, M. Auer, and R. S. Goody, *Biochemistry*, **28**, 4318–25 (1989).

60. I. R. Vetter, J. Reinstein, and P. Rösch, *Biochemistry*, **29**, 7459–67 (1990).

61. I. R. Vetter, M. Konrad, and P. Rösch, *Biochemistry*, **30**, 4137–42 (1991).

62. T. Dahnke, Z. Shi, H. Yan, R.-T. Jian, and M.-D. Tsai, *Biochemistry*, **31**, 6318–28 (1992).

63. U. Egner, A. G. Tomasselli, and G. E. Schulz, *J. Mol. Biol.*, **195**, 649–58 (1987).

64. G. E. Schulz, C. W. Müller, and K. Diederichs, *J. Mol. Biol.*, **213**, 627–30 (1990).

65. I. Bock-Möbius, M. Brune, A. Wittinghofer, H. Zimmermann, R. Leberman, M.-T. Dauergne *et al.*, *Biochem. J.*, **273**, 311–16 (1991).

66. I.-J. L. Byeon, H. Yan, A. S. Edison, E. S. Mooberry, F. Abildgaard, J. L. Markeley, *et al.*, *Biochemistry*, in press (1993).

67. B. D. Nageswara Rao, F. J. Kayne, and M. Cohn, *J. Biol. Chem.*, **254**, 2689 (1979).

68. R. J. Gupta, *J. Biol. Chem.*, **252**, 5183 (1977).

69. A. S. Mildvan, *Acc. Chem. Res.*, **10**, 246–52 (1977).

70. H. Muirhead, D. A. Clayden, D. Barford, C. G. Lorimer, L. A. Fothergill-Gilmore, E. Schiltz *et al.*, *EMBO J.*, **5**, 475–81 (1986).

71. D. G. Gorenstein (ed.), *Phosphorus-31 NMR*, Academic Press, London (1984).

72. S. Meshitsuka, G. M. Smith, and A. S. Mildvan, *J. Biol. Chem.*, **256**, 4460 (1981).

73. W. A. Blaettler and J. R. Knowles, *Biochemistry*, **18**, 3927 (1979).

74. A. Hassett, W. Blaettler, and J. R. Knowles, *Biochemistry*, **21**, 6335 (1982).

75. W. C. Hutton, E. M. Stephens, and C. M. Grisham, *Arch. Biochem. Biophys.*, **184**, 166 (1977).

76. D. E. Ash, F. J. Kayne, and G. H. Reed, *Arch. Biochem. Biophys.*, **190**, 571 (1978).

77. H. Hoving, B. Crysell, and P. F. Leadley, *Biochemistry*, **24**, 6163–9 (1985).

78. J. A. Stubbe and G. L. Kenyon, *Biochemistry*, **11**, 238–45 (1972).

79. T. Duffy, H. J. Saz, and T. Nowak, *Biochemistry*, **21**, 132–9 (1982).

80. R. Keck, J. Huus, and J. Rétey, *FEBS Lett.*, **114**, 287–90 (1980).

81. W. E. Bondinell and D. B. Sprinson, *Biochem. Biophys. Res. Commun.*, **40**, 1464 (1970).

82. J. A. Stubbe and G. L. Kenyon, *Biochemistry*, **10**, 2669–76 (1971).

83. I. A. Rose, *J. Biol. Chem.*, **245**, 6052–6 (1970).

84. J. J. Burbaum and J. R. Knowles, *Bioorg. Chem.*, **17**, 359–71 (1989).

85. B. D. Nageswara Rao and M. Cohn, *J. Biol. Chem.*, **256**, 1716 (1981).

86. E. J. Milner-White and D. S. Rycroft, *Biochem. J.*, **167**, 827 (1977).

87. G. K. Jakori, B. D. Ray, and B. D. Nageswara Rao, *Biochemistry*, **24**, 3487 (1985).

88. J. Boyd, K. M. Brindle, I. D. Campbell, and G. K. Radda, *J. Magn. Reson.*, **60**, 149 (1984).

89. D. E. Hansen and J. R. Knowles, *J. Biol. Chem.*, **256**, 5967 (1981).

90. G. Lowe and B. S. Sproat, *J. Biol. Chem.*, **255**, 3944 (1980).

91. T. L. James and M. Cohn, *J. Biol. Chem.*, **249**, 2599 (1974).

92. T. L. James, *Biochemistry*, **15**, 4724–30 (1976).

93. M. Vasak, K. Nagayama, K. Wüthrich, M. L. Mertens, and J. H. R. Ktigi, *Biochemistry*, **18**, 5050 (1979).

94. P. R. Rosevar, P. Desmeules, G. L. Kenyon, and A. S. Mildvan, *Biochemistry*, **20**, 6155 (1981).

95. K. R. Huskins, S. A. Berhard, and F. W. Dahlquist, *Biochemistry*, **21**, 4180–8 (1982).

96. B. D. Nageswara Rao, M. Cohn, and R. K. Scopes, *J. Biol. Chem.*, **253**, 8056 (1978).

97. B. D. Ray and B. D. Nageswara Rao, *Biochemistry*, **27**, 5574–8 (1988).

98. B. D. Ray, J. M. Moore, and B. D. Nageswara Rao, *J. Inorg. Biochem.*, **40**, 47–57 (1990).

99. B. D. Ray and B. D. Nageswara Rao, *Biochemistry*, **27**, 5579–85 (1988).

100. R. D. Banks, C. C. F. Blake, P. R. Evans, R. Haser, D. W. Rice, G. W. Hardy *et al.*, *Nature*, **279**, 773–7 (1979).

101. C. C. F. Blake and D. W. Rice, *Phil. Trans. R. Soc. Lond. A*, **293**, 93–104 (1981).

102. P. Tanswell, E. W. Westhead, and R. J. P. Williams, *Eur J. Biochem.*, **63**, 249 (1976).

103. M. A. Sherman, W. J. Fairbrother, and M. T. Mas, *Protein Sci.*, **1**, 752–60 (1992).

104. W. J. Fairbrother, P. Minard, L. Hall, J.-M. Betton, D. Missiakas, J. M. Yon, and R. J. P. Williams, *Protein Engin.*, **3**, 5–11 (1989).

105. H. C. Graham, R. J. P. Williams, J. A. Litlechild, and H. C. Watson, *Eur. J. Biochem.*, **196**, 261–9 (1991).

106. H. Joao, N. Taddei, and R. J. P. Williams, *Eur J. Biochem.*, **205**, 93–104 (1992).

107. W. J. Fairbrother, H. C. Graham, and R. J. P. Williams *Eur. J. Biochem.*, **190**, 1611–169 (1990).

108. J. R. Knowles, *Annu. Rev. Biochem.*, **49**, 877 (1980).

109. G. S. Begley, D. E. Hansen, G. R. Jacobson, and J. R. Knowles, *Biochemistry*, **21**, 5552 (1982).

110. O. Schrecker, R. Stein, W. Hengstenberg, M. Gassner, and D. Stehlik, *FEBS Lett.*, **51**, 309 (1975).

111. M. Gassner, D. Stehlik, O. Schrecker, W. Hengstenberg, W. Maurer, and H. Rüterjans, *Eur. J. Biochem.*, **75**, 287 (1977).

112. H. R. Kalbitzer and P. Rösch, *Org. Magn. Reson.*, **17**, 88 (1982).

113. A. A. van Dijk, L. C. M de Lange, W. W. Bachovchin, and G. T. Robillard, *Biochemistry*, **29**, 8164–71 (1990).

114. M. Wittekind, J. Reizer, J. Deutscher, M. H. Saier, and R. E. Klevit, *Biochemistry*, **28**, 9908–12 (1989).

115. O. A. L. El-Kabbani, B. E. Waygood, and L. T. J. Delbaere, *J. Biol. Chem.*, **262**, 12926–9 (1987).

116. R. E. Klevit and B. E. Waygood, *Biochemistry*, **25**, 7774–81 (1986).

117. P. K. Hammen, B. E. Waygood, and R. E. Klevit, *Biochemistry*, **30**, 11842–50 (1991).

118. N. A. J van Nuland, A. A. van Dijk, K. Dijkstra, F. H. J. van Hoesel, R. M. Scheek, and G. T. Robillard, *Eur. J. Biochem.*, **203**, 483–91 (1992).

119. O. Herzberg, P. Reddy, S. Sutrina, M. H. Saier, J. Reizer, and G. Kapafia, *Proc. Natl. Acad. Sci. USA*, **89**, 2499–503 (1992).

120. M. Wittekind, P. Rajagopal, B. R. Branchini, J. Reizer, M. H. Saier Jr, and R. E. Klevit, *Protein Sci.*, **1**, 1363–76 (1992).

121. H. R. Kalbitzer, K.-P. Neidig, and W. Hengstenberg, *Biochemistry*, **30**, 11186–92 (1991).

122. U. Finkeldei, H. R. Kalbitzer, R. Eiserman, G. C. Stewart, W. Hengstenberg, *Protein Engineer.*, **4**, 469–73 (1991).

123. J. G. Pelton, D. A. Torchia, N. D. Meadow, C.-Y. Wong, and S. Roseman, *Proc. Natl. Acad. Sci.*, **88**, 3479–83 (1991).

124. H. H. Pas, G. H. Meyer, W. H. Kruizinga, K. S. Tamminga, R. P. van Weeghel, and G. T. Robillard, *J. Biol. Chem.*, **266**, 6690–2 (1991).

125. N. A. J. van Nuland, G. J. A. Kroon, K. Dijkstra, G. K. Wolters, R. M. Scheek, and G. T. Robillard, *FEBS Lett.*, **315**, 11–15 (1993).

126. W. J. Fairbrother, A. G. Palmer III, M. Rance, J. Reizer, M. H. Saier Jr, and P. E. Wright, *Biochemistry*, **31**, 4413–25 (1992).

127. W. J. Fairbrother, G. P. Gippert, J. Reizer, M. H. Saier Jr, and P. E. Wright, *FEBS Lett.*, **296**, 148–52 (1992).

128. Y. Chen. J. Reizer, M. H. Saier Jr, W. J. Fairbrother, and P. E. Wright *Biochemistry*, **32**, 32–7 (1993).

129. J. E. Coleman and P. Gettins, *Adv. Enzymol.*, **55**, 381 (1983).

130. J. L. Bock and B. Sheard, *Biochem. Biophys. Res. Comm.*, **66**, 24 (1975).

131. J. F. Chlebowski, I. M. Armitage, P. P. Tusa, and J. E. Coleman, *J. Biol. Chem.*, **251**, 1207 (1976).

132. W. E. Hull, S. E. Halford, H. Gutfreund, and B. D. Sykes, *Biochemistry*, **15**, 1547 (1976).

133. J. D. Otvos, J. R. Alger, J. E. Coleman, and I. M. Armitage, *J. Biol. Chem.*, **254**, 1778 (1979).

134. S. R. Jones, L. A. Kindman, and J. R. Knowles, *Nature*, **257**, 564 (1978).

135. E. T. Fossel, R. L. Post, D. S. O'Hara, and T. H. Smith, *Biochemistry*, **20**, 7215 (1981).

136. C. M. Grisham, *Methods in Enzymol.*, **156**, 353–71 (1988).

137. G. M. Clore, A. M. Gronenborn, C. Mitchinson, and N. M. Green, *Eur. J. Biochem.*, **128**, 113 (1982).

138. M. R. Klemens and C. M. Grisham, *Biochem. Biophys. Res. Comm.*, **155**, 236–42 (1988).

139. L. D. Faller and G. A. Elgavish, *Biochemistry*, **23**, 6584 (1984).

140. J. Gairn, P. V. Vignais, A. M. Gronenborn, G. M. Clore, Z. Gao, and E. Baeuerlein, *FEBS Lett.*, **242**, 178–82 (1988).

141. C. C. Devlin and C. M. Grisham, *Biochemistry*, **29**, 6192–203 (1990).

142. A. E. Haddy and R. R. Sharp, *Biochemistry*, **28**, 3656–64 (1989).

143. A. E. Haddy, W. D. Frasch, and R. R. Sharp, *Biochemistry*, **28**, 3664–9 (1989).

144. M. D. Carr, D. Mulvey, A. Willis, S. J. Ferguson, and I. D. Campbell, *Biochim. Biophys. Acta*, **1015**, 79–86 (1990).

145. T. J. Norwood, D. A. Crawford, M. E. Steventon, P. C. Driscoll, and I. D. Campbell, *Biochemistry*, **31**, 6285–90 (1992).

146. M. Rico, J. Santoro, C. Gonzalez, M. Bruix, J. L. Neira, J. L. Nieto, and J. Herranz, *J. Biomol. NMR*, **1**, 283–98 (1991).

147. J. Santoro, C. Gonzalez, M. Bruix, J. L. Neira, J. L. Nieto, J. Herranz *et al.*, *J. Mol. Biol.*, **229**, 722–34 (1993).

148. F. M. Richards and H. W. Wyckoff, *Enzymes*, 3rd edn, **4**, 647–806 (1971).

149. T. Nakai, W. Yoshikawa, H. Nakamura, and H. Yoshida, *Eur. J. Biochem.*, **208**, 41–51 (1992).

150. K. Nagayama, T. Yamazaki, M. Yoshida, S. Kanaya, and H. Nakamura, *NATO ASI Ser.*, *Ser. A*, **225**, (Comput. Aspects Study Biol. Macromol. Nucl. Magn. Reson. Spectrosc.), 445–50 (1991).

151. I. Shimada and F. Inagaki, *Biochemistry*, **29**, 757–64 (1990).

152. D. H. Meadows, G. C. K. Roberts, and G. C. K. Roberts, *J. Mol. Biol.*, **45**, 491 (1969).

153. N. Haar, J. C. Thompson, W. Mauver, and H. Ruterjans, *Eur. J. Biochem.*, **40**, 259–66 (1973).

154. D. G. Gorenstein, A. M. Wyrwicz, and J. Bode, *J. Am. Chem. Soc.*, **98**, 2308–14 (1976).

155. U. Hahn, R. Desai-Hahn, and H. Rüterjans, *Eur. J. Biochem.*, **146**, 705–12 (1985).

156. B. Borah, C.-W. Chan, W. Egan, M. Miller. A. Wlodawer, and J. S. Cohen, *Biochemistry*, **24**, 2058–67 (1985).

157. C. M. Dobson and L.-Y. Lian, *FEBS Lett.*, **225**, 183–7 (1987).

158. H. Knoblauch, H. Rüterjans, W. Bloemhoff, and K. E. T. Kerling, *Eur. J. Biochem.*, **172**, 485–97 (1988).

159. J. L. Markley, I. Putter, and O. Jardetzky, *Science*, **161**, 1249–51 (1968).

160. D. A. Torchia, S. W. Sparks, and A. Bax, *Biochemistry*, **28**, 5509–24 (1989).

161. C. B. Grissom and J. L. Markley, *Biochemistry*, **28**, 2116–24 (1989).

162. J. Wang, A. P. Hinck, S. N. Loh, D. M. LeMaster, and J. L. Markley, *Biochemistry*, **31**, 921–36 (1992).

163. L. E. Kay, D. A. Torchia, and A. Bax, *Biochemistry*, **28**, 8972–9 (1989).

164. D. J. Weber, G. P. Mullen, and A. S. Mildvan, *Biochemistry*, **30**, 7425–37 (1991).

165. D. J. Weber, A. G. Gittis, G. P. Mullen, C. Abeygunawardana, E. E. Lattman, and A. S. Mildvan, *Proteins : Struct., Funct. and Genet.*, **13**, 275–87 (1992).

166. P. J. Loll and E. E. Lattman, *Proteins : Struct., Funct. and Genet.*, **5**, 183–201 (1989).

167. S. Mehdi and J. A. Gerlt, *J. Am. Chem. Soc.*, **104**, 3223–5 (1982).

8 · Other enzymes

In this chapter, we will consider a limited number of examples of enzymes which either involve unusual chemistry, or have been the subject of particularly intensive or elegant applications of NMR spectroscopy. Glycosyl transfer enzymes catalyse reactions of crucial importance to primary metabolism, and have been structurally well-characterized. Oxidoreductases, of which dihydrofolate reductase is a particularly fine example, are also of great importance in the cell, and have been the subject of intensive NMR investigations. Triosephosphate isomerase, the ultimate enzyme, is considered because of a particularly elegant application of NMR methods to address some mechanistic questions. Finally, three examples of carbon–carbon bond forming/breaking enzymes are presented, partly because the author has more than a passing interest in them, and partly because just about every trick of the trade in biomolecular NMR spectroscopy has been applied to them at one time or another.

8.1 GLYCOSYL TRANSFER ENZYMES—GLYCOGEN PHOSPHORYLASE

Glycogen phosphorylase (GP, EC 2.4.1.1, dimer with subunit $M_r = 97$ kDa, 842 residues) is an allosterically regulated enzyme which catalyses the breakdown of glycogen to glucose-1-phosphate in the presence of inorganic phosphate according to the mechanism outlined in Fig. 8.1. The enzyme is pyridoxal-phosphate (PLP) dependent, and, unlike other enzymes containing this co-factor, reduction with sodium borohydride does not result in reduction of a Schiff-base linkage. However, the X-ray structures both glycogen phosphorylase a determined by Robert Fletterick and co-workers[1] and phosphorylase b determined by Louise Johnson and co-workers[2] show that the phosphate group of the coenzyme points towards the substrate, suggesting a role for the phosphorus of PLP in catalysis. The two forms of the enzyme are structurally distinct, although quite similar, with phosphorylase a existing as an active tetramer or dimer with phosphorylated subunits. The phosphorylation is carried out by a phosphorylase kinase, which functions as a regulatory component. The dimeric form of phosphorylase b is unphosphorylated and may be active (R-state) or inactive (the T-state). In the phosphorylase b X-ray crystal structure, the PLP is attached to Lys-679 via a Schiff base linkage, and Ser-14 is the site for the covalent phosphorylation which results in the formation of phosphorylase b. The early ^{31}P NMR studies by Richards, Radda, and co-workers[3] and by Feldmann and Helmreich[4] demonstrated that PLP could be observed, although the resonance occurs in the same region of the spectrum as those of the product phosphate, the substrate glucose phosphate, the activator AMP, and the regulatory, covalently phosphorylated Ser-14 moiety. A number of tricks were used to simplify this region of the ^{31}P spectrum, such as replacement of phosphate with arsenate,[5] AMPS (thio-AMP) substituted for AMP,[6] and thiophosphorylation of the phosphoserine moiety which confers activity on the enzyme similar to phosphorylation.[7] Glucose-1-phosphate can be replaced by its 1,2-cyclic phosphate analogue, whose chemical shift differs due to bond strain.[8] Also the PLP can be replaced with analogues such as pyridoxal 5′-deoxymethylenephosphonate,[9] pyridoxal pyrophosphate,[10] and 6-fluoropyridoxal phosphate.[11]

The ^{31}P chemical shifts observed for the enzyme in its inactive (T-state) and active (R-state) conformations suggest that the phosphate moiety exists as a dianion in the active R-state form and as a monoanion in the inactive T-state form.[5] This suggests that the activation of the enzyme, which is accomplished by binding of the nucleotide activator AMP or its thiophosphate analogue, AMPS, is accompanied by a deprotonation of the coenzyme phosphate. Similar changes were observed as a consequence of covalent activation of the enzyme via phosphorylation, generating phosphorylase *a*. Activation is accompanied by an upfield ^{31}P shift of ≈ 3 ppm, which is essentially identical to the shift change observed upon deprotonation of the free PLP monoanion In these studies, it was tacitly assumed that differences in the ^{31}P NMR chemical shifts of the coenzyme phosphate between the T- and R-states are entirely a consequence of the influence of a protonation/deprotonation event. In the solution state, the isotropic shifts of phosphates are essentially governed by the π-bond order, the electronegativity of the substituents,

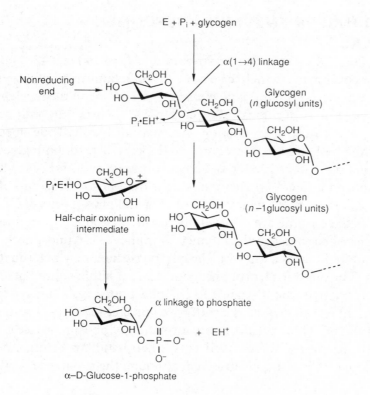

Fig. 8.1 The reaction mechanism of glycogen phosphorylase.

(a)

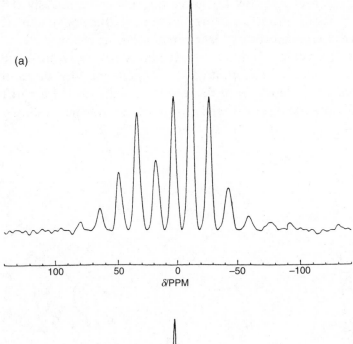

δ/PPM

(b)

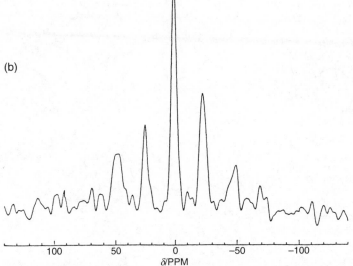

δ/PPM

Fig. 8.2 ^{31}P CPMAS solid-state NMR spectra of the crystalline glycogen phosphorylase *b* in (a) the T-state (tetragonal form) with $\sigma_{iso} = 3.4$ ppm and (b) the R-state (monoclinic) with $\sigma_{iso} = 2.0$ ppm. (Reprinted from ref. 21 with permission.)

and the O—P—O bond angles.[12–14] While the relative importance of such terms has been the subject of some debate,[15–16] it is clear that any geometrical differences that may result on conversion from the T- to the R-state conformation would be expected to influence the chemical shifts. An analysis of the ^{31}P shielding tensors of a wide variety of phosphate compounds[17] has established a linear relationship between the principal values of the shielding tensor and the O—P—O bond angles and bond lengths. Also, any changes that result in charge redistribution at the active site might also be expected to lead to a change in the ^{31}P isotropic shift. Indeed, differences in the charge density at the phosphate moiety in the T- and R-states are expected since the crystallographic studies of the T- to R-state conversion indicate that the side chain of Arg-569 displaces the side chain of Asp-283 upon activation.[10,18,19] This could result either in deprotonation of the phosphate (if previously protonated) to preserve charge neutrality, or in chemical-shift changes due to the changes in local charge density. This matter has been resolved recently by a solid-state ^{31}P NMR study by Withers and co-workers,[20] and confirmed independently by Taguchi and co-workers,[21] who both provided evidence from the analysis of the chemical-shift tensors that the PLP is dianionic in both the R- and T-state forms of the enzyme, as shown in Fig. 8.2. The relationship between these two states is shown in Fig. 8.3.

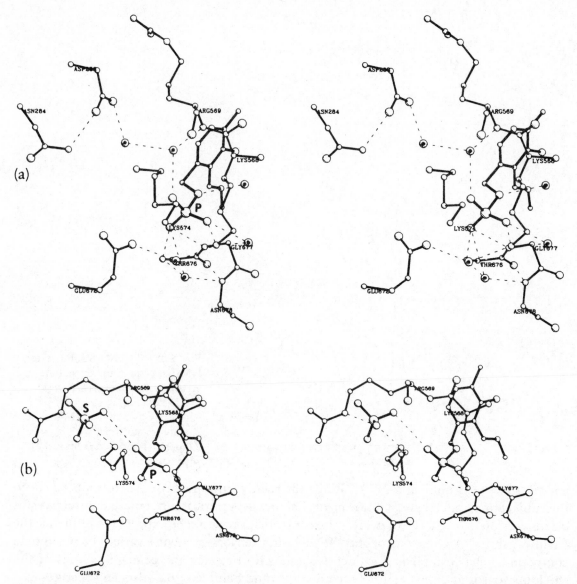

Fig. 8.3 Stereoviews from the crystal structures of the environment of the PLP phosphate group in (a) the T-state at 1.9 Å resolution; and (b) the R-state at 2.8 Å resolution. The phosphate and sulphate groups are labelled P and S respectively. Four water molecules are also indicated for the T-state. (Reprinted from ref. 21 with permission.)

The role of PLP is still not yet clear, despite the recent elegant experiments from Hajdu and Johnson using time-resolved Laue X-ray crystallography of glycogen phosphorylase *b* catalysis of the conversion of heptenitol to heptulose-2-phosphate at 2.7 Å resolution.[22,23] It has been suggested[24] that the dianionic phosphate of PLP might serve to act as an electrophile for the substrate phosphate oxyanion electrons, thereby promoting S_N1 cleavage of the substrate C—O bond. This has been questioned by Klein and co-workers, who suggest that the phosphate group probably plays a role in general acid–base catalysis, but not as an electrophilic catalyst, based on ^{31}P NMR results on heptulose-2-phosphate binding. However, these latter interpretations incorrectly interpreted the T-state as being monoanionic. Other proposed roles for the PLP phosphate relate to effects on the local dynamics[25] and on the local protein conformation and orientation of functional residues.[26,27]

8.2 OXIDOREDUCTASE ENZYMES—DIHYDROFOLATE REDUCTASE

Dihydrofolate reductase (DHFR, EC 1.5.1.3, *Lactobacillus casei*: 162 residues, $M_r =$ 18.3 kDa; *E.coli*: 159 residues, $M_r = 18$ kDa; Human DHFR: 186 residues, $M_r =$ 21.5 kDa) catalyses the NADPH-dependent reduction of 7,8-dihydrofolate to 5,6,7,8-tetrahydrofolate (see structures **8.1–8.4**). Since tetrahydrofolate (THF) or its derivatives are essential for the biosynthesis of purine nucleotides, thymidylate, and several amino acids, failure to maintain adequate levels of THF result in decreased nucleic acid synthesis. Therefore DHFR has become an important target for folate antagonists, including the anticancer drug methotrexate (MTX, structure **8.1**), and the antibacterial agent trimethoprim (TMP, structure **8.2**).

X-ray crystallography has been used to determine the structure of DHFR with and

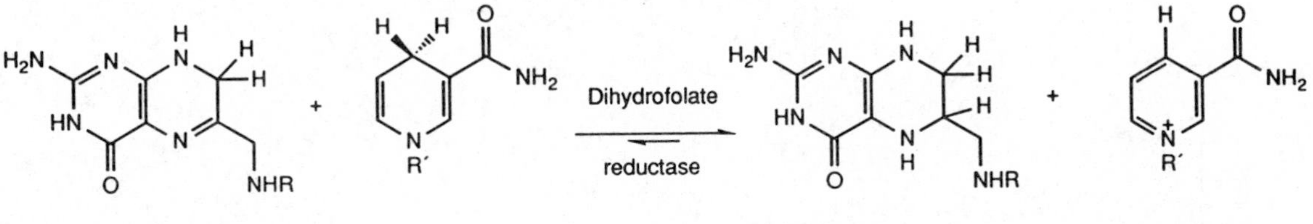

Dihydrofolate, DHF, **8.1**   NADPH

Dihydrofolate reductase

Tetrahydrofolate, THF, **8.2**   NADP+

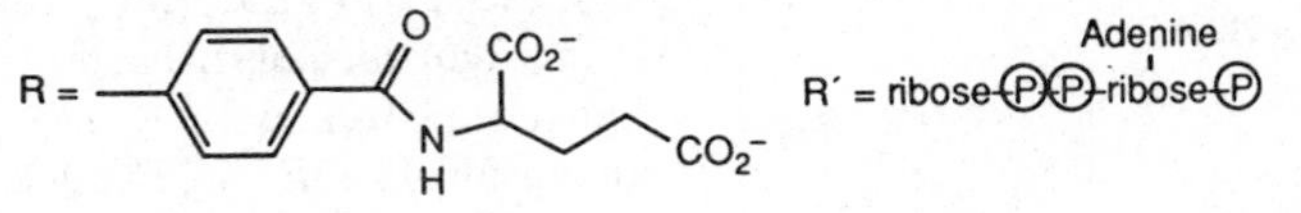

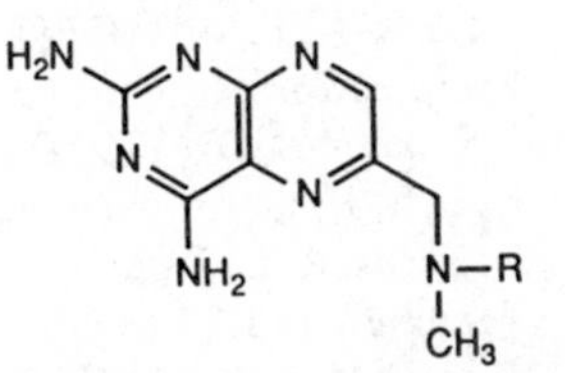

$$R = \qquad R' = \text{ribose-(P)(P)-ribose-(P)}$$

Adenine

Methotrexate, MTX, **8.3**

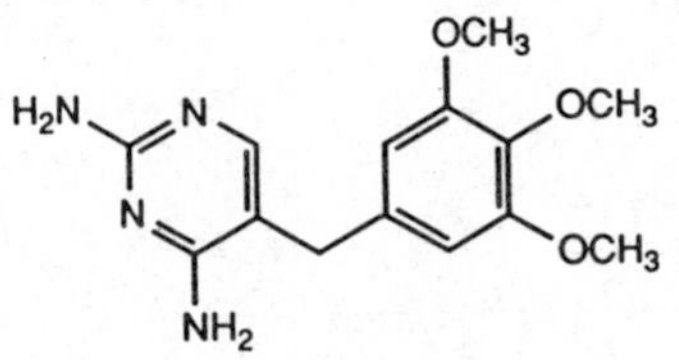

Trimethoprim, TMP, **8.4**

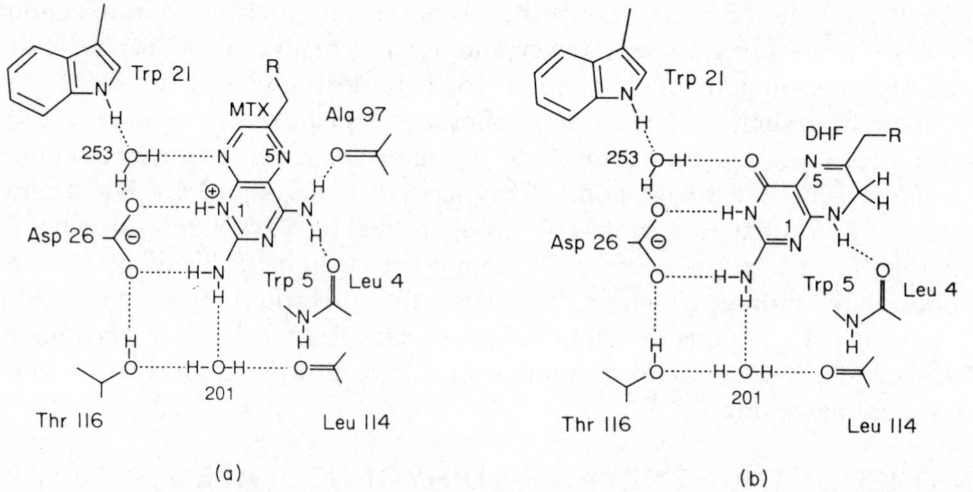

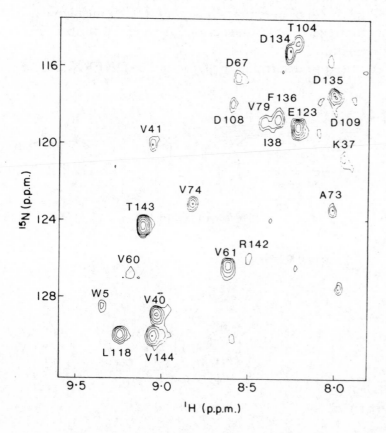

Fig. 8.4 Schematic diagrams of (a) the methrotrexate binding site of the *L.casei* DHFR–MTX–NADPH complex as determined by X-ray crystallography; and (b) proposed dihydrofolate (DHF) binding site. (Reprinted from I. P. Gerothanassis, B. Birdsall, C. J. Bauer, T. A. Frenkiel, and J. Feeney, *J. Mol. Biol.*, **226**, 549–54 (1992) with permission.)

Fig. 8.5 Part of the F_2F_3 plane at the frequency of the water signal in F_1 from a three-dimensional ROESY–HMQC ^{15}N—^{1}H experiment on the DHFR–MTX–NADPH complex formed with uniformly ^{15}N-labelled DHFR. (Reprinted from I. P. Gerothanassis, B. Birdsall, C. J. Bauer, T. A. Frenkiel, and J. Feeney, *J. Mol. Biol.*, **226**, 549–54 (1992) with permission.)

without ligands from a variety of sources, including *Lactobacillus casei*,[28] *E.coli*,[29,30] chicken liver,[31] mouse L1210 cells,[32] and human cells.[33,34] Solution-state NMR spectroscopy has been used to study ligand binding interactions of DHFR isolated from *L. casei*,[35,36] *E.coli*,[37,38] and human cells.[39,40] Although a complete three-dimensional structure from the NMR data was not available at the time of writing, three-dimensional heteronuclear NMR methods have been used to obtain the sequential assignments and the secondary structure of the enzymes complexed to methotrexate from all three sources. An example of the X-ray structure for MTX and DHF bound to the *L.casei* enzyme–NADPH complex is shown in Fig. 8.4. The relationship of the waters has been confirmed in solution by three-dimensional ROESY–HMQC NMR on uniformly ^{15}N-labelled DHFR (see Fig. 8.5).

Many of the early studies of DHFR were carried out by Roberts, Feeney, and co-workers,[41] and concentrated on the binding of ligands to the protein. ^{31}P, ^{1}H, ^{13}C, and ^{15}N NMR spectroscopy were used to investigate the binding of NADP$^+$, NADPH, and methotrexate/trimethoprim in binary and ternary complexes with DHFR. Rather than discuss the vast literature on these early experiments in detail, we will summarize their findings. The most important result was the discovery that with the enzyme from *L.casei* there are two distinct conformations of the DHFR–trimethoprim–NADP$^+$ ternary complex.[42] Furthermore, there are also two conformations of the binary *L.casei* DHFR–folate complex,[43] whereas Wright and co-workers found that there is only a single conformation for the *E.coli* DHFR–folate complex.[37] Huang and co-workers[44] used ^{15}N and ^{31}P one-dimensional NMR studies to investigate the conformation of TMP with *E.coli* DHFR, and found that a single conformation exists for each of the DHFR—TMP, DHFR—NADP$^+$, DHFR—NADPH, DHFR—NADPH—TMP complexes. Interestingly, the DHFR—NADP$^+$—TMP exhibited two conformations in the ternary complex, although because the conformations of the bound ligand do not correlate, there is the possibility of four conformations being present. Wright and co-workers[38] found that there are two protein conformations in the *E.coli* DHFR—MTX complex, and Wagner and co-workers[39] demonstrated that with human DHFR, MTX binds only in a non-productive orientation, with only one conformation detected.

Site-directed mutagenesis has been used in conjunction with NMR spectroscopy in order to assess the consequences of site mutants. For example, the D26N mutant of *L.casei* DHFR[45] shows that any structural changes in the DHFR—MTX complex are very small and local, whereas in the case of the T63Q substitution, the conformational effects are transmitted by as much as 25 Å through the protein.[46] Wright, Benkovic, and co-workers constructed a deletion mutant of *E.coli* DHFR missing Met–16—Ala–19, which is a highly mobile loop identified in the apoprotein and binary complex by X-ray crystallography and known to fold over both the folate and NADPH moieties in the ternary complex. ^{1}H NMR of the wild-type enzyme revealed that there is slow exchange in the residues in the vicinity of the loop at room temperature. In the mutant the dynamics of the same region alter substantially, not showing any EXSY cross-peaks. The mobility of this loop was postulated to provide a mechanism for recruiting hydrophobic residues which can properly align the nicotinamide and pteridine rings for the hydride transfer, so as such it provides a form of transition-state stabilization.

8.3 ISOMERASE ENZYMES—TRIOSEPHOSPHATE ISOMERASE

Triosephosphate isomerase (TIM, EC 5.3.1.1, 247 residues, $M_r \sim 27$ kDa) is a glycolytic enzyme which catalyses the interconversion of (R)-glyceraldehyde-3-phosphate (GAP) and dihydroxyacetone phosphate (DHAP). Fig. 8.6 shows that a single catalytic base, Glu-165, is responsible for proton abstraction from the carbon,[47–49] while His-95 is believed to be the catalytic acid.[50,51]

Campbell and co-workers[52] identified His-95 from the downfield ^{1}H NMR resonances of TIM in the absence and presence of the inhibitor 2-phosphoglycollate, a so-called transition-state analogue,[53] which shifted in the presence of the inhibitor. From ^{31}P NMR studies of the same inhibitor, it was found that the bound ligand is fully deprotonated and that a proton is accepted by an enzyme functional group such as Glu-165, which was proposed to form a hydrogen bond with the carboxylate of phosphoglycollate. In a more recent study along similar lines, Gracy and co-workers[54,55] found evidence from ^{31}P NMR studies for two different conformations of the enzyme when incubated with the affinity inhibitor chloroacetol phosphate, which covalently modifies Glu-165. The implication is that the binding of the inhibitor at each of the two active sites in the homodimer loosens the interaction between the two subunits, leading to slightly different ^{31}P chemical shifts. It was also found that if only one site was modified, then, from ^{31}P NMR, 2-phosphoglycollate bound to the vacant site in exactly the same way as to the unmodified enzyme, suggesting that the two catalytic sites act wholly independently.

The role of His-95 in the catalytic mechanism has been the subject of some speculation.[56] As outlined in Fig. 8.7, His-95 might act as an electrophile to polarize the carbonyl of the substrate (Fig. 8.7(a)), or it might act to stabilize developing oxygen negative charge on the proposed enediol intermediate (Fig. 8.7(b)). The difference between these two possibilities can be distinguished through the protonation state of the histidine imidazole ring. In one case (a) the protonated imidazolium ion becomes deprotonated, whereas in the other (b), the neutral imidazole becomes deprotonated to give the imidazolate ion. As we saw with α-lytic protease (see Section 7.1.1), the ^{15}N chemical shifts of the histidine are highly diagnostic not only of the protonation state of the two nitrogens, but also their involvement in hydrogen bonding. Lodi and Knowles, in a lovely set of experiments,[56] generated fully active TIM in which two of the three histidine residues had been mutated to glutamine (H103Q·H185Q), and examined the double mutant [^{15}N-His-95]-labelled enzyme by ^{15}N NMR in the absence and presence of the reaction intermediate analogue phosphoglycohydroxamate. The results are shown in Fig. 8.8, and illustrate that the imadazole is neutral over the entire pH range of isomerase activity, between pH 5 and 9.9.

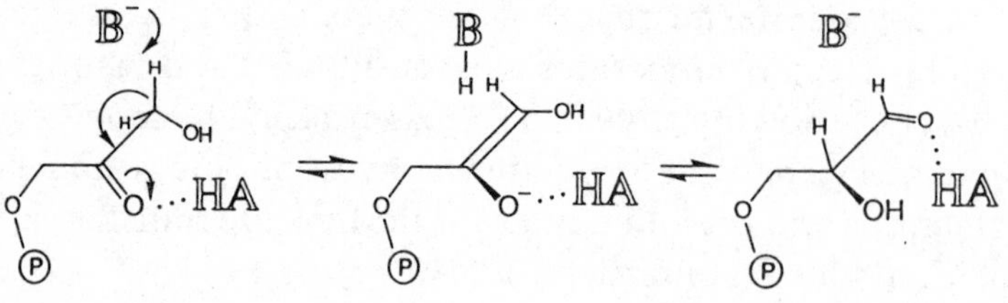

Fig. 8.6 An overview of the reaction catalysed by triosephosphate isomerase. B⁻ is the catalytic base (Glu-165), and HA is the catalytic acid (His-95). (Reprinted from ref. 56 with permission.)

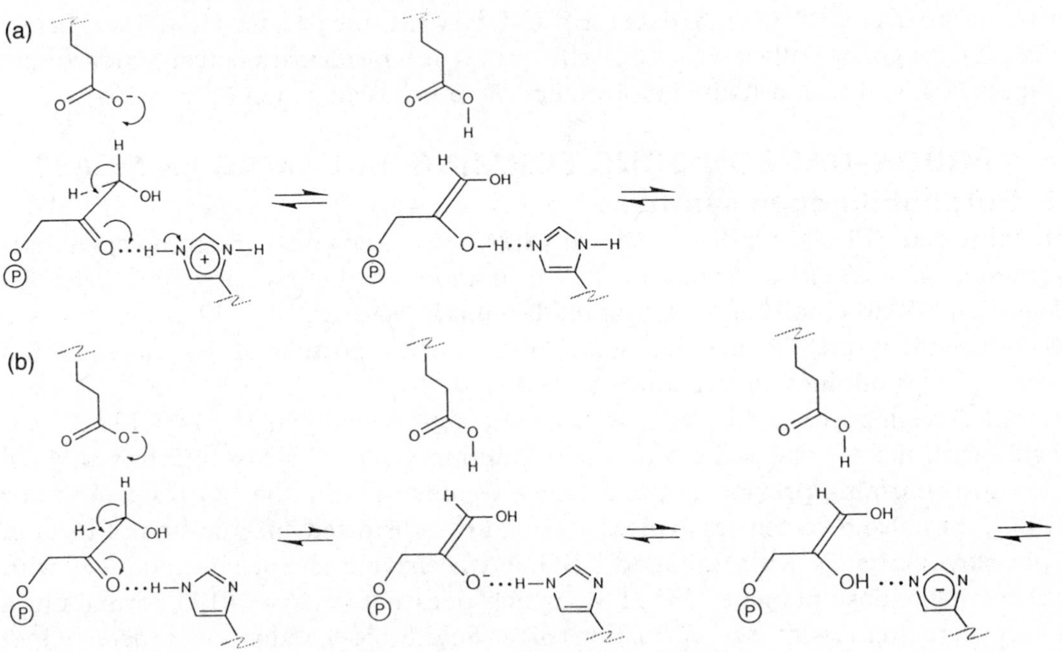

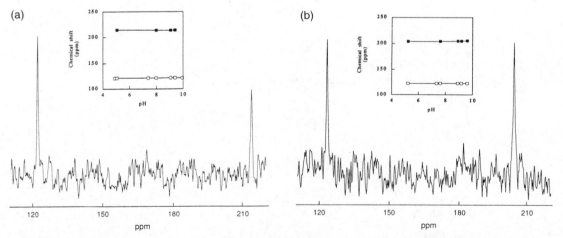

Fig. 8.7 Two possible roles for His95 in the mechanism of triosephosphate isomerase. Mechanism (a) involves the imidazolium of His95 that acts as a general acid, initially forming an imidazole–enediol pair. Mechanism (b) involves the neutral imidazole of His95 that acts as a general acid, initially forming an imidazole–enediol*ate* pair that (to accommodate the remaining proton movements) may form a transient imidazol*ate*–enediol pair. (Reprinted from ref. 56 with permission.)

Fig. 8.8 ^{15}N NMR spectra of H103Q·H185Q [^{15}N]histidine-labelled triosephosphate isomerase (a) unliganded at pH 9.4; and (b) the same sample as (a) in the presence of phosphoglycohydroxamate at pH 7.6. The insets show the pH titrations of $N^{\epsilon 2}$ (■) and $N^{\delta 1}$ (□) of His95 under their respective conditions. (Reprinted from ref. 56 with permission.)

Furthermore, from ^{13}C titration data for [^{13}C-His]TIM, the pK_a for His-95 was firmly established to be abnormally low (<4.5), which rules out the formation of an imidazolium ion (Fig. 8.7(a)) and supports the involvement of an imidazolate ion (Fig. 8.7(b)).

8.4 CARBON–CARBON BOND FORMING/BREAKING ENZYMES
8.4.1 Porphobilinogen synthase

Porphobilinogen (PBG) synthase (EC 4.2.1.24, also known as 5-aminolaevulinate dehydratase, $M_r = 280$ kDa) catalyses the formation of the monopyrrole PBG by condensation of two molecules of 5-aminolaevulinic acid (ALA). The enzyme is a zinc(II)-dependent metalloenzyme, whose mechanism, postulated by Shemin and co-workers,[57,58] is outlined in structures **8.5–8.6**.

In an interesting study by Jaffe, Markham, and co-workers,[59] [4-^{13}C]ALA was incubated with the enzyme and examined by solution-state ^{13}C NMR spectroscopy. In addition to confirming previously published assignments[60] for the ^{13}C resonances for [^{13}C]PBG, they also detected resonances assigned to a distorted enzyme-bound form of PBG, in slow exchange with unbound PBG. With enzyme chemically modified with methyl methanethiosulphonate (MMTS), which does not catalyse PBG formation, a resonance consistent with the formation of a Schiff base adduct was detected at 166.5 ppm, whose linewidth was 50 Hz, and in slow exchange with free ALA (see

5-aminolaevulinic acid, ALA, **8.5**

Porphobilinogen, PBG, **8.6**

Fig. 8.9). The linewidth is consistent with the expected rotational correlation time for a protein of this molecular weight.

In a subsequent study,[61] purely on the basis of interpreting chemical shifts by comparison with model compounds, the Schiff base intermediate was determined to represent an imine, as opposed to an eneamine, and the stereochemistry of the intermediate is shown in Fig. 8.10. This was tested[62] further by ^{15}N NMR of the [^{15}N]ALA-bound species, as shown in Fig. 8.11. In addition to exogenous [15]NH$_4$ at -354 ppm and free ALA at -349.9 ppm, a third, broad resonance is apparent whose chemical shift is at -351.1 ppm and which was assigned to the enzyme-bound [^{15}N]ALA. On the basis of model compounds, this latter chemical shift was interpreted as being consistent with being half-way between the protonated and non-protonated forms (which would be in fast exchange). The chemical shift for the enzyme-bound [3-^{13}C]ALA provided further evidence that the imine stereochemistry is likely to be E, as shown in Fig. 8.10. Overall, these results demonstrate the possibilities for detection of enzyme-substrate and stable enzyme-intermediate complexes with relatively large protiens.

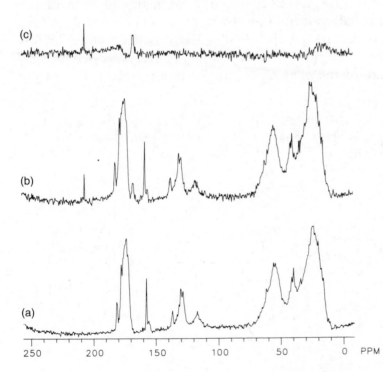

(c)

(b)

(a)

| 250 | 200 | 150 | 100 | 50 | 0 PPM |

Fig. 8.9 ^{13}C NMR spectra of methyl methanethiosulphonate (MMTS) modified PGB synthase (a) apo-enzyme; (b) sample of (a) after addition of [4-^{13}C]ALA; and (c) difference spectrum. (Reprinted from ref. 59 with permission.)

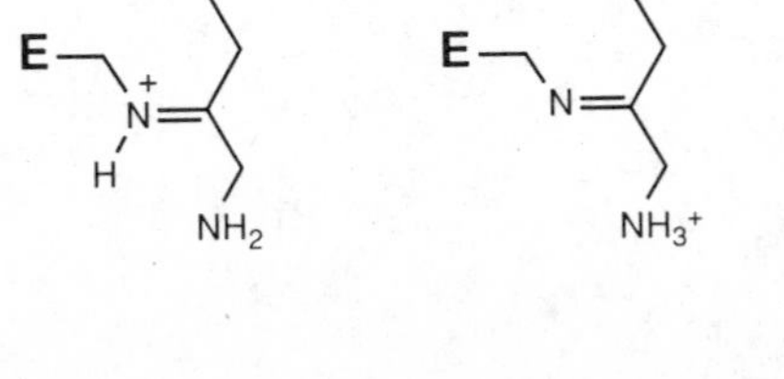

Fig. 8.10 Stereochemistry and possible protonation states of the P-side Schiff base intermediate. The C$_5$ of ALA is trans to the lysine methylene group; only one nitrogen carries a positive charge.

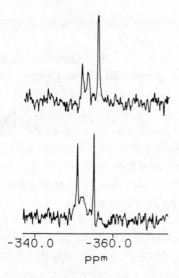

Fig. 8.11 ^{15}N NMR spectra of the enzyme-bound Schiff base complex of [^{15}N]ALA and MMTS-modified PBG synthase. The lower spectrum shows enzymes containing 2 mM active sites in H_2O, and the upper spectrum, 1.3 mM active sites in D_2O. In both cases the farthest upfield resonance is attributed to the ammonium ion, the centre resonance is due to the enzyme-bound Schiff base intermediate, and the downfield resonances from free ALA. (Reprinted from ref. 62 with permission.)

8.4.2 Porphobilinogen deaminase

Porphobilinogen deaminase (EC 4.3.1.8, also known as hydroxymethylbilane synthase, $M_r = 33$ kDa) catalyses the head-to-tail condensation of four porphobilinoen molecules to give hydroxymethylbilane (see structure group **8.6–8.7**). Early NMR studies on this enzyme from Scott and co-workers,[63–65] prior to the availability of overexpressed protein, attempted to identify the nature of the enzyme functional group (referred to as the

HYDROXYMETHYLBILANE (8.7)

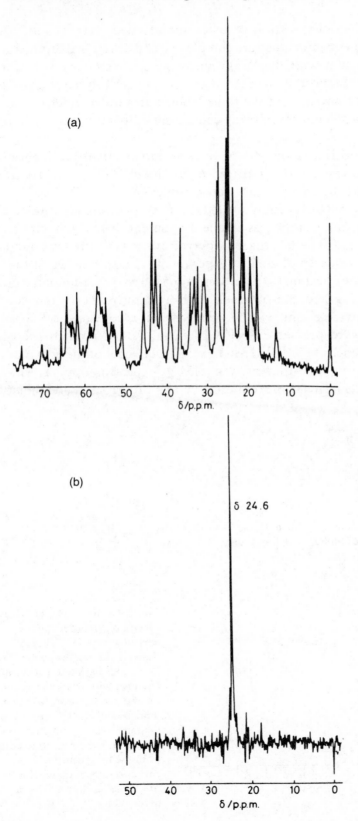

Fig. 8.12 [1]H-decoupled 100.6 MHz ^{13}C NMR spectra of (a) [11-^{13}C]PBG$_1$ 'mono' complex with PBG deaminase, pH 12; and (b) difference spectrum of (a) minus the same complex prepared with unlabelled PBG. (Reprinted from ref. 70 with permission.)

'X'-group) to which the substrate was known[66] to become attached. Both ³H and ¹³C NMR was carried out on isolated enzyme-substrate complexes, and a ³H NMR resonance was interpreted as being consistent with the 'X' group being a cysteine thiol. In contrast, Battersby and co-workers[67,68] reported that the 'X' group was probably a lysine residue. In fact, both laboratories were wrong, and the same laboratories independently identified that the substrate binds to a holoenzyme already containing a dipyrrole unit (see structure group **8.6–8.7**).

Jordan and Warren,[69] and Battersby and co-workers independently,[70] identified the dipyrrole in enzyme overexpressed from *E.coli* by incubating the enzyme with [11-¹³C]PBG and examining the enzyme-substrate complex by ¹³C NMR (see Fig. 8.12). In the difference spectrum, in which the unlabelled complex spectrum is subtracted from the labelled complex spectrum, a single resonance is apparent, consistent with a -CH₂- linkage to another pyrrole ring. The fact that two pyrrole units were already bound to the enzyme was implied from radioactive labelling experiments. Thus if the 'X'-group was a dipyrrole, this then raised the question as to the identity of the enzyme functional group to which the dipyrrole unit is attached, the 'Y' group, and also confirmation was needed as to the precise number of pyrrole units. This came from nice work by Scott and co-workers,[71,72] who used overexpressed *E.coli* [¹³C]-labelled PBG deaminase prepared by growth of *E.coli* in the presence of [5-¹³C]ALA. The ¹³C resonances of the enzyme-bound cofactor were detected using one-dimensional HMQC spectroscopy, as shown in Fig. 8.13. Although the chemical shifts are indicative of a dipyrromethane moiety, and

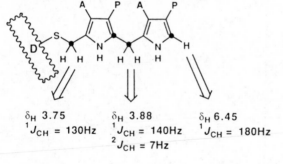

$$\delta_H\ 3.75 \qquad \delta_H\ 3.88 \qquad \delta_H\ 6.45$$
$$^1J_{CH} = 130Hz \qquad ^1J_{CH} = 140Hz \qquad ^1J_{CH} = 180Hz$$
$$^2J_{CH} = 7Hz$$

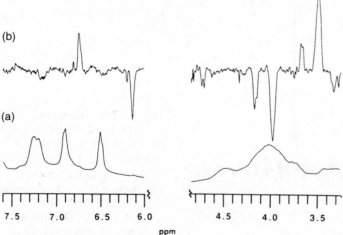

Fig. 8.13 300 MHz ¹H NMR spectra of PBG deaminase enriched from [5-¹³C]ALA showing the relevant regions of (a) the normal proton spectrum obtained with pre-saturation of the water resonance; (b) the one-dimensional HMQC spectrum of the same region containing the three proton signals originating from the biosynthetically enriched dipyrrole cofactor. (Reprinted from ref. 72 with permission.)

suggestive that the 'Y' linkage is sulphur, the latter point was proved by site-directed mutagenesis of two conserved cysteines. The Cys242Ser mutant enzyme was found to be inactive and not possible to label with [^{14}C]ALA.

8.4.3 5-Enolpyruvyl-shikimate-3-phosphate synthase
Background

5-Enolpyruvylshikimate-3-phosphate (EPSP) synthase (EC 2.5.1.19, $M_r = 46$ kDa) catalyses the penultimate step in the biosynthetic pathway in higher plants and bacteria that leads to chorismic acid, the common precursor for the three aromatic amino acids phenylalanine, tyrosine, and tryptophan. EPSP (structure **8.10**) is formed from shikimate-3-phosphate (S3P, structure **8.8**) and phosphoenolpyruvate (PEP, structure **8.9**). The enzyme is a monomer with no metal or co-factor requirements, and the cloned *E.coli* gene has been used to generate a hyperexpressing strain,[73] so that the

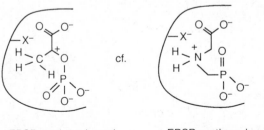

bacterial enzyme is available in gram quantities. Furthermore, EPSP synthase is the primary site of action of the herbicide glyphosate,[74] or *N*-phosphonomethylglycine. This is a broad spectrum post-emergence herbicide with worldwide applications in agriculture and horticulture. Glyphosate is thought to act as a transition-state analogue for a PEP-derived carbocation which might represent a transition-state structure *en route* to intermediate (although this point is in dispute—*vide infra*). This structure is shown in Fig. 8.14.

Two contrasting mechanisms have been proposed for the EPSP synthase reaction, according to what is thought to be the nucleophilic group that attacks C-2 of PEP. While both require the transient formation of an intermediate with a methyl group at C-3, and with tetrahedral geometry at the C-2 position, the 'shikimate nucleophile' mechanism[75,76] requires the attack on C-2 by the 5-hydroxyl of S3P, (Mechanism A), while the 'enzyme nucleophile' mechanism[77,78] requires the attack on C-2 by a nucleophilic group on the

EPSP synthase bound
PEP-derived carbocation

EPSP synthase bound
Glyphosate

Fig. 8.14 Postulated transition-state nature of the inhibitor glyphosate compared with a possible PEP-derived carbocation.

enzyme (Mechanism B). The involvement of a concerted reaction with a carbocation-like transition-state has also been proposed.[79] Anton *et al.*[77] have shown that incubation of ddS3P and PEP with EPSP synthase in tritiated water resulted in incorporation of tritium into PEP, thereby suggesting that the 5-hydroxyl of S3P is not the nucleophile, and that a covalent enzyme-PEP is formed (Mechanism B). Work by Anderson *et al.*[80,81] has provided evidence for the shikimate nucleophile mechanism, by isolation and character-ization of an enzyme-free species under rapid chemical-quench conditions whose structure is consistent with (**8.11**). The direct observation of the enzyme-intermediate (E·I) complex was first reported by our own laboratory,[82,83] and later confirmed by another laboratory.[84] Our work used solution-state ^{13}C and two-dimensional HMQC NMR spectroscopy has established that the same enzyme-free species detected by Anderson *et al.* is present in the enzyme-intermediate complex, and is not an isolation artefact. However, an enzymatic mechanism cannot be considered to be settled until the enzyme-bound species has been detected and characterized, since the intermediate (**8.11**) could be formed from intermediate (**8.13**) as a result of nucleophilic attack by inorganic phosphate.

Recently, the Monsanto group have reported[85] the X-ray crystal structure for the unliganded enzyme at 2.4 Å. Although the X-ray coordinates are not in the public domain, and a stereo picture was not provided in their publication, the protein shows two approximately equal globular domains connected by two segments of the main backbone which cross over. The protein has an open structure, suggesting that it may close up upon substrate binding. This implies that a large conformational change accompanies substrate binding, presumably PEP or glyphosate, since although the Monsanto group have succeeded in obtaining crystals of the S3P—EPSP synthase complex, despite repeated attempts, they have not managed to crystallize the ternary complex.

Mechanism A ("Shikimate Nucleophile")

Mechanism B ("Enzyme Nucleophile")

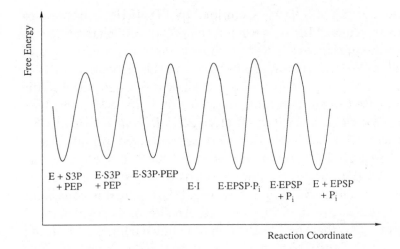

Fig. 8.15 Free energy profile of the reaction catalysed by EPSP synthase.

To date, the active site of EPSP synthase has been probed through a number of chemical modification and mutagenesis studies. For example, Glu-418 and His-385 were thought to form a diad in which the basicity of the His-385 imidazole ring is affected by the proximity of Glu-418.[86] The implication from this and other data is that His-385 is the likely base which donates (or abstracts) the proton to C-3 of PEP. More recent data argue that, while possibly being in close proximity, His-385 is probably not the principal base. Furthermore, there may be more than one base involved in the mechanism.[87]

There are only a handful of enzymes for which the full kinetic and thermodynamic profiles have been determined (see Fig. 8.15), and EPSP synthase would appear to be one of this select group as a result of widely acclaimed work by Anderson and Johnson.[80] One of the more remarkable results of this work is the prediction that the E·I complex of this enzyme is so unusually stable. It was suggested that EPSP synthase actually destabilizes the intermediate and selectively interacts more tightly with the substrates and products. However, implicit in this detailed kinetic analysis is an assumption as to the number of enzyme-substrate and enzyme-intermediate complexes (since none were detected directly, and only one indirectly). Also, Anderson and Johnson postulated an equilibrium sequential ordered mechanism, whereas recent evidence suggests that a random kinetic mechanism is operating,[88] which was confirmed by the observation that PEP can bind to the enzyme in the absence of S3P.[89] This then casts some doubt over the Anderson and Johnson kinetic analysis of EPSP synthase. Having said this, it still remains a particularly attactive system for developing novel NMR techniques for the elucidation of enzyme-intermediate complexes.

NMR studies of the enzyme-inhibitor complex

The ionization state of S3P and *N*-phosphonomethylglycine (glyphosate) when bound to EPSP synthase have been determined by examining the pH dependence of the resonances of glyphosate and [3-^{13}C, ^{15}N]glyphosate by ^{31}P, ^{13}C, and ^{15}N NMR spectroscopy.[90] The phosphonate group titrates with a pK_a of <1 and 5.7. The amino function titrates with a pK_a of 10.5, which is close to the value (10.6) reported previously by Sprankle *et al.*[91] The pH dependence of the phosphate group of S3P was examined by ^{31}P NMR, and found to titrate with a pK_a of 6.3. In the presence of EPSP synthase, the pK_a of S3P changes to 6.7, and the resonance broadens. In the presence of EPSP synthase and glyphosate, the bound S3P phosphate no longer titrates over the pH range 6.0–10.5, and the resonance shifts to −2.0 ppm, which is consistent with a monoanionic species. Since it does not titrate, this implies the proximity of a protonated enzyme amino acid side chain in the S3P phosphate binding site of EPSP synthase. One candidate is Lys-22 (Lys-23 in *Petunia hybrida* EPSP synthase) on the basis of chemical modification[92,93] and site-directed mutagenesis[94] studies. Another is Arg-27 (Arg-28 in *Petunia hybrida*) on the basis of chemical modification studies.[95] The bound glyphosate phosphonate also no longer titrates over the same pH range, and the resonance shifts to 7.7 ppm (see Fig. 8.16), which is consistent with a dianionic species with the amino group $\approx 50\%$ ionized. An alternative explanation, suggested by the Monsanto group,[96] is that the C—P—O bond angle changes on binding to the enzyme. The chemical shift of the ^{15}N resonance of bound [^{15}N]glyphosate falls outside the range in which the 50% ionized species resides, and does not titrate, implying that the change in bond angle is the likely explanation for shift (see Fig. 8.17). However, ^{15}N chemical shifts are susceptible to hydrogen bonding and electric field effects, so that the interpretation of this data is controversial.

Implicit in our discussion of the ^{31}P chemical shifts of the enzyme-bound species of EPSP synthase is the assumption that the ^{31}P chemical shifts of alkyl phosphate and

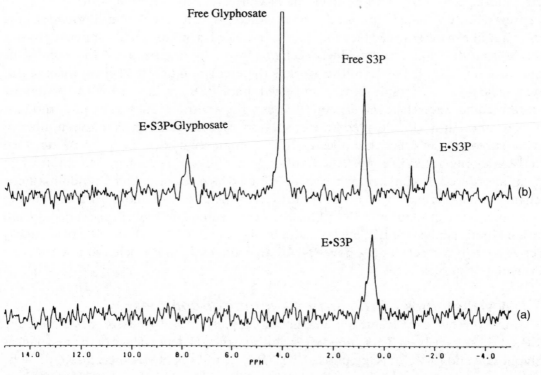

Fig. 8.16 ^{1}H-decoupled ^{31}P NMR spectrum of (a) EPSP synthase–S3P complex and (b) EPSP synthase–S3P–glyphosate complex, all at pH 7.9. (Reprinted from ref. 83 with permission.)

phosphonate resonances do not alter significantly on binding to an enzyme, and therefore the change in chemical shift reflects the ionization state of the group, rather than the local effects of the enzyme active site. There is a large body of literature to suggest that association of hydrogen-bonding donors (in active site of an enzyme) generally has little effect on the ^{31}P chemical shift other than that explained by a change in the pK_a.[97] For example, work on the 3′-CMP·RNase complex has shown that the proximity of positively charged groups in the active site to a phosphate group does not give rise to significant changes in ^{31}P chemical shifts (see Section 7.2.8), which is at most 0.3 ppm as compared to 4 ppm for phosphate ionization.[98–100] Although the rationalization for this latter shift is still the subject of some debate,[90] it may be attributed to a combination of the change in electron density upon ionization and the change in O—P—O bond angles. Other examples in which substrate binding was accompanied by changes in ^{31}P chemical shift of no more than 2 ppm, and often <1 ppm, come from work with pyruvate kinase,[101] 3-phosphoglycerate kinase,[102] methionyl tRNA synthetase,[103] dihydrofolate reductase,[104] G-actin,[105] glycogen phosphorylase,[106–108] and triose phosphate isomerase.[109] Indeed, all the known counterexamples in the literature in which large changes in ^{31}P chemical shift

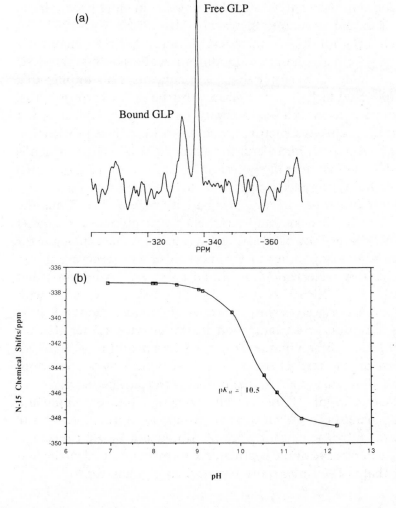

Fig. 8.17 (a) ^{1}H-coupled (without NOE) ^{15}N NMR spectrum of EPSP synthase–S3P–[3–^{13}C,^{15}N]–glyphosate complex. (b) pH titration of free [3-^{13}C,^{15}N]-glyphosate by ^{1}H-decoupled (with full NOE) ^{15}N NMR spectroscopy. (Reprinted from ref. 83 with permission.)

have been associated with the local environment have been exclusively with proteins containing a divalent metal ion sequestered at the active site.[110–113]

Work on the binding of S3P to EPSP synthase by Castellino *et al.*[96] has suggested, by modelling and ^{31}P NMR, that the small change in the S3P phosphate chemical shift of 0.3 ppm may be attributed either to ionization state or increased O—P—O bond angle (as opposed to protein active site effects). Similarly, these workers studied the binding of glyphosate to EPSP synthase–S3P complex and have suggested that the relatively large change in the phosphonate chemical shift of 3.6 ppm may be attributed to similar effects, and they favour the interpretation of major changes in C—P—O bond angle. Glyphosate has the opportunity to adopt a wider range of conformations in solution than alkyl phosphates, which may involve charge zwitterionic interactions, but which should be largely averaged. Binding to the enzyme might then allow a single unique conformation (and state of hydration) to be stabilized. However, there is little evidence to date that such a large change in chemical shift as observed for enzyme-bound glyphosate can be attributed wholly to conformational effects rather than electric field effects resulting from ionization.

The fact that the nitrogen of enzyme-bound glyphosate does not titrate is consistent with an amino acid functional group in the glyphosate binding site in close proximity to the amino group of glyphosate. Possible candidates are His-385,[114] Cys-408,[115] or Glu-418.[85] However, although 5,5′-dithiobis-(2-nitrobenzoic acid) (DTNB) inactivates the enzyme in the presence of S3P and glyphosate by modification of Cys-408, cyanolysis of this DTNB-inactivated enzyme leads to active thiocyano-enzyme. This implies that the smaller thiocyano group does not block the active site as does the DTNB group. Thus, while Cys-408 is undoubtedly located in close proximity to the active site, it is not required for catalysis. Both mechanisms in structure group **8.8–8.13** require protonation at C-3 of PEP, presumably via some protonated enzyme amino acid side chain which is either basic or nucleophilic. His-385 is most likely to fulfil this rôle, the pK_a of which appears to be 6.8.[92] The rôle of Glu-418 has been suggested[94] to be either involved directly in catalysis, or interacting with and thus affecting the basicity of His-385. Thus the suggestion by Anton *et al.*[77] and Steinrücken and Amrhein[74] that glyphosate inhibits EPSP synthase with a positively charged nitrogen acting as a transition-state analogue for the PEP-derived carbocation implied by either of the mechanisms in structure group **8.8–8.13** is open to question. The observation by Steinrücken and Amrhein[74] that inhibition of *Klebsiella pneumoniae* EPSP synthase by glyphosate increases with increasing pH, with an optimum between ≈6.8 (at 2 mM glyphosate) and ≈8.4 (at 0.1 mM glyphosate), is probably due to an ionization in the enzyme rather than in glyphosate. A likely rationalization for glyphosate's efficacy as an inhibitor of EPSP synthase lies in consideration of its stereochemistry. The structure of the enzyme-intermediate complex is tetrahedral (*vide infra*) so that probably the tetrahedral nature of the nitrogen is an important factor in inhibition, as well as having a phosph(on)ate and carboxylate with the correct separation between the two groups for optimal active site binding. This has been illustrated by the elegant work of Alberg and Bartlett recently, who synthesized a phosphonate derivative of the postulated intermediate which is the first rationally designed inhibitor that is 100 times more potent than glyphosate.[116]

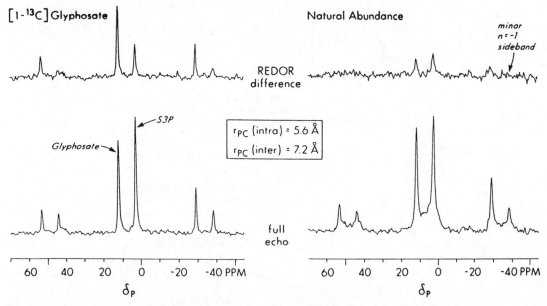

Fig. 8.18 121.3 MHz solid-state REDOR ^{31}P NMR of S3P–EPSP synthase to which [1-^{13}C]glyphosate (left) and unlabelled glyphosate (right) are bound. (Reprinted from ref. 117 with permission.)

Schaefer and co-workers[117] have used ^{31}P—^{13}C solid-state REDOR NMR experiments (see Section 3.3.3) to investigate the nature of the bound inhibitor glyphosate in the ternary complex. This is particularly significant, since all attempts to crystallize the ternary complex have failed. Fig. 8.18 shows the REDOR results for the ternary complex with [1-^{13}C]glyphosate compared to the same complex with unlabelled glyphosate. An intermolecular ^{31}P—^{13}C distance of 7.2 Å was measured between the phosphate of S3P and the labelled carbon of glyphosate. This means that S3P and glyphosate are relatively close in the enzyme-binding site. An intramolecular ^{31}P—^{13}C distance of 5.6 Å was measured between the phosphonate ^{31}P and the labelled carbon of glyphosate. This distance implies that glyphosate is completely extended when bound to the enzyme (although additional distance constraints are needed to confirm this), which is in principle consistent with an 'opening' of the C—P—O bond angle suggested by solution-state ^{31}P NMR studies. A proposed model structure for the conformation of the bound substrates in the ternary complex is shown in Fig. 8.19.

NMR studies of an enzyme-intermediate complex

Our own laboratory has carried out a detailed study of the enzyme-bound intermediate of EPSP synthase using ^{13}C and ^{1}H NMR spectroscopy. The interaction of EPSP synthase with S3P and [2-^{13}C]- or [3-^{13}C]- or [2,3-^{13}C$_2$]PEP was examined by ^{13}C NMR spectroscopy (see Fig. 8.20). From ^{13}C at C2 of PEP, in addition to the resonances due to C2 of PEP itself and [8-^{13}C]EPSP, new resonances appeared at 164.8, 110.9, and 107.2 ppm. The resonance at 164.8 ppm has been assigned to enzyme-bound EPSP. The resonance at 110.9 ppm has been assigned to C-8 of an enzyme-free tetrahedral

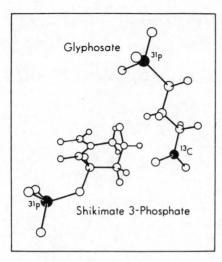

Fig. 8.19 Possible structure of S3P-glyphosate complex bound to EPSP synthase. The intramolecular ^{31}P—^{13}C distance for this structure is 5.6 Å, and the intermolecular ^{31}P—^{13}C distance is 7.2 Å. A hydrogen bond between the glyphosate NH proton and the C5 oxygen of S3P is assumed. (Reprinted from ref. 117 with permission.)

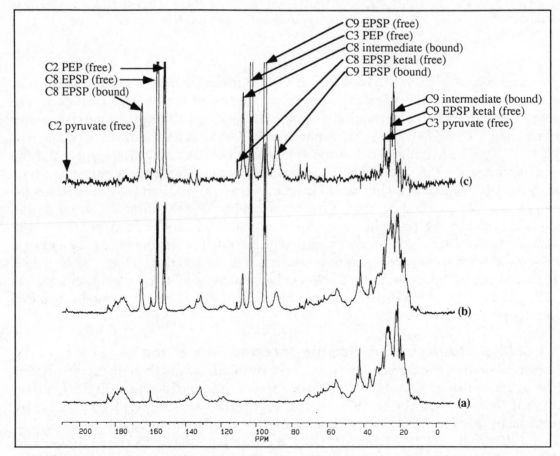

Fig. 8.20 ^1H-decoupled ^{13}C NMR spectra of EPSP synthase plus S3P and [2,3-$^{13}C_2$]PEP: (a) enzyme plus S3P; (b) sample in (a) plus [2,3-$^{13}C_2$]PEP; (c) difference spectrum (b) minus (a).

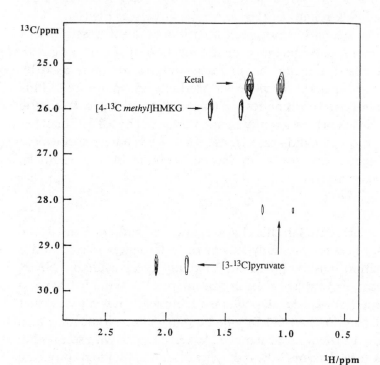

8.14

intermediate (structure **8.14**) of the sort originally proposed by Levin and Sprinson[75] and has recently been observed independently by Anderson *et al.*[84]

The resonance at 107.2 ppm has been assigned to an enzyme-bound intermediate whose structure is closely related to the tetrahedral intermediate. From ^{13}C at C3 of PEP, new resonances appeared at 88.9, 26.2, 25.5, and 24.5 ppm. The resonance at 88.9 ppm has been assigned to enzyme-bound EPSP. The resonance at 26.2 ppm, which was found to correlate with 1.48 ppm by isotope-edited multiple-quantum coherence ^1H NMR spectroscopy (see Fig. 8.21), has been assigned to the methyl group of 4-hydroxy-4-methyl-ketoglutarate. The resonance at 25.5 ppm, which was found to correlate with a ^1H resonance at 1.16 ppm, could be substantially reduced by pre-incubating the enzyme with the inhibitor glyphosate, and could be isolated from the protein by ultrafiltration. Resonances were detected in the ultrafiltrate at 26 ppm and 1.16 ppm, and these arise from the enzyme-free species. This was originally interpreted in terms of the same enzyme-free tetrahedral intermediate that gave rise to the resonance at 110.9 ppm with [2-^{13}C]PEP and that was proposed by Levin and Sprinson. However,

Fig. 8.21 Two-dimensional HMQC isotope-edited $^1H\{^{13}C\}$ spectrum of EPSP synthase plus S3P and [3-^{13}C]PEP. (Reprinted from ref. 82 with permission.)

more recent kinetic evidence obtained in our own laboratory suggested that this species is unlikely to be an intermediate, and is more likely to be a by-product of the breakdown of the enzyme-bound intermediate. Further investigations using one-dimensional NOE difference spectroscopy and two-dimensional ROESY have suggested that this species is an intramolecular ketal (structure **8.15**).

8.15

The above finding is consistent with the work of the Monsanto group, who reported the isolation of the same species.[118] No ^{1}H correlation has been detected for the ^{13}C resonance at 24.5 ppm due to its large linewidth (100 Hz), but it has been interpreted as arising from the same enzyme-bound intermediate which gave rise to the resonance at 107.2 ppm with [2-^{13}C]PEP. Furthermore, attempts to detect this enzyme-bound species by ^{31}P NMR have failed to date, presumably due to the inordinate anisotropy of ^{31}P at this field strength (11.75 T). The laboratory of Anderson *et al.*[84] confirmed our results with their own direct observation of the enzyme-bound intermediate complex of EPSP synthase.

As we discussed in Section 6.4.2, EPSP synthase was used for the development of the new technique of time-resolved solid-state NMR spectroscopy (see Fig. 6.4). One advantage that this method offers, which has not been mentioned, is the possibility of obtaining the pre-steady-state kinetic rate constants for the reaction by monitoring real, observable species. This is not possible using the chemical-quench methods used by Anderson and co-workers, since, in general, breakdown products are monitored. This introduces the added complication of the kinetics of chemical breakdown from the postulated intermediate species to the species actually being monitored by HPLC or other analytical methods. With time-resolved solid-state NMR, the intermediate species are monitored directly and their turnover can be measured without having to make assumptions about any of the rate constants.

8.5 OUTLOOK

As can be seen from the examples presented in this chapter, and in Chapters 6 and 7, the ways in which NMR can be applied to the study of enzymes is hugely varied and a considerable amount of information may be obtained. Virtually every available NMR method is necessary in some cases, as we have seen in this chapter. Although I have to admit significant bias, since the method was developed in my laboratory, it is my view that time-resolved solid-state NMR (see Section 6.4.2) will be particularly important as a general method for characterizing transient intermediates. When coupled with solid-state distance measurement methods (see Section 3.3) and Laue X-ray crystallography, the technique will be eminently suited to unravelling the molecular details of enzymatic mechanisms and to providing a glimpse of the intimate relationship between protein structure and function.

REFERENCES

1. R. J. Fletterick and N. B. Madsen, *Ann. Rev. Biochem.*, **49**, 31–61 (1980).

2. P. J. McLaughlin, D. I. Stuart, H. W. Klein, N. G. Oikonomakos, and L. N. Johnson, *Biochemistry*, **23**, 5865 (1984).

3. S. J. W. Busby, D. G. Gadian, G. K. Radda, R. E. Richards, and P. J. Seeley, *FEBS Lett.*, **55**, 14–17 (1975).

4. K. Feldmann and E. J. M. Helmreich, *Biochemistry*, **15**, 2394–401 (1976).

5. K. Feldmann and W. E. Hull, *Proc. Natl. Acad. Sci. USA*, **74**, 856–60 (1977).

6. S. G. Withers, B. D. Sykes, N. B. Madsen, and P. J. Kasvinsky, *Biochemistry*, **18**, 5342–8 (1979).

7. D. Gratecos and E. H. Fischer, *Biochim. Biophys. Acta*, **58**, 960–7 (1974).

8. S. G. Withers, N. B. Madsen, and B. D. Sykes, *Biochemistry*, **20**, 1748–56 (1981).

9. M. Hoerl, K. Feldmann, K. D. Schnackerz, and H. J. C. Helmreich, *Biochemistry*, **18**, 2457–64 (1979).

10. S. G. Withers, N. B. Madsen, and B. D. Sykes, *Biochemistry*, **21**, 6716–22 (1982).

11. Y. C. Chang and D. J. Graves, *J. Biol. Chem.*, **260**, 2709–14 (1985).

12. J. H. Lechter and J. R. van Wazer, *Top. Phosphorus Chem.*, 75–226 (1965).

13. J. H. Lechter and J. R. van Wazer, *J. Chem. Phys.*, **70**, 3300–16 (1965).

14. A. J. R. Constella, T. Glonek, and J. R. van Wazer, *Inorg. Chem.*, **15**, 972–4 (1976).

15. D. G. Gorenstein, *J. Am. Chem. Soc.*, **97**, 898–900 (1975).

16. D. G. Gorenstein, in *Phosphorus-31 NMR: Principles and Applications* (ed. D. G. Gorenstein), Academic Press, New York, pp. 7–36 (1984).

17. S. Un and M. P. Klein, *J. Am. Chem. Soc.*, **111**, 5119–24 (1989).

18. D. Barford, J. R. Schwabe, N. G. Oikonomakos, K. R. Acharya, J. Hajdu, A. C. Papageorgiou, *et al.*, *Biochemistry*, **27**, 6733–41 (1988).

19. L. N. Johnson and D. Barford, *J. Biol. Chem.*, **265**, 2409–12 (1990).

20. R. Challoner, C. A. McDowell, W. Stirtan, and S. G. Withers, *Biophys. J.*, **64**, 484–91 (1993).

21. J. E. Taguchi, S. J. Heyes, D. Barford, L. N. Johnson, and C. M. Dobson, *Biophys. J.*, **64**, 492–501 (1993).

22. J. Hajdu, P. A. Machin, J. W. Campbell, T. J. Greenhough, I. J. Clifton, S. Zurek, *et al.*, *Nature*, **329**, 178–81 (1987).

23. J. Hajdu, K. R. Acharya, D. I. Stuart, P. J. McLaughlin, D. Barford, N. G. Oikonomakos, *et al.*, *EMBO. J.*, **6**, 539–46 (1987).

24. S. G. Withers, N. B. Madsen, B. D. Sykes, M. Takagi, S. Shimomura, and T. Fukui, *J. Biol. Chem.*, **256**, 10759–62 (1981).

25. S. G. Withers, N. B. Madsen, and B. D. Sykes, *Biophys. J.*, **48**, 1019–26 (1985).

26. Y.-C. Chang, R. D. Scott, and D. J. Graves, *Biochemistry*, **25**, 1932–9 (1986).

27. Y.-C. Chang, R. D. Scott, and D. J. Graves, *Biochemistry*, **26**, 360–7 (1987).

28. J. T. Bolin, D. J. Filman, D. A. Matthews, R. C. Hamlin, and J. Kraut, *J. Biol. Chem.*, **257**, 13650–62 (1982).

29. C. Bystroff, S. J. Oatley, and J. Kraut, *Biochemistry*, **29**, 3263–77 (1990).

30. C. Bystroff and J. Kraut, *Biochemistry*, **30**, 2227–39 (1991).

31. D. A. Matthews, J. T. Bolin, J. M. Burridge, D. J. Filman, *et al.*, *J. Biol. Chem.*, **260**, 381–91; 392–399 (1985).

32. D. K. Stammers, J. N. Champness, C. R. Beddell, J. G. Dann, E. Eliopoulos, A. J. Geddes, *et al.*, *FEBS Lett.*, **218**, 178–84 (1987).

33. C. Oefner, A. D'Arcy, and F. K. Winkler, *Eur J. Biochem.*, **174**, 377–85 (1988).

34. J. F. Davies, T. J. Delcamp, N. J. Prendergast, V. A. Ashford, J. H. Freisham, and J. Kraut, *Biochemistry*, **29**, 9467–79 (1990).

35. B. Birdsall, J. Feeney, S. J. B. Tendler, S. J.

Hammond, and G. C. K. Roberts, *Biochemistry*, **28**, 2297–305 (1989).

36. M. D. Carr, B. Birdsall, T. A. Frenkiel, C. J. Bauer, J. Jimenez-Barbero, V. I. Polshakov, *et al.*, *Biochemistry*, **30**, 6330–41 (1991).

37. C. J. Falzone, S. J. Benkovic, and P. E. Wright, *Biochemistry*, **29**, 9667–77 (1990).

38. C. J. Falzone, P. E. Wright, and S. J. Benkovic, *Biochemistry*, **30**, 2184–91 (1991).

39. B. J. Stockman, N. R. Nirmala, G. Wagner, T. J. Delcamp, M. T. DeYarman, and J. H. Freisheim, *FEBS Lett.*, **283**, 267–9 (1991).

40. B. J. Stockman, N. R. Nirmala, G. Wagner, T. J. Delcamp, M. T. DeYarman, and J. H. Freisheim, *Biochemistry*, **31**, 218–29 (1992).

41. G. C. K. Roberts, *NATO ASI Ser., Ser. A*, **107**, (NMR Life Sci.), 73–86 (1986) and references cited therein.

42. A. Gronenborn, B. Birdsall, E. I. Hyde, G. C. K. Roberts, J. Feeney, and A. S. V. Burgen, *Nature*, **290**, 273–4 (1981).

43. B. Birdsall, J. DeGraw, J. Feeney, S. Hammond, M. S. Searle, G. C. K. Roberts, *et al.*, *FEBS Lett.*, **217**, 106–10 (1987).

44. F.-Y. Huang, Q.-X. Yang, T.-H. Huang, L. Gelbaum, and L. F. Kuyper, *FEBS Lett.*, **283**, 44–6 (1991).

45. M. A. Jimenez, J. R. P. Arnold, J. Andrews, J. A. Thomas, G. C. K. Roberts, B. Birdsall, and J. Feeney, *Prot. Engineer.*, **2**, 627 (1989).

46. G. C. K. Roberts, *Spec. Publ. R. Soc. Chem.* (*Spec. Biol. Mol.*), **94**, 219–22 (1991).

47. S. G. Waley, J. C. Miller, I. A. Rose, and E. L. O'Connell, *Nature*, **227**, 181 (1970).

48. E. Lolis, T. Alber, R. C. Davenport, D. Rose, F. C. Hartman, and G. A. Petsko, *Biochemistry*, **29**, 6619–25 (1990) and references cited therein.

49. D. W. Banner, A. C. Bloomer, G. A. Petsko, D. C. Phillips, C. I. Pogson, and I. A. Wilson, *Nature*, **255**, 609–14 (1975).

50. J. G. Belasco and J. R. Knowles, *Biochemistry*, **19**, 472–7 (1980).

51. E. A. Komives, L. C. Chang, E. Lolis, R. F. Tilton, G. A. Petsko, and J. R. Knowles, *Biochemistry*, **30**, 3011–19 (1991) and references cited therein.

52. I. D. Campbell, R. B. Jones, P. A. Kiener, and S. G. Waley, *Biochem. J.*, **179**, 607–21 (1979).

53. R. Wolfenden, *Biochemistry*, **9**, 3404–7 (1970).

54. K. D. Schnackerz, T. K. Kuan, W. J. Goux, and R. W. Gracy, *Biochem. Biophys. Res. Comm.*, **173**, 736–40 (1990).

55. K. D. Schnackerz and R. W. Gracy, *Eur. J. Biochem.*, **199**, 231–8 (1991).

56. P. J. Lodi and J. R. Knowles, *Biochemistry*, **30**, 6948–56 (1991) and references cited therein.

57. D. Shemin, *Phil. Trans. R. Soc. Lond. B*, **273**, 109 (1976).

58. P. M. Jordan and J. S. Seehra, *FEBS Lett.*, **114**, 283–6 (1980).

59. E. K. Jaffe and G. D. Markham, *Biochemistry*, **26**, 4258–64 (1987).

60. J. N. S. Evans, P. E. Fagerness, N. E. Mackenzie, and A. I. Scott. *Magn. Reson. Chem.*, **23**, 939–44 (1985).

61. E. K. Jaffe and G. D. Markham, *Biochemistry*, **27**, 4475–81 (1987).

62. E. K. Jaffe, G. D. Markham, and J. S. Rajagopalan, *Biochemistry*, **29**, 8345–50 (1990).

63. J. N. S. Evans, P. E. Fagerness, N. E. Mackenzie, and A. I. Scott, *J. Am. Chem. Soc.*, **106**, 5738–40 (1984).

64. J. N. S. Evans, R. C. Davies, A. S. F. Boyd, I. Ichinose, N. E. Mackenzie, A. I. Scott, *et al.*, , **25**, 896–904 (1986).

65. J. N. S. Evans, G. Burton, P. E. Fagerness, N. E. Mackenzie, and A. I. Scott, *Biochemistry*, **25**, 905–12 (1986).

66. P. M. Jordan and A. Berry, *Biochem. J.*, **195**, 177–81 (1981).

67. A. R. Battersby, C. J. R. Fookes, G. W. J. Matcham, E. McDonald, and R. Hollenstein, *J. Chem. Soc., Perkin I*, 3031–40 (1983).

68. A. R. Battersby, C. J. R. Fookes, G. Hart, G. W. J. Matcham, and P. S. Pandey, *J. Chem. Soc., Perkin I*, 3041–7 (1983).

69. P. M. Jordan and M. J. Warren, *FEBS Lett.*, **225**, 87–92 (1987).

70. G. J. Hart, A. D. Miller, F. J. Leeper, and A. R. Battersby, *J. Chem. Soc. Chem. Comm.*, 1762–5 (1987).

71. A. I. Scott, N. J. Stolowich, H. J. Williams, M. D. Gonzalez, C. A. Roessner, S. K. Grant, *et al.*, *J. Am. Chem. Soc.*, **110**, 5898–900 (1988).

72. A. I. Scott, C. A. Roessner, N. J. Stolowich, P. Karuso, H. J. Williams, S. K. Grant, *et al.*, *Biochemistry*, **27**, 7984–90 (1988).

73. W. A. Shuttleworth, C. D. Hough, K. P. Bertrand, and J. N. S. Evans, *Protein Engineering*, **5**, 461–6 (1992).

74. H. C. Steinrucken and N. Amrhein, *Eur. J. Biochem.*, **143**, 351 (1984).

75. J. G. Levin and D. B. Spirnson, *J. Biol. Chem.*, **239**, 1142 (1964).

76. C. E. Grimshaw, S. G. Sogo, S. D. Copley, and J. R. Knowles, *J. Amer. Chem. Soc.*, **106**, 2699 (1984).

77. D. L. Anton, L. Hedstrom, S. M. Fish, and R. H. Abeles, *Biochemistry*, **22**, 5903 (1983).

78. B. Ganem, *Tetrahedron*, **34**, 3353 (1978).

79. C. E. Grimshaw, S. G. Sogo, and J. R. Knowles, *J. Biol. Chem.*, **257**, 596 (1982).

80. K. S. Anderson, J. A. Sikorski, and K. A. Johnson, *Biochemistry*, **27**, 7395 (1988).

81. K. S. Anderson, J. A. Sikorski, A. J. Benesi, and K. A. Johnson, *J. Amer. Chem. Soc.*, **110**, 6577 (1988).

82. P. N. Barlow, R. J. Appleyard, B. J. O. Wilson, and J. N. S. Evans, *Biochemistry*, **28**, 7985 and 10093 (1989).

83. J. N. S. Evans, NMR and Enzymes, in *Pulsed Magnetic Resonance: NMR, ESR and Optics (A Recognition of E. L. Hahn)*, (ed. D. Bagguley), Oxford University Press, pp. 123–73 (1992).

84. K. S. Anderson, R. D. Sammons, G. C. Leo, J. A. Sikorski, A. J. Benesi, and K. A. Johnson, *Biochemistry*, **29**, 1460–5 (1990).

85. W. C. Stallings, S. Abdel-Meguid, L. W. Lim, H. S. Shieh, H. E. Dayringer, N. K. Leimgruber, *et al.*, *Proc. Natl. Acad. Sci. USA*, **88**, 5046–50 (1991).

86. Q. K. Huynh, *J. Biol. Chem.*, **263**, 11631–5 (1988).

87. W. A. Shuttleworth and J. N. S. Evans, *Biochemistry*, **33**, 7062–8 (1994).

88. K. J. Gruys, M. C. Walker, and J. A. Sikorski, *Biochemistry*, **31**, 5534–44 (1992).

89. J. E. Ream, H. K. Yuen, R. B. Frazier, and J. A. Sikorksi, *Biochemistry*, **31**, 5528–34 (1992).

90. R. J. Appleyard, P. N. Barlow, B. J. O. Wilson, and J. N. S. Evans, unpublished results.

91. P. Sprankle, W. F. Megitt, and D. Penner, *Weed Sci.*, **23**, 299–34 (1975).

92. Q. K. Huynh, G. M. Kishore, and G. S. Bild, *J. Biol. Chem.*, **263**, 735–9 (1988).

93. Q. K. Huynh, *J. Biol. Chem.*, **265**, 6700–4 (1990).

94. Q. K. Huynh, S. C. Bauer, G. S. Bild, G. M. Kishore, and J. R. Borgmeyer, *J. Biol. Chem.*, **263**, 11636–9 (1988).

95. S. R. Padgette, C. Smith, Q. K. Huynh, and G. M. Kishore, *Arch. Biochem. Biophys.*, **266**, 254–62 (1988).

96. S. Castellino, G. C. Leo, R. D. Sammons, and J. A. Sikorski, *Biochemistry*, **28**, 3856–68 (1989).

97. D. G. Gorenstein in *Phosphorus-31 NMR: Principles and Applications* (ed. D. G. Gorenstein), Academic Press (New York), p. 31 (1984).

98. W. Haar, J. C. Thompson, W. Maurer, and H. Rüterjans, *Eur. J. Biochem.*, **40**, 259–66 (1973).

99. D. G. Gorenstein and A. M. Wyrwicz, *Biochem. Biophys. Res. Commun.*, **54**, 976–82 (1973).

100. D. G. Gorenstein, A. M. Wyrwicz, and J. Bode, *J. Am. Chem. Soc.*, **98**, 2308–14 (1976).

101. B. D. Nageswara Rao, F. J. Kayne, and M. Cohn, *J. Biol. Chem.*, **254**, 2689–96 (1979).

102. B. D. Nageswara Rao, M. Cohn, and R. K. Scopes, *J. Biol. Chem.*, **253**, 8056–60 (1978).

103. G. Fayat, S. Blanquet, B. D. Nageswara Rao, and M. Cohn, *J. Biol. Chem.*, **255**, 8164–9 (1980).

104. E. I. Hyde, B. Birdsall, G. C. K. Roberts, J. Feeney, and A. S. V. Burgen, *Biochemistry*, **19**, 3738–54 (1980).

105. M. Brauer and B. D. Sykes, *Biochemistry*, **21**, 5934–9 (1982) and references cited therein.

106. K. Feldmann and W. E. Hull, *Proc. Natl. Acad. Sci. USA*, **74**, 856–60 (1977).

107. M. Hoerl, K. Feldmann, K. D. Schnackerz, and E. J. M. Helmreich, *Biochemistry*, **18**, 2457–64 (1979).

108. S. G. Withers, N. B. Masden, and B. D. Sykes, *Biochemistry*, **20**, 1748–56 (1981).

109. I. D. Campbell, R. B. Jones, P. A. Kiener, and S. G. Waley, *Biochem. J.*, **179**, 607–21 (1979).

110. J. L. Bock and B. Sheard, *Biochem. Biophys. Res. Commun.*, **66**, 24–30 (1975).

111. W. E. Hull, S. E. Halford, H. Gutfreund, and B. D. Sykes, *Biochemistry*, **15**, 1547–61 (1976).

112. J. F. Chlebowski, I. M. Armitage, P. P. Tusa, and J. E. Coleman, *J. Biol. Chem.*, **254**, 1207–16 (1976).

113. B. D. Nageswara Rao and M. Cohn, *Proc. Natl. Acad. Sci. USA*, **74**, 5355–7 (1977).

114. Q. K. Huynh, *Arch Biochem. Biophys.*, **258**, 233–9 (1987).

115. S. R. Padgette, Q. K. Huynh, S. Aykent, R. D. Sammons, J. A. Sikorski, and G. M. Kishore, *J. Biol. Chem.*, **263**, 1798–802 (1988).

116. D. G. Alberg and P. A. Bartlett, *J. Am. Chem. Soc.*, **111**, 2337 (1989).

117. A. M. Christensen and J. Schaefer, *Biochemistry*, **32**, 2868–73 (1993).

118. G. C. Leo, J. A. Sikorski, and R. D. Sammons, *J. Am. Chem. Soc.*, **112**, 1653–4 (1990).

Part IV · *Nucleic acids and carbohydrates*

9 · Nucleic acids and carbohydrates

9.1 OLIGONUCLEOTIDE STRUCTURE

The first application of NMR to DNA[1] took place one year after Watson and Crick's historic discovery[2] of the double helical structure in DNA fibres. Unlocking the molecular details of DNA has spawned the new field of molecular biology, and has resulted in a shift from the dominance of physics in the early part of this century to a clear dominance (intellectually, if not in terms of funding!) of biology in the late twentieth century. One of the attractions of NMR spectroscopy is that it is the experimental application of the principles of quantum mechanics, a field which inspired so many physicists in the 1920s–1940s, to the chemistry of biological systems, and it is now being used in part to sustain the molecular biology revolution through determining biomolecular structure and function, of which DNA is of course of central importance. Although structural genes are far too large (200–800 kDa) for solution-state NMR studies, the upstream control sequences in DNA, such as promoters and operators, consist of only 10–20 base pairs, and are therefore readily studied by multidimensional solution-state NMR methods. Promoters are the sites at which RNA polymerase begins to copy a gene, and operators are DNA sequences to which repressor proteins bind, thus controlling the occurrence of transcription.

NMR studies of oligonucleotide structures have lagged behind protein studies, primarily because of sample limitations. With the advent of solid-phase oligonucleotide synthesis, this situation has changed dramatically, and the wealth of different types of secondary structure that can occur in oligonucleotides is beginning to be investigated. In general, almost all the NMR studies that have been reported to date have been with synthetic oligonucleotides, and it has been possible to study single strands, stacked helical single strands, hairpins, regular duplexes formed between complementary strands, triplexes and quadruplexes, and a variety of structures involving bulges, bends, or other distorted shapes. Furthermore, the interaction of these DNAs with ligands such as drugs and small molecules, and also with DNA-binding proteins can be studied by NMR, and the structure of such complexes can be determined. Finally, the secondary structure of RNA can be investigated, for which there is a paucity of X-ray crystal structures, and for which the solution-state structure is crucial in understanding the nature of the interactions of RNA with other biomolecules.

9.1.1 Sequential assignments

There are only five bases (including uracil), compared with the 20 amino acids present in proteins, which renders the assignment of ¹H NMR spectra of oligonucleotides easier. The structure of DNA depends critically on the base–base interactions as outlined in Fig. 9.1, which shows the common Watson–Crick base pairing, together with some less common alternatives present in RNA (see Section 9.6).

Deoxyribose and ribose protons are designated 1′ through to 5′ and 5″, and the chain direction is defined as going from the 3′ of one nucleotide through phosphorus to the 5′ of the next (see Fig. 9.2(a)). The five-membered ribose ring is puckered and there are a number of conformations that it can adopt. The two sugar conformations most commonly found in oligonucleotides are 3′- and 2′-endo, abbreviated ³E and ²E, respectively, and these are shown in Fig. 9.2.

The E/endo nomenclature denotes the atom which is out of plane, endo meaning that the out of plane atom is on the same side of the plane as C5 and D-ribofuranose. The glycosidic bond is the C1′—N bond, and its conformation is described by the angle χ, which here is taken as the angle O4′-C1′-N9-C4 for purines and O4′-C1′-N1-C2 for

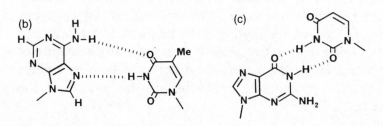

Fig. 9.1 Nomenclature and structure of (a) GC, AT base pairs, and U; (b) AT Hoogsteen pair; and (c) GU wobble pair. (Reprinted from ref. 22, Ch. 1 with permission.)

pyrimidines. Some earlier conventions have χ differing from this by $180°$; χ angles of $0 \pm 90°$ are described as *syn* and $180 \pm 90°$ as *anti*, as shown in Fig. 9.3.

A standard notation used for distances in nucleic acids similar to that employed with proteins has been suggested by Wüthrich,[3] although it does not appear to have been widely adopted in the literature. We will mention it here because it is logically consistent with the conventions in protein NMR: $d_i(A;B)$, where d refers to the distance for residue i between protons A and B in the 3′ to 5′ direction, so that, for example, the intranucleotide distance between the base H8 and the sugar H1′ would be $d_i(8;1')$; for sequential distances, $d_s(1';M)$, for example, refers to the distance between H8 and the methyl protons of thymidine of the next residue; within a base pair, interstrand distances are denoted $d_{pi}(A+;B-)$ and between labile base protons or A2H in adjoining stacked base pairs, the distance is denoted $d_{ps}(A+;B-)$, where $+$ and $-$ are used to distinguish between the two strands.

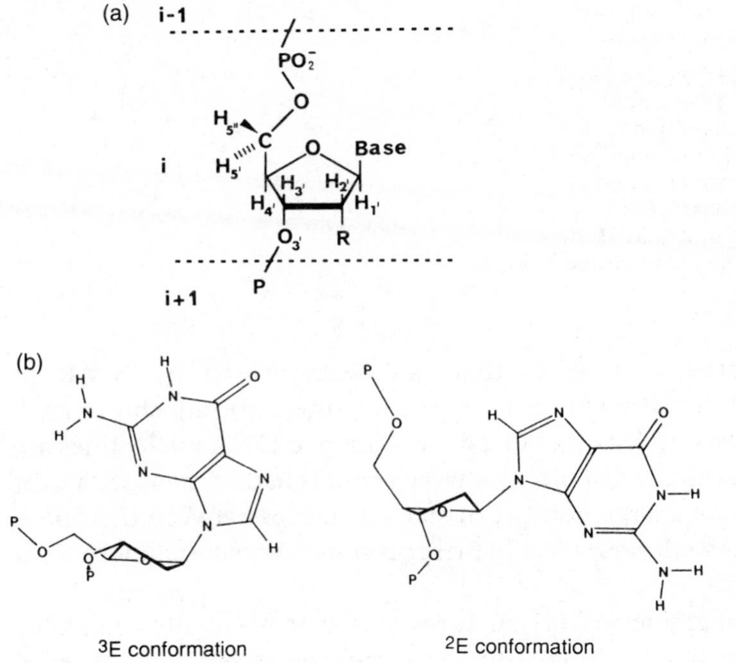

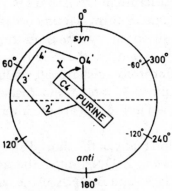

(a)

i-1

i

i+1

(b)

3E conformation

2E conformation

Fig. 9.2 (a) Nomenclature of sugar protons. For deoxyribose, R=H2″; for ribose, R=OH. (b) Glycosidic conformations and ring puckers, with the two most commonly encountered in oligonucleotides shown. The ^{3}E (or 3′-endo) has a *syn* conformation about the glycosidic bond, and the ^{2}E (2′-endo) has an *anti* glycosidic conformation. (Reprinted from ref. 22, Ch. 1 with permission.)

Fig. 9.3 Definition of the glycosidic angle χ for a purine in the *syn* conformation. (Reprinted from ref. 22, Ch. 1 with permission.)

Table 9.1 Short covalent ^1H—^1H distances in the common DNA nucleotides[a]

Distance	Value (Å)
$d_i(2';2'')$	1.8
$d_i(5';5'')$	1.8
$d_i(A1;6NH_2)$	4.4, 5.2
$d_i(A6NH;6NH)$	1.7
$d_i(A6NH_2;8)$	4.8, <u>6.1</u>
$d_i(G1;2NH_2)$	<u>2.3</u>, 3.4
$d_i(C4NH_2;5)$	2.4, <u>3.6</u>
$d_i(C4NH_2;6)$	4.6, <u>5.3</u>
$d_i(C5;6)$	2.4
$d_i(T3;M)$	4.9
$d_i(TM;6)$	2.9

[a]Data from Ref. 3. Computed from the atom coordinates in a standard B-DNA double helix. Identical distances occurring in several nucleotides are listed only once. For NH_2, the distance to the hydrogen-bonded proton in the Watson–Crick base pairs (Fig. 9.1) is underlined. M denotes $5CH_3$ of T, and the distances are to the centre of the three methyl protons.

Unlike proteins, the interproton distances that can be measured by NMR in oligonucleotides are restricted to adjacent base pairs (i.e. they are all short-range distances). The typical short covalent distances that are common in DNA nucleotides are given in Table 9.1. This includes all the distances between protons in the furanose ring. It is interesting to note that there are a large number of short distances between the ribose ring and the base, which is just as well, since usually these protons cannot be connected via J coupling.

With the exception of RNA, oligonucleotide structures consist of essentially straight or slightly curved helices, and nearly all solution-state structures have either A, B, or Z helices. The B-form of the helix is the most common structure in DNA and DNA/RNA hybrids, while the A-form is common only in RNA, and the Z-form is found occasionally in DNA, especially with substituents at the 5-position of cytidine, in high salt concentrations, and in alternating purine–pyrimidine sequences. B-DNA is a regular right-handed double helix, and some relevant short interproton distances for non-labile protons are given in Table 9.2, labile protons in Table 9.3, interstrand distances in Table 9.4, and depicted pictorially in Fig. 9.4. Also shown in Fig. 9.4 are the interproton distances in A-DNA, and in Fig. 9.5 those in Z-DNA.

A-DNA is also a regular right-handed double helix, and many of the short distances are similar to B-DNA, as is shown in Tables 9.2, 9.3, and 9.4. Z-DNA differs because the repeat unit is dinucleotide, and the helix is left-handed. In A- and B-DNA all the bases are

Table 9.2 Short sequential distances between non-labile protons in three standard DNA conformations

Distance[a]		A-DNA[b]	B-DNA[b]	Z-DNA,CG[b]	Z-DNA,GC[b]
d_s	(1';5")	4.6 Å	3.0 Å	3.7 Å	—
	2';1'	4.0	—	—	—
	2';3'	4.1	—	—	—
	2';5"	2.7	—	5.0	—
	2";2'	—	3.7	—	—
	2",5"	2.4	4.0	4.6	—
	3';2'	—	—	—	33 Å
	3';3'	—	—	—	4.5
	4';5"	4.1	3.4	2.0	—
	1';5	—	4.3		4.8
	1';6	4.6	3.5		—
	1';8	4.6	3.6	—	
	2';5	3.5	3.6		2.8
	2';M	3.8	3.4		
	2';6	2.0	4.0		3.0
	2';8	2.1	3.8	—	
	2";5	—	2.7		4.2
	2";M	—	2.9		
	2";6	3.7	2.2		4.2
	2";8	3.9	2.1	—	
	3';5	3.3	—		4.2
	3';M	3.0	—	—	
	3';6	3.1	4.6		3.4
	3';8	3.0	4.6	—	
	2 ;1'	4.1	—		
	5;5	3.7	4.7		
	5;M	3.3	4.7		
	M;5	4.3	—		
	M;M	3.7	5.0		
	6;5	3.9	3.9		
	6;M	3.5	3.8		
	6;8	4.5	4.8	—	
	2;2	4.8	3.6		
	8;5	3.8	3.9		4.9
	8;M	3.4	3.8		
	8;8	4.6	5.0		

[a]Distances are listed if they are shorter than 4.5 Å in at least one conformation, except that those involving 4'H, 5'H, or 5"H are only included if they are shorter than 3.0 Å at least once. Then, values up to 5.0 Å are listed, and a dash indicates values longer than 5.0 Å.
[b]Data from Ref. 3.

anti; in the Z-DNA fragment of $d(CG)_n$, the cytidines are all *anti* and ^2E as in B-DNA, while the guanosines are *syn* and ^3E. As is clear from Fig. 9.5, there are many fewer short distances in Z-DNA than in A- or B-DNA. The characteristics of the three types of helix are outlined in Table 9.5.

The first step in carrying out the sequential assignments of oligonucleotide spectra involve identification of the spin systems. This can be done in part on the basis of chemical

Table 9.3 Short sequential ^{1}H—^{1}H distances with labile base protons in three standard DNA conformations

Distance[a]		A-DNA	B-DNA	Z-DNA
d_{ps}	$(A1+,2;A1+,2)$[b]	4.8 Å	3.6 Å	
	A1+,6'A2+,6	3.0	3.7	
	A1+,2;T2+,3	4.3	4.0	
	A1+,6;T2+,3	3.9	4.5	
	A1+,2;G2+,1	—	4.3	
	A1+,6;G2+,2	—	4.2	
	A1+,6;G2+,1	4.0	3.7	
	A1+,2;C2+,4	3.0	4.2	
	T1+,3;T2+,3	3.8	3.5	
	T1+,3;A2+,3	—	4.4	
	T1+,3;A2+,1	4.5	4.0	
	T1+,3;G2+,2	4.8	3.8	
	T1+,3;C2+,1	3.8	3.9	
	G1+,1;G2+,1	3.8	3.4	
	G1+,1;G2+,2	4.8	3.7	
	G1+,2;A2+,2	4.3	3.7	
	G1+,2;A2+,6	4.2	4.5	
	G1+,1;A2+,2	4.2	3.5	
	G1+,1;T2+,3	3.4	3.9	
	G1+,2;T2+,3	4.3	4.5	
	G1+,2;C2+,4	4.0	4.7	4.0
	C1+,4;C2+,4	3.0	3.7	
	C1+,4;A2+,6	3.3	3.4	
	C1+,4;T2+,3	4.1	4.3	
	C1+,4;G2+,1	4.6	4.0	—
	C1+,6;G2+,1	—	—	3.8
	X1+,1';G2+,1	—	—	3.5
	X1+,1';G2+,2	—	—	3.5

[a]Data from Ref. 3. Distances are included if they are shorter than 4.5 Å in at least one conformation. Then, values up to 5.0 Å are listed, and a dash indicates values longer than 5.0 Å. The first nucleotide is arbitrarily placed in sequence position 1. For NH_2 groups the distance to the hydrogen-bonded proton is given. In the last two entries, X stands for any of the four nucleotides.

shift as outlined in Appendix 3. Furthermore, the nucleotides display characteristic cross-peak patterns and connectivities in DQF–COSY and TOCSY experiments, as outlined for the four common DNA deoxyribonucleoside monophosphates in Appendix 3. These can all be carried out in D_2O, since there are no scalar couplings involving labile protons.

There are two independent scalar coupling networks within the basic nucleotide unit, one comprising the sugar and the other the base. The coupling networks are best identified by means of TOCSY experiments which show direct and relayed through-bond connectivities along the H1′→H2′/H2″→H3′→H4′→H5′/H5″ pathway in each sugar unit, and between the H6 and H5 protons of cytosine and the H6 and CH_3 protons of thymine

Table 9.4 Short interstrand ^{1}H—^{1}H distances in three standard DNA conformations

Distance[a]		A-DNA	B-DNA	Z-DNA
d_{pi}	(A1;T3)	2.8 Å	2.8 Å	
	A6;T3	2.5	2.5	
	G1;C4	2.5	2.5	2.5 Å
d_{ps}	(A2+,2;A1−,2)	4.2 Å	2.9 Å	
	A2+,2′A3−,2	—	4.4	
	A2+,6;A1−,6	3.2	4.0	
	A2+,6;A3−,6	2.7	3.3	
	A2+,2;T1−,3	4.0	3.8	
	A2+,6;T1−,3	4.1	4.8	
	A2+,6;T3−.3	4.1	3.5	
	A2+,2;G1−,1	4.7	3.8	
	A2+,2;G1−,2	4.9	3.4	
	A2+,6;G1−,1	4.3	4.2	
	A2+,2;G3−,1	4.7	4.1	
	A2+,2;G3−,2	4.7	4.1	
	A2+,6;G3−,1	3.9	4.1	
	A2+,6;C1−,4	3.2	4.5	
	A2+,6;C3−,4	3.0	2.9	
	T2+,3;T1−,3	3.6	4.3	
	T2+,3;T3−,3	—	4.1	
	T2+,3;G1−,1	3.5	3.5	
	T2+,3;G1−,2	4.3	3.5	
	T2+,3;G3−,1	4.3	3.6	
	T2+,3;C1−,4	4.2	—	
	T2+,3;C3−,4	4.9	3.9	
	G2+,1;G1−,1	4.3	3.4	—
	G2+,1;G1−,2	—	4.1	—
	G2+,2;G1−,1	—	4.1	—
	G2+,2;G1−,2	—	4.1	—
	G2+,1;G3−,1	3.6	3.6	3.6
	G2+,1;G3−,2	4.5	4.3	4.3
	G2+,2;G3−,1	4.5	4.3	4.3
	G2+,2;G3−,2	4.3	4.0	4.3
	G2+,1;C1−,4	3.9	4.2	
	G2+,1;C3−,4	4.4	4.2	
	C2+,4;C1−,4	3.5	—	4.2
	C2+,4;C3−,4	3.6	2.9	4.9
	C2+,5;C3−,5	—	—	4.5
	A2+,2;X1−,1′	4.1	—	

[a]Data from Ref. 3. Distances are included if they are shorter than 4.5 Å in at least one conformation. Then, values up to 5.0 Å are listed, and a dash indicates values exceeding 5.0 Å. The first base is positioned arbitrarily at 2+ and the bases 1− and 3− are paired with the bases 1+ and 3+ respectively. X denotes any base.

Table 9.5 Mean double-helix parameters for A-DNA, B-DNA, and Z-DNA[a]

Helix parameter	A-DNA	B-DNA	Z-DNA	
Handedness	right	right	left	
Bases per turn	10.9	10.0		12.0
Height per base	2.0 Å	3.4 Å	GC	3.5 Å
			CG	4.1 Å
Pitch	31.6 Å	34.0 Å		45.6 Å
Glycosyl angle	anti	anti	C	anti
			G	syn
Sugar pucker endo	3'C endo	2'C endo	C	2'C
Base pairing	Watson–Crick	Watson–Crick	Watson–Crick	
Repeating helix unit	1 base pair	1 base pair	2 base pairs	
Twist per base pair	33°	36°	GC	−51°
			CG	−9°
Axis displacement[b]	4 Å	0	−3 Å	
Major groove	very deep	deep, wide	shallow	
Minor groove	shallow	deep, narrow	very deep	
Base inclination	13°	−2°	9°	
Base roll	6°	−1°	3°	
Propeller twist	15°	12°	4°	

[a]Data Ref. 3.
[b]Positive numbers indicate a displacement of the helix axis toward the major groove edge of the base pair.

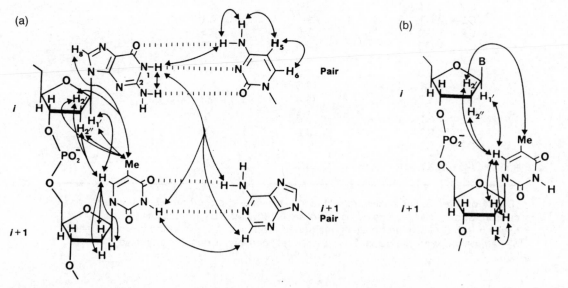

Fig. 9.4 Interproton contacts of less than approximately 4 Å involving base, 1′, 2′, and 2″ protons for (a) B-DNA and (b) A-DNA. Note the differences in the magnitude of the distances in the two different types of DNA. (Reprinted from ref. 22, Ch. 1 with permission.)

within the pyrimidine bases. The base and sugar moieties of the same nucleotide are connected by means of NOESY experiments via the intraresidue H1'/H2'(*i*)—H8/H6(*i*) cross-relaxation pathway, while through-space connectivities along the H1'/H2'/H2″ (*i*-1)→H8/H6(*i*)→H1'/H2'(*i*) and the H8/H6(*i*)→H5/CH$_3$(*i*+1) pathways yield sequential assignments along one strand. In virtually all cases for small oligonucleotides (not exceeding about 25 base pairs), complete assignments can be obtained by means of NOESY experiments alone, without recourse to scalar coupling experiments. Having said this, TOCSY and NOESY spectra are frequently used to complete the sequential assignments for oligonucleotides. Assignment of the imino protons as well as the adenine H2 protons is also carried out sequentially on the basis of NOESY experiments in H$_2$O. Also ^{1}H—^{31}P $\mathcal{J}$ couplings may be used to carry out sequential assignments using one of a

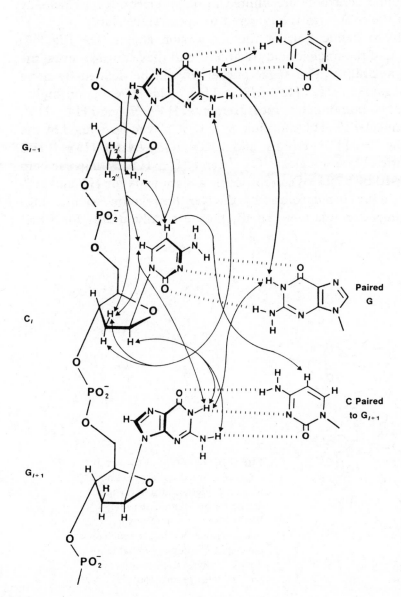

Fig. 9.5 Interproton contacts of less than approximately 4 Å involving base, 1', 2', and 2″ protons for Z-DNA. (Reprinted from ref. 22, Ch. 1 with permission.)

number of experiments, such as the ³¹P-detected heteronuclear COSY (HETCOR),[4-6] ¹H-detected HMQC,[7,8] ³¹P-detected heteronuclear TOCSY,[9] or, more recently, ¹H-detected heteronuclear TOCSY,[10] which promises to be the experiment of choice for this approach to making sequential connectivities in oligonucleotides. All the experiments permit the determination of a number of sequential $\mathcal{J}_{HP}$ couplings, in particular from H3′, H4′ and H5′, H5″ and relayed from H1′ and H2′, H2″ to the ³¹P. This is particularly important for larger oligonucleotides (>25 bp), where other heteronuclei from isotopic labelling can also be employed (see Section 9.1.4).

9.1.2 $\mathcal{J}$ coupling constants

Restraints based on torsion angles (see Fig. 9.6 for a definition of oligonucleotide torsion angles) derived from $\mathcal{J}$ coupling constants are limited in oligonucleotides, principally because the measurement of the couplings is hampered by spectral overlap.

Scalar couplings can help to define four of the six torsion angles (see Fig. 9.7) determining the conformation of the nucleic acid backbone, and they can also reveal the presence of conformational flexibility. The sugar pucker (δ) can be defined by three homonuclear ¹H—¹H couplings: H1′—H2′, H2′—H3′ and H3′—H4′. The torsion angle γ (C5′—C4′) can be defined by two homonuclear measurements: H4′—H5′ and H4′—H5″, and by the four-bond heteronuclear P—H4′ coupling. β (O5′—C5′) can be defined by the P—H5′ and P—H5″, plus the P—H4′ couplings, and ε (C3′—O3′) by the H3′—P and H2′—P sequential connectivities. Heteronuclear ¹³C—³¹P and ¹³C—¹H couplings can help to overcome ambiguities caused by a lack of stereospecific assignments for H5′ and H5″ protons, and by uncertainties in the correct solution to the Karplus relation (see Fig. 9.8), although since these couplings are relatively small, they are only useful for small oligonucleotides.

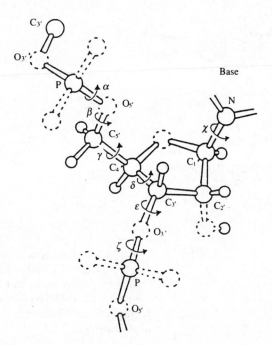

Fig. 9.6 Definition of the seven degrees of freedom (six backbone torsion angles and the glycosidic angle) for an oligonucleotide. With the exception of oxygen (dashed nuclei), all other nuclei have spin-$\frac{1}{2}$ isotopes amenable to NMR studies. (Reprinted from ref. 157 with permission.)

The way in which homonuclear couplings are measured is very similar to that already described for extracting homonuclear couplings from DQF–COSY experiments with proteins (see Sections 3.2 and 4.1.2), except that, to date, experiments such as the PE.COSY do not appear to have been employed as frequently as simulations of the DQF–COSY cross-peak lineshapes.[11–14] These approaches appear to work well for DNA but are less relevant for RNA, in which the absence of the H2″ protons gives rise to only two couplings involving H1′, H2′, and H3′, provided ^{31}P is decoupled. [31]P decoupling is worth carrying out in all attempts to measure homonuclear couplings in oligonucleotides, since cross-peaks involving H1′, H2′, and H3′ protons are simplified by removal of the heteronuclear coupling, with considerable improvement in sensitivity and sometimes a removal of spectral overlap. For the H4′, H5′, and H5″ protons, the generally small [31]P—[1]H heteronuclear couplings (2–4 Hz) increase the apparent linewidth and lead to less precise measurements. In this region, cross-peaks between resonances separated by only 0.03 ppm can be observed in favourable cases.

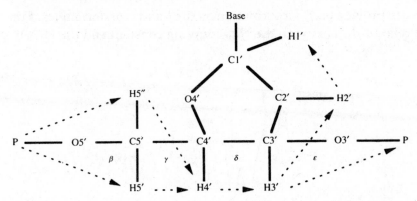

Fig. 9.7 Three-bond scalar couplings involving proton and phosphorus define four out of six backbone angles, β, γ, δ, and ε, as indicated by dashed lines. [13]C—[31]P and [13]C—[1]H 3-bond couplings can also be used in conformational analysis. (Reprinted from ref. 157 with permission.)

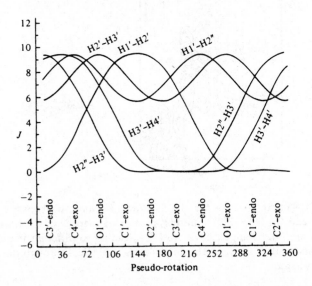

Fig. 9.8 Variation of the vicinal coupling constants in the D-ribose ring as a function of the ring geometry. The curves are based on the Karplus relation (Equation (1.39)) with $A = 10.2$ and $B = 0.8$. (Reprinted from ref. 58 with permission.)

The homonuclear coupling most easily measured is between H1′ and H2′ protons. In principle, this can be measured using the E.COSY or PE.COSY (see Section 3.2) experiments, although it is rare in the nucleic acid NMR literature. This is much more difficult for the H2′—H3′ and H3′—H4′ couplings, where spectral overlap can be severe. One trick for estimating these makes use of the differences between the active and passive couplings in the cross-peak for each coupling partner (see Fig. 9.9).

The method relies on the partial cancellation of the active coupling antiphase components of the H2′—H3′ cross-peak leading to an overestimation of the H2′—H3′ coupling, and the partial summing of the passive coupling in-phase components of the H3′—H4′ cross-peak leading to an underestimation of the H2′—H3′ coupling. The mean of the two estimates lies close to the true multiplet separation. For the couplings between H4′ and H5′, H5″ protons, which define the backbone torsion angle γ, the values can be 2–3 Hz for the common *gauche*$^+$ conformer of γ, and the cross-peaks are too weak to detect. The H4′—H5′ coupling can be measured under favourable circumstances from the H5′—H5″ cross-peak.

The ^{31}P—^{1}H couplings can provide useful information on the β and ε torsion angles. For example, a Karplus-type relationship between the $^3\mathcal{J}_{HP}$ coupling constant and the HCOP

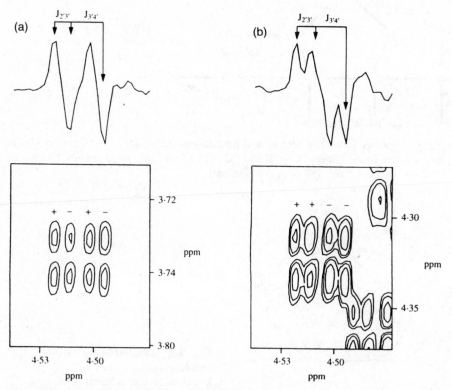

Fig. 9.9 Comparison between DQF–COSY cross-peaks for the N-type (3′-*endo*) sugar pucker with phosphorus decoupling for (a) H2′—H3′ and (b) H4′—H3′. The H3′ proton resonances in the horizontal dimension. Active and passive couplings are interchanged in (a) and (b), leading to different patterns of positive and negative multiplet components, as illustrated by the one-dimensional slices through the cross-peaks. (Reprinted from ref. 157 with permission.)

dihedral angle has been derived,[15] which has the form:

$$^3\mathcal{J}_{HP} = 15.3 \cos^2 \phi - 6.1 \cos \phi + 1.6 \qquad (9.1)$$

Using a selective version of the ^{1}H-detected HMQC experiment, $^1\mathcal{J}_{C3'H,P}$ coupling constants for oligonucleotides can be measured accurately.[16] For example, in the case of the DNA dodecamer 5'-d(CGCGAATTCGCG)$_2$, values for ε backbone torsion angle were derived in the bases of $^3\mathcal{J}_{C3'H,P}$ coupling constants using the simple relationship $\varepsilon = 120° + \phi$. The resulting values of ε were in good agreement with the values obtained from the single-crystal X-ray structure of the same oligonucleotide.[17] With the exception of one residue, G10, where the value of ε in the X-ray structure was in the *gauche*$^+$ conformation rather than the usual *trans* conformation, all the residues had a *trans* conformation for ε and the average RMSD between the solution and X-ray values was only $8 \pm 6°$, which is well within the errors of the two techniques.

Other heteronuclear coupling constants such as ^{1}H—^{13}C and ^{1}H—^{15}N can also be used, although to date there have been no measurements and application to structure refinement.

9.1.3 Imine exchange rates

The labile imine protons are usually in fast exchange with the solvent. In the earliest studies of this by von Hippel and co-workers,[18] and Englander and co-workers,[19] evidence for an 'open state' of the oligonucleotide was discovered by employing tritium and, later, deuterium exchange experiments. Clearly, as long as the imino and amino protons are involved in a hydrogen bond, they cannot be labile. Therefore the notion was advanced that the structure must open up or 'breathe' a little, and there is transient loss of the hydrogen bond.[20] It was therefore assumed that exchange is opening limited, although Guéron and co-workers have shown that exchange in both DNA and tRNA is not opening limited except at high buffer concentrations.[21] It appears that the buffer acts as a catalyst in the exchange process, and that hydrogen exchange correlates with base pair opening only at infinite buffer concentrations. More recently, in regions of secondary structure, Guéron and co-workers have found[22] that there is a correlation between exchange rate and structure. This is shown in Fig. 9.10, in which magnetization transfer experiments were used to measure imino proton exchange lifetimes.

In B-DNA, it was found that imino protons in an A·T pair have shorter lifetimes (1–7 ms at 15°C) than have G·C pairs (7–40 ms), irrespective of the identity of the

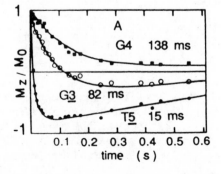

Fig. 9.10 Magnetization transfer experiment on the 5'-d(CGCGATCGCG) decamer. The magnetization of the imino protons is plotted versus the delay between selective inversion of the water and acquisition. (Reprinted from ref. 22 with permission.)

neighbouring pairs. Interestingly, however, in DNA containing A·T tracts, which are known to be curved as a result of their anomalous electrophoretic mobilities, longer lifetimes were observed. This observation might also lend support to the notion that A·T-rich DNA deviates substantially from B-DNA. Similar correlations of imino proton exchange with functional regions of operator sequences have been observed for the *trp* operator[23] and the *lac* operator.[24] As outlined in Fig. 9.11, the available lifetimes for base pairs in tRNA[Phe] in the absence of Mg^{2+} also show a dependence on secondary structure, with the imino protons of GU4 and UA6 exhibiting shorter lifetimes than the rest of the acceptor stem.[25]

9.1.4 Isotopic labelling of nucleic acids

Although at a much earlier stage of development than with proteins, the use of isotopic labelling promises to unlock the door to a variety of multidimensional multinuclear solution- and solid-state NMR experiments which will provide critical information for unambiguous assignments and long-range distance correlations. This will be particularly important for oligonucleotides in the 300–100mer range ($M_r = 10$–30 kDa). There are currently two approaches that have been adopted for the production of ^2H-, ^{13}C-, and ^{15}N-labelled oligonucleotides. The first approach is the synthesis of the oligonucleotide with labelled nucleoside triphosphates, which can be prepared through total synthesis, or, in the case of deuterated nucleosides,[26] by catalytic exchange. The second approach, which has principally been employed in labelling RNA, involves the growth of a suitable auxotrophic strain of *E.coli* on minimal media containing $^{15}NH_4Cl$ or [$^{13}C_6$]glucose, and isolation of the labelled RNA. This latter approach is exactly analogous to the isotopic labelling procedure for proteins already outlined in Section 4.2.2, and has been employed by a number of laboratories to prepare natural RNAs, including 5S rRNA fragments,[27,28]

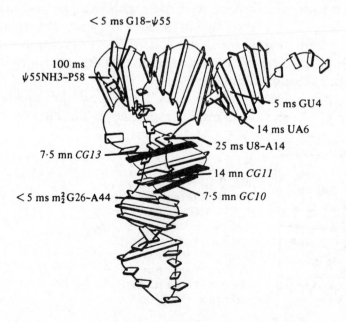

Fig. 9.11 A summary of the available base-pair lifetimes of yeast tRNA[Phe] in 0.1 M NaCl with no added Mg^{2+}. (Reprinted from ref. 58 with permission.)

and DNA.[29] Selective labelling has also been achieved through administering synthetic bases with specific labels into the growth media.[30–33]

Pardi and co-workers[34,35] recently prepared isotopically labelled oligonucleotides by *in vitro* RNA labelling, which involved growth of *E.coli* on minimal media containing $^{15}NH_4Cl$ or $[^{13}C_6]$glucose, and then isolation of the nucleic acids, which were degraded to their mononucleotides. These were enzymatically converted to their ribonucleoside triphosphates, and used in *in vitro* transcription reactions catalysed by T7 RNA polymerase to prepare labelled RNA. In an interesting variation, Williamson and co-workers[36] used *E.coli* for ^{15}N labelling, but *Methylophilus methylotrophus* for ^{13}C labelling, since this organism grows on methanol. $[^{13}C]$methanol is considerably cheaper than $[^{13}C]$glucose. King and co-workers have used the same approach, but separated the ribonucleoside triphosphates by HPLC, thereby enabling labelling by residue type.[37] Although this has been applied so far to RNAs, primarily because the DNA polymerases available are far less efficient than RNA polymerase, this approach will no doubt be developed soon.

9.1.5 Solid-state NMR of DNA

Both small oligonucleotides and larger DNA structures (even up to supercoiled DNA) are amenable to solid-state NMR studies. The approach has been the subject of an excellent review recently,[38] so we will confine our discussion here to the salient features. Some of the earliest solid-state NMR studies of DNA probed the conformation and dynamics of the phosphodiester backbone. For example, Shindo and co-workers,[39–41] and, independently, Waugh and co-workers[42] used the chemical-shift anisotropy of ^{31}P to study backbone conformation and dynamics of highly oriented Li/Na-DNA fibres in the A- and B-forms. They observed that the ^{31}P lineshape changes as a function of the orientation of the fibre axis with respect to the magnetic field, as illustrated for the A-form of DNA in Fig. 9.12.

At parallel orientations, the spectrum consists of a single resonance about 50 ppm wide (for both A- and B-forms), and at perpendicular orientations the A-form displays a broad resonance, whereas the B-form displays a narrower resonance (see Fig. 9.13). At intermediate orientations, the spectrum of the A-form is reminiscent of a powder pattern (see Section 1.2.5), while the B-form displays a broad, featureless resonance. These lineshapes can be simulated provided the following principal assumptions are made: (i) there is no motional averaging of the ^{31}P chemical shifts; (ii) the orientation of the ^{31}P chemical shift anisotropy tensor from model systems is assumed;[43–45] and (iii) the samples of oriented DNA are considered to be composed of crystallites which are oriented relative to one another. These observations have also been corroborated by Harbison and co-workers,[46] who used magic-angle spinning techniques to determine the orientational distribution function of the ^{31}P chemical shift anisotropy tensor, in agreement with Shindo and Waugh's results.

The early studies of the dynamics of DNA were carried out by Opella and co-workers,[47] who used ^2H and ^{31}P NMR to study high molecular weight DNA. For example, they studied high molecular weight calf thymus DNA, deuterated by exchange of the purine H8 protons, in the B-form as a function of temperature. They found that, although ^{31}P NMR showed non-axial averaging of the ^{31}P CSA tensor at rates $> 10^4$ Hz, little or no averaging of the ^{2}H powder pattern occurred. The conclusion was that large-amplitude motions of the bases do not occur on nanosecond timescales.

In addition to the use of ^{31}P NMR, ^{13}C and ^{15}N NMR have been used to study DNA structure and dynamics. Opella and co-workers have used ^{15}N CP–MAS NMR to study [U-^{15}N]DNA (6800 bp), and found 15 well-resolved lines from the chemically distinct nitrogen nuclei of DNA. Two-dimensional ^{15}N—^{15}N solid-state EXSY experiments, performed with mixing times $\tau_m = 2$, 4, and 6 s, showed variations in cross-peak intensity, and the spectra were simplified further by using multiple-quantum filtration of the ^{1}H—^{15}N heteronuclear dipolar coupling to select only those ^{15}N lacking an attached proton[29] (see Fig. 9.14). With this approach it should be possible to study the structure of DNA fragments much larger than the DNA oligomers currently under investigation by solution-state NMR methods (see Section 9.2.1).

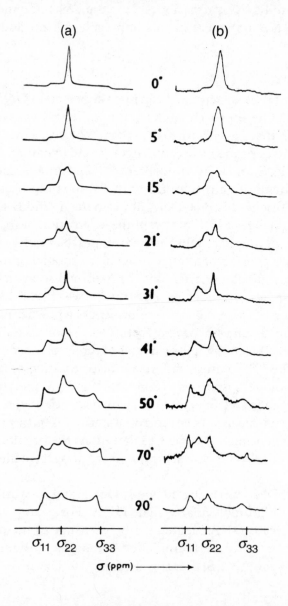

Fig. 9.12 Simulated (a) and observed (b) ^{1}H-decoupled ^{31}P NMR spectra for an aligned sample of A-form DNA at the indicated angles between the fibre direction and the magnetic field direction. (Reprinted from ref. 42 with permission.)

Harbison and co-workers[46] have used solid-state ^{13}C NMR to determine the sugar ring puckering in DNA. By correlating ^{13}C chemical shifts with conformations of model compounds, they found that 3′-endo conformers have chemical shifts for 3′-C and 5′-C 5–10 ppm upfield from the comparable 3′-exo and 2′-endo conformers. These trends are maintained in high molecular weight calf thymus DNA, with ^{13}C chemical shifts confirming that A-form DNA is primarily 3′-endo, and B-form DNA is largely 2′-endo.

The dynamics of DNA has been extensively studied by ^{2}H NMR, initially by the groups of James,[48] and Kearns and Vold.[49] This work involved ^{2}H NMR studies of

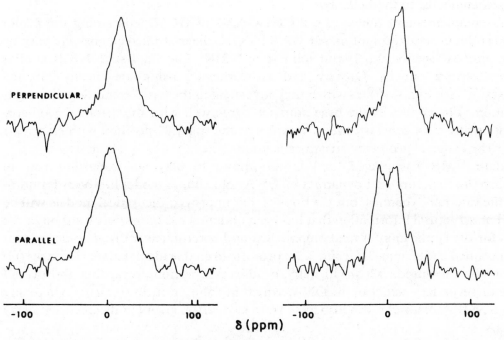

Fig. 9.13 24.7 MHz CP, ^{1}H-decoupled ^{31}P NMR spectra from A-form DNA, poly(dAdt)·poly(dAdt) fibres (right) and B-form DNA, salmon sperm DNA (left) at 98% relative humidity. (Reprinted from ref. 39 with permission.)

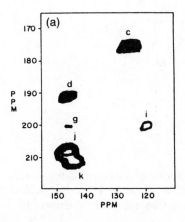
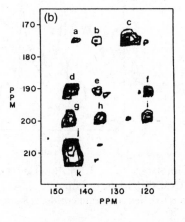

Fig. 9.14 Expanded regions of two-dimensional solid-state CPMAS ^{15}N—^{15}N EXSY experiments (a) with a selective ^{1}H—^{15}N dipolar coupling filter; and (b) standard 4 s EXSY. (Reprinted from ref. 29 with permission.)

hydrated samples of randomly oriented calf thymus DNA,[50] as well as ^{2}H NMR studies of oriented DNA fibres.[51] The quadrupolar coupling constant and quadrupolar relaxation of [^{2}H]DNA was studied as a function of relative humidity (RH), which results in a change from A- form to B-form between 75 and 92% RH. Such a study can be used to estimate the librational amplitudes of motion according to a variety of different models. Similar studies by Shindo, Torchia, and co-workers[52] estimated motional amplitudes in a similar regime. They also determined, on the basis of a lineshape analysis, that the purine base planes of the A-form DNA were tilted about 70% relative to the helix axis, and although the orientation of purine base planes in B-form DNA varied significantly, they were on average perpendicular to the helix axis.

The self-complementary dodecamer d(CGCGAATTCGCG)$_2$ containing the *Eco*RI restriction endonuclease recognition site GAATTC has been studied extensively both by solution-state (see Section 9.2.2) and solid-state NMR. The solid-state NMR studies have been carried out by Drobny and co-workers,[38] using specifically labelled dodecamers[53,54] (see Fig. 9.15), in which they investigated the base orientation and sugar conformation. The results of the base domains agree well with the distance geometry solution structure.[55] In contrast, the orientation of the sugar 2″-position with either the X-ray[56] or the distance geometry structure is poor.

Solid-state NMR has played, and is continuing to play, an important rôle in determining the structure and dynamics of DNA. No simple model has been found to simulate the internal dynamics, but the hope is that improved theoretical models will be devised. The structural information that has been obtained has been useful, although the prospects for the application of the homonuclear and heteronuclear distance-dependent methods outlined in Sections 3.3.2 and 3.3.3 promise to enable much longer distances (up to 12 Å) to be measured. This will help to address questions regarding the overall curvature or secondary structure of DNA, which in solution measurements where any long-range NOEs are absent, can also arise from systematic errors in distance geometry calculations.

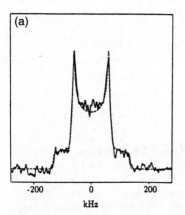
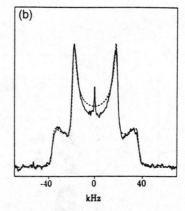

Fig. 9.15 Experimental (——) and simulated (. . . . . .) 76.75 MHz ^{2}H quadrupole echo (see Section 1.4.4) spectra of labelled [d-(CGCGAATTCGCG)]$_2$ in the oriented liquid-crystal phase. (a) [8-^{2}H]purine labelled and (b) [*methyl*-^{2}H]-2′-deoxythymidine labelled oligonucleotides. (Reprinted from ref. 38 with permission.)

9.2   EXAMPLES OF DNA STRUCTURES

An increasing number of structures of DNA in solution have been determined in recent years, principally in the laboratories of Reid, James, Patel, Clore and Gronenborn, Kaptein, and Wüthrich. Two excellent reviews have also appeared recently.[57,58] In this section we will consider only early study, which gives an example of the sequential assignment procedure for the *lac* operator, and then one more sophisticated study in which the DNA structure is completely refined.

9.2.1   The *lac* operator

Although the *lac* operator DNA structure was not determined properly in the absence of its DNA-binding protein headpiece, we will consider it here as an example of the sequential assignment procedure for oligonucleotides, and consider the structure of the complex in Section 9.5. The *lac* operator sequence is a region 20–25 bp long with a pseudo-dyad axis going through CG11 (see Fig. 9.16).

Based on the hypothesis that the *lac* repressor binds with two headpieces, thereby preserving the pseudo-C_2 symmetry in the repressor–operator complex, the 14bp half-operator fragment was chosen for studies by NMR with single headpieces by Kaptein and co-workers.[59] In right-handed DNA, as we discussed in Section 9.1.1, short distances prevail from base H6 or H8 protons (for pyrimidines and purines respectively) to ribose H1′, H2′, and H2″ protons, both within a nucleotide unit and to ribose protons on the neighbouring nucleotide at the 5′ side (and not the 3′ side). This provides the basis for sequential assignment,[60,61] although the handedness of the helix is confirmed or rejected at a later stage in the process. Fig. 9.17 outlines regions of the NOESY experiment showing the cross-relaxation pathway for H1′, H2′, and H2″ and the base H6/H8 protons for the 14 bp *lac* operator. The lines drawn in the fingerprint region *a* connect intranucleotide H1—H6/H8 cross-peaks with internucleotide ones. Since several different cross-relaxation pathways also exist involving ribose H3′, and H4′ and H5 (cytosine), and 5-methyl protons (thymine), the consistency of the assignments needs to be checked very carefully. For the 14 bp operator fragment all non-exchangeable protons were assigned in this way, except for some H5′ and H5″ protons, whose resonances tend to overlap.

When two-dimensional NOESY spectra of the oligonucleotide are recorded in H_2O, then the assignments of the imino and amino protons can be carried out.[62] Since the amino protons are close to non-exchangeable protons such as H5 for cytosine and H2 for adenine, networks of NOE cross-peaks can be traced out, connecting them to those protons which were assigned initially in D_2O. Cytosine amino protons appear as sharp doublets, but those of guanine and adenine are, under normal conditions, collapsed to broad singlets due to exchange processes, and are therefore less useful. These assignments

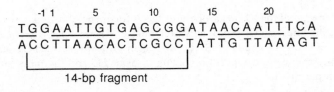

Fig. 9.16 Sequence of the *lac* operator, including the 14 bp DNA fragment.

were used in assigning the NMR spectrum of the *lac* headpiece-operator complex (see Section 9.5).

9.2.2 The *Eco*RI recognition dodecamer

The *Eco*RI restriction site DNA dodecamer [d(CGCGAATTCGCG)$_2$] has attracted much attention since it is one of the few DNA duplexes to crystallize in the B-form.[56] Since then, the structure of the *Eco*RI endonuclease complexed to its cognate hexanucleotide sequence GAATTC has been solved,[63] and supports the notion that unique structural variations within this sequence are responsible for its recognition by the endonuclease. However, crystallization of DNA takes place under mildly dehydrating conditions, which, as we mentioned in Section 9.1.5, are known to promote B→A or B→Z transitions,[64,65] and therefore there is the potential that crystallization can perturb the structure substantially. NMR assignments were carried out in Reid's laboratory,[55,56] and the imino protons were assigned by Patel and co-workers,[67] although the relevant NOESY assignments are shown in Fig. 9.18. Time-dependent NOE spectra were used to obtain the initial cross-relaxation rates between 155 pairs of protons. These initial cross-

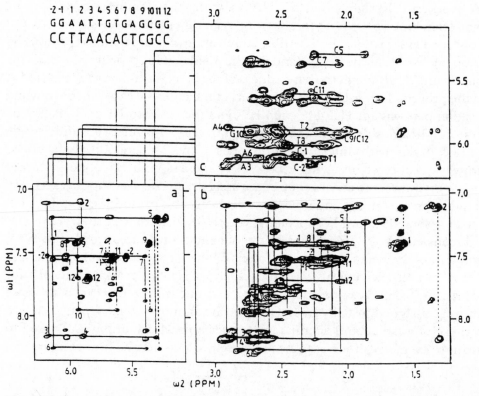

Fig. 9.17 Two-dimensional NOESY spectra of the 14 bp *lac* operator DNA fragment, showing networks of NOEs for the bottom strand, in which solid lines connect intra- and inter-nucleotide cross-peaks between H6/H8 and H1′ (in region (a)), between H6/H8 and H2′ and H6/H8-H2″ (in region (b)), and between H1′ and H2′ and H2″ (in region (c)), and dashed lines connect cross-peaks involving cytosine H5 and thymine 5-methyl protons. (Reprinted from ref. 155 with permission.)

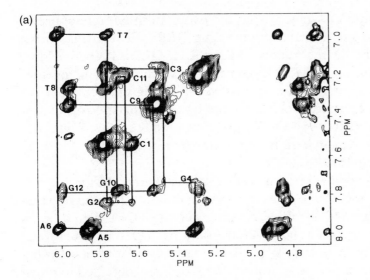

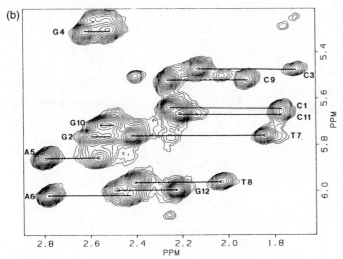

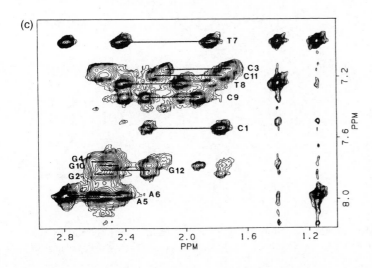

Fig. 9.18 NOESY spectra (200 ms) of the [d(CGCGAATTCGCG)]₂ dodecamer showing (a) H8/H6-1′H/3′H region; (b) 1′H-2′H,2″H region; and (c) H8/H6-2′H,2″H region. Connectivities are indicated by solid lines. (Reprinted from ref. 55 with permission.)

relaxation rates were converted into distances (using the two-spin initial rate approximation) and used in a distance geometry calculation (using DSPACE) starting with a family of random structures. The resultant structures had an RMSD of ≈ 2 Å per residue, and differed significantly from 'normal DNA'. The consistency within the data, together with a test of the distance geometry algorithm against fictional data derived from the X-ray structure, argued against the structure being derived by a systematic error. One of the structures was used to back calculate the NOESY spectrum, and the volume integrals were used to refine the DG structure in an iterative relaxation matrix approach. The final structure obtained is shown in Fig. 9.19.

The structure, which generates good agreement between calculated and experimental NOESY spectra, displays kinks at the C3—G4 base step and at the A6—T7 base step that appear to be similar to those reported for *Eco*RI restriction site DNA bound to its endonuclease. The solution structure is not, however, the same as the X-ray structure of the DNA duplex alone. In a more recent study by Beveridge and co-workers,[68] using unrestrained molecular dynamics refinement in water, a structure was obtained in much closer agreement with the duplex crystal structure. On the other hand, in a study on 5-fluorouracil-containing dodecamer,[69] iterative relaxation matrix refinement following

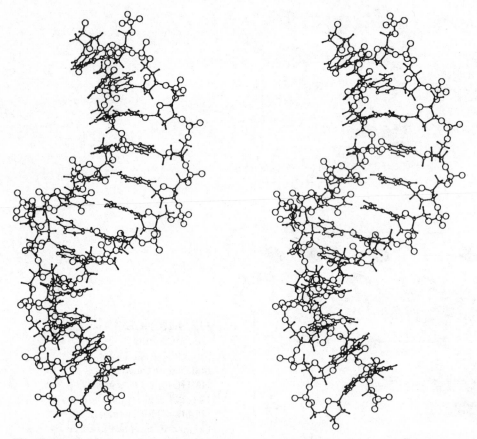

Fig. 9.19 Stereoview of the final structure of the [d(CGCGAATTCGCG)]$_2$ dodecamer refined by back calculation. The two kinks are between A6 and T7, and C3 and G4. (Reprinted from ref. 55 with permission.)

restrained molecular dynamics yielded a structure which had more in common with Beveridge's structure than with Reid's in that the kinks at the A·T and C·G base pairs were absent, although some minor groove narrowing was observed.

There has been some discussion in the DNA literature as to the relative merits and demerits of the use of iterative relaxation matrix approaches (IRMA) as opposed to restrained molecular dynamics (RMD). Clore and Gronenborn have argued[70,71] that restrained molecular dynamics is more efficient at finding the global minimum, and is essentially independent of starting structure, be it A-form or B-form DNA. In contrast, Reid argues[55] that both IRMA and RMD suffer from exactly the same spin diffusion effects, but RMD satisfies more incorrect distances and energy potentials. This was reinforced by an investigation[72] of the local motions in the *Eco*RI dodecamer, where it was found that the sugar and bases have the same correlation time. This was in contrast to the work of Clore and Gronenborn, who estimated a correlation time for the sugar moieties in DNA that was three-fold shorter, leading to systematic errors in all their DNA structure determinations which were not revealed using RMD. There would appear to be a consensus that IRMA is a better approach than RMD alone for determining oligonucleotide structures.[73–76]

9.2.3 Higher-order and larger DNA structures

One of the advantages of NMR spectroscopy is its ability to probe and characterize secondary structures hitherto undiscovered. The polymorphism of DNA has been known for some time, and recently NMR has been used to characterize hairpins,[77–80] as shown in Fig. 9.20. The rôle of the hairpin loop is to negotiate a complete turn to allow base pairing of a single strand with itself.

Another secondary structural element characterized by NMR is the DNA triplex, illustrated in Fig. 9.21, which was first structurally characterized by Arnott and co-workers,[81] and its rôle in sequence-specific probed recently by Dervan and co-workers.[82] The triplex was first studied by NMR in the laboratories of Feigon and co-workers[83] and independently by Patel and co-workers.[84] Since then a number of papers have appeared on triplexes, including a three-dimensional NMR study.[85] An example of such a structure from the work of Feigon and co-workers[86] is shown in Fig. 9.22.

Of course, the logical extension to the studies on triplex DNA are on quadruplex DNA, and, following pilot studies by Patel and co-workers,[87] two independent studies were reported, one from Feigon and co-workers,[88] and one from Lilley and co-workers.[89] More recently, Guéron and co-workers[90] have also reported a quadruplex structure, and an illustration for their model structure is shown in Fig. 9.23.

There have also been a large number of studies on a variety of bends, distortions, bulges, and junctions in DNA structures. Attempts have also been made to determine some structural elements of supercoiled or chromatin DNA in the nucleosome core by NMR, for example by Kallenbach and co-workers,[91] who argued on the basis of a ^{31}P lineshape analysis that there is a single average state of DNA which is not kinked in any regular fashion in the nucleosome core. Also, van Holde and co-workers[92] used ^{1}H NMR of the exchangeable imino protons to provide evidence for a histone-induced DNA

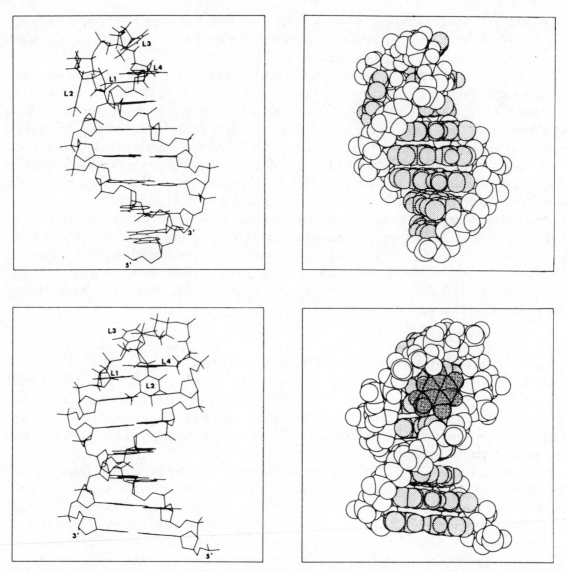

Fig. 9.20 A model for the hairpin DNA structure based on NMR data and a B-type helix. (Reprinted from ref. 80 with permission.)

conformational change which involves localized alteration of the base pairing in the core particle.

9.3 DRUG–DNA INTERACTIONS

The interaction of antitumour drugs with DNA is an area where NMR has been particularly useful in determining the molecular details of the interaction. There have been a number of studies involving a variety of drugs, many of which bind to the minor groove of DNA. Among the better known of these came from Wemmer and co-workers,[93,94] who studied the binding of distamycin A to an A·T-rich undecamer DNA fragment. Intramolecular NOEs were used to model the structure of the complex through

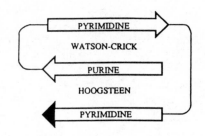

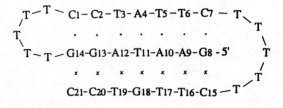

Fig. 9.21 AN example of the arrangement of interactions in a triplex DNA structure. (Reprinted from ref. 85 with permission.)

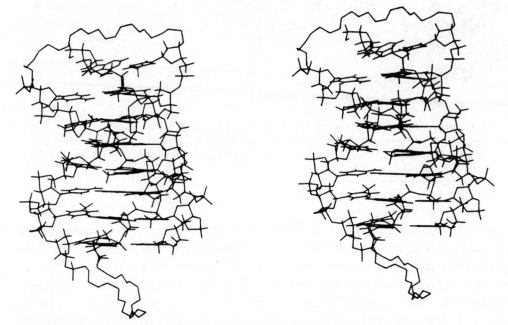

Fig. 9.22 Stereoview of an NMR model of triplex DNA. (Reprinted from ref. 86 with permission.)

manual docking and using the AMBER molecular mechanics program. Using a different fragment,[95,96] Wemmer and co-workers identified a 2:1 complex distamycin A:oligo-nucleotide, obtained by AMBER calculations, shown in Fig. 9.24, in which widening of the minor groove is apparent.

Other drugs interacting with DNA which have been studied by solution-state NMR methods include psoralen,[97] actinomycin,[98] nogalomycin,[99] luzopeptin,[100] echinomycin,[101] Hoechst 33258,[102] chromomycin,[103,104] berenil,[105] 1-methylimidazole-2-carboxamide netropsin,[106] and SN-6999.[107] In an interesting study by Harbison and co-workers,[108] [13]C

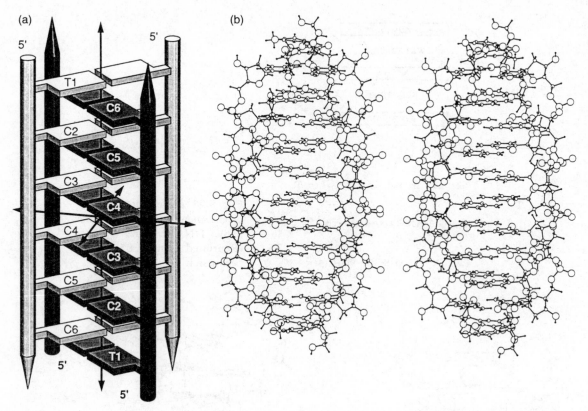

Fig. 9.23 Proposed model for the d(TC$_5$) quadruplex DNA deduced from NMR. (a) Intercalation scheme of the two parallel duplexes, with arrows showing the three axes of 2-fold symmetry; (b) the stereoview represents the energy-minimized structure calculated from 70 NOE distance and ten dihedral restraints. (Reprinted from ref. 90 with permission.)

and ^{31}P one- and two-dimensional solid-state CP–MAS NMR was used to show that the drug proflavine is stacked within the aromatic ring plane perpendicular to the fibre axis and that it is essentially immobile. ^{13}C NMR showed that proflavine binding does not change the puckering of the deoxyribose ring. But ^{31}P NMR spectra showed that some of the phosphodiesters tilted almost parallel to the helix axis, and a second set was perpendicular. This was interpreted as the first set of phosphodiesters spanning the intercalation sites, whereas the second set compensated for the unwinding of the DNA by the intercalator.

9.4 STRUCTURES OF DNA-BINDING PROTEINS

Although this chapter is concerned with DNA, RNA, and carbohydrates, it is appropriate here to provide a brief summary of the structures known from NMR studies of DNA-binding proteins. We have already considered one example—the *lac* repressor head-piece—in Section 4.3.1. We will discuss the details of protein–DNA interactions in Section 9.5.

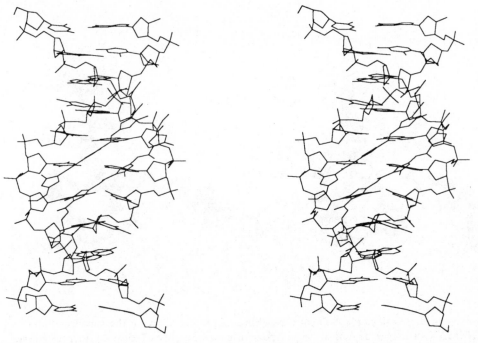

Fig. 9.24 Stereoview of the NMR structure of the complex of distamycin A and d(CGCAAATTGGC)$_2$ after energy minimization. (Reprinted from ref. 95 with permission.)

9.4.1 Helix–turn–helix DNA-binding proteins

The helix–turn–helix motif is perhaps the most familiar DNA recognition element, and in this section we will very briefly review the structures of those helix–turn–helix proteins whose structure has been determined by NMR. The helix–turn–helix involves recognition of a specific base sequence of the operator by insertion of the second helix into the major groove of the operator DNA, in contact with the specific bases leading to its name, the recognition helix. Many of these proteins are too large for current solution-state NMR methods, although the sequential assignments[109] (using fractional deuteration—see Section 4.2.3) and low-resolution structure[110] for the *trp*-repressor ($M_r = 25$ kDa) per dimer) has been carried out by Jardetsky and co-workers. One important finding from this work was that the *trp*-repressor recognition helix is more flexible than was originally thought from the X-ray structure for the protein-DNA complex determine by Sigler and co-workers.[111] Indeed, in the X-ray crystal structure of the *trp*-repressor-DNA complex proposed by Sigler and co-workers, it is suggested that the principal protein-base contacts are mediated through structured water. This structure, which has already made it into the textbooks of biochemistry,[112] has been suggested to represent a non-specific complex which results from the crystallization conditions.[113,114] Indeed, it has been suggested that the *trp* repressor may bind to the operator in a tandem fashion,[115,116] as has been suggested for the *met*J repressor.[117,118] These criticisms have been addressed by Sigler's laboratory,[119] in which it is argued that the crystalline protein-DNA complex does not change lattice form at the higher salt concentrations normally used for specific protein-

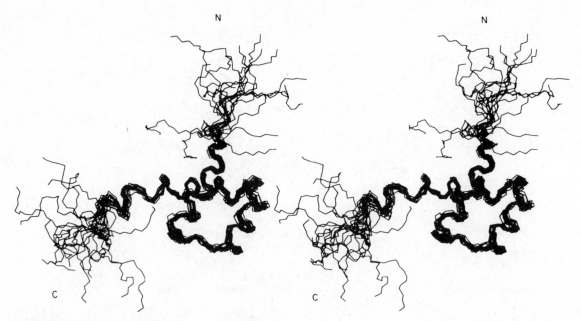

Fig. 9.25 Stereoview of the *Ant*P homeodomain DNA-binding protein showing the superposition of 19 energy-refined DISMAN structures depicting the backbone from residues 0 to 67. N and C identify the start and end of the polypeptide chain. (Reprinted from ref. 123 with permission.)

DNA complex formation (up to 200 mM NaCl), and the structure is consistent with the mutagenesis studies of Bass and co-workers.[120]

The structure of the *Antennapedia* homeodomain from *Drosophila melanogaster* (68 residues, $M_r = 7500$) has been determined by Wüthrich and co-workers,[121–123] using angle space distance geometry (DISMAN) and restrained energy minimization with AMBER. A family of 19 structures had an RMSD for the backbone atoms relative to the mean of 0.6 Å. The helix–turn–helix motif occurs from residues 28 to 52, with a somewhat flexible fourth helix from residues 53 to 59, which essentially forms an extension of the presumed recognition helix (see Fig. 9.25).

The protein-DNA complex structure was determined with a 14 bp DNA fragment[124] ($M_r = 17.8$ kDa). Interestingly, evidence was obtained that the loop immediately preceding the helix–turn–helix motif reaches with Arg-5 into the minor groove of the operator DNA, where in the free protein, residues 0–6 were disordered. More recently, the structure of a 69–residue fragment of the closely related 434 repressor has been determined,[125] using ^{15}N-labelling and fractional ^{13}C-labelling, and it corresponded closely to the X-ray structure.[126]

9.4.2 Zinc fingers

The zinc finger motif has been identified as a nucleic acid-binding domain. These motifs have been further categorized into two classes, where the Zn^{2+} ion in the classical zinc finger is coordinated by two Cys and two His residues, and in the second class the ligands are four Cys residues. Three-dimensional structures have been described for four synthetic peptides, each corresponding to a classical single zinc finger, but from different sources: the *Xenopus* protein Xfin-31, corresponding to the 31 st zinc finger of this transcription factor, determined by Wright and co-workers;[127,128] the yeast transcription activator ADR1, determined by Klevit and co-workers;[129] the human enhancer-binding domain (HEBD—see Fig. 9.26), determined by Clore, Gronenborn, and co-workers;[130] and the mouse Krupell-like gene mKr2, determined by Rösch and co-workers.[131] With the exception of the latter protein, the samples were prepared with at least a stoichiometric amount of zinc. All the proteins show similarities to each other, although there are also differences. The Xfin-31 finger starts with two β strands arranged in a hairpin structure. This structure is not observed in the other zinc fingers, where an irregular, antiparallel β-sheet connected in an unusual turn is observed (see Fig. 9.26). In each finger, a helix of varying length is observed, between 7 and 13 residues, starting at the same residue, with the exception of mKr2 which starts four residues later. The shorter helix for mKr2 may arise because of the lack of zinc, however. In the ADR1 and Xfin-1, but not in the mKr2 and HEBd fingers, the helix includes both His ligands. In ADR1, mKr2, and HEBD the helix is all α-helix, whereas in Xfin-31 the helix starts as α-helix, and after the first His ligand finishes as a 3_{10}-helix. The NOE data for Xfin-31, ADR1, and HEBD proteins are consistent with the N^{ε} atom of the two His ligates to the zinc atom.

An example of the second class of zinc fingers is the segment between Cys-440 and Arg-510 of the rat glucocorticoid receptor,[132,133] and the DNA-binding domain of the oestrogen receptor,[134] which both contain two fingers. The structures for these fingers differ significantly from the classical fingers in that no helical regions are found within the finger. For the glucocorticoid receptor fragment, a small piece of antiparallel β-sheet is found in the N-terminal region of the first finger, and a type I and type II turn encompasses the N-terminal Cys ligands of the second finger. There are α-helices and several portions of extended structure on the C-terminal side of each finger, so that the motif 'finger-helix-extended chain' is repeated. NOE data show a number of long-range interactions between the two fingers and between the two helical domains. In the three-dimensional structure, calculations show that the two helices pack perpendicular to each other with hydrophilic faces exposed to solvent. One finger contacts with the two helices, and the second finger extends out from the protein containing the two zincs, some 13 Å apart. In this second finger, there is an exposed region, Ala-477 to Asp-481, proposed to be involved in protein interactions between monomers of the glucocorticoid receptor, and thus considered important for cooperative binding. This is also consistent with the recent X-ray crystal structure of the complex with DNA.[135] The structure calculations of the oestrogen receptor DNA-binding domain show similar structures to those described for the glucocorticoid receptor, except that a nine-residue loop appears to bond to the minor groove.

These studies show differences that have implications for the model for DNA-binding. The class 2 zinc fingers appear to bind to the DNA major groove, whereas the model[136] for the binding of the class 1 (or 'classical') fingers is that successive fingers bind on one face of the major groove, so that successive minor grooves are crossed. The classical fingers that have been discussed are single isolated fingers, and because the proteins from which these were derived contain multiple fingers, it is important to establish the relationship between individual fingers. The transcription factor SW15 contains three fingers near its

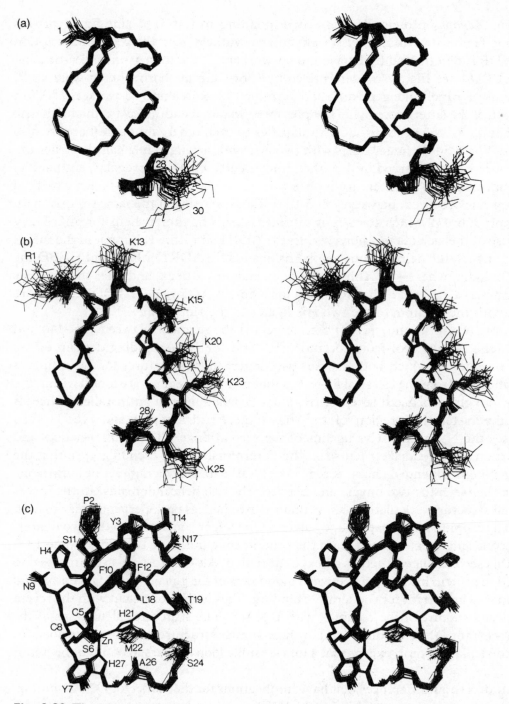

Fig. 9.26 The structure of human enhancer binding protein, showing the best-fit superposition of the backbone (N, C^{α}, C′) atoms of 40 simulated annealing structures (a) residues 1–30; (b) plus arginine and lysine side chains for residues 1–28; and (c) for the remaining side chains for residues 1–27. For residues 1–28, the RMSD was 0.4 Å for the backbone atoms, and 0.8 Å for all atoms, or 0.41 Å for all atoms excluding lysine and arginine side chains, which are disordered. The solution structure of the zinc finger consists of two irregular antiparallel β-strands connected by an atypical turn (residues 3–12) and a classical α-helix (residues 14–24). (Reprinted from ref. 130 with permission.)

C-terminus, and NMR studies by Neuhaus and co-workers[137] on constructs containing the first two fingers also suggest that the fingers are modular. The recent X-ray structure by Pavletich and Pabo[138] of the Zif268 protein-DNA complex containing three fingers also suggests that the isolated zinc finger domains are identical with the single classical fingers. In more recent studies Weiss and co-workers[139,140] have shown that the human male associated protein ZFY, which contains two fingers, has a structure which is similar to the classical single fingers.

9.4.3 Leucine zippers and assorted DNA-binding proteins

Strictly speaking, leucine zipper proteins are not DNA-binding proteins, since they principally interact with other proteins which bind to DNA. We mention them here because they can contain small DNA-binding contact regions, and because some NMR studies have been carried out on the transcription factors GCN4, *fos*, and *jun*,[141,142] and a histone DNA-binding protein.[143] The leucine zipper contains a helical region of 30 residues with a periodic repeat of Leu every seven residues. From the NMR study on GCN4, the peptide forms a symmetric coiled coil dimer,[144] and is not interdigitated, as had been supposed, with α-helix running almost the entire length of the peptide. There is thought to be a DNA-binding basic region in addition to the leucine zipper. The DNA-binding region was found to be relatively poorly defined by NOEs, suggesting some conformational flexibility, which was confirmed in a later study of the amide exchange rates[145] and by circular dichroism spectroscopy.[146] The amide exchange data also demonstrated an interesting periodicity indicative of supercoiling. These results were consistent with the X-ray structure.[147]

A number of other DNA-binding protein structures have been determined by NMR. For instance, the Arc repressor[148] is probably a β-sheet DNA-binding protein similar to the *Met*J repressor. The GAL4 transcription factor, while containing two zinc ions ligated to Cys and His residues, turns out to be a binuclear metal cluster rather than a zinc finger. This was initially shown by ^{113}Cd NMR studies,[149] and has since been completely characterized by NMR in four different laboratories[150–153] and by X-ray crystallography.[154]

9.5 PROTEIN–DNA INTERACTIONS—THE *LAC* REPRESSOR–OPERATOR COMPLEX

The interaction of the 14 bp operator, whose sequential assignments were considered in Section 9.2.1, with the *lac*-headpiece (whose assignments were considered in Section 4.3.1), was examined by NOESY experiments (see Fig. 9.27) by Kaptein and co-workers.[155] The complex $M_r = 14$ kDa represented the limit of the two-dimensional NMR approach employed in 1983. A low-resolution structure was determined on the basis of 23 observed intermolecular NOE interactions, which are summarized in Table 9.6.

In order to facilitate the assignments the NOESY spectra were recorded with rather long mixing times of 100 and 250 ms. They therefore contain cross-peaks resulting from spin diffusion. For instance, it is unlikely that Leu-6 lies in the DNA binding site and protein–DNA NOEs involving this residue are weak and probably transmitted through Tyr-7. For the same reason the distances corresponding to those NOEs may be up to 6 Å. In Table 9.6, the distinction is made between NOEs that are unambiguous because they involve protons with unique resonance positions and those that are probable. The latter protons occur in crowded regions where overlap of resonances may occur. They were assigned on the basis of a pattern recognition procedure, which involves the following reasoning. Suppose a cross-section of the NOESY showing a headpiece proton exhibits NOEs to a set of other protons of the same amino acid residue. Then if a cross-section of the NOESY showing a DNA proton exhibits cross-peaks at the same frequencies and at least one of these can be uniquely assigned to a headpiece proton, then the assignment of the other cross-peaks in the set is extremely likely. Consider the case of His-29. In the NOESY spectrum, a horizontal line at the C2 proton frequency of His-29 (8.52 ppm) shows a set of four cross-peaks in the DNA ribose region which are in a crowded region of the spectrum. A similar set of cross-peaks is observed at the line of H8 of adenine 2, and,

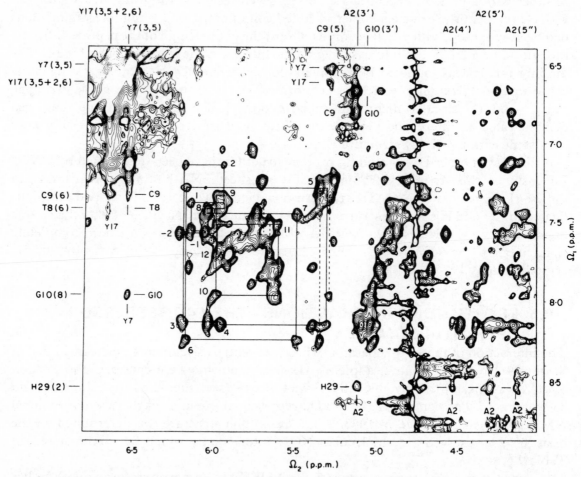

Fig. 9.27 Part of the 500 MHz NOESY spectrum of the *lac* headpiece-14 bp operator complex. Sequential NOEs of the same strand as in Fig. 9.17 are indicated by connecting lines. The protein-DNA NOE cross-peaks present in this part of the spectrum are also indicated. (Reprinted from ref. 155 with permission.)

Table 9.6 NOEs between *lac* repressor headpiece 56 and a 14 bp *lac* operator fragment

Protein	DNA	Protein	DNA
Unambiguous[a]		Probable[a]	
Tyr7 H3,5	−G10 H8	Thr5 CδH$_3$	−G10 H8
Tyr7 H3,5	−G10 H1′	Thr5 CδH$_3$	−G10 H3′
Tyr7 H3,5	−G10 H3′	Leu6 CδH$_3$	−C9 H5
Tyr7 H3,5	−C9 H5	Leu6 CδH$_3$	−C9 H6
Tyr7 H3,5	−C9 H6	Leu6 CδH$_3$	−C9 H3′
Leu6 CδH$_3$	−C9 H5	Leu6 CδH$_3$	−T8 H6
Tyr17 H3,5+H2,6	−C9 H5	Tyr17 H3,5+H2,6	−T8 CH$_3$
Tyr17 H3,5+H2,6	−C9 H6	Ser21 CαH	−T8 CH$_3$
Tyr17 H3,5+H2,6	−T8 H6	His29 H2	−A2 H3′
His29 H2	−A2 H8	His19 H2	−A2 H4′
		His29 H2	−A2 H5′[b]
		His29 H2	−A2 H5″[b]

[a]The unambiguous NOEs were assigned at unique resonance frequencies, while the probable NOEs were from resonances which could overlap with resonances of other protons (see text for further discussion).
[b]H5′ and H5″ protons were only pairwise assigned.

moreover, a very weak cross-peak is observed at the crossing of this line and that of His-29, which both have unique resonance positions. Hence, the weak NOE between H8 of A2 and the His-29 C2 proton is listed as 'unambiguous' in Table 9.6, although it is undoubtedly the result of spin diffusion. The other NOEs of His-29 with the ribose protons of A2 are stronger and represent shorter distances, but occur in a region of the spectrum with much overlap and are therefore indicated as probable.

This generates a low-resolution model for the protein–DNA complex (since the distance constraints can be up to 6 Å as a result of limited spin diffusion). With the DNA kept in the standard B-DNA conformation, the headpiece backbone conformation was held rigid as in the protein conformation alone, although some of the side chains in the DNA binding site were allowed to change their conformations. The model that was obtained then satisfied all the NOE constraints. Energy minimization of this model was carried out to ensure that it has reasonable non-bonded interactions (see Fig. 9.28).

The fact that all NOE constraints were satisfied simultaneously is significant, because it means that the model must represent the major specific complex. Other, non-specific complexes that are undoubtedly formed at the high concentration (5 mM) of the NMR experiments apparently do not lead to inconsistent NOEs either because they do not involve short proton–proton distances or because each of them does not have a long enough lifetime to allow build-up of NOEs. A 1:1 mixture of two different complexes would have led to inconsistent NOEs, and can also be excluded.

The most surprising feature of the model is that the orientation of the second or 'recognition' helix in the major groove of DNA with respect to the dyad axis at GC11 is opposite to that found in all other models of repressor–operator interactions, either from X-ray (434 repressor) or from model building (CAP, 1, and *cro* repressor). In these

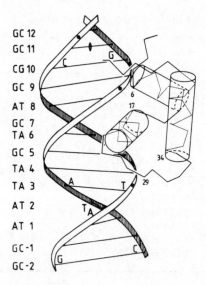

GC 12
GC 11
CG 10
GC 9
AT 8
GC 7
TA 6
GC 5
TA 4
TA 3
AT 2
AT 1
GC -1
GC -2

Fig. 9.28 Model of the *lac headpiece-lac* operator complex. The orientation of the recognition helix in the major groove of the DNA is shown schematically, using a standard B-DNA conformation. The positioning of the protein with respect to the DNA was carried out to satisfy the NOE restraints. The black dots indicate positions of the phosphates where ethylation with ethylnitrosourea interferes with lac repressor binding. (Reprinted from ref. 155 with permission.)

models the first helix would be away from the dyad axis, while in the NMR structure it is close to it. However, this structure accounts for the phosphate ethylation interference experiments indicated, and also for a functional contact between Gln 18 and GC 7 found by a genetic 'loss of contact' study involving mutants of both the repressor and operator. Although the whole *lac* repressor may bind to the operator differently, NMR cannot address this question, given its $M_r = 154$ kDa.

The complex of the *lac* repressor headpiece and a longer (22mer) DNA operator has been investigated by Kaptein and co-workers,[156] when it was found that two headpieces bound to the longer operator in an exactly analogous way to the binding of the one headpiece to the 14 bp operator.

9.6 EXAMPLES OF RNA STRUCTURE

The investigation of RNA structure by NMR spectroscopy has been reviewed recently.[58,157] Since the aspects of sequential assignments discussed for DNA in Section 9.1.1 are directly applicable to RNA, we will confine our discussion to the structural information that has been obtained for a small selection of RNAs.

The way in which the structure and stability of nucleic acid hairpins were initially characterized was developed from tRNA X-ray crystal structures and thermodynamic considerations.[158] For the helical stem of the RNA hairpin a comparison of the backbone torsion angles and helical parameters is given in Table 9.7. In general the folding of a hairpin takes place in such a way as to maximize the stacking interactions, so that in tRNA loops containing 7–8 unpaired nucleotides, for example, stacking is continued on the 5′-side of the stem, leaving only two nucleotides unstacked. Many of the NMR studies (reviewed in ref. 58) have been on the hairpins of yeast tRNA[Phe] and *E.coli* tRNA[Val], as shown in Fig. 9.29.

The pioneering one-dimensional NMR studies of the imino protons of tRNAs were carried out in the laboratories of Redfield and co-workers,[159] Kearns, Patel, and Shulman,[160] and Reid and co-workers,[161] and considerable difficulty was encountered in

Table 9.7 Helical parameters and backbone torsion angles for the helical stem of the RNA hairpin 5'GGAC(UUCG)GUCC[a]

	NMR	X-ray
Torsion angles		
α (P—O_5')	-75 ± 5	-61
β (O_5'—C_5')	-171 ± 4	-191
γ (C_5'—C_4')	47 ± 4	53
δ (C_4'—C_3')	86 ± 2	80
ϵ (C_3'—O_3')	-159 ± 5	-148
ζ (O_3'—P)	-68 ± 5	-79
X (C_1'—N_1/N_9)	-156 ± 4	-159
Helical parameters and groove widths		
Twist angle	31.3°	33.1°
Rise per residue (Å)	3.05	2.79
Tilt	$-0.7°$	16.7°
Propeller twist	6.5°	18.6°
Displacement (Å)	3.6	3.6
Minor groove width (Å)	9.0	10.2
Major groove width (Å)	—	3.7

[a](Values of the backbone torsion angles have been averaged over ten structures generated using distance geometry without any structural assumption, followed by restrained energy minimization; the helical parameters are from a single representative structure. Values for a duplex RNA in a single-crystal X-ray diffraction study are shown for comparison.)

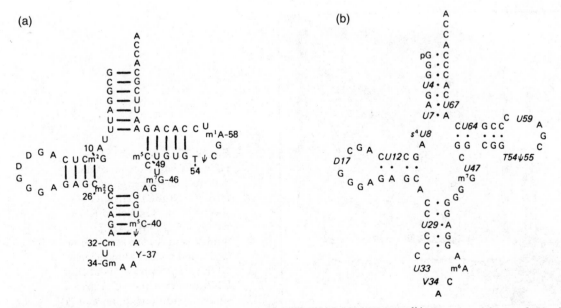

Fig. 9.29 The secondary structure of (a) yeast tRNA[Phe]; and (c) *E.coli* tRNA[Val]. (Reprinted from ref. 58 with permission.)

those studies making the correct assignments. With the use of two-dimensional NMR and ^{15}N-labelling, this has become much easier. For example, Tinoco and co-workers[162] have determined the three-dimensional structure of a RNA hairpin containing the common tetraloop UUCG, which is unusually stable. The unpaired nucleotides in the loop were found to adopt an extended *2'-endo* conformation as opposed to the usual *3'-endo*

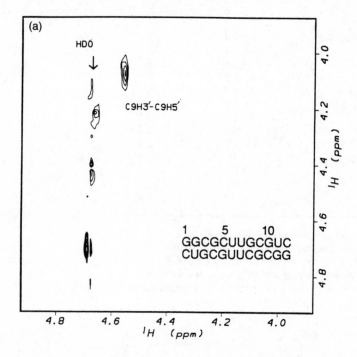

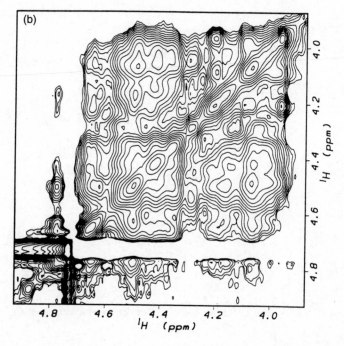

Fig. 9.30 A plot of (a) the H2'—H5' sugar region of the 400-ms four-dimensional HMQC–NOESY–HMQC spectrum of ^{13}C-labelled RNA r(GGCGCUUGCGUC)$_2$ duplex, and (b) the same region of the 400 ms two-dimensional NOESY spectrum of unlabelled RNA duplex acquired under similar conditions to those of (a). (Reprinted from ref. 164 with permission.)

conformation. Furthermore, instead of the normal *gauche⁻ gauche⁻* conformation, the two angles next to phosphorus, ζ and α adopt the *gauche⁻ trans* conformer found in the π-turn of the anticodon loop of tRNAs. The mismatched nucleotides stack, presumably to compensate for these two thermodynamically unfavourable conformational changes.

One of the problems with assigning RNA spectra is the problem of spectral overlap. Pardi and co-workers[163] have used three-dimensional and, more recently,[164] four-dimensional NMR methods to simplify the assignment procedure. Fig. 9.30 shows the dramatic improvement in spectral resolution obtainable using a four-dimensional HMQC–NOESY–HMQC experiment on uniformly ^{13}C-labelled RNA duplex. The three-dimensional experiments employed uniformly ^{13}C/^{15}N-labelled RNA, which promises to extend the structure determination possibilities to include hammerhead catalytic RNA ($M_r = 15$ kDa) and yeast tRNA ($M_r = 25$ kDa).

An example of an unusually stable RNA tetraplex structure is shown in Fig. 9.31 for $(UG_4U)_4$ (24 nucleotides, $M_r > 13$ kDa) determined by Moore and co-workers.[165] The assignments were carried out using NOESY, DQF–COSY, ^{1}H—^{31}P heteronuclear COSY and TOCSY experiments. A total of 428 distance restraints was derived, with 212

(a)

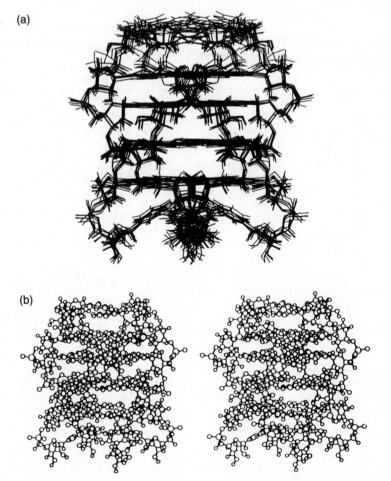

(b)

Fig. 9.31 The fully refined NMR model for UG_4G, showing (a) the stereoview of the superposition of eight structures, and (b) the ball-and-stick model. (Reprinted from ref. 165 with permission.)

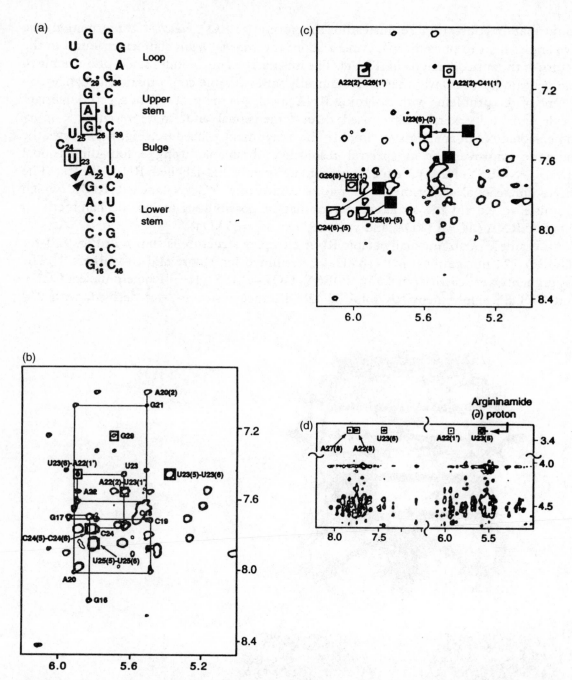

Fig. 9.32 (a) Secondary structure of TAR RNA, with phosphates whose ethylation interferes with binding of Tat peptides or arginine shown with arrows. (b) Two-dimensional NOESY showing H8/H6/H2-pyrimidine H5/ribose H1′ NOEs and sequential connectivities indicated for G16 to C24. (c) Same region of NOESY as (b) but in the presence of 6 mM argininamide, with those NOE cross-peaks which are boxed, indicating a conformational change. (d) Portion of the NOESY spectrum showing NOEs between argininamide C$^\delta$H protons and RNA proton. (Reprinted from ref. 166 with permission.)

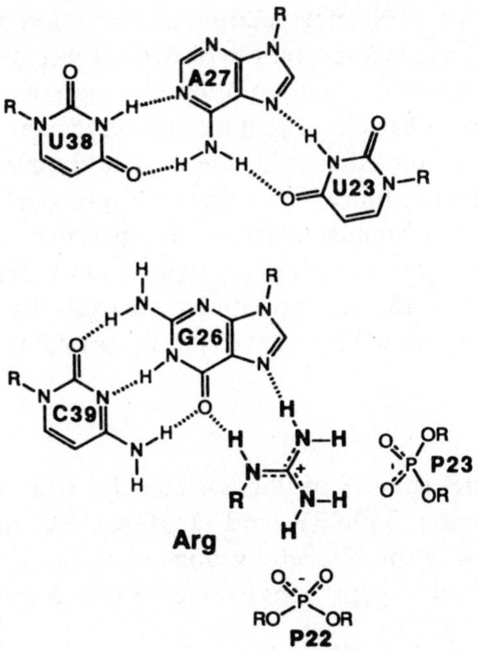

Fig. 9.33 Schematic representation of the proposed base triple between U23 and A27·U38 and the interaction of arginine with G26 and two phosphate groups. (Reprinted from ref. 166 with permission.)

intranucleotide NOEs, 112 internucleotide NOEs, 36 hydrogen bonds, eight exchangeable proton NOEs, 60 backbone torsion angles, and 104 additional torsion angles. The structure was calculated using distance geometry with eight structures, restrained energy minimization, restrained molecular dynamics, energy minimization again, and refinement by the iterative relaxation matrix approach, all using the program X-PLOR. The final RMSD was 0.71 ± 0.16 Å for all the atoms.

A recent interesting example of a hairpin structure involved in protein binding is that of the TAR RNA-arginine complex reported by Williamson and co-workers.[166] The predicted secondary structure involves two stem regions separated by a bulge, and a loop of six nucleotides (see Fig. 9.32). The loop region is not involved in binding to the HIV Tat protein, but is important in transcriptional activation, and binding of Tat to TAR is known to be mediated through a single arginine. Using NOESY experiments and chemical shift changes, Williamson and co-workers determined that arginine binds at a base triple between U23 and A27–U38, as depicted in Fig. 9.33.

Although there have not been many reports of RNA-protein complex structures, Müller and co-workers[167] examined the interaction of uniformly ^{13}C/^{15}N hnRNP C protein (93 residues, $M_r \approx 10$ kDa) with r(U)$_8$ RNA. However, changes in the protein structure were monitored, rather than in the RNA or in both simultaneously.

9.7 CARBOHYDRATES

The field of glycobiology is relatively young, and yet oligosaccharides are involved in many important biological structures, both in proteins and membranes. We will defer a discussion of glycolipids to Chapter 10, and here concentrate on the ways in which NMR spectroscopy is beginning to help to define the structure and shape of more complex carbohydrates, principally in solution. This type of structural information is often impossible to obtain by any other technique, and is important in biological recognition.[168]

NMR spectroscopy has been able to determine the configuration, position of linkages, branching, and microheterogeneity of oligosaccharides. Complex carbohydrates are also flexible to some extent, mainly in the glycosidic linkages. The pyranoid rings of the constituent glycosyl residues are generally far more rigid than the glycosyl bonds. Branched oligosaccharides can usually adopt a smaller number of conformations than linear oligosaccharides. With their unusual combination of rigidity and flexibility, oligosaccharides present a difficult challenge for conformational analysis by NMR, sometimes in conjunction with molecular dynamics. There have been a number of reviews recently on this field.[169–171]

9.7.1 Sequential assignments

The strategy for assignment of the ^{1}H NMR spectra of oligosaccharides is as follows: (i) assignment of aliphatic resonances using TOCSY and TQF–COSY in D_2O; (ii) assignment of amide and methyl protons of the N-acetyl groups in amino sugars by ROESY in H_2O; and (iii) assignment of the hydroxyl protons by one- or two-dimensional pre-steady-state NOESY in H_2O.

As an example the TOCSY and TQF–COSY data for the diantennary oligosaccharide known as M8 is shown in Fig. 9.34. Illustrated on the TOCSY spectrum is the tracing for

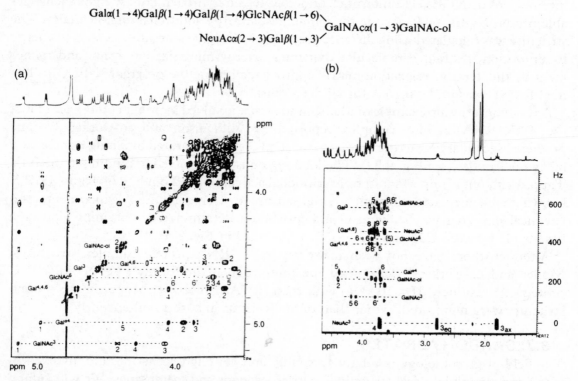

Fig. 9.34 The M8 oligosaccharide and its 500 MHz (a) TOCSY spectrum recorded in D_2O/acetone (9:1) with horizontal lines connecting protons within a particular glycosyl residue; and (b) TQF–COSY spectrum recorded in D_2O/acetone. (Reprinted from ref. 169 with permission.)

a number of individual monosaccharide units, separated in F_1 by their anomeric resonances. Subspectra along the dotted lines indicated provide complete spectra for a single monosaccharide, such as GlcNac, and so on. Complete connectivities can be obtained for GalNAc[3] H1→H6′ and for NeuAc H3→H9′. The GalNAc[3] residue shows exceptionally large dispersion of its seven ring protons. Similarly, the situation is favourable for the NeuAc H7, H8, H9, and H9′ protons because no strong coupling is present to complicate the spectrum. Unfortunately H6 and H6′ of all four Gal residues, and H1/H1′ and H6/H6′ of the GalNAc-ol residue, show higher-order patterns not easily unravelled by TOCSY. However, mutually coupled three-spin systems can be detected in TQF–COSY spectra, where the dotted horizontal lines connect three-spin systems such as Gal H5—H6—H6′. Since H5 has been assigned in the TOCSY experiment, the individual residue assignments can be made, as indicated in Fig. 9.34(b).

The next step in the procedure is the assignment of the exchangeable protons. The NH protons in amino sugars can be correlated to their *N*-acetyl methyl resonances through ROEs. In the case of M8 the ROESY experiment is shown in Fig. 9.35, in which the connectivities between the NH protons and methyl resonances are correlated via H2 or H5 protons which were already assigned in the TOCSY experiment.

The hydroxyl protons can be assigned using one-dimensional NOE difference experiments or NOESY with selective excitation. Then two-dimensional ROESY and NOESY experiments can be used to identify the NOEs in the usual manner. Typical

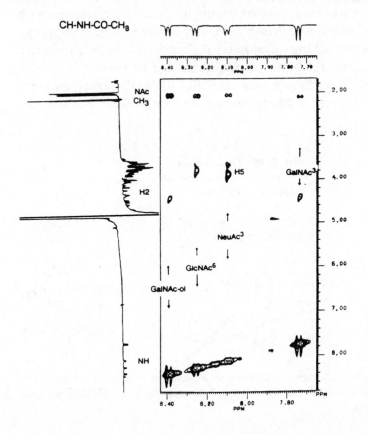

Fig. 9.35 The 500 MHz ROESY spectrum of the M8 oligosaccharide recorded in H$_2$O/acetone with the F_2 projection highlighting the NH signals. (Reprinted from ref. 169 with permission.)

NOEs are outlined schematically in Fig. 9.36. The essential step in determining the conformational preference of the (1→6)-branch is the assignment of the prochiral C6 protons of the branching unit, α-GalNAc in the case of M8. This can be solved by examination of the vicinal couplings and NOEs for H4→H5→ the two H6 protons. Using the following Karplus relations adapted for carbohydrates:

$$\mathcal{J}(H5,H6^{proR}) = 13.22 \cos^2 \theta - 0.99 \cos \theta - 6.4 \cos^2 (25.87 - \theta) \\ - 0.96 \cos^2 (7.96 + \theta) + 2.61$$

$$\mathcal{J}(H5,H6^{proS}) = 13.22 \cos^2 \theta - 0.99 \cos \theta - 0.96 \cos^2 (7.96 + \theta) \\ - 3.2 \cos^2 (25.87 - \theta) - 3.2 \cos^2 (25.87 + \theta) + 2.61 \quad (9.2)$$

the torsion angles can be calculated. The definition of the torsion angles is shown in Fig. 9.37. The conformations are expressed by the angles ϕ, ψ, and ω, where ϕ is the angle O5—C1—On'—Cn', ψ is the angle C1—On'—Cn'—C$(n-1)'$, and ω, which is needed only for 1→6 linkages, is the angle O6—C6—C5—C4. Angles are positive when the bond to the front, viewed along the central bond, must be rotated clockwise to eclipse the bond at the rear, so that the angle ω illustrated in Fig. 9.37 is $-60°$.

The problem with assigning oligosaccharide NMR spectra in general is that they are relatively crowded. A solution to this is the use of ^{13}C-labelling and HMQC or other heteronuclear correlated methods.[172] Carbohydrates may be labelled by growing cells in the presence of uniformly ^{13}C-labelled glucose, and isolating the oligosaccharides by standard procedures. Alternatively, small, simple oligosaccharides may be labelled by solid-phase synthesis (machines for which are just appearing on the market) starting with ^{13}C-labelled sugars. Having said this, it is not always necessary to label, since HMQC can be carried out with carbohydrates containing ^{13}C at natural abundance. As an alternative to heteronuclear NMR, homonuclear 3D NMR experiments can be used.[173] Fig. 9.38 shows the cross-sections through a three-dimensional NOESY–TOCSY experiment on a diantennary oligosaccharide, in which the diagonal shows the TOCSY correlations, the

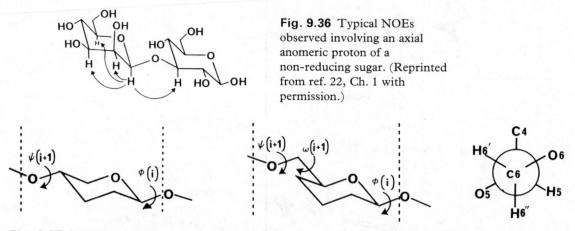

Fig. 9.36 Typical NOEs observed involving an axial anomeric proton of a non-reducing sugar. (Reprinted from ref. 22, Ch. 1 with permission.)

Fig. 9.37 Definition of the torsion angles in sugars. Also shown is a Newman projection depicting the nomenclature for H6′ and H6″ drawn looking down the C6—C5 bond. (Reprinted from ref. 22, Ch. 1 with permission.)

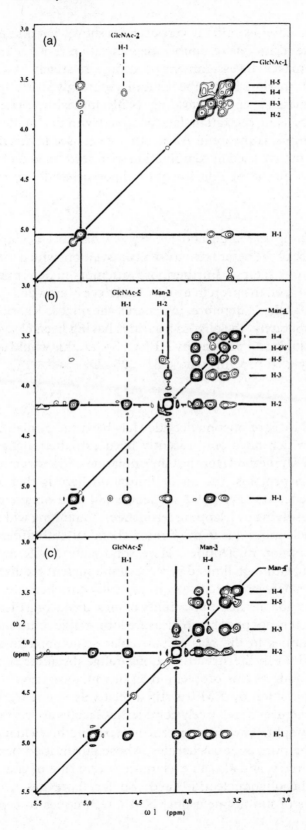

Fig. 9.38 Cross-sections perpendicular to F_3 of the three-dimensional NOESY–TOCSY spectrum of a diantennary asparagine-linked oligosaccharide. The cross-sections shown are at the F_3 positions of (a) GlcNAc-1 H1; (b) Man-4 H-2; and (c) Man-4′ H-2. (Reprinted from ref. 173 with permission.)

horizontal lines indicate NOE connectivities, and the vertical lines show what is known as the back transfer connectivity. The technique of double magnetization transfer ensures that there is minimal overlap and makes the assignment procedure relatively easy.

One of the problems with oligosaccharides is that the flexibility at the glycosidic linkage can lead to averaging of conformations and misleading NOE intensities. Here the importance of probing the dynamics using relaxation data, analysed with the 'model-free' approach (see Section 3.8), for example, is apparent. Although not yet used in this field to any extent, the iterative relaxation matrix used in combination with restrained molecular dynamics promises to be very powerful in the refinement of oligosaccharide structures.

9.7.2 Glycoproteins

The number of glycosylated proteins is very large, and yet very few have been completely characterized. This is in part because of the heterogeneity of glycosylation, which can vary from sample to sample and within one sample. Furthermore, attempts to determine the structure of the oligosaccharide *in situ* on the protein usually lead to very crowded spectra where even multidimensional NMR fails. Therefore, in general, the oligosaccharide and proteins segments are examined separately, the oligosaccharides having been cleaved off from the protein under mild conditions, and purified by HPLC. So far, this would appear to be the best approach for glycosylated proteins whose $M_r \leq 25$ kDa.

9.8 OUTLOOK

One of the limiting factors in the NMR of oligonucleotides has been the availability of isotopically labelled nucleotides, which have only recently been exploited along with multidimensional solution-state NMR methods for the determination of the structures of relatively short oligomers. As with proteins, the major limitation here is that as the biomolecule becomes large, coherence transfer pulse sequences fail, and no amount of higher dimensionality can help in resolving overlapping resonances, something which can be particularly severe with oligonucleotides due to the limited chemical-shift differences between the resonances of the common nucleotides. Here solid-state NMR distance measurement methods can be useful, although limited because of the current requirement to measure single pairwise distances between specifically isotopically enriched sites. One particular advantage of the latter method lies in the possibility of measuring long distances in short oligonucleotides, for which it is not possible to measure long-range solution-state distances, thus leading to uncertainties in the bias of particular structure refinement methods for generating spurious kinks in the structure. Long-range distances between key isotopically enriched sites can address this problem in an unequivocal way.

Oligosaccharide structure determination by NMR is still a relatively young field, and yet the prospects for solving the structures of relatively complex molecules are very good, again depending on the molecular weight. The problems encountered in this field include difficulties associated with obtaining pure oligosaccharides, whose purification becomes harder for more complex carbohydrates, and also an even more severe case of chemical-shift overlap problems than in the oligonucleotide field. So far there has been an insufficient number of applications of multidimensional NMR techniques to complex oligosaccharides to know with any certainty just what rôle NMR will play in structure determination.

REFERENCES

1. B. Jacobson, W. A. Anderson, and J. T. Arnold, *Nature*, **173**, 772–3 (1954).

2. J. D. Watson and F. H. C. Crick, *Nature*, **171**, 737 (1953).

3. K. Wüthrich, *NMR of Proteins and Nucleic Acids*, John Wiley and Sons, New York (1986).

4. A. Pardi, R. Walker, H. Rapoport, G. Wider, and K. Wüthrich, *J. Am. Chem. Soc.*, **105**, 1652 (1983).

5. S. A. Schroeder, J. M. Fu, C. R. Jones, and D. G. Gorenstein, *Biochemistry*, **26**, 3812 (1987).

6. J. M. Fu, S. A. Schroeder, C. R. Jones, R. Santini, and D. G. Gorenstein, *J. Magn. Reson.*, **77**, 577 (1988).

7. M. H. Frey, W. Leupin, O. W. Sørensen, W. A. Denney, R. R. Ernst, and K. Wüthrich, *Biopolymers*, **24**, 2371 (1985).

8. V. Sklenar, H. Miyashiro, G. Zon, H. T. Miles, and A. Bax, *FEBS Lett.*, **208**, 94 (1986).

9. M. G. Zagorski and D. G. Norman, *J. Magn. Reson.*, **83**, 167–72 (1989).

10. G. W. Kellog, *J. Magn. Reson.*, **98**, 176–82 (1992).

11. W. J. Chazin, K. Wüthrich, S. Hyberts, M. Rance, W. A. Denny, and W. Leupin, *J. Mol. Biol.*, **190**, 439–53 (1986).

12. H. Widmer and K. Wüthrich, *J. Magn. Reson.*, **74**, 316–36 (1987).

13. B. Celda, H. Widmer, W. Leupin, W. J. Chazin, W. A. Denny, and K. Wüthrich, *Biochemistry*, **28**, 1462–71 (1989).

14. M. Gochin and T. L. Kames, *Biochemistry*, **29**, 11172–80 (1990).

15. P. P. Lankhorst, C. A. G. Haasnoot, C. Erkelens, and C. Altona, *J. Biomol. Struct. Dyn.*, **1**, 1387 (1984).

16. V. Sklenar and A. Bax, *J. Am. Chem. Soc.*, **109**, 7525–6 (1987).

17. R. E. Dickerson and H. R. Drew, *J. Mol. Biol.*, **149**, 761 (1981).

18. M. P. Printz and P. H. von Hippel, *Proc. Natl. Acad. Sci. USA*, **53**, 363 (1965).

19. H. Teitelbaum and S. W. Englander, *J. Mol. Biol.*, **92**, 55, 79 (1975).

20. S. W. Englander and N. R. Kallenbach, *Quat. Rev. Biophys.*, **16**, 521–655 (1983).

21. J. L. Leroy, D. Broseta, and M. Guéron, *J. Mol. Biol.*, **184**, 165–78 (1985), and references cited therein.

22. M. Guéron, E. Charretier, M. Kochoyan, and J. L. Leroy, *Frontiers of NMR in Molecular Biology*, UCLA Symposia on Molecular and Cellular Biology (eds. D. Live. I. M. Armitage, and D. Patel), **109**, 225–38 (1990), and references cited therein.

23. J. F. Lefevre, A. N. Lane, and O. Jardetsky, *J. Mol. Biol.*, **185**, 689–99 (1985).

24. S. Cheung, K. Arndt, and P. Lu, *Proc. Natl. Acad. Sci. USA*, **81**, 3665–9 (1985).

25. J. L. Leroy, N. Bolo, N. Figueroa, P. Plateua, and M. Guéron, *J. Biomol. Str. Dyn.*, **2**, 915–39 (1985).

26. J. D. Puglisi, J. R. Wyatt, and I. Tinoco Jr, *J. Mol. Biol.*, **214**, 437–53 (1990).

27. M. J. Kime, *FEBS Lett.*, **173**, 342–6 (1984).

28. D. T. Gewirth, S. R. Abo, N. B. Leontis, and P. B. Moore, *Biochemistry*, **26**, 5213–20 (1987).

29. S. J. Opella and K. M. Morden, In *Dynamic Properties of Biomolecular Assemblies* (eds. S. E. Harding and A. J. Rowe), The Royal Society of Chemistry, Cambridge, p. 196 (1989).

30. S. Roy, M. Z. Papastavros, V. Sanchez, and A. G. Redfield, *Biochemistry*, **23**, 4395–400 (1984).

31. R. H. Griffey, D. Davis, Z. Yamaizumi, S. Nishimura, A. Bax, B. Hawkins *et al.*, *J. Biol. Chem.*, **260**, 9734–41 (1985).

32. S. Yokoyama, K. M. J. Usuki, Z. Yamaizumi, S. Nishimura, and T. Miyazawa, *FEBS Lett.*, **119**, 77–80 (1980).

33. J. I. Olsen, M. P. Schweizer, I. J. Walkiw, W. D. Hamill, W. J. Horton, and D. M. Grant, *Nucleic Acids Res.*, **10**, 4449–64 (1982).

34. E. P. Nikonowicz and A. Pardi, *Nature*, **355**, 184–6 (1992).

35. E. P. Nikonowicz and A. Pardi, *J. Am. Chem. Soc.*, **114**, 1082–3 (1992).

36. R. T. Batey, M. Inada, E. Kujawinski, J. D. Puglisi, and J. R. Williamson, *Nucleic Acids Res.*, **20**, 4515–23 (1992).

37. M. J. Michnicka, J. W. Harper, and G. C. King, *Biochemistry*, **32**, 395–400 (1993).

38. T. M. Alam and G. Drobny, *Chem. Rev.*, **91**, 1545–90 (1991).

39. H. Shindo and S. B. Zimmerman, *Nature*, **283**, 690–1 (1980).

40. H. Shindo, J. B. Wooten, B. H. Pheiffer, and S. B. Zimmerman, *Biochemistry*, **19**, 518 (1980).

41. H. Shindo, J. B. Wooten, and S. B. Zimmerman, *Biochemistry*, **20**, 745–50 (1981).

42. B. T. Nall, W. P. Rothwell, J. S. Waugh, and A. Rupprecht, *Biochemistry*, **20**, 1881–7 (1981).

43. R. Landridge, H. R. Wilson, C. W. Hooper, M. H. F. Wilkins, and L. D. Hamilton, *J. Mol. Biol.*, **3**, 19 (1960).

44. S. Arnott and D. W. L. Hukins, *Biochem. Biophys. Res. Commun.*, **47**, 1504 (1972).

45. S. J. Kohler and M. P. Klein, *Biochemistry*, **15**, 967 (1976).

46. P. Tang, R. A. Santos, and G. S. Harbison, *Adv. Magn. Reson.*, **13**, 225 (1990), and references cited therein.

47. J. A. DiVerdi and S. J. Opella, *J. Mol. Biol.*, **149**, 307 (1981).

48. P. Bendel, J. Murphy-Boesch, and T. L. James, *Biochem. Biophys. Acta*, **759**, 205 (1983).

49. R. R. Vold, R. Brandes, P. Tsang, D. R. Kearns, R. L. Vold, and A. Rupprecht, *J. Am. Chem. Soc.*, **108**, 302 (1986).

50. R. Brandes, R. R. Vold, R. L. Vold, and D. R. Kearns, *Biochemistry*, **25**, 7744 (1986).

51. R. Brandes, R. R. Vold, D. R. Kearns, and A. Rupprecht, *Biochemistry*, **29**, 1717 (1990), and references cited therein.

52. H. Shindo, Y. Hiyama, S. Roy, J. S. Cohen, and D. A. Torchia, *Bull. Chem. Soc. Japan*, **60**, 1631 (1987).

53. A. Kintanar, T. M. Alam, W. -C. Wang, D. C. Schindele, D. E. Wemmer, and G. P. Drobny, *J. Am. Chem. Soc.*, **110**, 6367 (1988).

54. T. M. Alam and G. P. Drobny, *J. Chem. Phys.*, **92**, 6840 (1990).

55. W. Nerdal, D. R. Hare, and B. R. Reid, *Biochemistry*, **28**, 10008 (1989).

56. R. Wing, H. Drew, T. Takano, C. Broka, S. Tanaka, K. Itakura *et al.*, *Nature*, **287**, 755–8 (1980).

57. B. R. Reid, *Quat. Rev. Biophys.*, **20**, 1–34 (1987).

58. D. J. Patel, L. Shapiro, and D. Hare, *Quat. Rev. Biophys.*, **20**, 35–112 (1987).

59. E. R. P. Zuiderweg, R. M. Scheek, G. Veeneman, J. H. van Boom, R. Kaptein, H. Rüterjans *et al.*, *Nucl. Acids Res.*, **9**, 6553–69 (1981).

60. R. M. Scheek, N. Russo, R. Boelens, and R. Kaptein, *J. Am. Chem. Soc.*, **105**, 2914–26 (1983).

61. R. M. Scheek, R. Boelens, N. Russo, J. H. van Boom, and R. Kaptein, *Biochemistry*, **23**, 1371–6 (1984).

62. R. Boelens, R. M. Scheek, K. Dijkstra, and R. Kaptein, *J. Magn. Reson.*, **62**, 378–86 (1985).

63. J. A. McClarin, C. A. Frederick, B. -C. Wang, P. Greene, H. W. Boyer, J. Grable *et al.*, *Science*, **234**, 1526–41 (1986).

64. J. Feigon, A. H. -J. Wang, G. A. van der Marel, J. H. van Boom, and A. Rich, *Science*, **230**, 82–4 (1985).

65. M. Coll, A. H. -J. Wang, G. A. van der Marel, J. H. van Boom, and A. Rich, *J. Biomol. Struct. Dyn.*, **4**, 157–72 (1986).

66. D. R. Hare, D. E. Wemmer, S. -H. Chou, G. Drobny, and B. R. Reid, *J. Mol. Biol.*, **171**, 319–36 (1983).

67. D. J. Patel, S. Ikuta, S. Kozlowski, and K. Itakura, *Proc. Natl. Acad. Sci. USA*, **80**, 2184–8 (1983).

68. S. Swaminathan, G. Ravishanker, and D. L. Beverisge, *J. Am. Chem. Soc.*, **113**, 5027–40 (1991).

69. R. Stolarski, W. Egan, and T. L. James, *Biochemistry*, **31**, 7027–942 (1992).

70. G. M. Clore and A. M. Gronenborn, *J. Magn. Reson.*, **84**, 398–409 (1989).

71. G. M. Clore and A. M. Gronenborn, *Crit. Rev. in Biochem. and Mol. Biol.*, **24**, 479–564 (1989).

72. B. R. Reid, K. Banks, P. Flynn, and W. Nerdal, *Biochemistry*, **28**, 10001–7 (1989), and references cited therein.

73. W. J. Metzler, C. Wang, D. B. Kitchen, R. M. Levy, and A. Pardi, *J. Mol. Biol.*, **214**, 711–36 (1990).

74. U. Schmitz, D. A. Pearlman, and T. L. James, *J. Mol. Biol.*, **221**, 271–92 (1991).

75. E. P. Nikonowicz and D. G. Gorenstein, *J. Am. Chem. Soc.*, **114**, 7494–503 (1992).

76. H. Robinson and A. H. -J. Wang, *Biochemistry*, **31**, 3542–33 (1992).

77. S. Ikuta, R. Chattopadhyaya, H. Ito, R. E.

Dickerson, and D. R. Kearns, *Biochemistry*, **25**, 4840–9 (1986).

78. D. R. Hare and B. R. Reid, *Biochemistry*, **25**, 5341–50 (1986).

79. J. R. Williamson and S. G. Boxer, *Biochemistry*, **28**, 2819–31, 2831–6 (1989).

80. M. J. J. Blommers, F. J. M. van de Ven, G. A. van der Marel, J. H. van Boom, and C. W. Hilbers, *Eur. J. Biochem.*, **201**, 33–51 (1991).

81. S. Arnott and E. Selsing, *J. Mol. Biol.*, **88**, 509–21 (1974).

82. H. E. Moser and P. B. Dervan, *Science*, **238**, 645–50 (1987).

83. P. Rajagopal and J. Feigon, *Nature*, **28**, 7859–70 (1989).

84. C. de los Santos, M. Rosen, and D. J. Patel. *Biochemistry*, **28**, 7282–9 (1989).

85. I. Radhakrishnan, D. J. Patel, and X. Gao, *Biochemistry*, **31**, 2514–23 (1992).

86. R. Macaya, E. Wong, P. Schultze, V. Sklenar, and J. Feigon, *J. Mol. Biol.*, **225**, 755–73 (1992).

87. Y. Wang, C. de los Santos, X. Gao, K. Greene, D. Live, and D. J. Patel, *J. Mol. Biol.*, **222**, 819–32 (1991).

88. F. W. Smith and J. Feigon, *Nature*, **356**, 164–6 (1992).

89. F. Aboul-ela, A. I. H. Murchie, and D. M. J. Lilley, *Nature*, **360**, 280–2 (1992).

90. K. Gehring, J. -L. Leroy, and M. Guéron, *Nature*, **363**, 561–5 (1993).

91. N. R. Kallenbach, D. W. Appleby, and C. H. Bradley, *Nature*, **272**, 134–8 (1978).

92. C. T. McMuray, K. E. van Holde, R. L. Jones, and W. D. Wilson, *Biochemistry*, **24**, 7037–44 (1985).

93. R. E. Klevit, D. E. Wemmer, and B. R. Reid, *Biochemistry*, **25**, 3296–303 (1986).

94. J. G. Pelton and D. E. Wemmer, *Biochemistry*, **27**, 8088–96 (1988).

95. J. G. Pelton and D. E. Wemmer, *Proc. Natl. Acad. Sci. USA*, **86**, 5723–7 (1989).

96. J. G. Pelton and D. E. Wemmer, *J. Am. Chem. Soc.*, **112**, 1393–9 (1990).

97. M. T. Tomic, D. E. Wemmer, and S. -H. Kim, *Science*, **238**, 1722–5 (1987).

98. R. L. Jones, E. V. Scott, G. Zon. L. G. Marzilli, and W. D. Wilson, *Biochemistry*, **27**, 6021–6 (1988).

99. H. Robinson, Y. -C. Liaw, G. A. van der Marel, J. H. van Boom, and A. H. -J. Wang, *Nucl. Acids Re.*, **18**, 4851–8 (1990).

100. X Zhang and D. J. Patel, *Biochemistry*, **30**, 4026–41 (1991).

101. D. E. Gilbert and J. Feigon, *Biochemistry*, **30**, 2483–94 (1991).

102. A. Fede, A. Labhardt, W. Bannwarth, and W. Leupin, *Biochemistry*, **30**, 11377–88 (1991).

103. X. Gao and D. J. Patel, *Biochemistry*, **28**, 751–62 (1989).

104. X. Gao, P. Mirau, and D. J. Patel, *J. Mol. Biol.*, **223**, 259–79 (1992).

105. S. Hu, K. Weisz, T. L. James, and R. H. Shafer, *Eur. J. Biochem.*, **204**, 31–8 (1992).

106. M. Mrksich, W. S. Wade, T. J. Dwyer, B. H. Geierstanger, D. E. Wemmer, and P. B. Dervan, *Proc. Natl. Acad. Sci. USA*, **89**, 7586–90 (1992).

107. S. -M. Chen, W. Leupin, M. Rance, and W. J. Chazin, *Biochemistry*, **31**, 4406–13 (1992).

108. P. Tang, C. -L. Jong, and G. S. Harbison, *Science*, **249**, 70–2 (1990).

109. C. H. Arrowsmith, R. Pachter, R. B. Altman, S. B. Iyer, and O. Jardetsky, *Biochemistry*, **29**, 6332–41 (1990).

110. C. H. Arrowsmith, R. Pachter, R. B. Altman, and O. Jardetsky, *Eur. J. Biochem.*, **202**, 53–66 (1991).

111. Z. Otwinowski, R. W. Schevitz, R. -G. Zhang, C. L. Lawson, A. Joachimiak, R. Q. Marmorstein, *et al.*, *Nature*, **335**, 321–9 (1988).

112. D. Voet and J. G. Voet, *Biochemistry*, John Wiley and Sons, New York, p. 873 (1990).

113. B. W. Matthews, *Nature*, **335**, 294–5 (1988).

114. R. G. Brennan and B. W. Matthews, *Trends in Biochem. Sci.*, **14**, 286–90 (1989).

115. A. A. Kumamoto, W. G. Miller, and R. P. Gunsalus, *Genes and Dev.*, **1**, 556–64 (1987).

116. D. Staake, B. Walter, B. Kisters-Woike, B. von Wilcken-Bergmann, and B. Müller-Hill, *EMBO J.*, **9**, 1963–7 and corrigendum p. 3023 (1990).

117. J. B. Rafferty, W. S. Somers, I. Saint-Girons, and S. E. V. Phillips, *Nature*, **341**, 705–10 (1989).

118. S. E. V. Phillips, I. Manfield, I. Parsons, B. E. Davison, J. B. Rafferty, W. S. Somers *et al.*, *Nature*, **341**, 711–15 (1989).

119. B. Luisi and P. B. Sigler, *Biochim. Biophys. Acta*, **1048**, 113–26 (1990).

120. S. Bass, V. Sorrels, and P. Youderian, *Science*, **242**, 240–5 (1988).

121. G. Otting, Y.-Q. Qian, M. Müller, M. Affolter, W. Gehring, and K. Wüthrich, *EMBO J.*, 7, 4305–9 (1988).

122. Y.-Q Qian, M. Billeter, G. Otting, M. Müller, W. J. Gehring, and K. Wüthrich, *Cell*, **59**, 573–80 (1989), and corrigendum, **61**, p. 548 (1990).

123. M. Billeter, Y.-Q. Qian, G. Otting, M. Müller, W. J. Gehring, and K. Wüthrich, *J. Mol. Biol.*, **214**, 183–97 (1990).

124. G. Otting, Y.-Q. Qian, M. Billeter, M. Müller, M. Affolter, W. J. Gehring *et al.*, *EMBO J.*, **9**, 3085–92 (1990).

125. D. Neri, M. Billeter, and K. Wüthrich, *J. Mol. Biol.*, **223**, 743–67 (1992).

126. A. Mondragon, S. Subbiah, S. C. Almo, M. Drottar, and S. C. Harrison, *J. Mol. Biol.*, **205**, 189–201 (1989).

127. M. S. Lee, J. Cavanagh, and P. E. Wright, *FEBS Lett.*, **254**, 159–64 (1989).

128. M. S. Lee, G. P. Gippert, K. V. Soman, D. A. Case, and P. E. Wright, *Science*, **245**, 635–7 (1989).

129. R. E. Klevit, J. R. Herriott, and S. J. Horvath, *Proteins: Struct., Funct. and Genet.*, 7, 215–26 (1990).

130. J. G. Omichinski, G. M. Clore, E. Apella, K. Sakaguchi, and A. M. Gronenborn, *Biochemistry*, **29**, 9324–34 (1990).

131. M. D. Carr, A. Pastore, H. Gausepohl, R. Frank, and P. Rösch, *Eur. J. Biochem.*, **188**, 455–61 (1990).

132. T. Haard, E. Kallenbach, R. Boelens, B. A. Maler, K. Dahlman *et al.*, *Science*, **249**, 157–60 (1990).

133. M. L. Remerowski, E. Kellenbach, R. Boelens, G. A. van der Marel, J. H. van Boom, B. A. Maler *et al.*, *Biochemistry*, **30**, 11620–4 (1991).

134. J. W. R. Schwabe, D. Neuhaus, and D. Rhodes, *Nature*, **348**, 458–61 (1990).

135. B. Luisi *et al.*, *Nature*, **352**, 487–505 (1991).

136. M. E. A. Churchill, T. D. Tullius, and A. Klug, *Proc. Natl. Acad. Sci. USA*, **87**, 5528–32 (1990).

137. D. Neuhaus, Y. Nakeseko, K. Nagai, and A. Klug, *FEBS Lett.*, **262**, 179–84 (1990).

138. N. P. Pavletich and C. O. Pabo, *Science*, **252**, 809–17 (1991).

139. M. Kochoyan, T. F. Havel, D. T. Nguyen, C. E. Dahl, H. T. Keutmann, and M. A. Weiss, *Biochemistry*, **30**, 3371–86 (1991).

140. X. Qian and M. A. Weiss, *Biochemistry*, **31**, 7463–76 (1992), and references cited therein.

141. E. K. O'Shea, R. Rutkowski, and P. S. Kim, *Science*, **243**, 538–42 (1989).

142. E. K. O'Shea, R. Rutkowski, W. F. Staford III, and P. S. Kim, *Science*, **245**, 646–8 (1989).

143. T. Tabate, H. Takase, S. Takayama, K. Mikami, A. Nakatsuka *et al.*, *Science*, **245**, 965–71 (1989).

144. T. G. Oas, L. P. McIntosh, E. K. O'Shea, F. W. Dahlquist, and P. S. Kim, *Biochemistry*, **29**, 2891–4 (1990).

145. E. M. Goodman and P. S. Kim, *Biochemistry*, **30**, 11615–20 (1991).

146. M. A. Weiss, T. Ellenberger, C. R. Wobbe, J. P. Lee, S. C. Harrison, and K. Struhl, *Nature*, **347**, 575–8 (1990).

147. E. K. O'Shea, J. D. Klemm, P. S. Kim, and T. Alber, *Science*, **254**, 539 (1991).

148. J. N. Breg, J. H. J. van Opheusden, M. J. M. Burgering, R. Boelens, and R. Kaptein, *Nature*, **346**, 586–9 (1990).

149. T. Pan and J. E. Coleman, *Proc. Natl. Acad. Sci. USA*, **87**, 2077 (1990).

150. T. Pan and J. E. Coleman, *Biochemistry*, **30**, 4212–22 (1991).

151. J. D. Baleja, R. Marmorstein, S. C. Harrison, and G. Wagner, *Nature*, **356**, 450–3 (1992).

152. P. Kraulis, A. R. C. Raine, P. L. Gadhavi, and E. D. Laue, *Nature*, **356**, 448–50 (1992).

153. M. Shirakawa, W. J. Fairbrother, Y. Serikawa, T. Ohkubo, Y. Kyogoku, and P. E. Wright, *Biochemistry*, **32**, 2144–53 (1993).

154. R. Marmorstein, M. Carey, M. Ptashne, and S. C. Harrison, *Nature*, **356**, 408–14 (1992).

155. R. Kaptein and R. Boelens, In *Nucleic Acids and Molecular Biology* (ed. F. Eckstein and D. M. J. Lilley), Vol. 2, Springer-Verlag, Berlin, pp. 167–87 (1988).

156. R. M. J. N. Lamerichs, R. Boelens, G. A. van der Marel, J. H. van Boom, R. Kaptein, F. Buck *et al.*, *Biochemistry*, **28**, 2985–91 (1989).

157. G. Varani and I. Tinoco Jr., *Quat. Rev. Biophys.*, **24**, 479–532 (1991).

158. C. A. G. Haasnoot, C. W. Hilbers, G. A. van der Marel, J. H. van Boom, U. C. Singh

N. Pattabiraman *et al.*, *J. Biomol. Struct. Dyn.*, **3**, 843–57 (1986).

159. P. R. Schimmel and A. G. Redfield, *Ann. Rev. Biophys. Bioeng.*, **9**, 181–221 (1980), and references cited therein.

160. D. R. Kearns, D. J. Patel, and R. G. Schulman, *Nature*, **229**, 338–9 (1971).

161. B. R. Reid, N. S. Ribeiro, L. McCollum, J. Abbate, and R. E. Hurd, *Biochemistry*, **16**, 2086–94 (1977).

162. C. Cheong, G. Varanu, and I. Tinoco Jr, *Nature*, **346**, 680–2 (1990).

163. E. P. Nikonowicz and A. Pardi, *Nature*, **355**, 184–6 (1992).

164. E. P. Nikonowicz and A. Pardi, *J. Am. Chem. Soc.*, **114**, 1083–4 (1992).

165. C. Cheong and P. B. Moore, *Biochemistry*, **31**, 8406–14 (1992).

166. J. D. Puglisi, R. Tan, B. J. Calnan, A. D. Frankel, and J. R. Williamson, *Science*, **257**, 76–80 (1992).

167. M. Görlach, M. Wittekind, R. A. Beckman, L. Müller, and G. Dreyfuss, *EMBO J.*, **11**, 3289–95 (1992).

168. T. W. Rademacher, R. B. Parekh, and R. A. Dwek, *Annu. Rev. Biochem.*, **57**, 785 (1988).

169. H. van Halbeck and L. Poppe, *Magn. Reson. Chem.*, Special Issue, **30**, S74–S86 (1992).

170. C. A. Bush and P. Cagas, In *Advances in Biophysical Chemistry* (ed. C. A. Bush), JAI Press, Greenwich, Connecticut, Vol. 2, p. 149 (1992).

171. S. W. Homans, *Prog. NMR Spectroscopy*, **22**, 55–81 (1990).

172. P. de Waard, B. R. Leeflang, J. F. G. Vliegnthart, R. Boelens, G. W. Vuister, and R. Kaptein, *J. Biomol. NMR*, **2**, 211–26 (1992).

173. G. W. Vuister, P. de Waard, R. Boelens, J. F. G. Vliegenthart, and R. Kaptein, *J. Am. Chem. Soc.*, **111**, 772–4 (1989).

Part V · *Membranes*

10 · Membranes and membrane proteins

Membranes provide a vital interface between a biological organism and its environment. Furthermore, they provide a means for the formation of compartments within a cell. In every sense, then, they are vital to living systems. Indeed, membranes are also the location where a large number of cellular processes occur, and comprise a complex mixture of lipids and proteins to which carbohydrates may be covalently attached (see Fig. 10.1),[1] known as the *fluid mosaic model*.[2] Since membranes and membrane proteins cannot easily be crystallized, NMR spectroscopy is a particularly powerful method for probing their structure and function.[3,4] In this chapter, we will outline the ways in which multinuclear NMR, in particular ^{2}H and ^{31}P, to a lesser extent ^{13}C, and rarely ^{23}Na or ^{14}N NMR, has been used to determine with some accuracy the structure of lipids in a bilayer. We will also consider some examples of proteins and peptides whose structures have been determined in the presence of micelles or bilayers.

10.1 MULTINUCLEAR NMR OF PHOSPHOLIPIDS

One of the key features of lipids in membranes is their dynamic behaviour. This can be probed using ^{2}H NMR, with which, as we saw in Section 1.4.3, a lineshape analysis can be very revealing regarding orientation with respect to the magnetic field, and molecular motion.[5,6] Fig. 10.2 shows the ^{2}H NMR spectra for a mixture of lipids deuterated in their headgroups, and dispersed in buffer. The bilayers align with respect to the magnetic field,[7,8] and display characteristic rotational symmetry in the spectrum. Upon vortexing the sample, and introducing heterogeneity, the spectrum looks much more like a static powder pattern, in which the bilayer isochromats are aligned randomly, and all possible orientations are detected in the NMR experiment. The ordering of the lipid with respect to the magnetic field can be achieved in one of two ways. First, lipid can be macroscopically oriented by pressing lipid-water dispersions between flat glass plates,[9] thus orienting the membrane microdomains by mechanical shearing forces[10] (see Fig. 10.3). Second, lipids that align spontaneously with respect to the magnetic field provide an even simpler way of orienting membranes. This macroscopic ordering occurs because the fatty acyl chains have a negative diamagnetic anisotropy leading to an orientation of the membrane microdomains such that the planes of the membrane are parallel to the magnetic field.

The quadrupolar splitting, Δv_Q, which is given by the separation of the two resonances

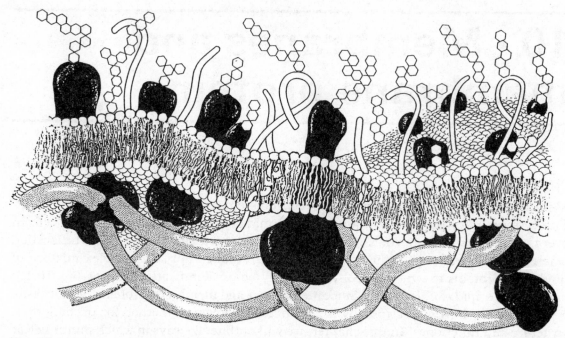

Fig. 10.1 Schematic illustration of a eukaryotic cell membrane, which highlights the membrane as a stratified composite of a fluid lipid bilayer sandwiched between the carbohydrate glycocalyx on the outside and the cytoskeleton on the inside. Integral membrane proteins and polypeptides are indicated as large molecular objects penetrating the lipid bilayer matrix. (Reprinted from ref. 1 with permission.)

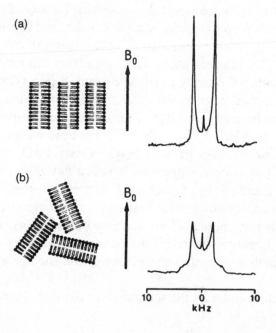

Fig. 10.2 ^{2}H NMR spectra of a mixture of synthetic lipids: 83% 1-palmitoyl-2-oleoyl-*sn*-glycero-3-phosphoethanolamine (POPE) and 17% [*sn*-2′-^{2}H]1-palmitoyl-2-oleoyl-*sn*-glycero-3-phosphoglycerol (POPG) dispersed in buffer with (a) sample oriented by magnetic field; and (b) sample (a) after vortexing. (Reprinted from ref. 5 with permission.)

in Fig. 10.2, can be expressed in terms of a deuterium order parameters, S_{CD}:

$$\Delta v_{Q} = \tfrac{3}{4}\left(\frac{e^2qQ}{h}\right)S_{CD} \tag{10.1}$$

where e^2qQ/h is the static quadrupolar coupling constant that we encountered in Section 1.2.5, typically 170 kHz for an aliphatic C—D bond. Equation (10.1) takes into account the membrane ordering depicted in Fig. 10.2, and the order parameter S_{CD} is an alternative way of expressing the angle θ of the C—D bond internuclear vector with respect to the magnetic field (see Equation (1.64)). Note that the value of the quadrupolar splitting does not change in the two spectra in Fig. 10.2, indicating that powder patterns, or Pake doublets as they are also called (see Section 1.2.5), also provide information on the order parameter S_{CD}.

The linewidth in ^{2}H spectra of oriented membranes is fairly small, indicating rapid rotational and translational motions of the lipid molecules. This can be analysed quantitatively through analysis of the transverse relaxation time T_1 which is related to the molecular rotational correlation time and the order parameter as follows:

$$(1/T_1) = \left(\frac{3\pi^2}{2}\right)\left(\frac{e^2qQ}{h}\right)^2[1 + \tfrac{1}{2}S_{CD} - \tfrac{3}{2}S_{CD}{}^2]\tau_c \tag{10.2}$$

Equation (10.2) is, to a first approximation, valid for the fast correlation time regime $(\tau_c < 10^{-8}$ s), assuming a single correlation time. The last term in parentheses contains the order parameter correction.[11]

Deuterium T_1 relaxation times are short because of the large quadrupole moment (see Table 1.2) which determines the relaxation. For pure D_2O, the $T_1 \approx 400$ ms, which in the presence of lipids reduces to 10–100 ms. The T_1 values for the aliphatic segments of fatty acyl chains vary between 5 and 200 ms.[3] However, in addition to the fast anisotropic motions of lipids, proteins in membranes are subject to slow motions and the lineshape

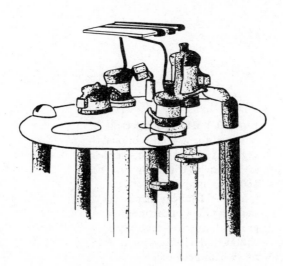

Fig. 10.3 Flat-coil probe designed for NMR of oriented bilayers. (Reprinted from ref. 9 with permission.)

can be analysed to infer the rate of any motion present.[12] Fig. 10.4 shows three different samples with differing rotational motions displaying the dramatic changes in lineshape in response to the motions. The quadrupolar splitting varies from a maximum value of 128 kHz to much smaller splittings as the motion goes from a two-site model, such as 180° ring flips,[13] to a continuous rotation.

Deuterium labelling can be carried out either in the fatty acid aliphatic chains, or in the headgroups, thereby providing a means to probe different motions and orientational dependencies. However, such specific labelling can prove to be relatively expensive, and ^{31}P of phospholipid headgroups, or, in some cases, specific ^{13}C-labelling, can provide alternative and additional sources of information. Since the natural abundance of ^{31}P is 100% and $I = \frac{1}{2}$, use of this nucleus is relatively simple, and its relatively large chemical-shift anisotropy and ^{1}H—^{31}P dipolar coupling can be used to determine motion and structure[14,15] (see Fig. 10.5). The effect of chemical-shift anisotropy on the phosphodiester headgroup is shown in Fig. 10.6.

From our consideration of shielding anisotropy in Section 1.2.4, we can define a reduced value from Equation (1.61) when in the presence of molecular motion:

$$\Delta\sigma = \sigma'_{\parallel} - \sigma'_{\perp} \tag{10.3}$$

where $\sigma'_{\parallel}$ and $\sigma'_{\perp}$ are the parallel and perpendicular components of the chemical-shift tensor in the presence of motion, and $\Delta\sigma$ is reduced in proportion to the amplitude of the motion. The effect of rotational diffusion $R_{\parallel}$ about an axis parallel to the magnetic field on the simulated ^{31}P NMR lineshape[16] is illustrated in Fig. 10.7. The simulations assume a Brownian diffusion model for axially symmetric motions, whereas other models are possible, such as free diffusion or discrete-jump diffusion.

The anisotropic motions of membranes reflect the symmetry of the membrane phases. Lamellar gel-to-fluid, lamellar fluid-to-hexagonal, and lamellar fluid-to-isotropic

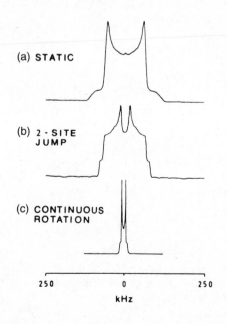

(a) STATIC

(b) 2-SITE JUMP

(c) CONTINUOUS ROTATION

250   0   250

kHz

Fig. 10.4 ^{2}H NMR spectra obtained for various motional regîmes, with (a) static sample of crystalline [$^{2}H_5$]tryptophan; (b) 180° ring flipping for [phenyl-$^{2}H_5$]phenylalanine; (c) rapid rotation of the methyl groups in methyl-deuterated choline iodide. (Reprinted from ref. 5 with permission.)

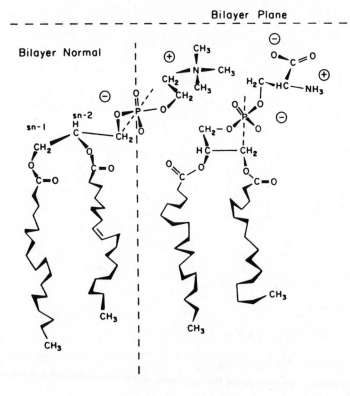

Fig. 10.5 Representation of some typical lipids found in biological membranes; the headgroups shown are phosphatidylcholine and phosphatidylserine. (Reprinted from ref. 15 with permission.)

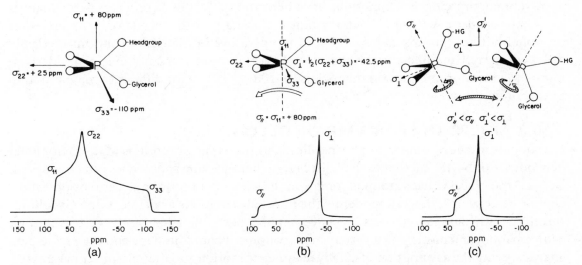

Fig. 10.6 Various possible motional states for the phosphodiester moety of a membrane lipid and the expected ^{31}P NMR spectra, with the phosphodiester (a) static; (b) ordered with rapid axial rotation; and (c) disordered with rapid axial rotation. In the case of motional averaging of the chemical-shift tensor due to axial symmetry, $\sigma_\parallel$ refers to the chemical shift for the external magnetic field parallel to the unique axis, and $\sigma_\perp$ to that for the field in the equatorial plane. (Reprinted from ref. 15 with permission.)

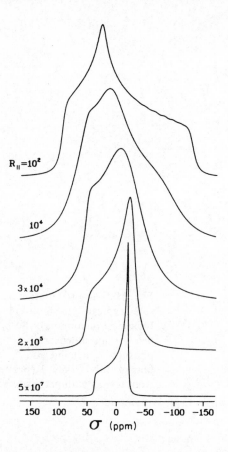

$R_\parallel = 10^2$

10^4

3×10^4

2×10^5

5×10^7

150 100 50 0 −50 −100 −150

σ (ppm)

Fig. 10.7 Simulated ^{31}P NMR spectra for anisotropic Brownian diffusion. The rates of diffusion $R_\parallel$ are given for each spectrum, $R_\perp = 15$ rad s^{-1}. (Reprinted from ref. 15 with permission.)

membrane transitions have been detected as a function of temperature and effectors, such as membrane proteins. Such motions have been the subject of a comprehensive analysis by Kothe and co-workers,[17] who studied lineshape and relaxation properties of phospholipid headgroups as a function of temperature in unoriented and magnetically oriented dimyristoylphosphatidylcholine (DMPC) model membranes, and derived motional activation energies (in the region 10–100 kJ mol^{-1}) for different types of motion in each lipid phase.

10.2 MULTINUCLEAR NMR OF GLYCOLIPIDS

Glycolipids are very similar to phospholipids in the types of motions and orientational dependence of NMR lineshapes they display, so that the same approaches can be adopted. Because the glycosyl headgroup is somewhat bulkier than phosphoglycerol or phosphocholine headgroups, this can moderate the molecular motions available to the glycolipid. Jarrell, Smith, and co-workers[18–21] have carried out a detailed study of orientation, conformation, and motion on a variety of glycolipids. From lineshape and relaxation data analysis, using the assumption of axially symmetric motion and a single-order parameter, the motion of the glycosyl ring was calculated. Although the assumption of a single-order parameter appears to be valid, Sanders and Prestegard[22] carried out an elegant study of the [^{2}H]- or [^{13}C]-labelled glycolipid analogue β-D-dodecyl glucopyranoside (BDOG) in a phospholipid bilayer comprising DMPC and 3-[(cholamidopropyl)dimethylammonio]-

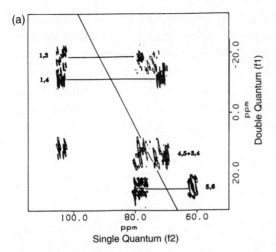 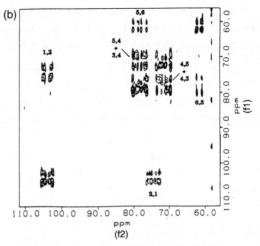

Fig. 10.8 ^{1}H-decoupled ^{13}C NMR experiments of ^{13}C$_6$-BDOG in 1:3 CHAPSO/DMPC by (a) double-quantum correlation (INADEQUATE); (b) DQF–COSY. Labelled cross-peaks indicate glucose carbons. (Reprinted from ref. 22 with permission.)

2-hydroxy-1-propanesulfonate (CHAPSO), employing a full generalized-order parameter matrix analysis. This bilayer forms a magnetically orientable membrane system, whose orientation relative to the magnetic field can be precisely controlled[23] by the ratio of CHAPSO:DMPC. Using ^{2}H and ^{13}C one- and two-dimensional NMR, the two quadrupolar coupling constants for the glucose ring deuterons, five ^{13}C—^{1}H dipolar splittings, and six ^{13}C—^{13}C dipolar splittings were determined. The latter two sets of splittings were measured first by assigning the ^{13}C resonances using a ^{1}H-decoupled $\{^{13}$C, ^{13}C$\}$ DQF–COSY (see Fig. 10.8(a)), and then using a double-quantum coherence pulse sequence (INADEQUATE) (see Fig. 10.8(b)). With the splittings that were measured, the generalized order matrix analysis was carried out followed by AMBER energy minimization. The final structure is shown in Fig. 10.9, which is gratifyingly similar to the simpler single-order parameter analyses. The limitation of this magnetically orientable membrane system for the study of membrane-protein interactions is that CHAPSO is a detergent which could cause partial unfolding of proteins. Sanders has applied the same approaches[24] as were used in the CHAPSO study to characterize mixtures of short-chain and long-chain phosphatidylcholine lipids (DMPC and DHPC). This builds on work by Gabriel and Roberts,[25] who characterized such mixtures below the gel-to-liquid crystalline phase transition.

The structural characterization of the oligosaccharide segment of glycolipids is carried out in a similar manner to oligosaccharides from glycoproteins (see Section 9.7). Usually they are cleaved from the lipid headgroup and analysed by the methods already outlined in Section 9.7.1, for example as outlined in the review by Inagaki.[26]

10.3 STRUCTURE OF MEMBRANE PROTEINS

In a casual survey of the 5000 or so known proteins, the average M_r is centred on about 100 kDa, with a substantial proportion being membrane-associated. Since only about 300 of those have been examined by X-ray crystallography, of which only a handful are membrane proteins, this means that we are basing our understanding on the relationship between protein structure and function on a sample size of less than 10%. This might be statistically significant if a *random* selection of those 3000 enzymes had been crystallized, but, generally, soluble globular proteins with well-defined structures are those which crystallize. Therefore we have only just begun to scratch at the surface of an understanding of the relationship between structure and function. NMR spectroscopy is unique in its capacity to bridge the structural gap between aqueous isotropic solution and

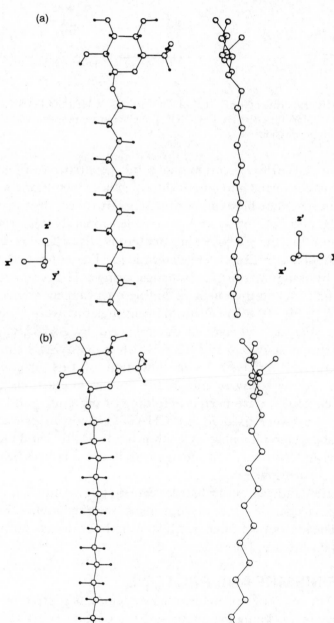

Fig. 10.9 Structures of BDOG generated with order matrix (a) and AMBER pseudo-energy (b) calculations. Hydrogen atoms have been omitted from the view along the approximate plane of the glucose ring for the sake of clarity. (Reprinted from ref. 22 with permission.)

anisotropic solid-like membranes. In the study of membrane proteins, both solution-state and solid-state NMR have been applied in Man's quest for structures at the interface.

There have been a few high-resolution solution-state NMR studies of membrane proteins solubilized in detergent micelles or organic solvents, and their structures have been determined. Examples include melittin,[27-30] δ-haemolysin,[31,32] lipopeptide,[33] murine epidermal growth factor,[34] M13 coat protein,[35] and the *c* subunit of the F_0 portion of F_1F_0 ATPase.[36] We will not consider these in detail in this chapter, since they represent conventional protein structure determinations (see Chapter 4), with the protein solubilized in a deuterated detergent such as sodium dodecyl sulphate (SDS) or deuterated solvents such as methanol. Once optimal conditions for the sample preparation have been achieved, conventional protein structure determination strategies can be adopted. Another study by Higashijima and co-workers[37] used transferred NOEs to measure the conformation of biologically active and inactive enkephalins when bound to a perdeuterated lipid bilayer in the form of liposomes. They found that the biologically inactive form adopted a completely different conformation in the presence of bilayers than did the active forms.

This section will concentrate on the currently less common solid-state NMR methods for the structural and dynamical investigation of peptides and proteins in lipid bilayers. Despite the large untapped resource of knowledge and understanding that will undoubtedly result from detailed studies of membrane proteins by solid-state NMR spectroscopy, this field has yet to catch on to the same extent as solution-state NMR. Advocates of the latter would be quick to point out that solid-state NMR has determined relatively few complete structures. The following examples demonstrate that the *potential* is there: it has just not been fully realized yet.

10.3.1 Effect of protein–membrane interactions on lipid bilayers

Lipids and proteins can interact by a number of mechanisms. (1) Proteins may bind electrostatically to the membrane surface, i.e. to the polar lipid headgroups. (2) Proteins may penetrate into the hydrophobic part of the membrane, influence the ordering and packing of the hydrocarbon chains, and vary the membrane thickness. (3) Proteins may modulate the long-range organization of lipids either by inducing non-bilayer types of structures or by stabilizing non-bilayer lipids into a bilayer structure. The precise details of these interactions are poorly understood, although it appears that the lipids at the interface between the membrane and the protein, the so-called boundary lipids, while thought to be 'immobilized' from EPR studies, display the absence of static lineshapes in the NMR spectra, leading to the hypothesis that the boundary lipids are in fast exchange with the bulk lipids of the bilayer.

The effect of added protein to membranes composed of either phospholipids or glycolipids can be monitored by examining the ^2H or ^{31}P NMR spectra, and analysing lineshape and relaxation parameters in a manner exactly analogous to the methods outlined in Sections 10.1 and 10.2. An example of this is illustrated in Fig. 10.10, in which the theoretical and experimental ^2H spectra of deuterated DMPC are compared with those obtained with added cytochrome *c* oxidase, sarcoplasmic reticulum ATPase, beef

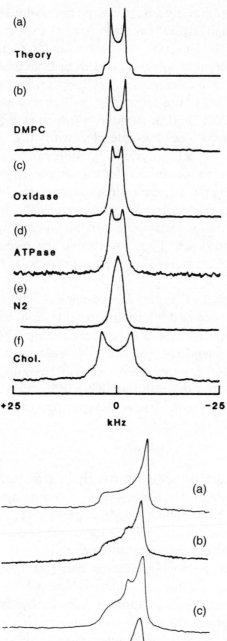

Fig. 10.10 Theoretical and experimental ^2H NMR quadrupole echo spectra of ^2H-labelled lipids, showing the effects of proteins and cholesterol: (a) theoretical ^2H powder pattern; (b) deuterated dimyristoylphosphatidylcholine (DMPC-d_3) in excess water; (c) sample (b) with 67% cytochrome c oxidase; (d) sample (b) with 67% sarcoplasmic reticulum ATPase; (e) sample (b) with 67% beef brain myelin proteolipid apoprotein (N2); and (f) sample (b) with 33% cholesterol. (Reprinted from ref. 6 with permission.)

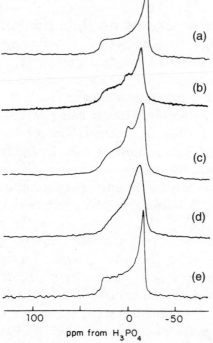

Fig. 10.11 ^1H-decoupled 60.7 MHz ^{31}P NMR spectra of pure dimyristoylphosphatidylcholine (DMPC) and of protein or cholesterol-containing complexes in excess water at room temperature: (a) pure DMPC; (b) sample (a) with 80% cytochrome c oxidase; (c) sample (a) with 70% sarcoplasmic reticulum ATPase; (d) sample (a) with 70% human lipophilin (N2 protein); and (e) sample (a) with 25% cholesterol. (Reprinted from ref. 38 with permission.)

brain myelin proteolipid apoprotein (N2), and the sterol cholesterol. A similar effect was seen on the lineshape of the ^{31}P NMR spectra (see Fig. 10.11).[38]

As we shall see shortly, the conformation of membrane proteins can be expected to be affected by the presence of the lipid bilayer. By the same token, the protein is also affected. There have been relatively few NMR studies attempting to correlate the interaction between protein and membrane by monitoring conformational and dynamical changes in both, but Marsh and co-workers[39] have carried out a nice study of cytochrome *c*-lipid interactions. They used ^{31}P NMR to follow the changes in an admixture of dioleoylglycerol or dioleoylphosphatidylcholine with dioleoylphosphatidylglycerol in the presence of cytochrome *c*, which was monitored by resonance Raman spectroscopy. It was found that conformational and coordination shifts in the protein accompany binding of cytochrome *c* to the lipid bilayer, and that the protein induces or increases the content of near isotropically diffusing lipid. When the lipid composition was changed with added dioleoylglycerol, the bilayer curvature of dioleoylphosphatidylglycerol increased, and the conformational change detected in the protein relaxed.

10.3.2 Gramicidin A

Gramicidin A is a small hydrophibic peptide composed of 15 alternating L and D amino acids ($M_r \approx 1600$).[40] The peptide dimerizes and forms channels that conduct monovalent cations. There are substantial differences between two X-ray crystal structures of the peptide[41,42] and the bilayer structure that has been proposed on the basis of circular dichroism measurements. Such a controversy is amenable to solution by NMR methods.

Cross and co-workers have conducted a nice set of experiments using orientation-dependent solid-state NMR studies to address the structure of gramicidin in membranes.[43–46] They examined, by ^{15}N and ^{13}C NMR, samples of gramicidin, specifically [^{15}N]- and [^{13}C]-labelled at particular residues, and oriented between glass plates in lipid bilayers. They succeeded in determining the structure by measurement of dipolar splittings and calculating ϕ and ψ torsion angles. Fig. 10.12 shows the spectra from gramicidin ^{15}N-labelled at Ala-3 and Leu-4. The measurements are of the Gly-2—Ala-3 and Ala-3—Leu-4 peptide bonds and the dipolar splittings (Δv) of the amide ^{15}N resonances resulting from the directly bonded carbonyl carbons. The top spectra are the spectra of oriented gramicidin singly ^{15}N-labelled at the amide nitrogens of Ala-3 (left) and Leu-4 (right), and the lower spectra show the dipolar splitting of the ^{15}N by ^{13}C when the peptide bonds are doubly labelled. The observed dipolar splittings are 670 Hz for the ^{13}C-Gly-2—^{15}N-Ala-3 bond and 820 Hz for the ^{13}C-Ala-3—^{15}N-Leu-4 bond. An accurate bond length is the only additional information needed to calculate the angle θ. A 1.34 Å C—N bond length taken from the crystal structure of alanylalanine translates into a dipolar coupling constant (D) of 1.26 kHz, so from Equation (3.2) these data define four possible orientations for each C—N bond because D is larger than the observed splittings.

To limit these possibilities, independent restraints on the peptide plane orientations are obtained from the ^{15}N–^{1}H dipolar couplings and the ^{15}N chemical shift tensor. The ^{15}N–^{1}H dipolar couplings are determined using separated local field experiments.[47–51] These experiments are designed to separate the dipolar couplings from the chemical

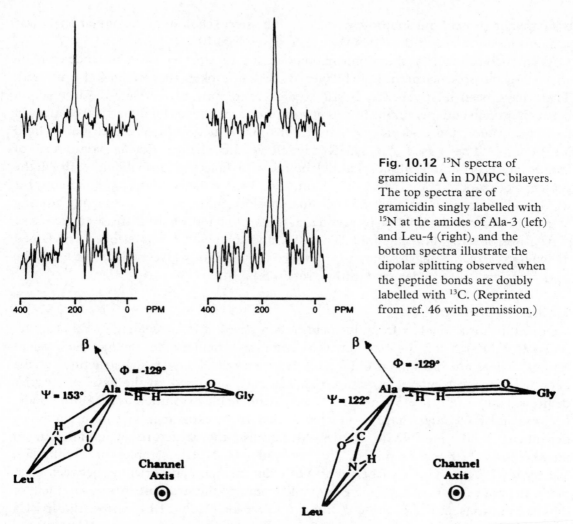

Fig. 10.12 ^{15}N spectra of gramicidin A in DMPC bilayers. The top spectra are of gramicidin singly labelled with ^{15}N at the amides of Ala-3 (left) and Leu-4 (right), and the bottom spectra illustrate the dipolar splitting observed when the peptide bonds are doubly labelled with ^{13}C. (Reprinted from ref. 46 with permission.)

Fig. 10.13 Two possible structures for the peptide planes at Ala-3 in gramicidin A. Both correspond to right-handed β-type helices but differ in the orientation of the Ala-3 carbonyl group, which points toward the gramicidin channel axis (left) or away from it (right). (Reprinted from ref. 46 with permission.)

shifts. They can also be used to relate the chemical shift and dipolar coupling tensors, and since the latter is readily related to the molecular coordinate system, the orientation of the principal components of the shift anisotropy can be determined directly. In gramicidin, the separated local field experiments yield only two possible orientations for each N—H bond because the ^{15}N—^{1}H dipolar coupling constant is smaller than the observed splittings. Since the orientation of the ^{15}N chemical-shift tensor relative to the molecular frame has been established independently by reference to model compounds[50,52–55] the observed ^{15}N chemical shifts provide a way to discriminate between several of the orientations of the N—H and N—C bonds implied by the dipolar couplings. Together these data, along with additional geometric restraints, define two possible sets of torsion angles for the Ala-3 position with errors in the ϕ and ψ angles of $\pm 6°$ and $\pm 5°$ respectively (see Fig. 10.13).

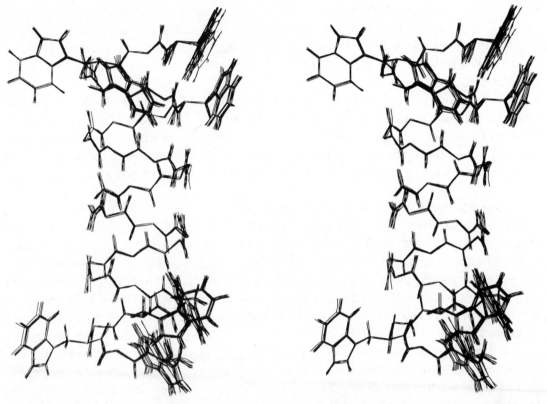

Fig. 10.14 Stereo side view of a set of ten refined solid-state NMR structures for gramicidin A in a membrane bilayer. The indole NH groups are clustered at the bilayer surface, and the amino terminus of each monomer is buried at the bilayer centre. (Reprinted from ref. 58 with permission.)

The two structures defined by the Ala-3 torsion angles are right-handed β-type helices. The difference between the two structures is the orientation of the carbonyl group with respect to the channel axis, which in the left-hand structure points towards the channel ($\psi = 153°$) and in the right-hand structure points away from it ($\psi = 122°$). The left structure is the favoured model because it is biologically sensible, since the selectivity of cations over anions by gramicidin might arise because of the partial negative charges of the backbone carbonyls, which could solvate the cations after thay have been stripped of water as they enter the channel.

An alternative approach has been adopted by Schaefer and co-workers,[56] who used two-dimensional REDOR on Val_1-$[1-^{13}C]Gly_2$-$[^{15}N]Ala_3$-gramicidin A in multilamellar dispersions of DMPC to measure dipolar couplings and infer from these four possible angles between the Gly_2—Ala_3 ^{13}C—^{15}N peptide bonds. This information supported a right-handed single-stranded helical dimer as a structure model for gramicidin in multilamellar dispersions.

We have given a detailed description of how torsion-angle restraints were derived for one set of dipolar splittings. Cross and co-workers[57,58] have carried out a systematic determination of the structure of the backbone and indole sidechains for gramicidin in a lipid environment using 144 orientational restraints derived from measurements of ^{15}N

chemical shift anisotropy, ^{15}N—^{1}H and ^{15}N—^{13}C dipolar splitting, and ^{2}H quadrupole splitting. The final structure (shown in Fig. 10.14) was calculated using an algorithm similar to simulated annealing on a set of ten random structures, and had RMSDs of ± 2–$3°$, which reflects the very high precision nature of the solid-state NMR measurements.

In addition to a structural study, Cross and co-workers[59,60] have made a detailed study of gramicidin dynamics in a lipid bilayer, principally by ^{2}H NMR. They found that the global motions of the peptide can be described by a correlation time in the microsecond regime, whereas it had been thought that the local peptide dynamics should play a role in the functional process of the channel, which operates in the nanosecond time regime.

10.3.3 Bacteriorhodopsin and rhodopsin

Bacteriorhodopsin (bR) and rhodopsin ($M_r \approx 26$ kDa) are both integral membrane proteins that contain the vitamin A aldehyde, retinal, as a photoreactive chromophore. Retinal is known to form a Schiff base with a lysine residue (see Figs. 10.15 and 10.17). Griffin and co-workers have studied bR and rhodopsin by ^{13}C CP–MAS and R^2 solid-state NMR. In these studies the retinal chromophore is removed and the apoprotein is regenerated with specifically ^{13}C-labelled retinal. In bR, the retinal functions as a light-driven proton–ion pump. Absorption of light initiates a photochemical reaction and the pigment passes through a series of intermediates. Solid-state ^{13}C NMR spectra of the dark-adapted pigment have yielded results on the structure of the retinal and specific protein charges in the retinal binding site.[61–63] These studies indicate that the two components of dark-adapted bR, bR_{568} and bR_{548}, contain all-*trans* and all-*cis* retinals, respectively, in agreement with other experiments. In addition, the spectra show that the C=N bond in bR_{568} is *anti* and in bR_{548} it is *syn*, and that in both components the C_6—C_7 bond is in the S-*trans* conformation.[60,64–66] Furthermore, spectra have been obtained of one of the photo-intermediates of bR, by blocking the thermal decay of the intermediate at $-23°C$. The MAS spectra[67] of bR and the photo-intermediate M_{412} regenerated with [13-^{13}C]retinal are shown in Fig. 10.15. The two components of dark-adapted bR are observed as two sharp resonances at 165.7 and 169.7 ppm (a), while the M_{412} resonance occurs at 146.7 ppm (b). The difference spectrum of (b)–(a) highlights the ^{13}C label (c), and the chemical shift indicates that the retinal chromophore has an unprotonated retinal-lysine Schiff base linkage. These studies illustrate the value of low-temperature solid-state NMR methods for obtaining spectra of reactive intermediates, coupled with difference methods[68,69] for removal of natural abundance backgrounds.

In a rotational resonance NMR study of bacteriorhodopsin, Griffin and co-workers[70] incorporated two specific ^{13}C labels into the retinal chromophore of the protein, one at C_1 on the ionone ring and the other at C_8 on the polyene chain. In order to remove the large contribution to the spectrum from the natural abundance ^{13}C, the intensities of the ^{13}C labels were obtained from difference spectra between bacteriorhodopsin regenerated with ^{13}C-labelled retinal and with unlabelled retinal. The distance between the labels depends on the conformation about the C_6—C_7 bond. Model compound studies showed that in the S-*cis* conformation, the labels are ≈ 3.1 Å apart, whereas in the S-*trans* conformation the distance is closer to 4.2 Å. In the protein the magnetization transfer rates agreed well with a separation of 4.2 Å (see Fig. 10.16), establishing that the retinal has the *S-trans* geometry.

Griffin and co-workers[71] have also reported rotational resonance measurements for the M_{412} intermediate, as shown in Fig. 10.17. These data measured the distance between the 14 retinal and the ε-lys$_{216}$ carbons in a sample of [14-^{13}C]retinal, [ε-^{13}C]lys-bR thermally trapped in the M state, which was found to be 3.9 ± 0.1 Å, corresponding to the *anti* C=N configuration (and not consistent with a *syn* configuration).

The geometry around the C_6—C_7 bond was also studied by Watts and co-workers[72] using ^{2}H and ^{31}P NMR. A uniaxially oriented sample of bacteriorhodopsin with retinal uniformly deuterated on the cyclohexene ring was oriented between glass plates and

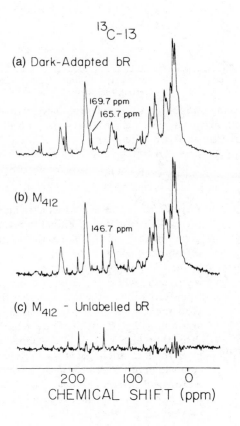

Fig. 10.15 ^{13}C CPMAS NMR spectra of bacteriorhodopsin containing [13-^{13}C]retinal. The spectrum (a) is of the dark-adapted pigment and exhibits two sharp resonances (and their associated rotational sidebands) corresponding to two distinct forms of the pigment. The spectrum in (b) is of the M_{412} photoreaction intermediate that has been trapped at 250 K. In (c) the difference spectrum (b) minus natural abundance protein is shown. (Reprinted from ref. 67 with permission.)

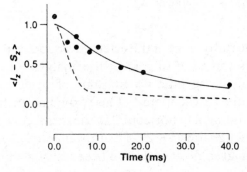

Fig. 10.16 Magnetization transfer data at $-30°$C measured for the [8,18-^{13}C$_2$]retinal-bR along with calculated curves based on known distances and geometries for the 6-S-*cis* (dashed line) and 6-S-*trans* (solid line). The data agree well with the 6-S-*trans* configuration and not with the 6-S-*cis*.

(a)

(b)

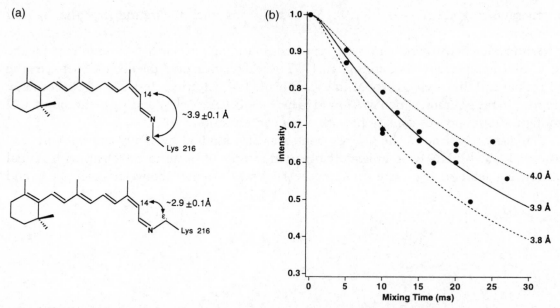

Fig. 10.17 (a) Comparison of the *anti* and *syn* C=N configurations for a 13-*cis*-deprotonated retinal Schiff base. (b) Observed (at $-60°C$) and simulated magnetization exchange between the ε-lysine$_{216}$ and 14-retinal carbons of bR in the M state under the $n=1$ rotational resonance condition. (Reprinted from ref. 71 with permission.)

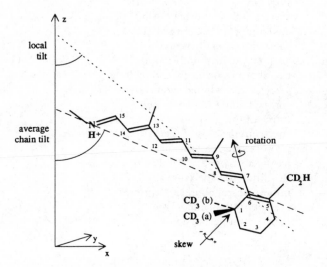

Fig. 10.18 Three-dimensional structure of the cyclohexene ring of retinal in bR as determined by ^{2}H NMR, relative to the membrane surface in the *xy*, plane. Analysis of the orientations of the three deuteromethyl groups on the puckered ring (skew around C_1—C_6) indicates that the chromophore has a (6*S*)-*trans* conformation around the C_6—C_7 bond. The reference axis defined by carbons C_4 and C_6 (dotted line) gives the local tilt angle at the cyclohexene ring which differs from that of the average end-to-end chromophore long axis (dashed line) due to an in-plane curvature of the polyene chain. (Reprinted from ref. 72 with permission.)

examined by ^{2}H NMR. In order to determine the quality of the oriented sample, ^{31}P NMR was also carried out, and lineshape analysis was used to correct the analyses of the ^{2}H quadrupole splittings. From the splittings, the specific orientations of three labelled methyl groups on the cyclohexene ring could be calculated. The two adjacent methyl groups on C_1 of retinal were found to lie approximately horizontal in the membrane and to make angles with the membrane normal of $94 \pm 2°$ and $75 \pm 2°$. The third methyl group, on C_5, points towards the cytoplasmic side with an angle of $46 \pm 3°$. These restraints indicate

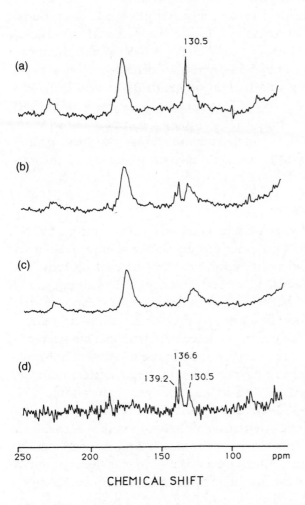

Rhodopsin

Isorhodopsin

Bathorhodopsin

Fig. 10.19 The structure and photochemical reaction of the retinal chromophore in rhodopsin. Rhodopsin has an 11-*cis*-retinal chromophore attached to the protein as a Schiff base. The isorhodopsin pigment has a 9-*cis* chromophore and can be generated by photolysis of the bathorhodopsin intermediate or by removal of the chromophore of rhodopsin and regeneration with exogenous 9-*cis*-retinal. Photolysis of both rhodopsin and isorhodopsin leads to a common intermediate, bathorhodopsin. The ^{13}C CPMAS NMR spectra are of (a) isorhodopsin; (b) a mixture of bathorhodopsin and isorhodopsin; (c) unlabelled isorhodopsin; and (d) difference spectrum of (b) minus (a). (Reprinted from ref. 74 with permission.)

that the cyclohexene lies approximately perpendicular to the membrane surface and that it has the S-*trans* conformation. From the estimated angle of tilt of the chromophore long axis, it was concluded that the polyene chain is slightly curved downward to the extracellular side of the membrane (see Fig. 10.18).

Solid-state NMR has also been used by Griffin and co-workers[73] to investigate rhodopsin, the visual pigment found in vertebrate rod cells, with ^{13}C-labelled chromophores. Unlike bacteriorhodopsin, rotational diffusion of the rhodopsin molecule in the membrane significantly broadens the ^{13}C protein resonances. Two different approaches for restricting protein motion were used in the solid-state work: low-temperature and lyophilization in diphytanolyglycerophosphocholine. These studies revealed the structure of the C_6—C_7 and C=N bonds of the chromophore. Furthermore, comparison of the chemical shifts of the chain positions with retinal model compounds have provided support for a negative charge near the C12–C13 region of the chromophore. A negative charge had been implicated in the mechanism for red-shifting the visible absorption band of rhodopsin. Furthermore, Smith and co-workers[74] examined the rhodopsin photo-intermediate bathorhodopsin trapped with specifically ^{13}C-labelled retinal at low temperature (<130 K) by CP–MAS NMR. When rhodopsin absorbs light, the first step is that the protein's 11-*cis*-retinylidene isomerizes to the 11-*trans* conformation (see Fig. 10.19), forming bathorhodopsin. By careful difference spectroscopy, Smith and co-workers were able to identify the resonances of the *all-trans*-retinal protonated Schiffs base (see Fig. 10.19), and, by comparison with the chemical-shift changes in rhodopsin-bound retinal, concluded that the local protein environment does not change significantly upon *cis*-trans isomerization of the chromophore.

10.3.4 Phage coat proteins

The filamentous bacteriophages are simple structures consisting of a strand of DNA protected by a surrounding coat of protein. This outer protein shell is composed of a few specialized cell-attachment proteins and several thousand copies of a coat protein. The coat proteins are generally small (≈ 50 residues, $M_r \approx 5500$) and have a C-terminal stretch of 20–25 hydrophobic amino acids that is α-helical.[75,76] They are assembled into the virus shell in a process that first involves incorporation into the membrane of the host cell.

The structure and dynamics of the phase coat proteins have been studied extensively by Opella and co-workers, in both oriented fibres[77] and in membrane bilayers.[78] Solution-state NMR studies[79,80] of the phage Pf1 coat protein show that the sequence contains two α-helical segments, a short helix from residues 6 to 13 and a longer hydrophobic helix from residues 19 to 42. Based on these data, ^{15}N chemical-shift measurements of oriented Pf1 coat protein have provided a picture of the orientation of these two helical segments in phospholipid bilayers. The ^{15}N chemical-shift tensor of the amide bond is nearly axially symmetric (with $\sigma_{11} = \sigma_{22} \neq \sigma_{33}$), resulting in a ^{15}N frequency close to σ_{33} when the amide N—H bond is nearly parallel to the z-axis of the magnetic field, and in a frequency near σ_{11} when the N—H bond is perpendicular to the applied field (see Equation (3.3) and Fig. 3.7). In an α-helix, the N—H bond is nearly parallel to the helix axis. Consequently, if the two helical segments are retained in bilayers, only a single measurement of the ^{15}N chemical shift from one amino acid in each helix is needed to define the helix orientation. Measurement of the ^{15}N resonance of Glu-9 in the N-terminal segment argues that this helix is oriented parallel to the membrane plane, and ^{15}N labelling of Tyr-25 and Tyr-40 shows that the second helix lies perpendicular to the bilayer plane. These data have lead to a model that uses the short N-terminal helix as an initiation site for phage assembly.[81]

10.3.5 Alamethacin

Alamethacin is a small membrane channel-forming peptide (20 residues), whose model N-terminal undecapeptide X-ray crystal structure has been solved to 0.9 Å resolution by Jung and co-workers, using crystals grown from dichloromethane solutions.[82] This peptide, whose sequence is Boc-L-Ala-MeA-Ala-MeA-Ala-Glu(OBzl)-Ala-MeA-Ala-MeA-Ala-OMe (where MeA = 2-methyl-alanine) forms a canonical α-helix, which Smith and co-workers[83,84] have studied by homonuclear rotational resonance and crystallization from methanol. The ^{13}C labels were placed on pairs of backbone carbonyl and alanine methyl carbons in the first five residues of the peptide such that the distances between them serve as markers for a helical structure. The rotational resonance transfer at $n = 1$ (see Section 3.3.2) was examined in crystals of the peptide in which the ^{13}C labels are 3.7, 4.8, and 6.8 Å apart (see Fig. 10.20(a)), and is shown in Fig. 10.20(b).

The ^{13}C-labelled carbonyl resonance at 175 ppm has been inverted selectively while spinning at rotational resonance with the ^{13}C methyl resonance at 20 ppm, and magnetization occurs as a function of mixing time. In the difference spectra shown, the increase in carbonyl intensity correlates with the loss in intensity of the methyl resonance intensity. Also, magnetization exchange is limited to the two rotationally coupled lines,

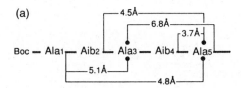

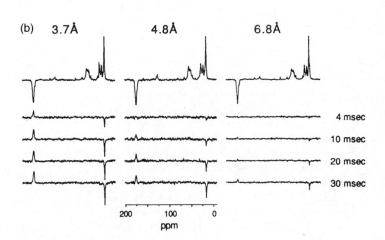

Fig. 10.20 (a) The N-terminal sequence of the alamethacin undecapeptide showing the positions and distances between the ^{13}C-labelled pairs. (b) ^{13}C rotational resonance NMR spectra of three microcystalline alamethacin undecapeptides at $n = 1$. Each peptide has two ^{13}C labels, one at the backbone carbonyl, and one at an alanine methyl group. The times indicated are the magnetization transfer mixing times. (Reprinted from ref. 84 with permission.)

with little transfer to the natural abundance background. Fig. 10.21 shows the transfer curves for five labelled crystalline peptides, which gives a sense of the range of distances which can be measured using this method, with 6.8 Å representing a probable upper limit. The structure of the peptide when bound to a membrane was examined in the same

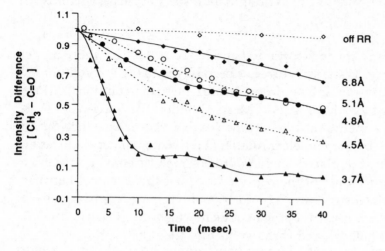

Fig. 10.21 Magnetization transfer curves for five microcrystalline alamethacin undecapeptides showing the correlation between internuclear distances and the rate of magnetization transfer. The off-rotational resonance data were obtained for the 6.8 Å sample spinning 541 Hz faster than the $n = 1$ condition. The line fits are not theoretical simulations. (Reprinted from ref. 84 with permission.)

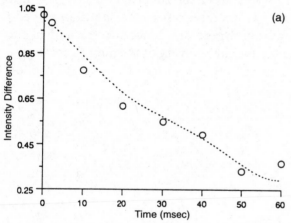

(a)

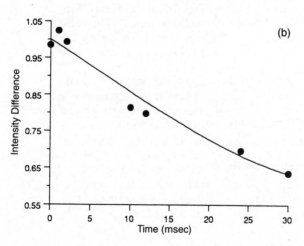

(b)

Fig. 10.22 Magnetization transfer curves for alamethacin peptidedipalmitoylphosphatidyl-choline (DPPC) complexes for (a) the 5.1 Å peptide (spinning at $n = 1$); and (b) the 4.8 Å peptide (spinning at $n = 2$). The lines correspond to the crystal data for each peptide and the symbols are experimental points. (Reprinted from ref. 84 with permission.)

study, in which ^{13}C-labelled peptides were reconstituted into lipid bilayers, and studied at $-40°C$ in order to reduce peptide motion. Fig. 10.22 shows a comparison of the magnetization transfer rates in the crystal and in the lipid for the 4.8 Å and 5.1 Å peptides, where the labels are one turn apart and across the helix from each other. The close correlation between the crystal data demonstrates that rotational resonance is a valid method for the investigation of the local secondary structure of membrane proteins.

10.4 OUTLOOK

NMR spectroscopy has already proved to be invaluable in the determination of the structures of phospholipids and glycolipids. Furthermore, it has been used with some success in solution to determine the structure of small membrane proteins in hydrophobic solvents or, in some cases, in the presence of lipid bilayers. Solid-state NMR has also been useful in determining the structures for small membrane proteins in bilayers. The current major limitation for both methods is the molecular weight limit (which in solids arises because of the impracticality of labelling every site in a large protein sequentially). X-ray crystallography, with the notable exception of the photo-synthetic reaction centre, has been applied with very limited success to solving the structures of either peripheral or integral membrane proteins. NMR spectroscopy will clearly be very important in the coming decade in helping to shape our understanding of the molecular details of interfaces between life and the environment.

REFERENCES

1. O. G. Mouritsen and M. Bloom, *Annu. Rev. Biophys. Biomol. Struct.*, **22**, p. 146 (1993).

2. S. J. Singer and G. L. Nicolson, *Science*, **175**, 720 (1972).

3. J. H. Davis, *Biochim. Biophys. Acta*, **737**, 117–71 (1983).

4. I. C. P. Smith, In *Structure and Properties of Cell Membranes*, Vol III (ed. G. Benga), CRC Press, Boca Raton, Florida (1985).

5. J. Seelig and P. M. Macdonald, *Acc. Chem. Res.*, **20**, 221–8 (1987).

6. R. L. Smith and E. Oldfield, *Science*, **225**, 280–8 (1984).

7. J. Seelig, F. Borle, and T. A. Cross, *Biochim. Biophys. Acta*, **814**, 195–8 (1985).

8. J. B. Speyer, P. K. Sripada, S. K. Das Gupta, G. G. Shipley, and R. G. Griffin, *Biophys. J.*, **51**, 687–91 (1987).

9. B. Bechinger and S. J. Opella, *J. Magn. Reson.*, **95**, 585 (1991).

10. J. J. de Vries and H. J. C. Berendsen, *Nature*, **221**, 1139–40 (1969).

11. M. F. Brown, J. Seeling, and U. Haeberlen, *J. Chem. Phys.*, **70**, 5045–53 (1979).

12. D. A. Torchia, *Annu. Rev. Biophys. Bioeng.*, **13**, 125–44 (1984).

13. R. G. Griffin, *Methods in Enzymol.*, **72**, 108–74 (1981).

14. J. Seelig, *Biochim. Biophys. Acta*, **515**, 105–40 (1978).

15. I. C. P. Smith and I. Ekiel, In *Phosphorus-31 NMR: Principles and Applications* (ed. D. G. Gorenstein), Academic Press, New York, pp. 447–75 (1984).

16. R. F. Campbell, E. Meirovitch, and J. H. Freed, *J. Phys. Chem.*, **83**, 525–33 (1979).

17. E. J. Dufourc, C. Mayer, J. Stohrer, G. Althoff, and G. Kothe, *Biophys. J.*, **61**, 42–57 (1992).

18. J. -P. Renou, J. B. Giziewicz, I. C. P. Smith, and H. C. Jarrell, *Biochemistry*, **28**, 1804–14 (1989).

19. H. C. Jarrell, A. J. Wand, J. B. Giziewicz, and I. C. P. Smith, *Biochim. Biophys. Acta*, **897**, 69–82 (1987).

20. M. Auger, D. Carrier, I. C. P. Smith, and H. C. Jarrell, *Biochemistry*, **112**, 1373–81 (1990).

21. I. C. P. Smith and H. C. Jarrell, *Pure and Appl. Chem.*, **63**, 529–34 (1991).

22. C. R. Sanders II and J. H. Prestegard, *J. Am. Chem. Soc.*, **113**, 1987–96 (1991).

23. C. R. Sanders II and J. H. Prestegard, *Biophys. J.*, **58**, 447–60 (1990).

24. C. R. Sanders II and J. P. Schwonek, *Biochemistry*, **31**, 889–905 (1992).

25. N. E. Gabriel, N. V. Agman, and M. F. Roberts, *Biochemistry*, **26**, 7409–18 (1987) and references cited therein.

26. F. Inagaki, *Magn. Reson. Chem.*, Special Issue, **30**, S125–S133 (1992).

27. L. R. Brown, A. Braun, A. Kumar, and K. Wüthrich, *Biophys. J.*, **37**, 319–28 (1982).

28. R. Bazzo, M. J. Tappin, A. Pastore, T. S. Harvey, J. A. Carver, and I. D. Campbell, *Eur. J. Biochem.*, **173**, 139–46 (1988).

29. C. E. Dempsey, *Biochim. Biophys. Acta*, **1031**, 143–61 (1990) and references cited therein.

30. C. E. Dempsey and G. S. Butler, *Biochemistry*, **31**, 11973–7 (1992).

31. K. H. Lee, J. E. Fitton, and K. Wüthrich, *Biochim. Biophys. Acta*, **911**, 144–53 (1987).

32. M. J. Tappin, A. Pastore, R. S. Norton, J. H. Freer, and I. D. Campbell, *Biochemistry*, **27**, 1643–7 (1988).

33. F. Macquaire, F. Baleaux, E. Giaccobi, T. Huynh-Dinh, J. -M. Neumans, and A. Sanson, *Biochemistry*, **31**, 2576–82 (1992).

34. D. Kohda and F. Inagaki, *Biochemistry*, **31**, 677–85 (1992).

35. G. D. Henry and B. D. Sykes, *Biochem. Cell. Biol.*, **68**, 318–29 (1990) and references cited therein.

36. M. F. Moody, P. T. Jones, J. A. Carver, J. Boyd, and I. D. Campbell, *J. Mol. Biol.*, **193**, 759–74 (1987).

37. A. Milon, T. Miyazawa, and T. Higashijima, *Biochemistry*, **29**, 65–75 (1990).

38. S. Rajan, S. -Y. Kang, H. S. Gutowsky, and E. Oldfield, *J. Biol. Chem.*, **256**, 1160–6 (1981).

39. T. Heimburg, P. Hildebrandt, and D. Marsh, *Biochemistry*, **30**, 9084–9 (1991).

40. B. A. Wallace, *Annu. Rev. Biophys. Biophys. Chem.*, **19**, 127–57 (1990).

41. D. A. Langs, *Science*, **241**, 188–91 (1988).

42. B. A. Wallace and K. Ravikumar, *Science*, **241**, 182–7 (1988).

43. F. Moll III and T. A. Cross, *Biophys. J.*, **57**, 351–62 (1990).

44. L. K. Nicholson and T. A. Cross, *Biochemistry*, **28**, 9379–85 (1989).

45. L. K. Nicholson, Q. Teng, and T. A. Cross, *J. Mol. Biol.*, **218**, 621–37 (1991).

46. Q. Teng, L. K. Nicholson, and T. A. Cross, *J. Mol. Biol.*, **218**, 607–19 (1991).

47. T. A. Cross and S. J. Opella, *J. Mol. Biol.*, **182**, 367–81 (1985).

48. A. C. Kolbert, M. H. Levitt, and R. G. Griffin, *J. Magn. Reson.*, **85**, 42–9 (1989).

49. M. Munowitz, W. P. Aue, and R. G. Griffin, *J. Chem. Phys.*, **77**, 1686–9 (1982).

50. M. Munowitz, R. G. Griffin, G. Bodenhausen, and T. H. Huang, *J. Am. Chem. Soc.*, **103**, 2529–33 (1981).

51. Q. Teng and T. A. Cross, *J. Magn. Reson.*, **85**, 439–47 (1989).

52. C. J. Hartzell, T. K. Pratum, and G. P. Drobny, *J. Chem. Phys.*, **87**, 4324–31 (1987).

53. C. J. Hartzell, M. Whitfield, T. G. Oas, and G. P. Drogby, *J. Am. Chem. Soc.*, **109**, 5966–9 (1987).

54. T. G. Oas, C. J. Hartzell, F. W. Dahlquist, and G. P. Drobny, *J. Am. Chem. Soc.*, **109**, 5962–6 (1987).

55. T. G. Oas, C. J. Hartzell, T. J. McMahon, G. P. Drobny, and F. W. Dahlquist, *J. Am. Chem. Soc.*, **109**, 5956–62 (1987).

56. A. W. Hing and J. Schaefer, *Biochemistry*, **32**, 7593–604 (1993).

57. W. Hu, K. -C. Lee and T. A. Cross, *Biochemistry*, **32**, 7035 (1993).

58. R. R. Ketchem, W. Hu, and T. A. Cross, *Science*, **261**, 1457–60 (1993).

59. C. L. North and T. A. Cross, *J. Magn. Reson.*, **101B**, 35 (1993).

60. K. -C. Lee, W. Hu and T. A. Cross, *Biophys. J.*, in press (1993).

61. G. S. Harbison, S. O. Smith, J. A. Pardoen, J. M. L. Courtin, J. Lugtenburg, J. Herzfeld *et al.*, *Biochemistry*, **24**, 6955–62 (1985).

62. G. S. Harbison, J. Herzfeld, and R. G. Griffin, *Biochemistry*, **22**, 1–5 (1983).

63. G. S. Harbison, S. O. Smith, J. A. Pardoen, C. Winkel, J. Lugtenburg, J. Herzfeld, *et al.*, *Proc. Natl. Acad. Sci. USA*, **81**, 1706–9 (1984).

64. R. van der Steen, P. L. Biesheuvel, R. A. Mathies, and J. Lugtenburg, *J. Am. Chem. Soc.*, **108**, 6410–11 (1986).

65. S. O. Smith, H. J. M. de Groot, R. Gebhard, J. M. L. Courtin, J. Lugtenburg, J. Herzfeld *et al.*, *Biochemistry*, **28**, 8897–904 (1989).

66. V. Copié, A. McDermott, K. Beshah, M. Spijker, J. Lugtenburg, J. Herzfeld *et al.*, *Biophys. J.*, **57**, 360a (1990).

67. S. O. Smith and R. G. Griffin, *Ann. Rev. Phys. Chem.*, **39**, 511–38 (1988).

68. H. J. M. de Groot, V. Copié, S. O. Smith, P. J. Allen, C. Winkel, J. Lugtenburg, J. Herzfeld, and R. G. Griffin, *J. Magn. Reson.*, **77**, 251–7 (1988).

69. L. C. P. J. Mollevanger, A. P. M. Kentgens, J. A. Pardoen, J. M. L. Courtin, W. S. Veeman, J. Lugtenburg *et al.*, *Eur. J. Biochem.*, **163**, 9–14 (1987).

70. F. Cruezet, A. McDermott, R. Gebhard, K. van der Hoef, M. B. Spijker-Assink, J. Herzfeld *et al.*, *Science*, **251**, 783–6 (1991).

71. K. V. Lakshmi, M. Auger, J. Raap, J. Lugtenburg, R. G. Griffin, and J. Herzfeld, *J. Am. Chem. Soc.*, **115**, 8515–16 (1993).

72. A. S. Ulrich, M. P. Heyn, and A. Watts, *Biochemistry*, **31**, 10390–9 (1992).

73. S. O. Smith, I. Palings, V. Copié, D. P. Raleigh, J. Courtin, J. A. Pardoen *et al.*, and *Biochemistry*, **26**, 1606–11 (1987).

74. S. O. Smith, J. Courtin, H. de Groot, R. Gebhard, and J. Lugtenberg, *Biochemistry*, **30**, 7409–15 (1991).

75. R. A. Schiksnis, M. J. Bogusky, P. Tsang, and S. J. Opella, *Biochemistry*, **26**, 1373–81 (1987).

76. W. Start, M. J. Glucksman, and L. Makowski, *J. Mol. Biol.*, **199**, 171–82 (1988).

77. T. A. Cross and S. J. Opella, *J. Mol. Biol.*, **182**, 367–81 (1985).

78. K. J. Shon, Y. Kim, L. A. Colnago, and S. J. Opella, *Science,*, **252**, 1303–5 (1991).

79. R. A. Schiksnis, M. J. Bogusky, P. Tsang, and S. J. Opella, *Biochemistry*, **26**, 1373–81 (1987).

80. R. A. Schiksnis, M. J. Bogusky, and S. J. Opella, *J. Mol. Biol.*, **200**, 741–3 (1988).

81. R. Nambudripad, W. Stark, S. J. Opella, and L. Makowski, *Science*, **252**, 1305–8 (1991).

82. R. Bosch, G. Jung, H. Schmitt, and W. Winter, *Biopolyners*, **24**, 961–78 (1985).

83. S. O. Smith and O. B. Peersen, *Annu. Rev. Biophys. Biomol. Struct.*, **21**, 25–47 (1992).

84. O. B. Peersen, S. Yoshimura, H. Hojo, S. Aimoto, and S. O. Smith, *J. Am. Chem. Soc.*, **114**, 4332–5 (1992).

Appendix 1 • Examples of product operator calculations

As an example of the use of product operators, we present here the complete product operator calculation for two pulse sequences: the COSY pulse sequence, and the REDOR pulse sequence. This illustrates the way in which these relatively simple calculations are carried out. If you can understand this, you should, in principle, be able to tackle any pulse sequence.

A1.1 THE COSY PULSE SEQUENCE

For the pulse sequence outlined in Section 2.1.2, consider two weakly coupled spins I_1 and I_2:

$$\sigma_0 = I_{1z} + I_{2z} \tag{A1.1}$$

After the first $90°$ pulse $(I_{1x} + I_{2x})$ (see Fig. A1.1):

$$\sigma_1 = -I_{1y} - I_{2y} \tag{A1.2}$$

Now during the evolution time, t_1, both the chemical shifts and the couplings evolve. We can consider the effects of these in any order. Here we will consider first (a) the effects of chemical shifts $(\Omega_1 t_1 I_{1z})$ on Equation (A1.2):

(a)
$$I_{1y} \cos \Omega_1 t_1 + I_{1x} \sin \Omega_1 t_1 - I_{2y} \cos \Omega_2 t_1 + I_{2x} \sin \Omega_2 t_1 \tag{A1.3}$$

Consider now (b) the effects of couplings $(\pi \mathcal{J}_{12} t_1 2 I_{1z} I_{2z})$ on each term in Equation (A1.2):

(b)
$$2 I_{1x} I_{2z} \cos \Omega_1 t_1 \sin \pi \mathcal{J}_{12} t_1 - I_{1y} \cos \Omega_1 t_1 \cos \pi \mathcal{J}_{12} t_1 + I_{1x} \sin \Omega t_1 \cos \pi \mathcal{J}_{12} t_1$$
$$+ 2 I_{1y} I_{2z} \sin \Omega_1 t_1 \sin \pi \mathcal{J}_{12} t_1 - I_{2y} \cos \Omega_2 t_1 \cos \pi \mathcal{J}_{12} t_1$$
$$+ 2 I_{1z} I_{2x} \cos \Omega_2 t_1 \sin \pi \mathcal{J}_{12} t_1$$
$$+ I_{2x} \sin \Omega_2 t_1 \cos \pi \mathcal{J}_{12} t_1 + 2 I_{1x} I_{2y} \sin \Omega_2 t_1 \sin \pi \mathcal{J}_{12} t_1 \tag{A1.4}$$

If we collect terms from Equations (A1.3) and (A1.4):

$$\sigma_2 = (-I_{1y} \cos \Omega_1 t_1 + I_{1x} \sin \Omega_1 t_1 - I_{2y} \cos \Omega_2 t_1 + I_{2x} \sin \Omega_2 t_1) \cos \pi \mathcal{J}_{12} t_1$$
$$+ (2 I_{1x} I_{2z} \cos \Omega_1 t_1 + 2 I_{1y} I_{2z} \sin \Omega_1 t_1 + 2 I_{1z} I_{2x} \cos \Omega_2 t_1$$
$$+ 2 I_{1z} I_{2y} \sin \Omega_2 t_1) \sin \pi \mathcal{J}_{12} t_1 \tag{A1.5}$$

After the second 90° pulse, the terms in Equation (A1.5) become:

$$(-I_{1z} \cos \Omega_1 t_1 + I_{1x} \sin \Omega_1 t_1 - I_{2z} \cos \Omega_2 t_1 + I_{2x} \sin \Omega_2 t_1) \cos \pi \mathcal{J}_{12} t_1$$
$$-(2I_{1x}I_{2y} \cos \Omega_1 t_1 + 2I_{1z}I_{2y} \sin \Omega_1 t_1 + 2I_{1y}I_{2x} \cos \Omega_2 t_1 + 2I_{1y}I_{2z} \sin \Omega_2 t_1) \sin \pi \mathcal{J}_{12} t_1$$

$$(A1.6)$$

Note that in Equation (A1.6), I_z magnetization, and product operators with both x and y magnetization (e.g. $2I_{1x}I_{2y}$) can be ignored when considering the observable magnetization, since they cannot be observed! Thus the observable magnetization is:

$$\sigma_3^{\text{obs}} = (I_{1x} \sin \Omega_1 t_1 + I_{2x} \sin \Omega_2 t_1) \cos \pi \mathcal{J}_{12} t_1 - (2I_{1y}I_{2z} \sin \Omega_2 t_1$$
$$+ 2I_{1z}I_{2y} \sin \Omega_1 t_1) \sin \pi \mathcal{J}_{12} t_1$$

$$(A1.7)$$

Of course, strictly speaking, the chemical shifts and couplings continue to evolve during t_2, but, nevertheless, the interpretation remains as that presented in Section 2.1.2.

A1.2 THE REDOR PULSE SEQUENCE

For an isolated $I - S$ spin pair (spin-$\frac{1}{2}$ nuclei), the dipolar transition frequencies of a static sample are given by:

$$\omega_D(\theta) = \pm \tfrac{1}{2} D (3 \cos^2 \theta - 1)$$

$$(A1.8)$$

Pulses

$$I_z \xrightarrow{\beta_y} I_z \cos \beta + I_x \sin \beta$$

$$I_z \xrightarrow{\beta_x} I_z \cos \beta - I_y \sin \beta$$

$$I_x \xrightarrow{\beta_x} I_x$$

$$I_y \xrightarrow{\beta_y} I_y$$

$$I_x \xrightarrow{\beta_y} I_x \cos \beta - I_z \sin \beta$$

$$I_y \xrightarrow{\beta_x} I_y \cos \beta + I_z \sin \beta$$

90° pulses

$$I_z \xrightarrow{90°_y} +I_x$$

$$I_z \xrightarrow{90°_x} -I_y$$

$$I_x \xrightarrow{90°_y} -I_z$$

$$I_y \xrightarrow{90°_x} +I_z$$

Scalar Coupling

$$I_z \xrightarrow{\pi J_{12} t \, I_{1z}I_{2z}} I_z$$

$$2I_{1x}I_{2y} \xrightarrow{\pi J_{12} t \, I_{1z}I_{2z}} 2I_{1x}I_{2y}$$

$$I_{1x} \xrightarrow{\pi J_{12} t \, I_{1z}I_{2z}} I_{1x} \cos\pi J_{12} t + 2 I_{1y}I_{2z} \sin\pi J_{12} t$$

$$I_{1y} \xrightarrow{\pi J_{12} t \, I_{1z}I_{2z}} I_{1y} \cos\pi J_{12} t - 2 I_{1x}I_{2z} \sin\pi J_{12} t$$

$$2I_{1x}I_{2z} \xrightarrow{\pi J_{12} t \, I_{1z}I_{2z}} 2I_{1x}I_{2y} \cos\pi J_{12} t + I_{1y} \sin\pi J_{12} t$$

$$2I_{1y}I_{2z} \xrightarrow{\pi J_{12} t \, I_{1z}I_{2z}} 2I_{1y}I_{2y} \cos\pi J_{12} t - I_{1x} \sin\pi J_{12} t$$

Chemical Shifts

$$I_z \xrightarrow{\Omega t \, I_z} I_z$$

$$I_x \xrightarrow{\Omega t \, I_z} I_x \cos\Omega t + I_y \sin\Omega t$$

$$I_y \xrightarrow{\Omega t \, I_z} I_y \cos\Omega t - I_x \sin\Omega t$$

Fig. A1.1 Summary of product operator transformations.

where $D = \gamma_I \gamma_S \hbar / r^3$, and $\hbar$ is Planck's constant divided by 2π, r is the internuclear distance, γ_I and γ_S are the gyromagnetic ratios for I and S respectively. Under MAS conditions, the dipolar transition frequencies become:

$$\omega_D(\alpha,\beta;t) = \pm \tfrac{1}{2}D[\sin^2 \beta \cos 2(\alpha+\omega_r t) - \sqrt{2} \sin 2\beta \cos(\alpha+\omega_r t)] \tag{A1.9}$$

where α, and β are defined according to Fig. A1.2.

The average of Equation (A1.9) over one rotor cycle is zero. However, if a single I-spin π pulse is applied at time t, after the start of the rotor cycle, ω_D averaged over one cycle is given by:

$$\overline{\omega_D}(\alpha,\beta;t) = \pm \frac{1}{T_r}\left[\int_0^{t_1} \omega_D(\alpha,\beta;t)\mathrm{d}t - \int_{t_1}^{T_r} \omega_D(\alpha,\beta;t)\mathrm{d}t \right] \neq 0 \tag{A1.10}$$

If we make the following loose analogy between the MAS experiment and two-dimensional $\mathcal{J}$ spectroscopy:

MAS		Two-dimensional $\mathcal{J}$ spectroscopy
$\overline{\overline{\omega_D}} \neq 0$	$\Leftrightarrow$	$\mathcal{J} \neq 0$

then we can discuss the REDOR experiment in terms of product operator formalism. Note that this analogy is not strictly correct, since the MAS frame of reference and the rotating frame of reference are not the same. In fact the dipolar coupling is zero when averaged over one rotor cycle. Thus this product operator analysis should be regarded as one simple way of looking at the REDOR experiment, but not to be taken too literally. Thus for chemical shifts:

$$I_z \xrightarrow{90^\circ_y} I_x \xrightarrow{t} I_x \cos \Omega_I t + I_y \sin \Omega_I t \xrightarrow{180^\circ_x} I_x \cos \Omega_I t - I_y \sin \Omega_I t \xrightarrow{t} I_x \tag{A1.11}$$

For heteronuclear coupling:

$$I_z \xrightarrow{90^\circ_y} I_x \xrightarrow{t} I_x \cos \pi \mathcal{J}_{IS} t + 2I_y S_z \sin \pi \mathcal{J}_{IS} t \xrightarrow{180^\circ_x(I)} I_x \cos \pi \mathcal{J}_{IS} t$$
$$- 2I_y S_z \sin \pi \mathcal{J}_{IS} t \xrightarrow{t} I_x \tag{A1.12}$$

However, if 180° pulses are applied both to the I and S channels, then:

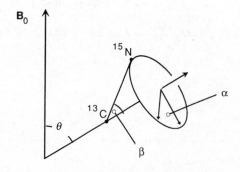

Fig. A1.2 Diagram showing the relationship between the spinning axis, the external magnetic field, and the heteronuclear dipolar vector.

$$I_z \xrightarrow{\ 90^\circ_y\ } I_x \xrightarrow{\ t\ } I_x \cos \pi \mathcal{J}_{IS} t + 2 I_y S_z \sin \pi \mathcal{J}_{IS} t \xrightarrow{\ 180^\circ_x (I,S)\ } I_x \cos \pi \mathcal{J}_{IS} t$$

$$+ 2 I_y S_z \sin \pi \mathcal{J}_{IS} t$$

$$\xrightarrow{\ t\ } I_x \cos \pi \mathcal{J}_{IS} 2t + 2 I_y S_z \sin \pi \mathcal{J}_{IS} 2t \tag{A1.13}$$

Comparing Equations (A1.12) and (A1.13), since $I_y S_z$ is unobservable, the intensity of I_x after Fourier transformation is: I_x for Equation (A1.12) and $I_x \cos \pi \mathcal{J}_{IS} 2t$ for Equation (A1.13). If we define S_0 for the signal when the 180° pulse is applied only to the I channel, and S for the signal when 180° pulses are applied to both I and S channels, then:

$$\Delta S = S_0 - S = I_x (1 - \cos \pi \mathcal{J}_{IS} 2t) \tag{A1.14}$$

and

$$\frac{\Delta S}{S} = \frac{I_x (1 - \cos \pi \mathcal{J}_{IS} 2t)}{I_x} = 1 - \cos \pi \mathcal{J}_{IS} 2t \tag{A1.15}$$

Recall that $2t$ is the time period between the initial 90° pulse and the start of the acquisition. Therefore, the following relationship exists between MAS and two-dimensional $\mathcal{J}$ spectroscopy experiments:

$$\pi \mathcal{J} \equiv \overline{\omega_D}$$

$$2t \equiv N T_r$$

where N is the number of rotor cycles, and T_r is the time to complete one rotor cycle. Thus in the REDOR experiment,

$$\frac{\Delta S}{S} = 1 - \cos \overline{\omega_D}\, N_r T_r \tag{A1.16}$$

Note that in a MAS experiment, Equation (A1.16) corresponds to one particular α and β value. Therefore, the real REDOR equation is:

$$\frac{\Delta S}{S} = \frac{\int_\alpha \int_\beta (1 - \cos \omega_D N_r T_r) \sin \beta \, d\alpha \, d\beta}{\int_\alpha \int_\beta \sin \beta \, d\alpha \, d\beta} \tag{A1.17}$$

which gives the curve depicted in Fig. 3.12. Thus, when $\overline{\omega_D}$ is small,

$$\frac{\Delta S}{S} = N_r T_r \overline{\omega_D}$$

$$= \alpha \cdot N_r T_r D \tag{A1.18}$$

where α is a universal constant. From Equation (A1.18), once $\Delta S/S$ is determined, the dipolar constant D can be calculated, and therefore the internuclear distance r can also be calculated.

Appendix 2 · Useful NMR data on amino acid residues

The characteristic ^{1}H chemical shifts for amino acid residues of random extended chain structures are given in Table A2.1. The characteristic ^{13}C chemical shifts for amino acid residues of random extended chain structures are given in Table A2.2.

The DQF–COSY and TOCSY spectra for 19 of the 20 amino acids obtained at 11.7 T (500 MHz ^{1}H) and pD 7.0 are shown below. Insets show expansions for selected cross-peaks. Glycine is not included, since in solution the α-methylene protons are equivalent and therefore show no cross-peaks in the two-dimensional spectra. However, it should be noted that, usually in proteins, these two proton resonances are not equivalent.

Table A2.1 Extended chain ^{1}H chemical shifts for the 20 common amino acid residues[a]

Residue	NH	αH	βH	Others	
Gly	8.39	3.97			
Ala	8.25	4.35	1.39		
Val	8.44	4.18	2.13	γCH_3	0.97, 0.94
Ile	8.19	4.23	1.9	γCH_2	1.48, 1.19
				γCH_3	0.95
				δCH_3	0.89
Leu	8.42	4.38	1.65, 1.65	γH 1.64	
				δCH_3	0.94, 0.90
Pro[b]		4.44	2.28, 2.02	γCH_2	2.03, 2.03
				δCH_2	3.68, 3.65
Ser	8.38	4.50	3.88, 3.88		
Thr	8.24	4.35	4.22	γCH_3	1.23
Asp	8.41	4.76	1.84, 1.75		
Glu	8.37	4.29	2.09, 1.97	γCH_2	2.31, 2.28
Lys	8.41	4.36	1.85, 1.76	γCH_2	1.45, 1.45
				δCH_2	1.70, 1.70
				ϵCH_2	3.02, 3.02
				ϵNH_3^+	7.52
Arg	8.27	4.38	1.89, 1.79	γCH_2	1.70 1.70
				δCH_2	3.32, 3.32
				NH	7.17, 6.62
Asn	8.75	4.75	2.83, 2.75	γNH_2	7.59, 6.91
Gln	8.41	4.37	2.13, 2.01	γCH_2	2.38, 2.38
				δNH_2	6.87, 7.59
Met	8.42	4.52	2.15, 2.01	γCH_2	2.64, 2.64
				ϵCH_3	2.13
Cys	8.31	4.69	3.28, 2.96		
Trp	8.09	4.70	3.32, 3.19	2H	7.24
				4H	7.65
				5H	7.17
				6H	7.24
				7H	7.50
				NH	10.22
Phe	8.23	4.66	3.22, 2.99	2,6H	7.30
				3,5H	7.39
				4H	7.34
Tyr	8.18	4.60	3.13, 2.92	2,6H	7.15
				3,5H	6.86
His	8.41	4.63	3.26, 3.20	2H	8.12
				4H	7.14

[a]Data for the non-terminal residues X in tetrapeptides GGXA, pH 7.0, 35°C from K. Wüthrich, *NMR of Proteins and Nucleic Acids*, John Wiley, 1986.
[b]Data for *trans*-Pro.

Table A2.2 [13]C Chemical shifts of the common amino acid residues in neutral D_2O^a

Amino acid residue	C=O	αC	βC	γC	δC	εC	ζC
Gly[a]	172.7	43.5					
Ala[b]	175.8	50.8	17.7				
Val[a]	174.9	60.7	30.8	19.3			
				18.5			
Ile[a]	174.8	59.6	36.9	25.4	11.3		
				15.7			
Leu[a]	175.9	53.6	40.5	25.2	23.1		
					21.6		
Ser[b]	172.6	56.6	62.3				
Thr[b]	172.7	60.2	68.3	20.0			
Pro[b,c] { trans	175.2	61.6	30.6	25.5	48.2		
cis		61.3	33.1	23.2	48.8		
Asp[b]	174.2	52.7	39.8	178.4			
Glu[b]	174.8	54.9	28.9	34.6	182.8		
Lys[b]	174.7	54.4	27.5	23.1	31.8	40.5	
Arg[a]	175.0	54.6	28.8	25.7	41.7		157.6
Asn[b]	173.1	51.5	37.7	175.6			
Gln[b]	174.0	54.1	28.1	32.2	179.0		
Met[a]	175.0	53.9	31.0	30.7		15.0	
Cys	175.7	57.9	26.0				

Ring carbon atoms

Trp	176.7	56.7	27.4	C2 } { 126.0			
				C5 } { 122.9			
				C3 108.4			
				C4 } { 120.3			
				C6 } { 119.3			
				C7 112.8			
				C8 137.3			
				C9 127.5			
Phe	176.0	57.4	37.0	C1 136.2			
				C2,6 130.3			
				C3,5 130.3			
				C4 128.6			
Tyr	176.0	57.4	37.0	C1 (128.0)			
				C2,6 130.0			
				C3,5 117.0			
				C4 156.0			
His[b]	172.6	53.7	28.0	C2 135.2			
				C4 118.7			
				C5 130.3			

[a]The resonance positions are given with respect to TMS.
[a]Data from K. Wüthrich, *NMR of Proteins and Nucleic Acids*, John Wiley, 1986. Measured in linear pentapeptides H-Gly-Gly-X-Gly-Gly-OH. To convert to the TMS scale, $\delta(CS_2) = 193.7$ ppm was used.
[b]Measured in protected linear tetrapeptides TFA-Gly-Gly-X-L-Ala-OCH$_3$. To convert to the TMS scale, δ (internal dioxane) = 67.8 ppm was used.
[c]In the prolyl peptide, *trans*- and *cis*-proline occurred with a relative abundance of approx. 4:1.

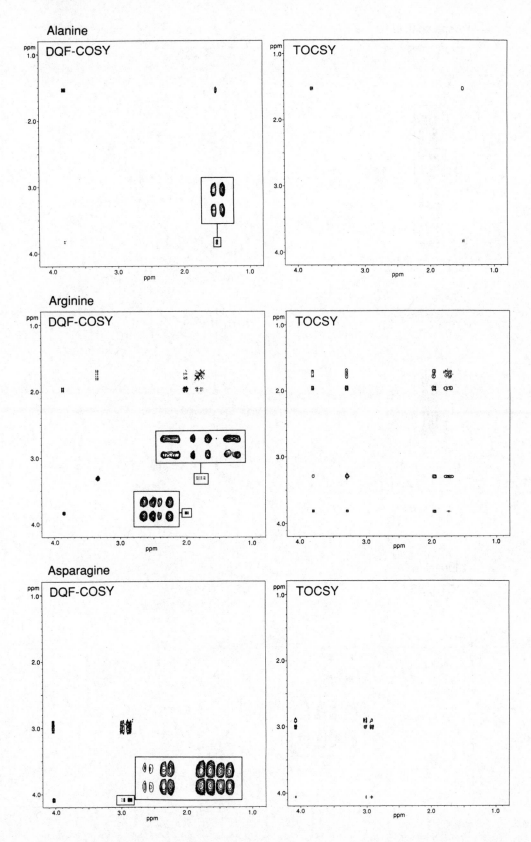

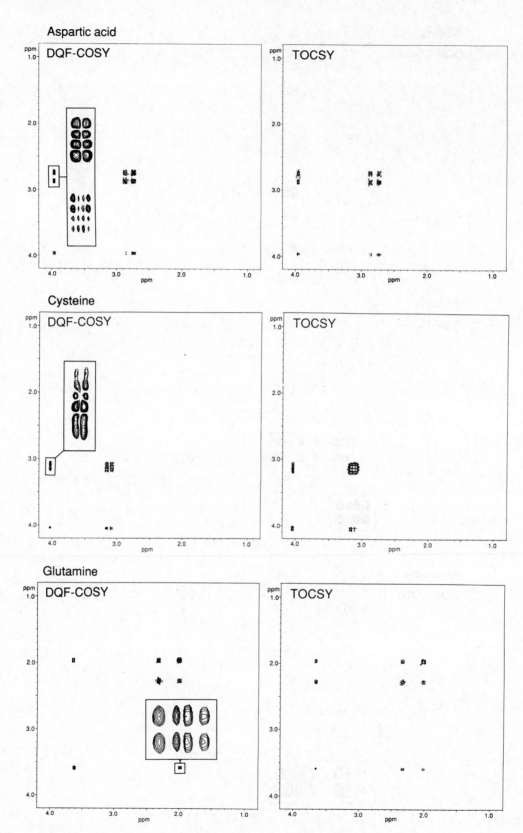

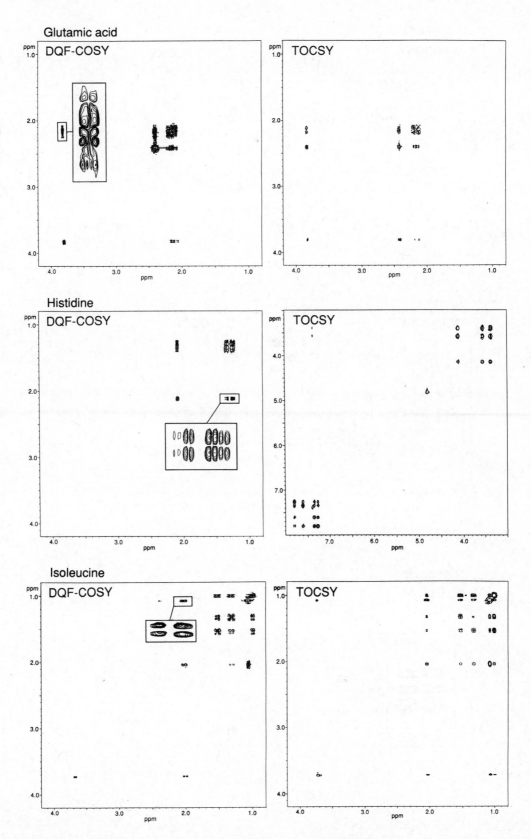

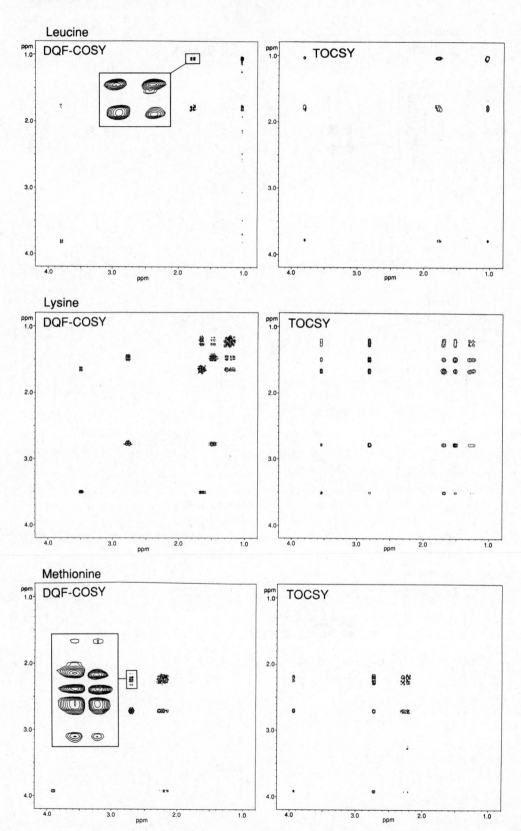

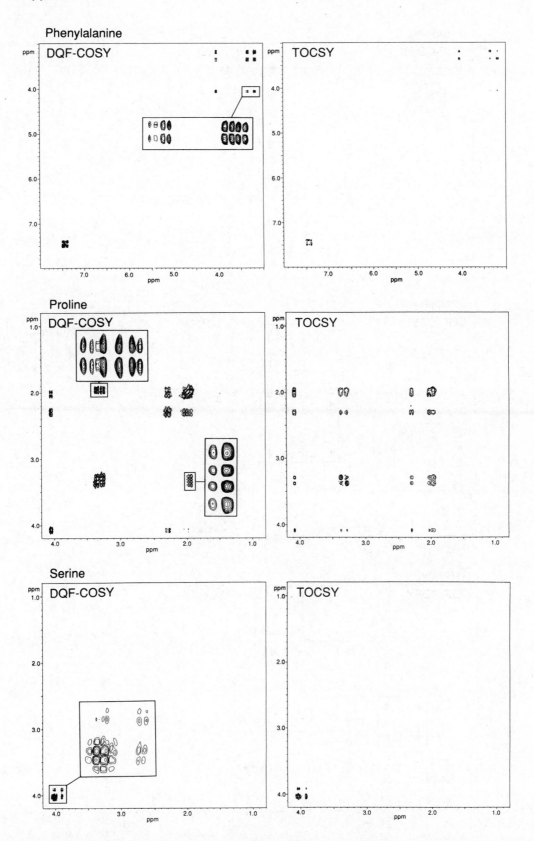

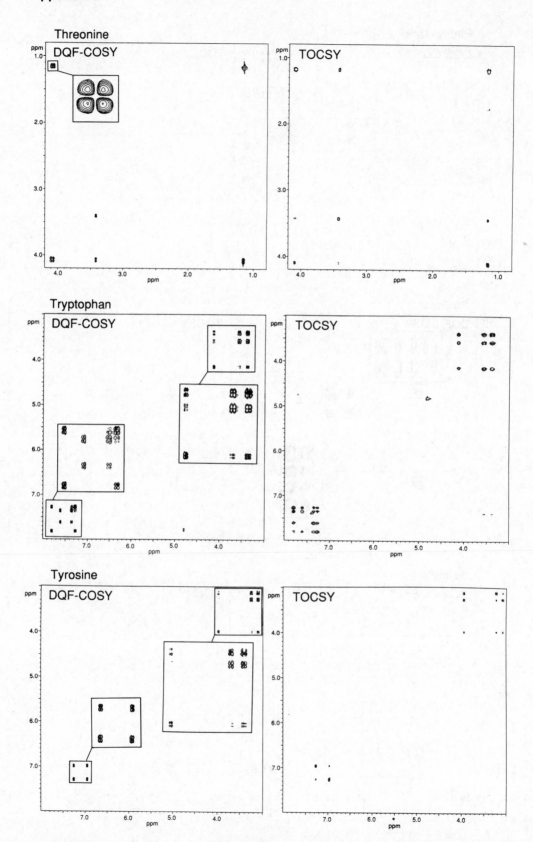

Valine

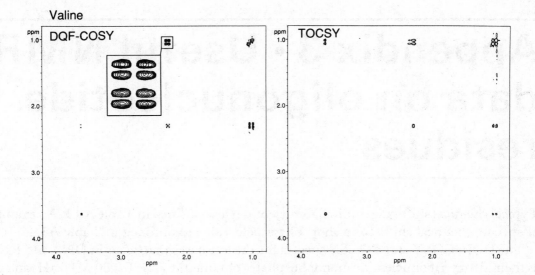

Appendix 3 · Useful NMR data on oligonucleotide residues

Typical chemical shift ranges for oligonucleotides are outlined in Table A3.1. An example of specific chemical shifts for a short oligonucleotide are outlined in Table A3.2.

The DQF–COSY and TOCSY spectra for the five bases that occur in DNA and RNA, as their (deoxy)ribonucleotide monophosphates obtained at 11.7 T (500 MHz ^{1}H) and pD 7.0 are shown below. These were carried out in D_2O, since there are no scalar couplings involving labile protons. Insets show expansions for selected cross-peaks.

Table A3.1 Approximate ^{1}H chemical-shift ranges in single stranded and duplex DNA and RNA fragments[a]

δ(ppm)	Assignment
1.8–3.0	2'H, 2"H in DNA
2.7–4.5	4'H, 5'H, 5"H in DNA
4.4–5.2	3'H in DNA
3.7–5.2	2'H, 3'H, 4'H, 5'H, 5"H in RNA
5.3–6.3	1'H
1.2–1.6	CH_3 of T
5.3–6.0	5H of C and U
7.1–7.6	6H of C, T and U
7.3–8.4	8H of A and G, 2H of A
6.6–9.0*	NH_2 of A, C and G
10–15*	Ring of NH of G, T and U

[a]Data from K. Wüthrich, *NMR of Proteins and Nucleic Acids*, John Wiley, 1986. The rather wide ranges result from the sequence effects on the shifts and for the labile protons (identified by *, observable only in H_2O) from hydrogen bonding.

Table A3.2 Chemical shifts of the non-labile protons of the first five residues in d(CGCGAATTCGCG)$_2^a$

Residue	Chemical Shifts (37°C, pD 7.0)							
	8H	6H	5H	1'H	2'H	2"H	3'H	4'H
C1		7.59	5.84	5.71	1.89	2.35	4.65	4.15
G2	7.92			5.84	2.59	2.68	4.91	4.31
C3		7.24	5.35	5.55	1.81	2.22	4.78	4.17
G4	7.83			5.42	2.63	2.74	4.98	4.29
A5	8.09			5.96	2.65	2.90	5.05	4.44

^{a}In parts per million.

Deoxyadenosine monophosphate

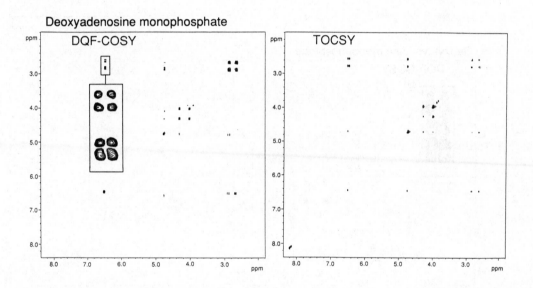

Deoxycytidine monophosphate

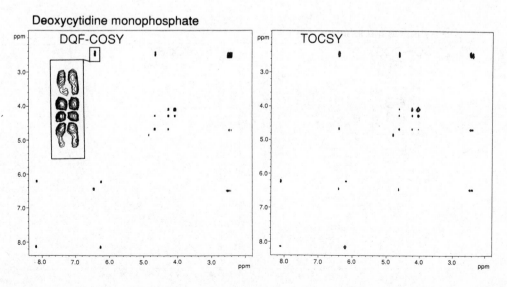

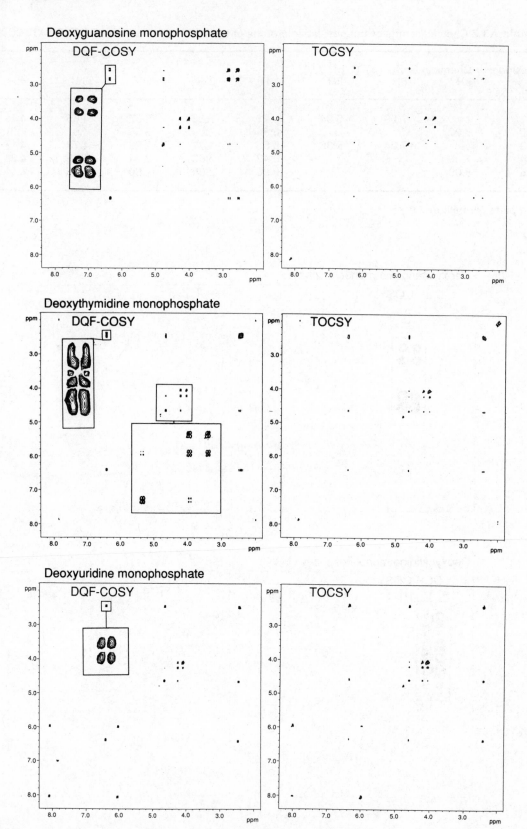

Deoxyguanosine monophosphate

Deoxythymidine monophosphate

Deoxyuridine monophosphate

Appendix 4 · Useful NMR data on common solvents

As a handy reference we present here the chemical shifts for common solvents. Table A4.1 shows the ^1H chemical shifts and properties for common NMR solvents. ^{13}C chemical shifts are shown in Table A4.2.

Table A4.1 Properties of some common NMR solvents[a]

Solvent	Structure	¹H chemical shift δ (rel. to TMS or DDS)	Exchange of labile protons with deuterium	Dielectric constant ε (25°)[f]	pK$_{a1}$[f]	pK$_{a2}$[f]
Water	H_2O	~4.8 (20°)[b]	no[c]	78.5	−1.8	15.3
Deuterium oxide	D_2O	~4.8 (20°)[b,d]	yes			
Methanol	CH_3OH	3.35	no[c]	32.6	−2.0	16.0
d$_4$-Methanol	CD_3OD	3.35[d,e]	yes			
Trifluoroacetic acid	CF_3COOH	~12.5	no[c]	8.4	−10.6	0.2
d-Trifluoroacetic acid	CF_3COOD	~12.5[d]	yes			
d$_6$-Dimethylsulfoxide	$(CD_3)_2-2=0$	2.58[d,e]	no	46.4	0.0	
d-Chloroform	$CDCl_3$	7.27[d]	no	4.9	−6.0	24.0
d$_7$-N-dimethylformamide	$(CD_3)_2NCD=O$	2.88, 2.97, 8.02[d]	no	36.7	0.0	
d$_5$-Pyridine	C_5D_5N	7.0, 7.6, 8.6[d,e]	no	12.3	5.2	
Trifluoroethanol	CF_3CH_2OH	~6.5, ~12.0	no[c]	26.7	−8.2	12.4
d$_3$-Trifluoroethanol	CF_3CD_2OD	~6.5, ~12.0[d,e]	yes			

[a] Indicated are the solvent structure, the ¹H-NMR spectral properties, the observability of labile protons in the dissolved materials, the dielectric constant, and the acid base character. The latter three properties are of particular interest for studies of peptides.

[b] The water resonance moves upfield at a higher temperature (at 50°, δH$_2$O ≈ 4.5 ppm).

[c] Labile protons of the solute may in certain cases not be observable as separate NMR signals because of rapid proton exchange with the solvent.

[d] NMR of residual protons in the deuterated solvent.

[e] The solvent spectrum may contain multiplet resonances arising from hetero-nuclear proton–deuterium spin–spin coupling.

[f] The dielectric constants, ε, pK$_{a1}$, which indicate the pH for half protonation of the neutral bases, and pK$_{a2}$, which indicate the pH for half protonation of the Brönsted acid, are from Ref. 14. The pK$_a$ values are meant only as a general indication of the acid–base character, which may be different in different solvent mixtures.

Table A4.2 Chemical shifts of some common solvents and standards for ^{13}C-NMR of biological molecules[a]

Solvent	Chemical shift (ppm from TMS)
DMSO	40.5
d_6-DMSO	39.6
p-Dioxane	(67.8)[b]
$CHCl_3$	77.2
$CDCl_3$	76.9
CH_3OH	49.3
CD_3OD	49.0
TFA	121, 166
CS_2	192.8
CS_2 ext. capillary	193.7

[a]Note that multiplet resonances will appear for the deuterated and fluorinated molecules even in ^{1}H noise-decoupled spectra.
[b]Often used as a secondary reference to convert data taken in aqueous solution to the TMS scale. The chemical shift depends somewhat on the concentration.

Index